CAMBRIDGE LIBRARY COLLECTION

Books of enduring scholarly value

Mathematical Sciences

From its pre-historic roots in simple counting to the algorithms powering modern desktop computers, from the genius of Archimedes to the genius of Einstein, advances in mathematical understanding and numerical techniques have been directly responsible for creating the modern world as we know it. This series will provide a library of the most influential publications and writers on mathematics in its broadest sense. As such, it will show not only the deep roots from which modern science and technology have grown, but also the astonishing breadth of application of mathematical techniques in the humanities and social sciences, and in everyday life.

The Dynamical Theory of Gases

Sir James Jeans (1877–1946) is regarded as one of the founders of British cosmology, and was the first to suggest (in 1928) the steady state theory, which assumes a continuous creation of matter in the universe. He made many major contributions over a wide area of mathematical physics, but was also well known as an accessible writer for the non-specialist. Jeans's primary aim with the first edition of this book, originally published in 1904, was to 'develop the theory of gases upon as exact a mathematical basis as possible'. Twenty years later those theories were being revolutionised by Quantum Theory. In this fourth edition of 1925, he does not attempt to avoid the discoveries of this new science, but rather exposes the many difficulties that classical theory was experiencing, and how those problems disappeared with Quantum Theory. This edition therefore offers a fascinating insight into a field of physics in transition between two great models of physical science.

The Dynamical Theory of Gases

JAMES JEANS

CAMBRIDGE UNIVERSITY PRESS

Cambridge, New York, Melbourne, Madrid, Cape Town, Singapore,
São Paolo, Delhi, Dubai, Tokyo

Published in the United States of America by Cambridge University Press, New York

www.cambridge.org
Information on this title: www.cambridge.org/9781108005647

This edition first published 1904
This digitally printed version 2009

ISBN 978-1-108-00564-7

THE DYNAMICAL THEORY OF GASES

THE DYNAMICAL THEORY OF GASES

BY

J. H. JEANS, D.Sc., LL.D., F.R.S.

FOURTH EDITION

CAMBRIDGE
AT THE UNIVERSITY PRESS
1925

CAMBRIDGE UNIVERSITY PRESS
Cambridge, New York, Melbourne, Madrid, Cape Town, Singapore, São Paulo, Delhi

Cambridge University Press
The Edinburgh Building, Cambridge CB2 8RU, UK

Published in the United States of America by Cambridge University Press, New York

www.cambridge.org
Information on this title: www.cambridge.org/9780521744782

First edition 1904
Second edition 1916
Third edition 1921
Fourth edition 1925
This digitally printed version 2008

A catalogue record for this publication is available from the British Library

ISBN 978-0-521-74478-2 paperback

PREFACE TO THE SECOND EDITION

MY primary aim in the first edition of this book was to develop the Theory of Gases upon as exact a mathematical basis as possible. This aim has not been forgotten in the preparation of a second edition, but has been combined with an attempt to make as much of the book as possible intelligible to the non-mathematical reader. I have adopted the plan, partially followed in the first edition, of dividing the book to a large extent into mathematical and physical chapters. The reader whose interest is mainly on the physical side will, it is hoped, get an intelligible account of the present state of the subject by reading the physical chapters I, VI, VII and XI to XVIII, and regarding the more mathematical chapters simply as material for reference. Apart from this, something is, I think, gained by clearing the ground by a full mathematical treatment before any physical discussion is attempted.

Since the first edition of this book appeared the position of the Kinetic Theory has been to some extent revolutionised by the growth and developments of the Quantum Theory, and it has been by no means easy to decide what exact amount of prominence ought to be given to the Quantum Theory in the arrangement of the book. The plan finally adopted has been to confine the Quantum Theory to the last chapter; the difficulties arising out of the classical treatment have been allowed to emerge in the earlier chapters, but have been left unsolved. The last chapter merely indicates how these difficulties disappear in the light of the new conceptions of the Quantum Theory: no attempt is made to give a full or balanced view of the whole theory. In the present status of the Quantum Theory this seemed to me the best procedure, but I anticipate that if the book is fortunate enough to run to a further edition, the Quantum Theory is likely to figure much more prominently there than in the present edition.

It will be found that the book has been very extensively rewritten since the first edition. A large amount of new matter has been inserted, and a good deal of labour has been expended in bringing the subject up to date, both on the theoretical and experimental sides. I hope the book in its new form will be of value to the large and increasing number of physicists who find a knowledge of the methods and results of the Kinetic Theory essential to their work, as well as to those who study the subject for its own sake.

J. H. JEANS.

LONDON,
January 1916.

PREFACE TO THE THIRD EDITION

IT will be found that this edition varies from its predecessor mainly in the greater prominence given to the Quantum Theory. The chapter on Radiation and the Quantum Theory which previously terminated the book has been added to and partially rewritten. To this I have further added an entirely new chapter on Quantum Dynamics, dealing mainly with the quite recent work of Ehrenfest, Sommerfeld, Epstein and others. This chapter can of necessity provide only a very brief introduction into the mysteries of Quantum Dynamics, but I hope it will be of value in stimulating the interest of English-speaking readers in a branch of science of which the development has so far been left mainly to other nations.

J. H. JEANS.

DORKING,
October 1920.

PREFACE TO THE FOURTH EDITION

IN preparing a fourth edition I have made only those few changes which were necessary to bring the book up to date.

J. H. JEANS.

DORKING,
June 1925.

CONTENTS

CHAPTER I

INTRODUCTION

1. THE Kinetic Theory of Matter rests essentially upon two closely related hypotheses; the first—that of the molecular structure of matter, the second—the hypothesis that heat is a manifestation of molecular motion.

The first of these hypotheses belongs to the domain of chemistry; indeed it forms the basis of modern chemical science. It is believed not only that all matter is composed of a great number of molecules, but also that all molecules of the same chemical substance are exactly similar as regards size, mass, etc. If this were not so, it would be possible to separate the molecules of different types by chemical processes of fractionisation, whereas Dalton found that the successively separated fractions of a substance were exactly similar. It is true that very modern research has thrown some doubt on whether the molecules of a substance are all as exactly identical as they were once thought to be, but it is also true that the hypothesis of exact similarity of molecules is now, as a broad truth, unassailable, and forms a suitable and convenient working hypothesis for the Theory of Matter.

The second hypothesis, the identification of heat with molecular motion, is that with which the Kinetic Theory of Matter is especially concerned. This hypothesis was for long regarded as pure conjecture, incapable of direct proof, and probable just in proportion to the number of phenomena which could be explained by its help. In recent years, however, the study of the Brownian movements has provided brilliant visual demonstration of the truth of this conjecture, and the actual heat-motion of molecules—or at least of particles which play a rôle exactly similar to that of molecules—may now be seen by anyone who can use a microscope.

The Three States of Matter.

2. One of the most striking and universal properties of the different kinds of matter is that of appearing in three distinct states—solid, liquid, and gaseous. Broadly speaking, the three states are associated with different ranges of temperature; as the temperature of a substance is raised, the substance passes through the solid, liquid, and gaseous states in succession. It is natural to conjecture that the three states of matter are associated with three different types or intensities of molecular motion, and it is not difficult to see how the necessity for these three different states arises.

We know that two bodies cannot occupy the same space; any attempt to compel them to do so brings into play a system of repulsive forces tending to keep the two bodies apart, and this system of forces can only be interpreted as the aggregate of the forces from individual molecules. It follows that molecules exert forces on one another, and that these forces are, in general, repulsive when the molecules are sufficiently close to one another. On the other hand, the phenomenon of cohesion shews that the force between two molecules may, under certain conditions, be one of attraction.

3. *The solid state.* The fact that a solid body, when in its natural state, resists both compression and dilatation, indicates that the force between molecules changes from one of repulsion at small distances to one of attraction at greater distances. This change from a repulsive to an attractive force suggests a position of stable equilibrium in which a pair of molecules can rest in proximity to one another. If we imagine a great number of molecules placed in proximity and at rest in an equilibrium configuration, we have, on the Kinetic Theory conception of matter, a mass of matter in the solid state, and, as there is no motion, this matter must be supposed, in accordance with the fundamental hypothesis of the theory, to be entirely devoid of heat.

The molecules of which the substance is formed will be capable of vibration about their positions of equilibrium, and when these vibrations occur, we say that the body possesses *heat.* As the vibrations become more vigorous we say that the *temperature* of the body increases.

For example, we may imagine the vibratory motion of the molecules to be set up in the first instance by rubbing the surface of the body against a surface of a similar body: here we have a case of heat generated by friction. The act of rubbing will consist in first placing the surfaces of the two bodies so near to one another that the molecules near the surface of one exert a perceptible action on the molecules near the surface of the other, and then in moving the surfaces over one another so as to disturb these surface molecules from their positions of equilibrium. At first the motion will be confined to the neighbourhood of the parts actually rubbed, but the motion of these parts will gradually induce motion in the adjoining regions, until ultimately the motion will have spread over the whole mass.

As a second example, we may imagine two masses, both devoid of internal motion, to impinge one upon the other. The impact will at first cause systems of waves to be set up in the masses, but after a sufficient time the wave character of the motion will have become obliterated, whilst motion of some kind must persist in order to account for the energy of the original motion. This original motion will, in fact, have become replaced by a small vibratory motion of the molecules about their positions of equilibrium—according to the Kinetic Theory, by heat-motion.

4. *The liquid state.* If the body under consideration continues to acquire heat in any way—if, that is, the energy of vibration is caused continually to increase—a stage will in time be reached in which some of the molecules are possessed of so much kinetic energy that the forces from the other molecules no longer suffice to hold them in position: they will, to borrow an astronomical term, escape from their orbits. When the body has reached a state such that this is true of a great number of molecules, it is clear that the application of even a small force, provided it is continued for a sufficient length of time, will, by taking advantage time after time, as opportunity occurs, of the weakness of the forces tending to retain individual molecules, cause the mass to change its shape. When this stage is reached, the body has assumed a plastic or liquid state.

When a molecule of a liquid escapes from its orbit it will in general wander about amongst the other molecules until it falls into a new orbit. If, however, it was initially near to the surface of the liquid, it may be possible for it to escape altogether from the attraction of the other molecules, just as it is possible for a projectile, if projected from the earth's surface with sufficient velocity, to escape from the earth altogether. When this happens the molecule will leave the liquid, so that the mass and volume of liquid will continually diminish owing to the loss of such molecules. Here we have the Kinetic Theory interpretation of the process of evaporation, the vapour being formed by the escaped molecules.

If the liquid is contained in a closed vessel the escaping molecules will impinge on the side of the vessel, and after a certain number of impacts, may fall back again into the liquid. When a state is reached in which the number of molecules which fall back in this way is exactly balanced by the number which escape, we have, according to the Kinetic Theory, a liquid in equilibrium with its own vapour.

5. *The gaseous state.* If we suppose the whole of the liquid transformed into vapour in this way, we have the Kinetic Theory conception of a gas. The molecules can no longer be said to describe orbits, but describe rectilinear paths with uniform velocity except when they encounter other molecules or the walls of the containing vessel. It is clear that this view of the nature of a gas will sufficiently explain the property which a gas possesses of

spreading throughout the whole of any closed space in which it is placed. It is not necessary to suppose, as was at one time done, that this expansive property of a gas is the result of repulsive forces between the molecules.

The fundamental accuracy of this conception of the gaseous state is finely illustrated by some experiments of Dunoyer*. A cylindrical tube was divided into three compartments by means of two partitions perpendicular to the axis of the tube, and these partitions were then pierced in their centres by small holes, so as to form diaphragms. The tube was fixed vertically and in its lowest compartment was placed a piece of some substance, such as sodium, which is in the solid state at ordinary temperatures. After exhausting the tube of all air, the sodium (or other substance) was heated to a sufficient temperature to vaporise it. As the molecules of the vapour are shot off, they move in various directions, and the majority strike on the walls of the lowest compartment of the tube and form a deposit there. Some however pass through the first diaphragm, and describe paths radiating out from the hole in this diaphragm. A few of the molecules pass through both diaphragms into the upper compartment of the tube. These do not collide, for their paths cannot intersect since they are rectilinear paths all radiating from the same point, namely the hole in the lower diaphragm. They accordingly form a deposit on the top of the tube, and this deposit is found to coincide exactly with the projection of the hole in the second diaphragm formed by radii drawn from the hole in the first diaphragm. If a small obstacle is placed in the path of these molecules, it will be found to form a "shadow" on the upper surface of the tube; it may even be that an umbra and penumbra will be discernible.

Mechanical Illustration of the Kinetic Theory of Gases.

6. The Kinetic Theory of Matter is best approached through a study of the Kinetic Theory of the gaseous state. Indeed, until very recently, the Kinetic Theory of Matter has been identical with the Kinetic Theory of Gases; there has not been sufficient evidence as to the conditions prevailing in the solid and liquid states to formulate a Kinetic Theory of these states. The requisite evidence is now rapidly accumulating, so that theories of the solid and liquid states are becoming possible, in outline at least, but it is still true that the theory of gases is much more developed and complete than the corresponding theories ot liquids and solids ean possibly be, and the earlier chapters of the present book are devoted especially to the consideration of the gaseous state.

7. It is important to form as clear an idea as possible of the conception of the gaseous state on which the Kinetic Theory is based, and this can best be done by considering a mechanical illustration.

* L. Dunoyer, *Comptes Rendus*, 152 (1911), p. 592, and *Le Radium*, VIII. (1911), p. 142.

Very little is known as to the structure or shape of actual molecules, or the way in which they react upon one another. Since, however, it is desirable to have as concrete a representation as possible before the mind, at least at the outset, we may (following a procedure which is very usual in the development of the Kinetic Theory) agree for the present to associate the idea of a molecule with that of a spherical body of great elasticity and rigidity—to make the picture quite definite, let us say a billiard-ball. The justification for this procedure lies in its success: it will be found that the behaviour of a gas in which the molecules are complex structures of the most general kind can, to a large extent, be predicted from the behaviour of a much simpler imaginary gas in which the molecules are of the type just described. In fact, one of the most striking features of the Kinetic Theory is the extent to which it is possible to predict the behaviour of a gas as a whole, while remaining in almost complete ignorance of the behaviour and properties of the molecules of which it is composed. Indeed, so many of the results of the theory are true for all kinds of molecules that they would remain true even if the molecules actually were billiard-balls.

As it is somewhat difficult to imagine in detail the motion of a large number of spheres flying about in three dimensions, we may conveniently confine ourselves to a consideration of the analogous motion in two dimensions. As the molecules of the gas are to be represented by billiard-balls, let us suppose the vessel in which the gas is enclosed to be represented by a large billiard-table. The walls of the vessel will of course be represented by the cushions of the table, and if the vessel to be represented is a closed one, the table must have no pockets. Finally, the materials of the table must be supposed of such ideal quality that a ball once set in motion will collide many thousands of times with the cushions before being brought to rest by the friction and the various other passive forces which tend to destroy its motion. A great number of the properties of gases can be illustrated with this imaginary apparatus.

If we take a very large number of balls, and start them at random on the table with random velocities, the resulting state of motion will give a representation of what is supposed to be the condition of matter in its gaseous state. Every ball will be continually colliding both with the other balls and with the cushions of the table. The velocities of the balls will be of the most varying kinds: at one instant a ball may be brought absolutely to rest, while at another instant, as the result of a succession of favourable collisions, it may possess a velocity far in excess of the average velocity of the other balls. One of the problems we shall have to solve will be to find how the velocities of the various balls are distributed about the mean velocity. We shall find that whatever the way in which the velocities are grouped at the outset, they will tend, after a sufficient number of collisions, to group themselves according to the so-called law of trial and error—the law which governs the grouping in position of shots fired at a target.

If the cushions of the table were not fixed in position, they would be driven back by the continued impacts of the balls. The force exerted on the cushions by the balls colliding with them accordingly represents the pressure exerted on the walls of the containing vessel by the gas. Let us imagine a moveable barrier placed initially against one of the cushions, and capable of motion parallel to this cushion. Moving this barrier forward is equivalent to decreasing the volume of the gas. If the barrier is moved forwards while the motion of the billiard-balls is in progress, the impacts both on the moveable barrier and on the three fixed cushions will of course become more frequent: here we have a representation of an increase of pressure accompanying a diminution of volume of a gas. We shall have to discuss how far the law connecting the pressure and density of a gas, constituted in the way imagined by the Kinetic Theory, is in agreement with that found by experiment for an actual gas.

Let us imagine the barrier on our supposed billiard-table to be moved half-way up the table. Let us suppose that the part of the table in front of the barrier is occupied by white balls moving on the average with a large velocity, while the part behind it is similarly occupied by red balls moving on the average with a much smaller velocity. Here we may imagine that we have divided our vessel into two separate chambers; the one is occupied by a gas of one kind at a high temperature, the other by a gas of a different kind at a lower temperature. Returning to the billiard-table, let the barrier suddenly be removed. The white balls will immediately invade the part which was formerly occupied only by red balls, and vice versa. Also the rapidly moving white balls will be continually losing energy by collision with the slower red balls, and the red of course gaining energy through impact with the white. After the motion has been in progress for a sufficient time the white and red balls will be equally distributed over the whole of the table, and the average velocities of the balls of the two colours will be the same. Here we have simple illustrations of the diffusion of gases, and of equalisation of temperature. The actual problem to be solved is, however, obviously more complex than that suggested by this analogy, for in nature the molecules of different gases differ by something more fundamental than mere colour.

One further question must be considered. No matter how elastic the billiard-balls and table may be, the motion cannot continue indefinitely. In time, the energy of this motion will be frittered away, partly perhaps by frictional forces, such as air-resistance, and partly by the vibrations set up in the balls by collisions. The energy dissipated by air-resistance becomes transformed into energy in the air; the energy dissipated by collisions is transformed into energy of internal vibrations of the billiard-balls. What, then, does this represent in the gas, and how is it that a gas, if constituted as we have supposed, does not, in a very short time, lose the energy of translational motion of its molecules, and replace it by energy of internal vibrations of these molecules, and energy in the surrounding ether?

The difficulties raised by this and similar questions formed a most serious hindrance to the progress of the Kinetic Theory for many years. Attention was drawn to them by Maxwell, but it was not until the introduction of the Quantum-theory by Planck and his followers in the early years of the present century, that it was possible to give anything like a satisfactory explanation. The explanation supplied by the Quantum-theory will have to be examined in detail in a later chapter of the present book. It is at best only partial, but must, so far as it goes, probably be regarded as satisfactory. The explanation is, in brief, that there is no true analogy between the two cases when we consider questions of internal vibrations and transfer of energy to the surrounding medium. For the motion of the billiard-balls is governed by the well-known Newtonian laws, whereas the internal motions of molecules, and their transfer of energy to the ether, are now believed to be governed by an entirely different system of dynamical laws. The procedure of this book will be to develop the Kinetic Theory as far as it can be developed without departure from the Newtonian laws, and then to examine what light can be thrown on the various outstanding phenomena by the new system of dynamical laws suggested by Planck.

Numerical Values.

8. The foregoing rough sketch will, it is hoped, have given some idea of the nature of the problems to be attacked. As a conclusion to this preliminary chapter, it may be useful to give some approximate numerical values. These will give an indication of the order of magnitude of the quantities with which we shall be dealing, and will make it easier to form a clear mental picture of the processes under consideration.

Number of molecules per cubic centimetre. In accordance with the law of Avogadro (see below, § 147), the number of molecules in a cubic centimetre of gas at standard temperature and pressure (0° C. and 1 atmosphere) is independent of the chemical composition of the gas. This number, which will be denoted by N_0, is frequently referred to as Avogadro's number, and its numerical evaluation is naturally of great importance for the Kinetic Theory of Matter. Unfortunately the number is extremely difficult to evaluate with any great accuracy: many of the uncertainties in the numerical values used in the Kinetic Theory may be referred ultimately to uncertainties in the estimation of this number. It is fortunate that there are a great number of methods available for the determination of N_0.

The best determination of N_0 is that deduced from the value of the charge on the electron. Millikan* determines this, with a probable error of about one part in a thousand, as

$$e = 4{\cdot}774 \times 10^{-10} \text{ electrostatic units}$$
$$= 1{\cdot}592 \times 10^{-20} \text{ electromagnetic units.}$$

Since $N_0 e$ is the constant of electrolysis, and therefore equal to 9649·6 electromagnetic units, the corresponding value of N_0 is

$$N_0 = 2{\cdot}705 \times 10^{19}.$$

The density of hydrogen (molecular weight 2·016) at standard temperature and pressure is ·00008987, so that, if N_1 is the number of molecules in one gram-molecule, we have

$$N_0 = {\cdot}00004458\, N_1.$$

Thus the above value of N_0 corresponds to

$$N_1 = 6{\cdot}062 \times 10^{23}.$$

A less accurate determination of N_0 is obtained from a study of black-body radiation. Richardson† obtains the value $N_0 = 2{\cdot}76 \times 10^{19}$, but a recalculation in which use is made of more recent observational data, shews that the value of N_0 must be very close to the value $N_0 = 2{\cdot}705 \times 10^{19}$ given above.

Another determination of N_1 can be made from observations on the Brownian Movements. The method was first developed by Perrin, who obtained values uniformly larger than Millikan's value given above. More recent observations by H. Fletcher‡ give the value $N_1 = 6{\cdot}03 \times 10^{23}$, with a probable error of about 2 per cent. of the whole.

For the purpose of calculations in the present book we shall adopt Millikan's values

$$N_0 = \mathbf{2{\cdot}705} \times \mathbf{10^{19}},$$
$$N_1 = \mathbf{6{\cdot}062} \times \mathbf{10^{23}}, \text{ the number of molecules in 1 gm.-mol.,}$$
$$e = \mathbf{4{\cdot}774} \times \mathbf{10^{-10}} \text{ electrostatic units.}$$

Thus at 0° C. and at a pressure of 1 atmosphere (760 mm.), there are taken to be $2{\cdot}705 \times 10^{19}$ molecules per cubic centimetre. Under other conditions the number is of course directly proportional to the density.

The average distance apart of adjacent molecules at atmospheric pressure will clearly be about $(2{\cdot}705 \times 10^{19})^{-\frac{1}{3}}$ cms., or $3{\cdot}33 \times 10^{-7}$ cms. At a pressure of one-millionth of an atmosphere (·00076 mm.) this distance is increased to $3{\cdot}33 \times 10^{-5}$ cms. (·00033 mm.).

* *Phil. Mag.* XXXIV. (1917), p. 1.

† *Electron Theory of Matter*, p. 356.

‡ *Phys. Rev.* IV. (1914), p. 440.

Mass of a molecule. Since we are taking $N_1 = 6{\cdot}062 \times 10^{23}$, the mass of the atom of unit atomic weight must be taken to be $(6{\cdot}062 \times 10^{23})^{-1}$ gms. or $1{\cdot}651 \times 10^{-24}$ gms. Thus the hydrogen atom, of atomic weight 1·008, will be of mass $1{\cdot}662 \times 10^{-24}$ gms., and the masses of molecules will be in proportion to their molecular weights; that of oxygen for instance is 52×10^{-24} gms.

Velocity. The velocity of the molecules does not depend on the evaluation of Avogadro's number, and is known with great accuracy (see below, § 150).

For air at a temperature of 15° C. the average velocity is about 459 metres per second. For hydrogen at 0° C. it is about 1694 metres per second. As regards velocities under other conditions it may be said that, roughly, the mean velocity of a molecule of molecular weight m, at a temperature of θ degrees Centigrade, is proportional to $\sqrt{273 + \theta}$, and is inversely proportional to $\sqrt{m}$, so that, for instance, the velocity of the oxygen molecule is approximately a quarter of that of the hydrogen molecule at the same temperature. The velocity is independent of the density of the gas. A table of velocities will be given later (§ 151).

Since the molecules of hydrogen at 0° C. move with an average velocity of 1694 metres a second, the total distance described by the $2{\cdot}705 \times 10^{19}$ molecules in a c.c. of hydrogen at 0° C. will be 458×10^{17} kilometres per second.

Size. It is a matter of some difficulty to determine or even to define the size of a molecule. The trouble arises primarily from our ignorance of the shape and other properties of the molecule. If the molecules were known to be elastic spheres the question would be simple enough, and the size of the molecule would be measured by the diameter of the sphere. If, however, the molecules are *assumed* as a first approximation to be elastic spheres, experiment leads to discordant results for the diameters of these spheres, shewing that the original assumption is unjustifiable. The divergences arise not only from the fact that the shape of the molecules is not spherical, but also from the fact that the molecules are surrounded by fields of force, and in most experiments it is the extension of this field of force, rather than that of the molecules themselves, with which we are concerned.

If, however, we agree to regard the molecules as roughly represented by elastic spheres, it is found that these spheres must be supposed in the case of hydrogen to have a radius of about $1{\cdot}36 \times 10^{-8}$ cms. The size ought strictly to be different for different molecules, and more exact figures will be given later, but as the difference in size is hardly more than comparable with the error introduced by the supposition that the molecules are elastic spheres, these differences need not be discussed here. A full discussion of the evidence of the Kinetic Theory as to the size of molecules will be found in Chapter XIV of the present book.

Number and frequency of collisions. Regarding the molecule of hydrogen as a sphere of radius $1{\cdot}36 \times 10^{-8}$ cms. the number of collisions per cubic centimetre of hydrogen at 0° C. is found to be about $2{\cdot}037 \times 10^{29}$ per second.

Free paths. Each collision is the termination of two free paths, hence the number of free paths described in the gas just considered is about $4{\cdot}07 \times 10^{29}$ per second. It has already been said that the total distance described—*i.e.* the aggregate of these free paths—is 458×10^{22} cms. Hence on division we see that the mean length of these free paths is $1{\cdot}125 \times 10^{-5}$ cms.

It is obvious that the mean free path, being a pure length, will depend only on the diameter of the molecules, and on the number of molecules per cubic centimetre; it will not depend on the velocities of motion of the molecules. Thus the values we have obtained for the mean free path are approximately true for all gases so long as the molecules are supposed uniformly to be spheres of radius $1{\cdot}36 \times 10^{-8}$ cms. The free path is, however, inversely proportional to the number of molecules per cubic centimetre of gas. For instance in a vacuum tube in which the pressure is that of half a millimetre of mercury, the density of gas is only 1 : 1520 of the normal density, and therefore the free path is roughly equal to one-sixth of a millimetre.

It appears from these figures that the mean free path of a molecule is about 400 times its diameter in a gas at normal pressure, and is about 600,000 times its diameter when the pressure is reduced to half a millimetre of mercury. There is therefore every justification for assuming, as a first approximation, that the linear dimensions of molecules are small in comparison with their free paths.

Comparing the values obtained for the length of the free path with the values previously given for the velocity of motion, we find that the mean time of describing a free path ranges from about 3×10^{-10} seconds in the case of air under normal conditions, to about 10^{-7} seconds in the case of hydrogen at 0° C. at a pressure equal to that of half a millimetre of mercury.

The principal lesson to be learned from the foregoing figures is that the mechanism of the Kinetic Theory is extremely "fine-grained" when measured by ordinary standards. Molecules are, in fact, not infinitely small, and neither is their motion infinitely rapid, but the units of space and time appropriate for the measurement of the motion of individual molecules are so small in comparison with even the smallest quantities which we can measure experimentally that the phenomena exhibited by a gas constituted in the way described will be indistinguishable, so far as experiment and human observation go, from those of a continuous medium.

There are two other fundamental quantities of which the numerical values will frequently be required in the present book, namely the mechanical equivalent of heat, and the absolute zero of temperature.

Mechanical equivalent of heat. The calorie is the number of heat units required to raise one gramme of water through 1° C. at a temperature which is usually taken to be 15° C. or some near temperature. A summary of experimental values will be found in Glazebrook's *Dictionary of Applied Physics* (I. p. 493); the best determinations range from $4{\cdot}182 \times 10^7$ to $4{\cdot}185 \times 10^7$. In the present book we take

$$J = 4{\cdot}184 \times 10^7,$$

this being the value adopted both by the *Recueil de Constantes Physiques* of the French Physical Society (1913) and by the *Smithsonian Physical Tables* (Washington, 1920).

Absolute zero of temperature. The value of the absolute zero of temperature T_0 given in the *Recueil de Constantes Physiques* as most probable is $-273{\cdot}09°$ C. From a lengthy investigation, Callendar* deduces $-273{\cdot}10°$ C. as the most probable value. In the present book we shall adopt the value

$$T_0 = -273{\cdot}1° \text{ C}$$

Historical Note.

9. The rise of the Kinetic Theory was of a gradual nature, and it is difficult to mention any time at which the theory may be said to have arisen, or any single name to whom honour of its establishment is due. Three stages in its development may be traced. There is first the stage of speculative opinion, unsupported by scientific evidence. Given that a great number of thinkers are speculating as to the structure of matter, it is only in accordance with the laws of probability that some of them should arrive fairly near to the truth. An opinion which turns out ultimately to be near the truth remains, however, of no greater value to the advancement of science than a more erroneous opinion, until scientific reasons can be given for supposing the former to be more accurate than the latter. When this point is reached the theory may be said to have entered upon the second stage of its development; the true and false opinions are still equally in the field, but the former is supplied with weapons for defeating the latter. In the third stage there is general agreement as to the main foundations of the theory and their truth, and labour is devoted no longer to defeating adverse opinion, but to the elaboration of the detail of the theory, and to attempts to extend its boundaries.

In its earliest stage the growth of the Kinetic Theory is hardly distinguishable from that of the atomic theory. The view that matter was to be

* *Phil. Mag.* v. (1903), p. 95.

regarded as an aggregation of hard, indivisible and similar parts was upheld by Lucretius, who appears to have taken his opinions from Democritus and Epicurus, who again had been guided by Leucippus. This theory was revived by Gassendi in the middle of the seventeenth century*. Apparently Gassendi was the first to suspect that the motion alone of the atoms was sufficient to account for a number of phenomena, without the introduction of adventitious hypotheses to account separately for these phenomena. Lasswitz† describes Gassendi's work as follows: "Following Democritus and Epicurus, Gassendi in the seventeenth century re-established and elaborated an atomic theory based upon the assumption that all material phenomena can be referred to the indestructible motion of atoms and can therefore be described as 'kinetic.' Gassendi's atoms are devoid of all qualities except absolute rigidity; they are similar in substance, but different in size and form, and move in all directions through empty space. On this basis Gassendi explains a number of physical processes, in particular the three states of matter and the transitions from one to another, in a way very little different from that of the modern kinetic theory." It is obvious, then, that with Gassendi the theory is entering upon the second stage of its existence.

Twenty years later ideas of the same nature seem to have occurred independently to Hooke, the recognition of whose work in the foundation of the Kinetic Theory is due to Professor Tait‡.

The next advance in the theory is due to Daniel Bernoulli§, who frequently is credited with having been the first to make the discoveries of Gassendi and Hooke. In his *Hydrodynamica*, published in 1738, he points out that the elasticity of a gas may be regarded as due to the impacts of particles on the boundary. He deduces Boyle's law for the relation between pressure and volume, and attempts to find a general relation between pressure and volume when the finite size of the molecules, supposed absolutely hard and spherical, is taken into account.

After Bernoulli, there is little to record for almost a century. Then we find that in rapid succession Herapath|| (1821), Waterston (1845), Joule¶ (1848), Krönig** (1856), and Clausius (1857) take up the subject. Waterston attempted to found a scientific mathematical theory of the subject; but

* *Syntagma Philosophicum*, 1658, Lugduni.

† "Der Verfall der kinetischen Atomistik im 17 Jahrhundert," *Pogg. Ann.* 153, p. 373 (1874).

‡ "Hooke's Anticipation of the Kinetic Theory," *Proc. Edin. Roy. Soc.* March 16, 1885. Tait's *Collected Works*, II. p. 122.

§ Daniel Bernoulli, *Hydrodynamica.* Argentoria, 1738. Sectio decima, "De affectionibus atque motibus fluidorum elasticorum, praecipue autem aeris."

|| *Annals of Philosophy*, [2], I. p. 273.

¶ *British Association Report*, 1848, Part II. p. 21; *Memoirs of the Manchester Literary and Philosophical Society*, [2], IX. p. 107.

** *Poggendorff's Annalen.*

his paper, which was presented to the Royal Society in 1845, contained certain inaccuracies, and was for this reason not published in the *Philosophical Transactions* until 1892*, when Lord Rayleigh had it published on account of its historical interest. Clausius, in his first paper†, calculates accurately the relation between temperature, pressure and volume, and also the value of the ratio of the two specific heats for a gas in which the energy of the molecules is wholly one of translation. In 1859, Clerk Maxwell was added to the number of contributors to the theory, reading a paper on the subject before the British Association at Aberdeen‡. It has been suggested that Maxwell was first led to take an interest in the subject by his investigations on the motion of Saturn's rings, which gained for him the Adams Prize in 1857§. In the hands of Clausius and Maxwell the theory developed with great rapidity, so that to write the history of the subject from this time would be hardly less than to give an account of the theory in its present form. Among the more prominent contributors to the theory in the interval between the time of Clausius and Maxwell and the end of the nineteenth century may be mentioned Boltzmann, Kirchhoff, Van der Waals and Lorentz on the continent, and in this country Tait and Lord Rayleigh.

In the interval just mentioned, there had gathered around the theory what Lord Kelvin|| called "Nineteenth-century Clouds over the Dynamical Theory of Heat." Lecturing in the last year of the century, Lord Kelvin said¶, "The beauty and clearness of the dynamical theory which asserts heat and light to be modes of motion, is at present obscured by two clouds." The first of these clouds had to do with the question of the constitution of the ether, and need not concern us here; the second cloud was thrown over the Theory of Gases by difficulties such as those referred to at the end of § 7. This cloud has already to a great extent been dissipated by the development of the Quantum-theory, a theory which will be explained in its proper place in the present book. It may be remarked here that since the Quantum-theory came into being in 1901, the growth of the Kinetic Theory has been almost exactly identical with the growth of the Quantum-theory.

* *Phil. Trans.* 183, p. 1.

† "Ueber die Art der Bewegung welche wir Wärme nennen," *Pogg. Annalen*, 100, p. 353.

‡ *Phil. Mag.* Jan and July, 1860; *Collected Works*, I. p. 377.

§ See W. D. Niven, preface to Maxwell's *Collected Works*, p. xv.

|| *Phil. Mag.* II. (1901), p. 1.

¶ Lecture delivered at the Royal Institution of Great Britain, Friday, April 27, 1900.

CHAPTER II

THE LAW OF DISTRIBUTION OF VELOCITIES

I. The Method of Collisions.

10. The mathematical difficulties of the subject commence when we attempt to discuss the law according to which the velocities of the molecules are grouped about their mean value. We are of course at liberty to consider an imaginary gas in which the velocities are grouped at the outset according to any law we please, but in general every collision which occurs will tend to change this law. The problem before us is to investigate whether there is any law which remains, on the whole, unchanged by collisions; and if so whether the velocities of the molecules of a gas, starting from some arbitrarily chosen law, will tend after a sufficient time to obey some definite law which is independent of the particular law from which the gas started.

There are two totally distinct methods of attacking these problems, and these are given in this chapter and the next, the relation between them being discussed in Chapter IV. The present chapter contains the classical method of which the development is due mainly to Clerk Maxwell and Boltzmann (see § 60 below).

The definition of Density.

11. There is no difficulty in defining the density of a continuous substance. If we take a small volume v, enclosing a given point P, and denote by m the mass of matter contained within this volume, then the assumption of continuity ensures that as the volume v shrinks until it is of infinitesimal size, while still enclosing the point P, then the ratio m/v will approach a definite limit ρ, and we define the density at the point P as being the value of the limit ρ.

Again, when, as in the Kinetic Theory, the matter is composed of discrete molecules, there is no difficulty in defining density if the matter is homogeneous and if also it can be supposed that there is an infinitely great quantity of it. In this case, we take a volume V and suppose M to be the mass enclosed within it. The homogeneity of the matter now ensures that as V is increased indefinitely, the ratio M/V will approximate to a definite limit ρ, and, as before, we define the density of the matter to be the value of the limit ρ.

The gas of the Kinetic Theory will, in general, be neither continuous nor homogeneous. It will therefore be impossible to frame a general definition upon the model of the two foregoing definitions, since to do this we should have to suppose the element of volume to become infinitely great and infinitely small at the same time. But with reference to the actual conditions of nature this objection is not serious. We can find an element of volume which may, without appreciable error, be supposed to be infinitely great in comparison with the distance between neighbouring molecules, and at the same time infinitely small compared with the scale of variation of density of the gas. For instance, the density of a gas may generally be supposed homogeneous throughout a cube of edge equal to one millimetre, while such a cube is large compared with the scale of molecular structure, containing, as has already been mentioned, about $2{\cdot}75 \times 10^{16}$ molecules in the case of a gas under normal conditions of pressure and temperature.

The ratio of the mass contained in an element of this kind, to the volume of the element, will give the *mass-density* of the gas. If we substitute "number of molecules whose centre is contained in" for "mass contained in," the definition gives the *molecular-density* of the gas. We shall find it convenient to denote the mass-density by ρ and the molecular-density by ν. If m is the mass of each molecule, we have

$$\rho = m\nu \quad \text{.................................}(1).$$

It will be seen that this definition of density is not logically perfect, but it will be admitted that it is adequate for practical use. The difficulty of obtaining a logically perfect definition has been discussed by Burbury*. A similar difficulty is of common occurrence in statistical work: consider, for instance, the statement "the density of population in parts of London is as high as 105 per acre."

12. If Ω is a volume throughout which the density is sensibly constant, the number of molecules of which the centres are contained within this volume would, if the foregoing definitions were logically perfect, be $\Omega\nu$. As the definitions are not perfect, we must examine within what limits the statement is true, that the number of molecules is $\Omega\nu$. It is certainly not literally true, for neither Ω nor ν will in general be integers, while the number of molecules whose centres are contained in the volume Ω must necessarily be integral. In the language of the theory of probability the statement may be taken to mean that the "expectation" of the number

* S. H. Burbury, *Kinetic Theory of Gases*, p. 3.

of molecules in the region in question is $\Omega\nu$. Any appeal to the theory of probability implies that a certain amount of knowledge is given, while we remain in ignorance of the remaining facts. In this particular case, what is known is that the molecular density throughout the region Ω is ν; what is not known is the position of the individual molecules of the gas.

With this understanding it will be permissible to say that the number of molecules in an element of volume $dxdydz$ *selected at random* is $\nu dxdydz$. What is meant is that the probability of finding the centre of a molecule inside this element of volume is $\nu dxdydz$.

The definition of the Law of Distribution of Velocities.

13. The difficulties of the last two sections recur when we attempt to define the law of distribution of velocities. In fact at present we may consider that a molecule possesses six coordinates—the coordinates in space of its centre of gravity which we denote by x, y, z, and the corresponding velocity components, which we shall denote by u, v, w. In the last two sections we were virtually discussing the law of grouping of the coordinates x, y, z; we now have to discuss the law of grouping of the velocities u, v, w.

Let us take some fixed imaginary point as origin, and draw from this point a system of lines to represent in magnitude and direction the velocities of the different molecules of the gas. Referred to orthogonal axes the coordinates of the extremity of any line will be u, v, w, the components of velocity of the corresponding molecule. A discussion of the law of distribution of velocities is exactly equivalent to a discussion of the law of density of these points.

Subject to the limitations mentioned above (§ 12), we can define the density of these points in the manner already explained. If we denote this density by τ, then, on our former understanding, we can say that the number of molecules of which the velocities lie between u and $u + du$, v and $v + dv$, w and $w + dw$, is $\tau dudvdw$, where τ is the "density of points at the point u, v, w." We shall find it convenient to replace τ by Nf, where N is the total number of molecules of which the velocities have been represented. When it is necessary to specify the point u, v, w at which f is measured, we shall write $f(u, v, w)$ instead of f.

To avoid the continual repetition of these limits, let us agree to say that a molecule of which the components lie between u and $u + du$, v and $v + dv$, w and $w + dw$ is a molecule of class A.

14. The total number of molecules of class A is therefore

$$Nf(u, v, w)\,dudvdw,$$

and since there are N molecules altogether, it follows that the probability that the velocity of a molecule selected at random shall have components lying between u and $u + du$, v and $v + dv$, w and $w + dw$ is $f(u, v, w)\, du\, dv\, dw$.

In accordance with the definition of § 12, we can say that the number of molecules belonging to class A which are found within the element of volume $dx\,dy\,dz$ *selected at random* is

$$\nu f(u, v, w)\, du\, dv\, dw\, dx\, dy\, dz \quad \text{.......................(2).}$$

Interpreted literally this statement is unintelligible for $du\,dv\,dw\,dx\,dy\,dz$ is a small quantity of the sixth order; interpreted in the sense already explained, no exception can be taken either to its intelligibility or truth.

The assumption of Molecular Chaos.

15. Let us imagine that instead of the element $dx\,dy\,dz$ having been selected at random, we had supposed it to be an element in the immediate neighbourhood of a second molecule of which the components of velocity were known to lie between u' and $u' + du'$, v' and $v' + dv'$, w' and $w' + dw'$, let us say a molecule of class B. We are no longer justified in saying that the probability of finding a molecule belonging to class A inside this element is given by expression (2). If all the molecules of class A were distributed at random, and then those of class B were independently distributed at random, the statement would be true enough. But if the gas is moving in accordance with the dynamical conditions of nature, it is quite conceivable that, for instance, molecules possessing nearly equal velocities might tend to flock together. If this were so the probability we are discussing would be greater than that given by expression (2) when the velocities of the two molecules of classes A and B were nearly equal; in general, it would depend on u', v', w'. as well as on u, v, w.

In the case which is discussed in the present chapter—that in which the molecules are hard elastic spheres—it is usual to *assume* that the molecules having velocity-components lying within any small specified limit are, at every instant throughout the motion of the gas, distributed at random, independently of the positions or velocities of the other molecules, provided only that two molecules do not occupy the same space. The legitimacy of this assumption is not self-evident. Indeed, nothing but a discussion of the dynamical equations which determine the motion of the molecules can decide whether the assumption is true or not. Such a discussion will be given in Chapter IV and the assumption will be proved to be justifiable; for the present we shall be content to make the assumption, without discussing its validity.

The changes produced by Collisions when the Molecules are Elastic Spheres.

16. The state of a gas is fully known, from the statistical point of view, when the density and the law of distribution of velocities at every point of the gas are known. The main problem of this chapter, which we now proceed to attack, is to search for a steady state: *i.e.* a state in which the density and law of distribution of velocities remain the same at every point of the gas throughout all time.

We begin by discussing the simplest case. Not only are the molecules supposed to be hard rigid spheres, but we suppose that the external physical conditions are the same at every point of space, and that the gas fills infinite space. The last of these assumptions is a temporary one, which enables us to consider separately the elements of the problem which are introduced by the presence of a containing vessel.

Under the conditions now postulated, we may clearly begin by assuming the gas to have the same molecular density ν and the same law of distribution of velocities f at every point of space. Since there is nothing to distinguish the different regions in space, this uniformity in space will obviously be maintained throughout all time, but the actual form of the function f will change with the time.

17. The first problem is to find an expression for the change in the number of molecules belonging to class A (defined on p. 16) which occurs during an interval of time dt. Since the motion of the molecules is one of uniform velocity except when collisions take place, it appears that molecules can only enter or leave class A through the occurrence of collisions. We begin by considering molecules which leave class A through collisions.

Let us consider a special kind of collision which we shall call a collision of class α. This is to be defined as a collision in which the three following conditions are satisfied:

(i) One of the two colliding molecules is to be a molecule of class A.

(ii) The second colliding molecule is to be of class B (defined on p. 17).

(iii) The direction of the line joining the centre of the former molecule to that of the latter at the moment of impact is to be such that a line drawn parallel to it from the centre of a fixed sphere of unit radius to the surface of this sphere meets the surface inside a small element of area $d\omega$, this element being such that the direction cosines of a line drawn to its centre from the centre of the sphere are l, m, n.

The number of molecules of class A is $\nu f(u, v, w)\, du dv dw$ per unit volume, and each of these is capable of taking part in a collision of class α. Let σ be the *diameter* of a molecule, and imagine a sphere of *radius* σ drawn round each molecule and concentric with it. As the molecule moves, the sphere is to move so as to remain concentric, but is not to rotate with the molecule. If a collision of class α occurs, the centre of the second molecule—that of class B—must lie on this sphere at the moment of impact, and further, since condition (iii) is to be satisfied, must lie within a small element of surface of area $\sigma^2 d\omega$. In figure 1, the sphere of radius σ is drawn thick. The other spheres represent the two molecules just before and at the instant of collision.

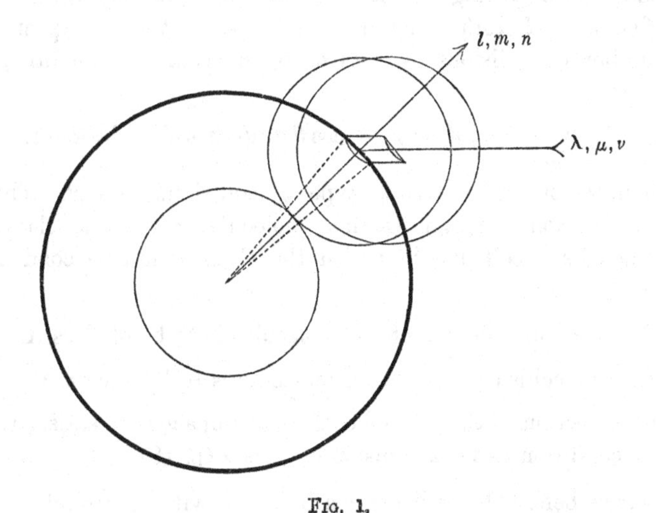

FIG. 1.

Supposing that a collision of class α takes place, we see that before collision the second molecule must have been moving relatively to the first with a velocity of which the components, except for infinitesimally small quantities, were $u' - u$, $v' - v$, $w' - w$; let us say a velocity V in a direction λ, μ, ν. Hence at an infinitesimally small time δt before collision, the centre of the second molecule must have been upon a small element of area $\sigma^2 d\omega$ obtained by moving the element $\sigma^2 d\omega$ from an initial position upon the surface of the sphere through a distance $V\delta t$ in a direction $-\lambda$, $-\mu$, $-\nu$. If, therefore, a collision is to take place within an interval dt the centre of the second molecule must, at the beginning of this interval, have been inside the cylinder which is described by moving the original element through a distance Vdt in this same direction.

The volume of this cylinder is equal to its base multiplied by its height. The former is $\sigma^2 d\omega$, the latter is $Vdt \cos\theta$, where θ is the angle between the axis of the cylinder and a perpendicular to the base. The direction cosines of the axis are $-\lambda$, $-\mu$, $-\nu$, those of the perpendicular to the base are of course l, m, n, so that

$$\cos\theta = -(l\lambda + m\mu + n\nu) \quad\text{.......................(3).}$$

The volume of the cylinder is therefore $V\sigma^2 \cos\theta\, d\omega\, dt$, so that for any single molecule of class A, the probability that the centre of a molecule of class B shall lie within this cylinder at the beginning of the interval dt is, in accordance with § 15,

$$\nu f(u', v', w')\, du'\, dv'\, dw'\, V\sigma^2 \cos\theta\, d\omega\, dt.$$

This, then, is the probability for each molecule of class A that a collision of class α shall occur during the interval dt. The number of molecules of class A is $\nu f(u, v, w)\, du dv dw$ per unit volume, so that the "expectation" of the total number of collisions of class α which occur in time dt per unit volume will be

$$\nu^2 f(u, v, w) f(u', v', w')\, V\sigma^2 \cos\theta\, du dv dw du' dv' dw' d\omega dt \quad\text{......(4).}$$

18. We now consider a second type of collision, class β. This is to be a type of collision through which a molecule enters into class A, and is to be defined as a collision in which the three following conditions are satisfied:

(i) After the collision, one of the molecules is to be of class A.

(ii) After the collision, the second molecule is to be of class B.

(iii) The direction of the line of centres at impact is to satisfy the same condition as for a collision of class α (p. 18).

The velocities before the collision can be found without trouble. For the relative velocity can be divided into two parts—the one in the common tangent plane through the point of contact of the spheres, and the other along the line of centres. Of these, the former remains unaltered by the collision, while the latter is reversed in direction, but remains unaltered in magnitude. Now the normal relative velocity *after* impact must, in virtue of the three conditions satisfied, be the same as for a collision of class α *before* impact. It must, therefore, be $V\cos\theta$. We have, by equation (3),

$$\begin{aligned} V\cos\theta &= -V(l\lambda + m\mu + n\nu) \\ &= l(u-u') + m(v-v') + n(w-w'). \end{aligned}$$

Let $\bar{u}$, $\bar{v}$, $\bar{w}$ and $\bar{u}'$, $\bar{v}'$, $\bar{w}'$ be the components of the velocities of two molecules such that after a collision along a line of centres having direction cosines l, m, n the velocities are u, v, w and u', v', w', then by what has just been said we must have

$$\bar{u} = u - lV\cos\theta = u - \{l^2(u-u') + lm(v-v') + ln(w-w')\} \quad\text{...(5),}$$

$$\bar{u}' = u' + lV\cos\theta = u' - \{l^2(u'-u) + lm(v'-v) + ln(w'-w)\} \quad\text{...(6).}$$

The number of collisions per unit volume, such that before collision the components of the velocities lie between $\bar{u}$ and $\bar{u}+d\bar{u}$, etc., and $\bar{u}'$ and $\bar{u}'+d\bar{u}'$, etc., and such that the third condition of p. 18 is satisfied by the line of centres at impact is, by comparison with expression (4), seen to be

$$\nu^2 f(\bar{u}, \bar{v}, \bar{w}) f(\bar{u}', \bar{v}', \bar{w}') V\sigma^2 \cos\theta \, d\bar{u} d\bar{v} d\bar{w} d\bar{u}' d\bar{v}' d\bar{w}' d\omega dt \quad \text{......(7).}$$

19. These collisions will all belong to class β, provided that the limits determined by $d\bar{u}$, $d\bar{v}$, etc., are such that the values of u, v given by equations (5) and (6) lie within the appropriate limits u and $u+du$, v and $v+dv$, etc. To obtain the whole number of collisions of class β we must integrate expression (7) over all values of $\bar{u}$, $\bar{v}$, etc., such that the values of u, v, etc., lie within these limits.

To do this we need only consider the ratio of the two products of differentials $du dv dw du' dv' dw'$ and $d\bar{u} d\bar{v} d\bar{w} d\bar{u}' d\bar{v}' d\bar{w}'$. We use Jacobi's theorem that

$$d\bar{u} d\bar{v} d\bar{w} d\bar{u}' d\bar{v}' d\bar{w}' = \pm \Delta \, du dv dw du' dv' dw' \quad \text{..............(8),}$$

where Δ denotes the determinant

$$\begin{vmatrix} \frac{\partial \bar{u}}{\partial u}, & \frac{\partial \bar{u}}{\partial v}, & \frac{\partial \bar{u}}{\partial w}, & \frac{\partial \bar{u}}{\partial u'}, & \frac{\partial \bar{u}}{\partial v'}, & \frac{\partial \bar{u}}{\partial w'} \\ \frac{\partial \bar{v}}{\partial u}, & \frac{\partial \bar{v}}{\partial v}, & \frac{\partial \bar{v}}{\partial w}, & \text{etc.,} & \cdots & \\ \cdots & \cdots & \cdots & \cdots & \cdots & \cdots \\ \cdots & \cdots & \cdots & \cdots & \cdots & \cdots \\ \cdots & \cdots & \cdots & \cdots & \cdots & \cdots \end{vmatrix}.$$

Using the values given by equations (5) and (6) we find without trouble that $\Delta = -1$. The numerical value of Δ, which is all that is required, may in fact be seen without actual calculation. For since equations (5) and (6) are linear as regards the velocities, the value of the above determinant cannot depend on the velocities. Also since the relation between the velocities before and after collision is, on account of the reversibility of the motion, a reciprocal relation, it is clear from equation (8) that the only possible values for Δ are ± 1.

Hence, since the ratio in question must necessarily be positive in sign, equation (8) becomes

$$d\bar{u} d\bar{v} d\bar{w} d\bar{u}' d\bar{v}' d\bar{w}' = du dv dw du' dv' dw' \quad \text{..................(9);}$$

and expression (7) may be written in the form

$$\nu^2 f(\bar{u}, \bar{v}, \bar{w}) f(\bar{u}', \bar{v}', \bar{w}') V\sigma^2 \cos\theta \, du dv dw du' dv' dw' d\omega dt \quad \text{......(10).}$$

If this number of collisions is exactly to include all of class β, the values of du, dv, dw, du', dv', dw' must be those which occur in the specification of a collision of class α (p. 18) and therefore those which occur in expression (4).

20. Suppose that expression (4) is summed over all possible classes of collisions which can occur to a molecule of class A. Or, what is the same thing, suppose that expression (4) is integrated over all possible values of u', v', w' and over all elements of spherical surface $d\omega$ for which a collision is possible. Obviously the quantity obtained in this way will represent the total number of molecules of class A which enter into collision in the interval dt*. So also expression (10) integrated through the same range of values gives the total number of molecules of class A which emerge from collision during the same interval.

The net gain to class A in the interval dt is therefore the difference of these two integrals, and this is

$$\nu^2 du\,dv\,dw\,dt \iiiint\!\!\int (\bar{f}\bar{f}' - ff')\, V\sigma^2 \cos\theta\, du'\,dv'\,dw'\,d\omega \quad \text{.........(11)},$$

in which $f, f', \bar{f}, \bar{f}'$ are written for $f(u, v, w)$, $f(u', v', w')$, $f(\bar{u}, \bar{v}, \bar{w})$ and $f(\bar{u}', \bar{v}', \bar{w}')$ respectively.

21. The number of molecules which belong to class A at the beginning of the interval dt is known to have been $\nu f du\,dv\,dw$ per unit volume, whilst the number at the end of this interval may be supposed to be

$$\nu\left(f + \frac{\partial f}{\partial t} dt\right) du\,dv\,dw.$$

The gain to class A is therefore

$$\nu \frac{\partial f}{\partial t} du\,dv\,dw\,dt.$$

Equating this quantity to that given by expression (11), we obtain the equation

$$\frac{\partial f}{\partial t} = \nu \iiiint\!\!\int (\bar{f}\bar{f}' - ff')\, V\sigma^2 \cos\theta\, du'\,dv'\,dw'\,d\omega \quad \text{...........(12)}.$$

The condition for a steady state is that $\partial f/\partial t$ shall vanish for all values of u, v and w. No progress can however be made by equating the right-hand side of equation (12) to zero: the problem of determining the steady state has to be attacked in a different manner.

The H-theorem.

22. Consider the quantity H defined by

$$H = \iiint f \log f\, du\,dv\,dw \quad \text{.......................(13)},$$

in which the integration is to extend over all possible values of u, v, w, so that H is a pure quantity and not a function of u, v, w. This quantity depends

* Not the total number of collisions in which a molecule initially of class A is involved, since collisions for which both molecules are of class A are counted twice.

solely upon the law of distribution of velocities and therefore remains unchanged so long as this law remains unchanged. Hence a necessary condition for a steady state is given by $dH/dt = 0$. We proceed to evaluate dH/dt in the general case.

After an interval dt the value of $f \log f$ corresponding to any specified values of u, v, w will of course have changed into

$$f \log f + \frac{\partial}{\partial t}(f \log f)\, dt,$$

or, what is the same thing, into

$$f \log f + (1 + \log f) \frac{\partial f}{\partial t}\, dt.$$

Hence the increase in H, which may be written $\frac{dH}{dt}\, dt$, will be given by

$$\frac{dH}{dt}\, dt = \left\{ \iiint (1 + \log f) \frac{\partial f}{\partial t}\, du\, dv\, dw \right\} dt \quad \text{..............(14)},$$

and if we substitute the value of $\partial f/\partial t$ from equation (12), this becomes

$$\frac{dH}{dt} = \nu \iiiint\!\!\!\iiiint (1 + \log f)(\bar{f}\bar{f}' - ff')\, V\sigma^2 \cos\theta\, du\, dv\, dw\, du'\, dv'\, dw'\, d\omega \text{...(15)}.$$

23. Equation (13) regards H as the sum of a number of contributions, one from each class of molecule, and in this equation class A is taken as the typical class. If we had chosen class B as the typical class, we might have written H in the form

$$H = \iiint f' \log f'\, du'\, dv'\, dw' \quad \text{.....................(16)},$$

and the increase in H, instead of being given by equation (14), would then have been given by

$$\frac{dH}{dt} = \iiint (1 + \log f') \frac{\partial f'}{\partial t}\, du'\, dv'\, dw' \quad \text{.................(17)}.$$

To evaluate the right-hand member of this equation we need to know the value of $\partial f'/\partial t$. Now equation (12) regards the change in f as the sum of a number of contributions, one from every class of collision in which either of the molecules either before or after impact is of class A, and the typical classes of collision are taken to be classes α and β. In a similar way we can express $\partial f'/\partial t$ as the sum of a number of contributions, one from every class of collision in which either of the molecules either before or after impact is of class B. The typical classes of collision may again be taken to be classes α and β, and if this is done we obtain for $\partial f'/\partial t$ an expression similar to that given for $\partial f/\partial t$ in equation (12), except that accented symbols replace unaccented, and vice versa. In fact molecules of classes A and B exchange rôles.

If we now substitute this value for $\partial f'/\partial t$ in equation (17) we obtain (cf. equation 15)

$$\frac{dH}{dt} = \nu \iiiint\!\!\!\iiiint (1 + \log f')(\bar{f}\bar{f}' - ff')\, V\sigma^2 \cos\theta\, du\,dv\,dw\,du'\,dv'\,dw'\,d\omega,$$

an equation which is of course the same as (15) except that accented and unaccented symbols are interchanged. If we add together the two values for dH/dt which have been obtained, we have

$$\frac{dH}{dt} = \tfrac{1}{2}\nu \iiiint\!\!\!\iiiint (2 + \log ff')(\bar{f}\bar{f}' - ff')\, V\sigma^2 \cos\theta\, du\,dv\,dw\,du'\,dv'\,dw'\,d\omega \quad \text{...(18)}.$$

This equation expresses dH/dt as the sum of a number of contributions, one from every possible class of collision. The typical class of collision is taken to be class α, in which

$$u,\ v,\ w,\ u',\ v',\ w'$$

become changed into

$$\bar{u},\ \bar{v},\ \bar{w},\ \bar{u}',\ \bar{v}',\ \bar{w}'.$$

If we use the same equation, but take as the typical collision one of class β, in which

$$\bar{u},\ \bar{v},\ \bar{w},\ \bar{u}',\ \bar{v}',\ \bar{w}'$$

become changed into

$$u,\ v,\ w,\ u',\ v',\ w',$$

we obtain, as a still different form for dH/dt,

$$\frac{dH}{dt} = \tfrac{1}{2}\nu \iiiint\!\!\!\iiiint (2 + \log \bar{f}\bar{f}')(ff' - \bar{f}\bar{f}')\, V\sigma^2 \cos\theta\, d\bar{u}\,d\bar{v}\,d\bar{w}\,d\bar{u}'\,d\bar{v}'\,d\bar{w}'\,d\omega \quad \text{...(19)}.$$

Equation (9) enables us to replace the product of the first six differentials on the right-hand of this equation by $du\,dv\,dw\,du'\,dv'\,dw'$, and if we add this modified value of dH/dt to that given by equation (18) we obtain

$$\frac{dH}{dt} = \tfrac{1}{4}\nu \iiiint\!\!\!\iiiint (\log ff' - \log \bar{f}\bar{f}')(\bar{f}\bar{f}' - ff')\, V\sigma^2 \cos\theta\, du\,dv\,dw\,du'\,dv'\,dw'\,d\omega \quad \text{.........(20)}.$$

Now $(\log ff' - \log \bar{f}\bar{f}')$ is positive or negative according as ff' is greater or is less than $\bar{f}\bar{f}'$ and is therefore always of the sign opposite to that of $\bar{f}\bar{f}' - ff$. Hence the product

$$(\log ff' - \log \bar{f}\bar{f}')(\bar{f}\bar{f}' - ff'),$$

if not zero, is necessarily negative. Since $V \cos\theta$, the relative velocity along the line of centres before impact, is necessarily positive for every type of collision, it follows that the integrand of equation (20) is always either negative or zero. Hence equation (20) shews that dH/dt is either negative or zero.

The Solution for a Steady State.

24. In order that the gas may be in a steady state, it is necessary, as has been already said, that dH/dt shall be zero. Now equation (20), as we have seen, expresses dH/dt as the sum of a number of contributions, one from every type of collision, and each contribution is either negative or zero. Hence for dH/dt to be zero, each contribution separately must be zero. In other words we must have

$$ff' = \bar{f}\bar{f}' \quad \text{.................................(21)}$$

for every type of collision.

This condition has been seen to be *necessary* for a steady state. Equation (12) shews that it is also *sufficient*, for if it is satisfied then $\partial f/\partial t = 0$ for every value of u, v and w. The problem of determining the steady state is therefore reduced to the problem of obtaining the solution of equation (21). We shall find it convenient to take logarithms of both sides, and write the equation in the form

$$\log f + \log f' = \log \bar{f} + \log \bar{f}' \quad \text{.....................(22).}$$

25. Let χ be a function of the velocities u, v, w, such that when two molecules collide, the sum of the χ's appropriate to the two molecules before impact is equal to the sum of the two χ's after impact. Since χ is, by hypothesis, a function only of u, v, w, the value of χ will remain the same for every molecule except when it is altered by collision. We may therefore say that χ is defined as being capable of exchange between molecules at a collision, but is indestructible; $\Sigma\chi$ remains the same throughout the motion, where Σ denotes summation which extends over all the molecules of the gas.

It is clear that a particular solution of equation (22) is

$$\log f = \chi.$$

Further it will be seen that the difference between χ and the most general solution of (22) is such as to satisfy the conditions postulated for χ. Let $\chi_1, \chi_2, \chi_3, \ldots$ be independent quantities, each satisfying these conditions, and let it be supposed that there are no other such quantities, then the most general solution of (22) must be

$$\log f = \alpha_1\chi_1 + \alpha_2\chi_2 + \alpha_3\chi_3 + \ldots \quad \text{.....................(23),}$$

in which $\alpha_1, \alpha_2, \alpha_3, \ldots$ are independent and, so far, arbitrary constants.

From the dynamics of a collision we know that there are four quantities which satisfy the condition in question: namely, the energy and the three components of linear momentum. These give four forms for χ: a fifth is obtained by taking χ equal to a constant, and it is obvious that there can be no others. For if there were any additional form possible for χ, there would be five equations giving $\bar{u}, \bar{v}, \bar{w}, \bar{u}', \bar{v}', \bar{w}'$ in terms of u, v, w, u', v', w', so that

$\bar{u}, \bar{v}, \bar{w}, \bar{u}', \bar{v}', \bar{w}'$ would be determined except for one unknown. There must however be two unknowns, as the direction of the line of centres is unknown.

The general solution of equation (22) is therefore seen to be

$$\log f = \alpha_1 m (u^2 + v^2 + w^2) + \alpha_2 mu + \alpha_3 mv + \alpha_4 mw + \alpha_5 \quad \text{......(24)}.$$

The constants $\alpha_2, \alpha_3, \alpha_4, \alpha_5$ may be replaced by new ones and the solution written in the form

$$\log f = \alpha_1 m [(u - u_0)^2 + (v - v_0)^2 + (w - w_0)^2] + \alpha_6,$$

or, if we still further change the constants,

$$f = Ae^{-hm[(u-u_0)^2+(v-v_0)^2+(w-w_0)^2]} \quad \text{.....................(25)},$$

in which A, h, u_0, v_0, w_0 are new arbitrary constants.

26. By giving different values to these five constants we obtain all the steady states which are possible for a gas. The different values of the constants depend upon the different values of $\Sigma\chi_1, \Sigma\chi_2, \Sigma\chi_3, \Sigma\chi_4, \Sigma\chi_5$, *i.e.*, upon the total energy, momentum and mass of the gas. We proceed to determine the relations between these constants and the corresponding physical quantities.

The value per unit volume of any quantity χ summed over all the molecules is given by

$$\Sigma\chi = \nu \iiint \chi A e^{-hm[(u-u_0)^2+(v-v_0)^2+(w-w_0)^2]} du\, dv\, dw \quad \text{......(26)}.$$

If we write

$$u - u_0 = \mathsf{u},$$
$$v - v_0 = \mathsf{v},$$
$$w - w_0 = \mathsf{w},$$

this becomes transformed into

$$\Sigma\chi = \nu \iiint \chi A e^{-hm(\mathsf{u}^2+\mathsf{v}^2+\mathsf{w}^2)} d\mathsf{u}\, d\mathsf{v}\, d\mathsf{w} \quad \text{..............(27)},$$

and if we further transform variables according to the scheme

$$\left.\begin{aligned} \mathsf{u} &= c \sin\theta \cos\phi \\ \mathsf{v} &= c \sin\theta \sin\phi \\ \mathsf{w} &= c \cos\theta \end{aligned}\right\} \quad \text{.......................(28)},$$

the equation becomes

$$\Sigma\chi = \nu \iiint \chi A e^{-hmc^2} c^2 \sin\theta \, d\theta \, d\phi \, dc \quad \text{.................(29)}.$$

If we take $\chi = 1$, $\Sigma\chi$ is the number of molecules per unit volume, and is therefore equal to ν. Equation (29) accordingly becomes

$$1 = 4\pi A \int_0^\infty e^{-hmc^2} c^2 dc,$$

and since the value of the integral is known* to be $\frac{1}{4}\sqrt{\frac{\pi}{h^3m^3}}$ this gives the relation

$$A = \sqrt{\frac{h^3m^3}{\pi^3}} \quad\text{.................................(30).}$$

Next put $\chi = \mathsf{u}$ in equation (27). We obtain

$$\Sigma\mathsf{u} = \nu A \int_{-\infty}^{+\infty} e^{-hm\mathsf{u}^2}\mathsf{u}\,d\mathsf{u} \int_{-\infty}^{+\infty} e^{-hm\mathsf{v}^2}d\mathsf{v} \int_{-\infty}^{+\infty} e^{-hm\mathsf{w}^2}d\mathsf{w},$$

and the right-hand vanishes, since the value of the first integral is zero. Hence $\Sigma\mathsf{u} = 0$, or, what is the same thing,

$$\Sigma u = \Sigma u_0 = \nu u_0.$$

Thus u_0 is the mean value of u, and is therefore the x-component of the mean-velocity of the gas. Similarly v_0, w_0 are the y and z components of this velocity.

Lastly, let us put $\chi = \mathsf{u}^2 + \mathsf{v}^2 + \mathsf{w}^2$. If we substitute this value in equation (29) we obtain

$$\Sigma(\mathsf{u}^2 + \mathsf{v}^2 + \mathsf{w}^2) = 4\pi\nu A \int_0^\infty e^{-hmc^2}c^4 dc.$$

The value of the integral on the right-hand is known* to be $\frac{3}{8}\sqrt{\frac{\pi}{h^5m^5}}$, and this leads to

$$\Sigma(\mathsf{u}^2 + \mathsf{v}^2 + \mathsf{w}^2) = \tfrac{3}{2}A\sqrt{\frac{\pi^3}{h^5m^5}}\,\nu,$$

or, on substituting the value of A from equation (30),

$$\Sigma(\mathsf{u}^2 + \mathsf{v}^2 + \mathsf{w}^2) = \frac{3}{2hm}\nu \quad\text{.........................(31).}$$

The kinetic energy per unit volume of the gas is $\Sigma\frac{1}{2}m(u^2 + v^2 + w^2)$, and since $\Sigma\mathsf{u} = \Sigma\mathsf{v} = \Sigma\mathsf{w} = 0$, we have

$$\begin{aligned}\Sigma\tfrac{1}{2}m(u^2 + v^2 + w^2) &= \tfrac{1}{2}m\Sigma((\mathsf{u} + u_0)^2 + (\mathsf{v} + v_0)^2 + (\mathsf{w} + w_0)^2)\\ &= \tfrac{1}{2}m\Sigma(\mathsf{u}^2 + \mathsf{v}^2 + \mathsf{w}^2 + u_0^2 + v_0^2 + w_0^2)\\ &= \tfrac{1}{2}m\nu\left(\frac{3}{2hm} + u_0^2 + v_0^2 + w_0^2\right)\\ &= \frac{3}{4h}\nu + \tfrac{1}{2}\rho(u_0^2 + v_0^2 + w_0^2) \quad\text{.........................(32),}\end{aligned}$$

where ρ is the mass-density of the gas, given by equation (1).

We have now determined five relations between the unknown constants and the density, kinetic energy and momenta of the gas. It appears that for

* For the value of this and similar integrals, see Appendix A at the end of the book.

given density, kinetic energy and momenta the values of the constants are unique, h being determined by equation (32), u_0, v_0, w_0 by the momenta, and A by equation (30). Hence there is only one steady state possible for given values of the density, energy and momenta.

Gas in a closed vessel.

27. This completes the determination of the steady states of an infinite mass of gas. We have next to consider the modifications introduced when the gas is confined in a closed vessel. Supposing the walls of this vessel to be absolutely rigid and elastic, we shall shew that the law of distribution already found in equation (25), namely

$$f = Ae^{-hm[(u-u_0)^2+(v-v_0)^2+(w-w_0)^2]} \quad \text{.....................(33)},$$

will still represent a steady state, independently of the shape of the containing vessel, provided that this vessel is moving with a velocity u_0, v_0, w_0.

To prove this we consider the collisions of molecules with a single small element of the wall of the containing vessel. Let this element be of area dS and let the direction cosines of a line drawn perpendicular to it be l, m, n. Consider the class of collisions such that the components of velocity of the colliding molecule before impact lie between

$$u \text{ and } u + du,\ v \text{ and } v + dv,\ w \text{ and } w + dw \quad \text{...........(34)}.$$

As before, let all such molecules be called molecules of class A. Let us, as on a former occasion (§ 13), take a fixed point O as origin and represent the velocities of the different molecules in magnitude and direction by a system of lines drawn from this point. All the molecules of class A will be represented by lines having their representative points inside a certain small rectangular parallelepiped—the rectangular parallelepiped of which the orthogonal coordinates lie within the limits (34) (see fig. 2 opposite).

Let P be any one of these points, so that OP represents the velocity of the corresponding molecule. Let OR represent the velocity u_0, v_0, w_0 of the vessel, then RP will represent the velocity of the molecule relatively to the vessel. After collision with the element dS, the normal component of this velocity will be reversed, while the tangential component will persist unaltered. Hence if TRS is a plane through R parallel to the element dS, the relative velocity after impact is RP', where P' is the image of P in the plane TRS.

The small parallelepiped in which P must lie if the corresponding molecule is to belong to class A will have as its image in the plane TRS a second parallelepiped which is obviously of the same volume as the former. Let us denote the two parallelepipeds by α, β, and when the velocity of a molecule is such that the line representing it has its end within β, let us say that the molecule is of class C. Then we have seen that a molecule of class A is changed by collision into a molecule of class C, and from symmetry it is obvious that the converse is true.

The number of molecules of class A which collide with the element dS in time dt is equal to the number of molecules of class A which lie within a certain cylinder at the beginning of the interval dt. Similarly the number of molecules of class C which collide with dS in the same time dt is equal to the number of molecules of class C which lie within a second cylinder at the same instant.

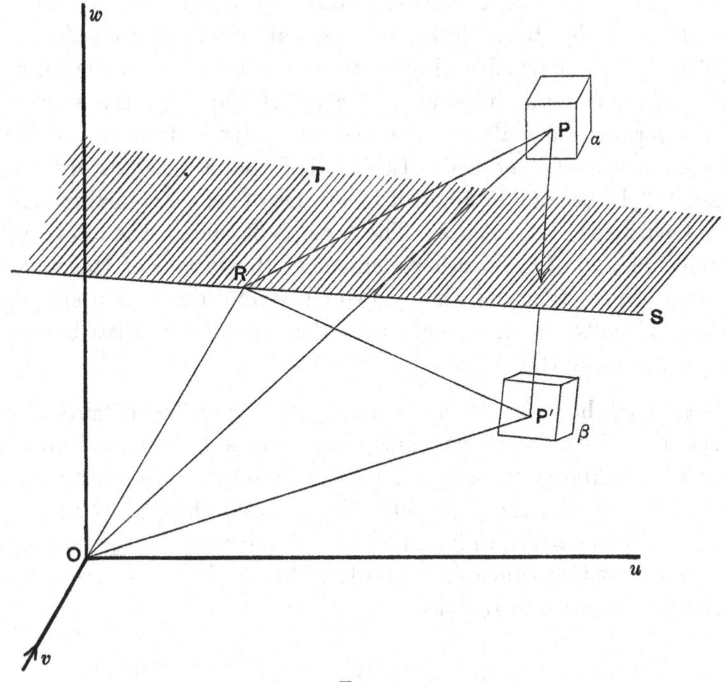

FIG. 2.

The former cylinder is of base dS and of a height equal to the distance described by a molecule of class A in time dt measured normal to the element dS. The second cylinder is of the same base dS, but of height equal to the distance described by a molecule of class C in time dt measured normal to dS. Since the normal velocity is the same for a molecule of class A as for one of class C, the heights of these two cylinders are the same, and since their bases are the same, their volumes will be the same.

The density of molecules of class A in the first cylinder is, in accordance with equation (33),

$$\nu A e^{-hm[(u-u_0)^2+(v-v_0)^2+(w-w_0)^2]}\,du\,dv\,dw,$$

and since in fig. 2 the coordinates of P are u, v, w while those of R are u_0, v_0, w_0, we have

$$RP^2 = (u - u_0)^2 + (v - v_0)^2 + (w - w_0)^2,$$

and the density of molecules of class A in the first cylinder is

$$\nu A e^{-hmRP^2} \times \text{volume of } \alpha \quad\quad (35).$$

Similarly the density of molecules of class B in the second cylinder is

$$\nu A e^{-hmRP'^2} \times \text{volume of } \beta \quad\quad (36).$$

Now $RP = RP'$, and the volume of α has been seen to be equal to the volume of β. Hence the two densities (35) and (36) are equal. Since the two cylinders have also been shewn to be equal it follows that the number of molecules of class A which collide with the element dS in the interval dt is equal to the number of molecules of class C which do the same thing. Each of the former molecules is changed by collision from a molecule of class A to one of class C, and each of the latter from a molecule of class C to one of class A. Hence the number of molecules of class A remains unaltered by collisions with the element dS. The same is of course true of every other class of molecule, and of every other element of the surface of the containing vessel, whence we see that the whole law of distribution is unaltered by the presence of the walls, or, in other words, that the law of distribution (33) represents a steady state.

It now appears, however, that there are only the two constants A and h at our disposal in the case of a gas enclosed in a vessel which is either at rest or moving with a known velocity u_0, v_0, w_0, and these two constants are of course connected by the relation (30). By varying these constants we are enabled to assign to our gas different values of the total energy, or, speaking physically, different temperatures. Similarly by varying ν we are enabled to assign different densities to the gas.

Mass Motion and Molecular Motion.

28. We have seen that the most general "steady state" possible consists of a motion compounded of a mass-motion and a molecular-motion. The mass-motion has velocity components u_0, v_0, w_0, the molecular-motion has velocity components $u - u_0$, $v - v_0$, $w - w_0$, which we have denoted (p. 26) by $\mathsf{u}, \mathsf{v}, \mathsf{w}$. The number of molecules having molecular velocities lying between u and $\mathsf{u} + d\mathsf{u}$, v and $\mathsf{v} + d\mathsf{v}$, w and $\mathsf{w} + d\mathsf{w}$ is the number of molecules having actual resultant velocities lying between u and $u + d\mathsf{u}$, etc., etc., and this by equation (25)

$$= A e^{-hm[(u-u_0)^2+(v-v_0)^2+(w-w_0)^2]}\, d\mathsf{u}\, d\mathsf{v} d\mathsf{w}$$

$$= A e^{-hm(\mathsf{u}^2+\mathsf{v}^2+\mathsf{w}^2)}\, d\mathsf{u}\, d\mathsf{v} d\mathsf{w}.$$

Hence we may suppose the molecular velocities distributed according to the law

$$Ae^{-hm(u^2+v^2+w^2)}\,du\,dv\,dw \quad\text{.......................(37).}$$

If we adopt the scheme of transformation (28) we may replace the velocity of which the components are u, v, w by a velocity of magnitude c, in a direction which makes an angle θ with the axis of z, and such that a plane through this direction and the axis of z makes an angle ϕ with the axis of x. The law of distribution (37) now becomes

$$Ae^{-hmc^2}c^2\sin\theta\,d\theta\,d\phi\,dc \quad\text{..........................(38).}$$

This shews that the velocities of molecular motion are distributed equally in

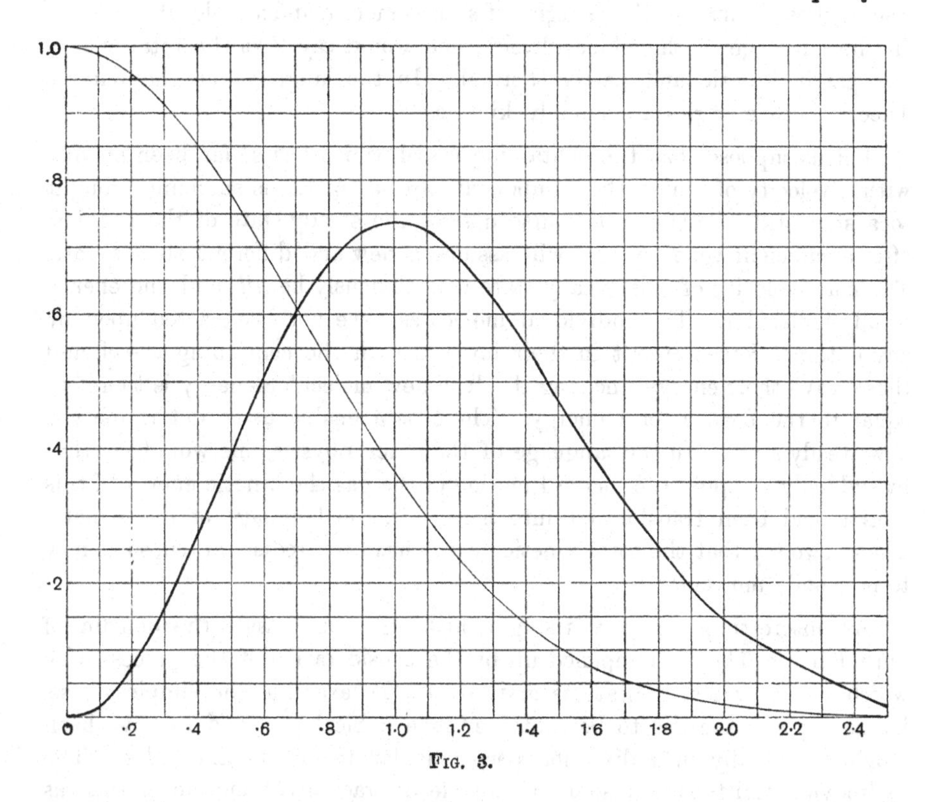

Fig. 3.

all directions in space. The law of distribution of the magnitudes of these velocities independently of their directions in space is found, upon integration of expression (38) with respect to θ and ϕ, to be

$$4\pi Ae^{-hmc^2}c^2dc \quad\text{...............................(39).}$$

The law indicated by expression (37) can also be written in the form

$$A\,[e^{-hmu^2}du]\,[e^{-hmv^2}dv]\,[e^{-hmw^2}dw] \quad\text{.................(40),}$$

shewing that the distributions of u, v, w are independent.

In fig. 3, the thick line is the curve

$$y = 2x^2 e^{-x^2},$$

while the thin line is the curve

$$y = e^{-x^2}.$$

The factor 2 is introduced in the former equation in order that the two curves may have the same area, namely $\frac{1}{2}\sqrt{\pi}$. The former curve shews the grouping of the magnitudes of the velocities independently of their directions in space, the latter that of the magnitudes of a single component.

29. Equation (32) expresses that the total energy of a gas may be regarded as the sum of the energies of a mass-motion and a molecular-motion. In the language of the older physics, one would say that the total energy was partly kinetic and partly thermal. In the language of the Kinetic Theory, both energies are equally kinetic.

Let us suppose that the containing vessel, which has so far been moving with a velocity of which the components are u_0, v_0, w_0, is suddenly brought to a standstill. This will of course destroy the steady state of the gas, but after a sufficient time, the gas will assume a new and different steady state. The mass-velocity of this steady state will obviously be *nil*, and the energy wholly molecular. The individual molecules have not been acted upon by any external forces except in their impacts with the containing vessel, and these leave their energy unchanged. The new molecular energy is therefore equal to the former total energy. These data enable us to determine the new steady state. In the language of the older physics, one would say that by suddenly stopping the forward motion of the gas the kinetic energy of this motion had been transformed into heat. In the language of the Kinetic Theory, we say that the total kinetic energy has been redistributed, so as now to be wholly molecular.

An interesting region of thought, although one outside the domain of pure Kinetic Theory, is opened up by the consideration of the processes by which this new steady state is arrived at. To examine the simplest case, let us suppose the gas to be contained in a cubical box, and to have been moving originally in a direction perpendicular to one of the sides. The hydrodynamical theory of sound is capable of tracing the motion of the gas throughout all time, subject of course to the assumptions on which the theory is based. The solution obtained to the problem from the hydrodynamical standpoint is that the original motion of the gas is perpetuated in the form of plane waves of sound in the gas, the wave fronts all being perpendicular to the original direction of motion. This solution is obviously very different from that arrived at by the Kinetic Theory. For instance, the solution of hydrodynamics indicates that the original direction of motion remains differentiated from other directions in space through all time, whereas the solution of the

Kinetic Theory indicates that a state is soon attained in which there is no differentiation between directions in space.

The explanation of the divergence of the two solutions is naturally to be looked for in the differences of the assumptions made. The conception of the perfect non-viscous fluid postulated by hydrodynamics is an abstract ideal which is logically inconsistent with the molecular constitution of matter postulated by the Kinetic Theory. Indeed we shall in a later part of the book be able to shew that the actual viscosity of gases is simply and fully accounted for by their molecular structure. If we introduce viscosity terms into the hydrodynamical equations, these equations will lead to a solution in which the ultimate state is one in which there is no mass-motion in the gas. On the hydrodynamical view, the energy of the original motion has been "dissipated" by viscosity. On the Kinetic Theory view, this energy has been converted into molecular motion. In fact the Kinetic Theory enables us to trace as molecular motion, energy which other theories are content to regard as lost from sight.

Gas devoid of Mass-motion.

30. In what follows we may be content to neglect the complication of mass-motion. We shall suppose that the whole motion of the molecules consists of its molecular motion of components u, v, w, and we shall put

$$u^2 + v^2 + w^2 = c^2 \qquad \text{(41)}.$$

Then we have seen (equations (25) and (30)) that out of N molecules under consideration, the number having a velocity with components between u and $u + du$, v and $v + dv$, w and $w + dw$ is

$$N \left(\frac{hm}{\pi}\right)^{\frac{3}{2}} e^{-hm(u^2+v^2+w^2)}\, du\, dv\, dw \qquad \text{(42)},$$

while the number having a resultant velocity between c and $c + dc$ is

$$4\pi N \left(\frac{hm}{\pi}\right)^{\frac{3}{2}} e^{-hmc^2} c^2 dc \qquad \text{(43)}.$$

The mean velocity of all the molecules will be the average value of c, and may be denoted by $\bar{c}$. It is given by

$$\bar{c} = 4\pi \left(\frac{hm}{\pi}\right)^{\frac{3}{2}} \int_0^\infty e^{-hmc^2} c^3 dc$$

$$= \frac{2}{\sqrt{(\pi hm)}} \qquad \text{(44)}.$$

It is convenient to introduce a velocity C, defined as being such that the mean value of c^2 is equal to C^2. The mean kinetic energy of translation of a molecule is then $\frac{1}{2}mC^2$, and the total kinetic energy per unit volume of the gas is $\frac{1}{2}\nu mC^2$ or $\frac{1}{2}\rho C^2$; it is the same as though the mass ρ were moving forward with a velocity C.

From equation (31) νC^2 is equal to $\frac{3}{2hm}\nu$, so that

$$C^2 = \frac{3}{2hm} \quad \text{.................................(45).}$$

In terms of C, the average velocity $\bar{c}$ is given by

$$\bar{c} = \frac{2}{\sqrt{(\pi hm)}} = \sqrt{\left(\frac{8}{3\pi}\right)} C = \cdot 921 C \quad \text{..................(46).}$$

Thus the kinetic energy (and, as we shall see, the pressure also) is the same as if each molecule had a velocity C equal to 1·086 times the average velocity.

31. It is frequently convenient, in obtaining rough approximations to the solution of a physical problem, to assume that all the molecules have exactly the same velocity, and in order that this velocity may be consistent with the actual values of the pressure and of the kinetic energy, this uniform velocity must be supposed to be C.

Some idea of the amount of error involved in this approximation may be obtained from a study of fig. 3. Since $C^2 = 3/(2hm)$, and since hm is taken equal to unity in drawing the curves of fig. 3, the approximation amounts to the assumption that the whole area of the curve is collected close to the abscissa

$$x = \sqrt{\frac{3}{2}} = 1 \cdot 225.$$

The approximation is thus seen to be a very rough one.

32. It is sometimes required to know the number of molecules which at any instant have a speed greater than a given speed c_0. Out of a total of N molecules the number which have a speed in excess of c_0 is, by formula (43),

$$4\pi N \left(\frac{hm}{\pi}\right)^{\frac{3}{2}} \int_{c_0}^{\infty} e^{-hmc^2} c^2 dc,$$

and, on integration by parts, this becomes

$$2N \left(\frac{hm}{\pi}\right)^{\frac{1}{2}} \left\{ \int_{c_0}^{\infty} e^{-hmc^2} dc + c_0 e^{-hmc_0^2} \right\}.$$

In terms of the probability integral, or error function defined by

$$\text{erf}\, x = \frac{2}{\sqrt{\pi}} \int_x^{\infty} e^{-x^2} dx,$$

this number becomes

$$N\left(\operatorname{erf} x + \frac{2}{\sqrt{\pi}} x e^{-x^2}\right) \quad\quad (47),$$

where

$$x = (hm)^{\frac{1}{2}} c_0 = \sqrt{\frac{3}{2}}\left(\frac{c_0}{C}\right).$$

From a table* of values of $\operatorname{erf} x$, it is easy to calculate the number of molecules either having velocity greater than any value c_0, or within any range c_0 to c_1. The general run of the numbers to be expected will be sufficiently seen by an inspection of the curves of fig. 3.

Number of Collisions, Mean Free Path, etc.

33. We shall now use the results which have been obtained to calculate the total number of collisions per unit volume of gas. The number of collisions is not affected by the mass-motion of the gas, so that we may continue to take this mass-motion to be zero.

In expression (4) we found the number of collisions of class α occurring per unit time to be

$$\nu^2 f(u, v, w) f(u', v', w') V\sigma^2 \cos\theta \, du\, dv\, dw\, du'\, dv'\, dw'\, d\omega \quad\quad (48),$$

and the problem of determining the total number of collisions amounts to integrating this expression over all values of the variables when $f(u, v, w)$ has the form appropriate to the steady state, *i.e.*, when

$$f(u, v, w) = \sqrt{\frac{h^3 m^3}{\pi^3}}\, e^{-hmc^2} \quad\quad (49).$$

In expression (48), V is the relative velocity and θ is the angle between this velocity and that of the line of centres. If ϕ is the azimuth of the line of centres referred to any definite plane through the direction of the relative velocity, we may, in expression (48), replace $d\omega$ by $\sin\theta\, d\theta\, d\phi$. Since collisions can occur for all values of ϕ and for all values of θ from 0 to $\pi/2$, we must integrate expression (48) from $\phi = 0$ to $\phi = 2\pi$ and from $\theta = 0$ to $\theta = \pi/2$. Performing the integrations, and substituting for $f(u, v, w)$ from equation (49), we obtain

$$\pi\nu^2\left(\frac{h^3 m^3}{\pi^3} e^{-hm(c^2+c'^2)}\right) V\sigma^2 du\, dv\, dw\, du'\, dv'\, dw' \quad\quad (50),$$

as the total number of collisions in which the molecules before collision have velocities lying within the usual limits $du\, dv\, dw\, du'\, dv'\, dw'$.

Let us now suppose that the variables are transformed to new variables, given by

$$\mathbf{u} = \tfrac{1}{2}(u + u'), \text{ etc.,} \quad \alpha = u' - u, \text{ etc.,}$$

* Values of $1 - \operatorname{erf} x$ are given in the fourth column of the table in Appendix B.

so that $\mathbf{u}, \mathbf{v}, \mathbf{w}$ are the components of the velocity of the centre of gravity of the two molecules, and α, β, γ are the components of the velocity of the second molecule relatively to the first. We have

$$\frac{\partial(\mathbf{u}, \alpha)}{\partial(u, u')} = \begin{vmatrix} \frac{1}{2}, & \frac{1}{2} \\ -1, & 1 \end{vmatrix} = 1,$$

so that
$$d\mathbf{u}\, d\mathbf{v}\, d\mathbf{w}\, d\alpha\, d\beta\, d\gamma = du\, dv\, dw\, du'\, dv'\, dw'.$$

Hence from expression (50) the number of collisions for which the new variables lie within a range $d\mathbf{u}\, d\mathbf{v}\, d\mathbf{w}\, d\alpha\, d\beta\, d\gamma$ is

$$\pi\nu^2 \left(\frac{h^3m^3}{\pi^3} e^{-hm(c^2+c'^2)}\right) V\sigma^2 d\mathbf{u}\, d\mathbf{v}\, d\mathbf{w}\, d\alpha\, d\beta\, d\gamma \quad \text{.........(51)},$$

in which we have

$$V^2 = (u'-u)^2 + (v'-v)^2 + (w'-w)^2 = \alpha^2 + \beta^2 + \gamma^2,$$

$$c^2 + c'^2 = u^2 + u'^2 + v^2 + v'^2 + w^2 + w'^2 = 2(\mathbf{u}^2 + \mathbf{v}^2 + \mathbf{w}^2) + \tfrac{1}{2}(\alpha^2 + \beta^2 + \gamma^2),$$

or, if we write $\mathbf{u}^2 + \mathbf{v}^2 + \mathbf{w}^2 = \mathbf{c}^2$,

$$c^2 + c'^2 = 2\mathbf{c}^2 + \tfrac{1}{2}V^2.$$

Let us again transform variables according to the schemes

$$\mathbf{u} = \mathbf{c}\sin\theta\cos\phi, \qquad \alpha = V\sin\psi\cos\chi,$$
$$\mathbf{v} = \mathbf{c}\sin\theta\sin\phi, \qquad \beta = V\sin\psi\sin\chi,$$
$$\mathbf{w} = \mathbf{c}\cos\theta, \qquad \gamma = V\cos\psi.$$

In order that $\mathbf{u}, \mathbf{v}, \mathbf{w}$ may have all possible values, θ must range from 0 to π, ϕ from 0 to 2π, and $\mathbf{c}$ from 0 to ∞. If, however, we give a similar range to the new variables in the second scheme of transformation, we shall be counting each collision twice over. For a collision in which α, β, γ have given values can, by merely changing the rôles of the two molecules, be regarded as a new collision in which the signs of α, β, γ are altered. This source of error can be eliminated by limiting the integration with respect to ψ from 0 to $\pi/2$, instead of from 0 to π. Hence we obtain, for the number for which $\mathbf{c}$ lies between $\mathbf{c}$ and $\mathbf{c} + d\mathbf{c}$, while V lies between V and $V + dV$,

$$8h^3m^3\nu^2\sigma^2 e^{-hm(2\mathbf{c}^2+\frac{1}{2}V^2)}\,\mathbf{c}^2 V^3 d\mathbf{c}\, dV.$$

Integrating with respect to $\mathbf{c}$ from 0 to ∞, the number of collisions for which V lies between V and $V + dV$ is

$$8h^3m^3\nu^2\sigma^2 \left[\int_0^\infty e^{-2hm\mathbf{c}^2}\mathbf{c}^2 d\mathbf{c}\right] e^{-\frac{1}{2}hmV^2} V^3 dV,$$

or
$$\nu^2\sigma^2\sqrt{\frac{\pi h^3m^3}{2}}\, e^{-\frac{1}{2}hmV^2} V^3 dV \quad \text{......................(52)},$$

a result which will be required later.

If we finally integrate this from $V=0$ to $V=\infty$, we obtain for the total number of collisions

$$\nu^2\sigma^2\sqrt{\frac{2\pi}{hm}} \quad\text{.................................(53),}$$

an expression which, on using relation (40), may be replaced by

$$\frac{\pi}{\sqrt{2}}\nu^2\sigma^2\bar{c}.$$

There are ν molecules per unit volume, and each collision terminates two free paths. Hence the ν molecules describe

$$\sqrt{2}\pi\nu^2\sigma^2\bar{c} \quad\text{.................................(54)}$$

free paths per unit time.

The average duration of a free path is accordingly

$$\frac{1}{\sqrt{2}\pi\nu\sigma^2\bar{c}} \quad\text{.................................(55).}$$

The distance described per unit time by the ν molecules occupying unit volume is $\nu\bar{c}$, and this distance is the aggregate of all the free paths, of which the number is given by expression (54). By division we find as the length of the mean free path

$$\frac{1}{\sqrt{2}\pi\nu\sigma^2}=\frac{\cdot 7071..}{\pi\nu\sigma^2} \quad\text{...........................(56).}$$

If the average is taken in any other way, the result is of course different. We might for instance average over all the free paths which are being described at a particular instant of time.

Tait* took a particular instant of time, and defined the mean free path as the average of the distances described by each molecule between this instant and the instant of its next collision. He calculated as the value of the mean free path defined in this way,

$$\frac{\cdot 677..}{\pi\nu\sigma^2} \quad\text{.................................(57),}$$

the factor ·677.. arising from an integral of which the value cannot be calculated in finite terms. We shall return later to the actual calculations by which this result is obtained.

From the results obtained in this section the numerical values given in § 8 can be calculated without trouble.

* *Royal Soc. Edin. Trans.* XXXIII. p. 74 (1886).

Apparent Irreversibility of Motion.

34. When a gas is not in a steady state, it follows from § 23 (p. 24) that dH/dt must be negative. Some writers have interpreted this to mean that H will continually decrease, until it reaches a minimum value, and will then retain that value for ever after. A motion of this kind would, however, be dynamically irreversible, and therefore inconsistent with the dynamical equations of motion from which it professes to have been deduced. As will appear later (cf. §§ 70—73), the truth is that we have at this point reached the limit within which the assumption of molecular chaos leads to accurate results. The motion is, in point of fact, strictly reversible, and the apparent irreversibility is merely an illusion introduced by the imperfections of the statistical method.

CHAPTER III

THE LAW OF DISTRIBUTION OF VELOCITIES (*continued*)

II. THE METHOD OF STATISTICAL MECHANICS.

The Conception of a Generalised Space.

35. IN the last chapter it was twice found convenient to represent the three velocity coordinates u, v, w of a molecule, by a point in space of which the coordinates referred to three rectangular axes were u, v, w. The principle involved is a useful one, capable of almost indefinite extension, and will be largely used both in the present chapter and elsewhere in the book.

The space of nature possesses three dimensions, but just as it is open for us to represent any two coordinates in an imaginary space of only two dimensions, so in the same way we may represent any four coordinates in an imaginary space of four dimensions. Similarly if a dynamical system is specified by any number n of coordinates, we can represent these coordinates in a space of n dimensions, and the various points in this space will correspond to the various configurations of the dynamical system.

In the present chapter, we attempt to find the law of distribution of velocities by a method which consists essentially in regarding the whole gas as a single dynamical system, and in representing its coordinates in a single imaginary space of the appropriate number of dimensions.

Let us suppose that the gas consists of a great number N of exactly similar molecules, enclosed in a vessel of volume Ω. At the outset we shall suppose these molecules to be elastic spheres of the kind already described. Each molecule will possess six coordinates, the three positional coordinates of its centre referred to three fixed rectangular axes in the containing vessel, and the three components of the velocity of its centre resolved parallel to these three axes. We shall denote the separate molecules by the letters A, B, C, etc., and the six coordinates of molecule A will be denoted by $x_a, y_a, z_a, u_a, v_a, w_a$. The whole gas may accordingly be regarded as a single dynamical system possessing $6N$ coordinates, namely,

$$x_a, y_a, z_a, u_a, v_a, w_a, x_b, y_b, z_b, u_b, v_b, w_b, x_c, \dots \text{ etc.} \quad \dots\dots(58).$$

We can suppose this dynamical system represented in a generalised space of $6N$ dimensions. The configuration of the system in which the coordinates are those given by (58) will be represented by a single point in this space, namely the point of which the coordinates referred to $6N$ rectangular axes are those given by (58).

36. In this way every possible configuration is represented by a point, but it does not follow that every point represents a possible configuration. For instance if as before the diameter of each molecule is σ, then a configuration in which

$$(x_a - x_b)^2 + (y_a - y_b)^2 + (z_a - z_b)^2 < \sigma^2 \quad \text{..............(59)}$$

is physically impossible because it represents a configuration in which the centres of molecules A and B are separated by a distance less than σ—a configuration, therefore, in which parts of these two molecules occupy the same space. We must therefore suppose the region in which the inequality (59) is satisfied to be excluded from our generalised space. If we shut off all such regions, found by substituting for a, b in (59) the suffixes corresponding to all possible pairs of molecules, we see that every point in the space which is left will represent a system which is physically possible in so far that no two molecules overlap.

There is still the boundary to be considered. For a configuration to be physically possible it is necessary that the centre of each molecule shall be at a normal distance from the boundary which is greater than the radius of the molecule. Thus all parts of the space must be excluded which do not satisfy equations of the form

$$\left.\begin{aligned}\phi(x_a, y_a, z_a) > \tfrac{1}{2}\sigma \ldots \text{etc.}\\ \phi(x_b, y_b, z_b) > \tfrac{1}{2}\sigma \ldots \text{etc.}\end{aligned}\right\} \quad \text{....................(60)},$$

where $\phi(x, y, z)$ is the shortest normal distance from x, y, z to the surface of the enclosing vessel.

37. If we exclude all the regions just indicated, it is clear that any point in the space which remains will represent a configuration of the system of molecules which is physically possible. In the course of the motion of the gas, this configuration will give place to other configurations, and by tracing out this series of configurations in the generalised space, we should obtain a "path" indicating the motion of the gas. By starting from a great number of points, and tracing the motion backwards as well as forwards, the whole space can be mapped out into paths in this way. Since the motion of the gas is completely determined when all the coordinates (58) are known, it follows that through any point there is one and only one path; two paths can never intersect. Also, of course, the paths are fixed in the generalised space;

the motion of a gas, starting from given values of velocity and position coordinates, is always the same. These paths are identical with the "trajectories" of abstract dynamics.

The Motion in the Generalised Space.

38. The general nature of these paths can be seen without trouble. A collision either of two molecules, or of a molecule and the boundary, occurs when, and only when, a path meets one of the surfaces of the regions excluded in § 36. Now between collisions every molecule moves with uniform velocity in a straight line. Thus if at time $t=0$, the coordinates of a system are

$$x_a', y_a', z_a', u_a', v_a', w_a', x_b', y_b', z_b', u_b', v_b' w_b' \quad \text{.........(61)},$$

the coordinates at time t, assuming that no collision has taken place in the interval, will be given by

$$x_a = x_a' + u_a't, \quad y_a = y_a' + v_a't, \text{ etc.,}$$
$$u_a = u_a', \quad v_a = v_a', \text{ etc.}$$

To find the equations of the path described by the representative point in the generalised space we eliminate t, and so obtain

$$\left.\begin{array}{l} \dfrac{x_a - x_a'}{u_a'} = \dfrac{y_a - y_a'}{v_a'} = \text{...... etc.} \\ u_a = u_a',\ v_a = v_a',\ w_a = w_a' \text{... etc.} \end{array}\right\} \quad \text{.................(62)},$$

and since these equations are linear they are of course the equations of a straight line. We therefore see that the paths in the generalised space are rectilinear except when they meet the excluded regions. Along the rectilinear parts of any paths, all the coordinates u_a, v_a, w_a, u_b ... etc., maintain constant values, and any series of paths for which these constant values are the same are all parallel. When a representative point, moving along one of these paths, meets a boundary of the excluded space—corresponding to a collision—it must be supposed to move along this boundary until it reaches the point of which the coordinates are those of the system after collision, and then to start from here and describe the new rectilinear path through this point.

39. Now in the gas of the Kinetic Theory, we do not know anything as to the coordinates of the individual molecules of the gas: the problem we have to attack is virtually that of finding as much as we can about the behaviour of a dynamical system, without knowing on which of the paths in our generalised space its representative point is moving.

Our method is therefore to start an infinite number of systems, each system being a complete gas of the kind already specified, so as to have systems starting from every conceivable configuration, and moving over every path. We then investigate, as far as possible, the motion of this series of systems, in the hope of finding features common to all. Or, what comes to exactly the same thing, we shall imagine our generalised space filled with a dust of points, so close together that they may be regarded as forming a continuous fluid; we shall suppose the different points of this dust or fluid to move along the paths in the generalised space as stream-lines, as directed by the dynamical equations of the gas, and we shall then examine the motion of this fluid.

It is obvious that the initial distribution of density of this fluid may be chosen quite arbitrarily. We therefore choose that the initial distribution shall be homogeneous. The advantage of this choice is that the fluid remains homogeneous throughout its subsequent motion. This result follows from a general theorem which will be proved later (§ 85), but we now proceed to give a separate proof for the special case at present under consideration.

40. It has been seen that throughout the motion which takes place between two collisions, all the velocity coordinates u_a, v_a, w_a, u_b ... etc., remain constant for any single path.

Consider a series of systems starting simultaneously with the same values of these velocity coordinates, but having positional coordinates lying between

$$x_a \text{ and } x_a + dx_a,\ y_a \text{ and } y_a + dy_a,\ z_a \text{ and } z_a + dz_a,$$

$$x_b \text{ and } x_b + dx_b \dots \text{ etc.} \qquad\qquad (63).$$

Let these systems move for a time dt, and let it be supposed that no collision occurs during this interval. Then it is clear that at the end of the interval the various positional coordinates will have values lying between

$$x_a + u_a dt \text{ and } x_a + u_a dt + dx_a \dots \text{ etc., etc.,}$$

while the velocity coordinates of course remain unaltered.

Hence the element of generalised space occupied by these systems remains unaltered in shape, size and orientation, but has in the course of the time dt moved parallel to itself a distance $u_a dt$ parallel to the axis of x_a, $v_a dt$ parallel to the axis of y_a, etc. It follows that the density of the fluid which occupies this element of generalised space has remained constant through this rectilinear motion.

Just as rectilinear motion leaves the velocity coordinates unchanged while altering the positional coordinates, so a collision leaves the positional coordinates unchanged while altering the velocity coordinates. There are two types of collisions to be discussed—collisions between molecules and the boundary, and collisions between pairs of molecules.

As a specimen of the former, consider a collision between molecule A and the boundary: this leaves all the coordinates unchanged except u_a, v_a, w_a. Considering a series of systems in which before collision all the coordinates except u_a, v_a, w_a have the same values for each member of the system, whilst u_a, v_a, w_a lie between u_a and $u_a + du_a$, v_a and $v_a + dv_a$, w_a and $w_a + dw_a$, we see that after collision all the coordinates will remain unaltered except u_a, v_a, w_a, while these will lie within a new set of limits. Now in fig. 2 (p. 29) we may suppose the former limits represented by the parallelepiped α, in which case the latter set will be represented by the parallelepiped β. These parallelepipeds have been shewn to be equal in size although in the present case the orientations are not the same. This proves that the size of the element of volume of generalised space occupied by the series of systems now under consideration is unaltered by a collision of molecule A with the boundary, and hence that the density of the fluid may be supposed to remain unaltered.

The case of a collision between a pair of molecules may be treated in the same way. If the molecules are A and B, all the coordinates remain unaltered except u_a, v_a, w_a, u_b, v_b, w_b, and the same result as before follows at once from equation (9) if we change the notation so as to replace u, v, w by u_a, v_a, w_a and u', v', w' by u_b, v_b, w_b. For this equation shews that $du_a dv_a dw_a du_b dv_b dw_b$ remains unchanged by the collision, so that the element of volume in the generalised space remains the same, and therefore also the density of the fluid inside it.

Examining the motion of any small element of fluid in our generalised space, we have now proved that the density of this element remains unchanged by steady motion and by collisions, *i.e.*, remains unchanged throughout the whole motion of the gas. It follows that if the whole generalised space is filled with fluid initially homogeneous, then this fluid will remain homogeneous throughout the entire motion. Or, again resolving the fluid into a dust of representative points, we have seen that there is no tendency for these points to crowd together, or to spread apart.

41. It has already been remarked that the stream-lines along which the fluid moves are permanently fixed in the generalised space. This fact, combined with the result just proved, shews that the motion of the fluid we are discussing is a "steady-motion" in the hydrodynamical sense.

One further feature of this motion must be noticed. If we denote the total kinetic energy of any system by E, so that

$$2E = m\left(u_a^2 + v_a^2 + w_a^2 + u_b^2 + \dots\right) \quad\text{.................(64)},$$

it is clear that E remains constant throughout the whole length of any stream-line. When E is a constant, equation (64), regarded as a relation between the Cartesian coordinates of a point in the generalised space, expresses that the point lies on a certain locus (of dimensions $6N-1$) in this space. It follows, then, that the motion of any element of the fluid is confined to that member of the family of loci $E = \text{constant}$, in which it started.

To obtain some idea of the disposition of this family of loci in our generalised space, we notice that $2E/m$ is the square of the perpendicular distance from

$$u_a = v_a = w_a = u_b = \ldots = 0,$$

or, what is the same thing, from $E = 0$. Hence the loci enclose one another, being in fact a system of tubular surfaces of which the cross-sections are spherical loci of $3N$ dimensions. The tubes do not extend to infinity along their length. For we pass along a generator of a tube by varying x_a, y_a, z_a, $x_b \ldots$ etc., and none of these coordinates can become infinite, because each molecule of the gas is supposed to be contained in a finite closed vessel. The surfaces $E =$ constant are therefore finite closed surfaces in the generalised space, the surface $E = \infty$ alone being, in the limit, infinite and enclosing all the others.

Hence the motion of the fluid in the generalised space is one of circulation in closed surfaces, and, in particular, there is no motion of the fluid across the boundary at infinity.

42. Similarly, if there were any other quantities $\chi_1, \chi_2 \ldots$, functions of the coordinates in the generalised space, which remained constant throughout the motion of the gas, then the motion of the fluid in the generalised space would be confined to the loci

$$\chi_1 = \text{constant}, \quad \chi_2 = \text{constant, etc.} \quad \ldots\ldots\ldots\ldots\ldots\ldots(65).$$

The only quantities of which we know, other than the energy, which remain constant over a collision between any two molecules, are the three components of linear momentum, the three moments of angular momentum and the number of molecules in the gas; of these the components of momentum both linear and angular are in general changed by a collision between a molecule and the boundary, and the number of molecules in the gas is not a function of the coordinates in the generalised space. Thus in general the energy is the only quantity of which we know, satisfying the conditions in question.

An exception to this may occur if the vessel containing the gas is a figure of revolution, having its interior surface perfectly smooth. For then there is always a component of momentum which is not changed by a collision between a molecule and the boundary; namely, that parallel to a tangent to the containing vessel at the point at which the collision takes place. In this case, then, the moment of momentum of the whole gas about the axis of figure of the containing vessel remains constant throughout the motion. It will, however, be convenient to defer the consideration of special cases of this type until Chapter V.

The Partition of the Generalised Space—Positional Coordinates.

43. We have supposed the volume of the containing vessel to be Ω. Let us now suppose the vessel divided up in a number n of "cells" each of the same volume ω, so that $n\omega = \Omega$. These cells will be referred to as cell 1, cell 2, ..., respectively. The different possible configurations of the gas may be classified according to the number of molecules of which the centres fall within the different cells. As a typical class, we consider a class such that a_1 molecules have their centres in cell 1, a_2 in cell 2, and so on. Let this class be referred to as class A. We proceed to examine what proportion of the whole of the generalised space represents systems of class A.

Let us, for the present, suppose that the radius of the molecules is vanishingly small, so that those parts of the generalised space excluded by the conditions of § 36 may be neglected. Then the representative points of systems which are such that the centre of the molecule A lies within a single specified cell,—*i.e.*, is restricted to a range ω out of the whole volume Ω of the containing vessel—will clearly occupy a fraction ω/Ω of the whole of the generalised space. Since $n\omega = \Omega$, this may be written n^{-1}. If two molecules A and B both lie within specified cells, the representative points occupy a fraction n^{-2} of the whole, and so on. Thus if each of the N molecules lies within a specified cell, the representative points will occupy a fraction n^{-N} of the whole of the generalised space.

Now the number of different ways in which the N molecules can be assigned to the n different cells, so that the system shall belong to class A, defined as above, is

$$\frac{N!}{a_1!\,a_2!\,a_3!\,\dots\,a_n!} \qquad (66),$$

in which, since the total number of molecules is N,

$$a_1 + a_2 + a_3 + \dots + a_n = N \qquad (67).$$

It follows that the representative points of systems of class A will occupy a fraction, say θ_a, of the whole of the generalised space, given by

$$\theta_a = \frac{N!\,n^{-N}}{a_1!\,a_2!\,a_3!\,\dots\,a_n!} \qquad (68).$$

Similarly the representative points of systems of any other class, say B, will occupy a fraction θ_b of the whole of the generalised space, given by

$$\theta_b = \frac{N!\,n^{-N}}{b_1!\,b_2!\,b_3!\,\dots\,b_n!}.$$

The sum of all such expressions must of course be equal to unity.

44. We have already supposed N to be great; we now suppose that $a_1, a_2, a_3 \dots a_n$ separately are very great. This enables us to express (68) in a simpler form, by using the well-known theorem of Stirling,

$$\underset{p=\infty}{\mathrm{L^t}}\ p! = \sqrt{2p\pi}\left(\frac{p}{e}\right)^p \qquad (69).$$

On taking logarithms of both sides, this becomes

$$\underset{p=\infty}{\mathrm{Lt}} \log p! = \tfrac{1}{2}\log 2\pi + (p+\tfrac{1}{2})\log p - p \quad\text{..............(70).}$$

Taking logarithms of both sides of equation (68),

$$\log \theta_a = \log N! - \sum_{s=1}^{s=n} \log a_s! - N\log n,$$

and in the limit, when $a_1, a_2 \ldots a_n, N$ are all infinite, this may, in virtue of equation (70), be replaced by

$$\log \theta_a = \tfrac{1}{2}\log 2\pi + (N+\tfrac{1}{2})\log N - N$$
$$- \sum_{s=1}^{s=n} \{\tfrac{1}{2}\log 2\pi + (a_s+\tfrac{1}{2})\log a_s - a_s\} - N\log n.$$

Now $\Sigma a_s = N$, so that $\Sigma(a_s+\tfrac{1}{2}) = N + \tfrac{1}{2}n$, and hence it will be found that the foregoing equation may be transformed into

$$\log \theta_a = \frac{n}{2}\log n - \frac{n-1}{2}\log 2\pi N - \sum_{s=1}^{s=n}(a_s+\tfrac{1}{2})\log\frac{na_s}{N} \quad\text{......(71).}$$

It is convenient to write

$$K_a = \frac{1}{N}\sum_{s=1}^{s=n}(a_s+\tfrac{1}{2})\log\frac{na_s}{N} \quad\text{......................(72),}$$

so that θ_a, the fraction of the generalised space which represents systems of class A, is given by

$$\theta_a = n^{\frac{1}{2}n}(2\pi N)^{-\frac{1}{2}(n-1)}e^{-NK_a} \quad\text{......................(73),}$$

an expression which, it will be noticed, involves the a's only through the term K_a in the exponential.

45. To examine the way in which θ_a varies for the different classes of configurations (A, B, etc.) it is sufficient to study the variations of K_a given by equation (72).

For the moment, let $a_1, a_2, \ldots$ be regarded no longer as integers, but as continuous variables, subject only to condition (67),

$$a_1 + a_2 + \ldots = N \quad\text{........ ....(74),}$$

and let K_a be treated as a continuous function of these variables, defined by equation (72). We search first for maximum or minimum values of K. By variation of equations (72) and (74),

$$\delta K = \frac{1}{N}\sum_{s=1}^{s=n}\left(\log\frac{na_s}{N} + 1 + \frac{1}{2a_s}\right)\delta a_s \quad\text{..................(75),}$$

$$0 = \sum_{s=1}^{s=n}\delta a_s \quad\text{...(76).}$$

The values of $a_1, a_2, \ldots$ for which K has stationary values will be those for which δK vanishes for all values of $\delta a_1, \delta a_2, \ldots$ which satisfy relation (76). The vanishing of δK therefore requires that the coefficients of $\delta a_1, \delta a_2, \ldots$ in equation (75) shall all be equal, and this in turn requires that $a_1, a_2, \ldots$ shall all be equal.

Thus there is only one set of values of the a's for which K is stationary, and it is given by

$$a_1 = a_2 = \ldots = \frac{N}{n} \quad \text{.........................(77)},$$

the quantity N/n being obtained from relation (74).

We next examine how K varies for values of a near to these values. An adjacent set of values will be

$$a_1 = \frac{N}{n} + \alpha_1, \quad a_2 = \frac{N}{n} + \alpha_2, \text{ etc.} \quad \text{.....................(78)},$$

where the α's are small compared with N/n, and

$$\alpha_1 + \alpha_2 + \ldots = 0 \quad \text{..............................(79)},$$

in order that equation (74) may still be satisfied.

For this distribution of molecules in the different cells, we find for the value of K_a, from equation (72),

$$K_a = \frac{1}{N} \sum_{s=1}^{s=n} \left(\frac{N}{n} + \alpha_s + \frac{1}{2}\right) \log\left(1 + \frac{n\alpha_s}{N}\right).$$

Since N and $\frac{N}{n}$ are both very great, and $\log\left(1 + \frac{n\alpha_s}{N}\right)$ is very small, we may replace $\frac{N}{n} + \frac{1}{2}$ by $\frac{N}{n}$, and so obtain

$$K_a = \frac{1}{n} \sum_{s=1}^{s=n} \left(1 + \frac{n\alpha_s}{N}\right) \log\left(1 + \frac{n\alpha_s}{N}\right) \quad \text{...(80)}$$

$$= \frac{1}{N}\left\{\frac{1}{2}\frac{n}{N}\sum_1^n \alpha_s^2 - \frac{1}{6}\left(\frac{n}{N}\right)^2 \sum_1^n \alpha_s^3 + \ldots + \frac{(-1)^{p+1}}{p(p+1)}\left(\frac{n}{N}\right)^p \sum_1^n \alpha_s^{p+1} + \ldots\right\} \quad (81).$$

From this it is clear that K vanishes when $\alpha_1 = \alpha_2 = \ldots = 0$, while for small values of the α's, K is invariably positive. Thus the stationary value which has been found for K is a true minimum; further, as it was seen to be the only stationary value for K, it follows that K increases steadily as we recede from it, and so must be everywhere positive except at this minimum.

In the expansion (81), the ratio of each term to the preceding is of the order of $\frac{n\alpha}{N}$, which is small so long as α is small compared with N/n. Thus for small values of α, K_a is represented by the first term of (81). When α becomes comparable with N/n, the terms in (81) all become of the same order of magnitude, but NK is now of the order of magnitude of N/n, and so is very great. For such values of α, therefore, e^{-NK_a} is vanishingly small, so that θ_a, given by equation (73), is also vanishingly small. Thus arrangements of molecules for which the α's are comparable with N/n only occupy an infinitesimally small part of the generalised space. For all the remainder, the α's are small compared with N/n, and K_a is given by

$$NK_a = \frac{1}{2}\frac{n}{N}\sum_1^n \alpha_s^2 \quad \text{.........................(82).}$$

46. The number of distributions of molecules for which NK is less than some assigned value NK_0, which is itself not infinite, will be equal to the number of sets of integral values of $\alpha_1, \alpha_2, \ldots$, such that (cf. equation (82))

$$\alpha_1^2 + \alpha_2^2 + \ldots \alpha_n^2 < \frac{2N^2K_0}{n} \quad \text{.........................(83),}$$

while the sum of the α's vanishes.

Imagine $\alpha_1, \alpha_2, \ldots \alpha_n$ to be Cartesian coordinates in a space of n dimensions, then integral values of the α's occur at the rate of one per unit volume. The sets of values for which $\alpha_1 + \alpha_2 + \ldots \alpha_n = 0$, all occur in the plane of which this is the equation. It is a plane through the origin having direction cosines $n^{-\frac{1}{2}}, n^{-\frac{1}{2}}, \ldots n^{-\frac{1}{2}}$. Thus sets of integral values in this plane occur at the rate of $n^{-\frac{1}{2}}$ per unit content of this plane. If A is the content of that part of this plane in which inequality (83) is satisfied, then the number of sets of values for which we are in search is $n^{-\frac{1}{2}}A$.

The quantity A is (cf. equation (83)) clearly the content of a sphere (or circle) of $(n-1)$ dimensions and of radius $\left(\frac{2N^2K_0}{n}\right)^{\frac{1}{2}}$. Its value is accordingly

$$A = \frac{\pi^{\frac{1}{2}(n-1)}}{\Gamma\left(\frac{n+1}{2}\right)}\left(\frac{2N^2K_0}{n}\right)^{\frac{1}{2}(n-1)}$$

The number sought, namely the number of distributions for which K is less than K_0, is $n^{-\frac{1}{2}}A$ or

$$\frac{\pi^{\frac{1}{2}(n-1)}}{\Gamma\left(\frac{n+1}{2}\right)} n^{-\frac{1}{2}}\left(\frac{2N^2K_0}{n}\right)^{\frac{1}{2}(n-1)} \quad \text{.....................(84).}$$

On differentiating with respect to K, the number of distributions for which K lies between K_0 and $K_0 + dK$ is found to be

$$\frac{\pi^{\frac{1}{2}(n-1)}}{\Gamma\left(\frac{n+1}{2}\right)} N^2 n^{-\frac{3}{2}}(n-1)\left(\frac{2N^2K_0}{n}\right)^{\frac{1}{2}(n-3)} dK \quad \text{...........(85).}$$

47. Multiplying this by θ_a, given by equation (73), we find as the fraction of the original generalised space which represents distributions of molecules for which K lies between K_0 and $K_0 + dK$,

$$\frac{N^{\frac{1}{2}(n-1)}}{\Gamma\left(\frac{n-1}{2}\right)} e^{-NK_0} K_0^{\frac{1}{2}(n-3)} dK \quad \text{.....................(86).}$$

It is easily verified that the integral of this from $K=0$ to $K=\infty$ is equal to unity, as of course it ought to be (cf. § 43). Thus expression (86) gives the law of distribution of values of K in the generalised space, or, to put the same thing in another way, it expresses the chance that a system selected at random from all the representative points in the generalised space shall have a value for K intermediate between K_0 and K_0+dK.

If x is written for NK, the law of distribution (86) becomes

$$\frac{1}{\Gamma\left(\frac{n-1}{2}\right)} e^{-x} x^{\frac{1}{2}(n-3)} dx \quad \text{.....................(87),}$$

while x is from equation (82) given by

$$x=\frac{1}{2}\frac{n}{N}\sum_1^n \alpha_s^2 \quad \text{.............................(88).}$$

48. Clearly expression (87) only has appreciable values when x is finite: the contribution to the whole integral which is supplied by infinite values of x is infinitesimal. The mean value of x is

$$\frac{1}{\Gamma\left(\frac{n-1}{2}\right)} \int e^{-x} x^{\frac{1}{2}(n-1)} dx = \tfrac{1}{2}(n-1) \quad \text{..............(89).}$$

Thus for all except an infinitesimal fraction of the systems represented in the generalised space x, or NK, is finite, so that K is zero. Also the mean value of x for all systems is $\frac{1}{2}(n-1)$, so that the mean value of K is $\frac{1}{2}(n-1)/N$.

Hence for all except an infinitesimal fraction of the systems $\Sigma\alpha_s^2$ is comparable with $\frac{N}{n}$, so that each α^2 is comparable with $\frac{N}{n^2}$, and the mean value of any single α^2 is $\frac{N(n-1)}{n^2}$.

If ρ is the mean density in the gas, and if in any cell the density is $\rho(1+\delta)$, then $\delta=\alpha n/N$ from equations (78). Thus in all except an infinitesimal fraction of all the systems δ^2 is comparable with $\frac{1}{N}$, and the mean value of δ^2 is $\frac{n-1}{N}$.

To sum up, we have discovered a very strong tendency towards an equal distribution of density. In all except an infinitesimal fraction of the systems, the variations of density in the different cells will only be of the order of $1/N$ of the whole. Since we are supposing N to be so great that $1/N$ may be neglected, we may say that in all except an infinitesimal fraction of the systems there is uniform density throughout the gas.

49. Let the cells, each of volume ω, now become infinitesimal and coincide with the different elements of volume $dxdydz$ in the gas. The mean molecular density ν_0 in the whole gas is equal to N/Ω, and the molecular density ν_s in the sth cell is such that

$$a_s = \nu_s \, dxdydz.$$

As in § 45, we may neglect the difference between a_s and $a_s + \frac{1}{2}$, and may accordingly replace equation (72) by

$$\begin{aligned} K &= \frac{1}{N} \sum_1^n a_s \log \frac{na_s}{N} \\ &= \frac{1}{\Omega} \sum_1^n \frac{\nu_s}{\nu_0} \log \left(\frac{\nu_s}{\nu_0}\right) dxdydz \\ &= \frac{1}{\Omega} \iiint \left(\frac{\nu}{\nu_0}\right) \log \left(\frac{\nu}{\nu_0}\right) dxdydz \quad \text{.....................(90)}, \end{aligned}$$

where the integral is taken through the whole gas.

Thus K is the mean value of $\frac{\nu}{\nu_0} \log \left(\frac{\nu}{\nu_0}\right)$ averaged through the gas. When the variations from uniform density are small, so that ν may be replaced by $\nu_0 + \delta\nu$, the value of K becomes

$$K = \frac{1}{2\Omega} \iiint \left(\frac{\delta\nu}{\nu_0}\right)^2 dxdydz \quad \text{.......................(91)},$$

as could also be seen from equation (82). Thus K is equal to the mean value of $\left(\frac{\delta\nu}{\nu_0}\right)^2$ averaged through the gas, and so is a quantity which might naturally be taken to measure divergence from uniform density.

The Partition of the Generalised Space—Velocity Coordinates.

50. We can investigate the partition of the velocity coordinates in a way similar to that used for the partition of positional coordinates.

As in § 13, we take a three-dimensional space, and represent the velocity of each molecule by a point such that its coordinates in this space are u, v, w, the components of velocity of the molecule. In this way we get N points in this three-dimensional space, and their positions determine the velocities of all the molecules in the gas.

The space now under consideration will extend to infinity in all directions, for the values of u, v, w for any molecule can vary from $-\infty$ to $+\infty$. We shall, however, find it unnecessary to take the regions at infinity into account. We shall ultimately be concerned only with systems in which the total kinetic energy of the N molecules has an assigned value E, and for these systems there is an upper limit to the values of u, v, w, fixed by the circumstance that the energy of any one coordinate, $\frac{1}{2}mu^2$, $\frac{1}{2}mv^2$ or $\frac{1}{2}mw^2$, cannot be greater than E. Thus if we put $l^2 = \frac{2E}{m}$, we can suppose the three-dimensional space in which the velocity coordinates are represented, to be limited by the planes $u = \pm l$, $v = \pm l$, $w = \pm l$. And equally the original $6N$ dimensional space can be considered limited by the planes $u_a = \pm l$, $v_a = \pm l$, $w_a = \pm l$, $u_b = \pm l$, etc. With these limitations, it is clear that the three-dimensional space is exactly suited to represent the velocity coordinates of all systems in the $6N$-dimensional space which are of the assigned energy E.

51. The three-dimensional space, of volume $8l^3$, may now be thought of as divided into a very great number n of very small cells, each of content ω or $du\,dv\,dw$. We consider a definite partition of velocities, such that the number of points in cell 1 is a_1, in cell 2 is a_2, and so on, where of course

$$a_1 + a_2 + \ldots + a_n = N \quad\ldots\ldots\ldots\ldots(92).$$

This will be called a partition of class A.

We can shew, exactly as in § 43, that the systems of class A will occupy a fraction θ_a of the original generalised space of $6N$ dimensions, where (cf. equations (68) and (73))

$$\theta_a = \frac{N!\,n^{-N}}{a_1!\,a_2!\ldots a_n!} \quad\ldots\ldots\ldots\ldots(93)$$

$$= n^{\frac{1}{2}n}(2\pi N)^{-\frac{1}{2}(n-1)}\,e^{-NK_a} \quad\ldots\ldots\ldots\ldots(94),$$

where

$$K_a = \frac{1}{N}\sum_{s=1}^{s=n} a_s \log\frac{na_s}{N} \quad\ldots\ldots\ldots\ldots(95),$$

since the term $\frac{1}{2}$ in equation (72) may be neglected in comparison with a_s. And, as before, the partitions which are commonest are those for which K_a is least.

52. If we proceeded to find which partitions were commonest in the whole $6N$-dimensional space, we should naturally be led to the same result as in § 45. But what we now want is to find which partitions are commonest, not for all the systems in the space, but only for those systems of which the energy is E.

When a molecule has its velocity components represented by a point in cell 1, let us suppose its kinetic energy to be ϵ_1; when the point is in cell 2 let the energy be ϵ_2, and so on. Then the total kinetic energy of the N molecules will be $a_1\epsilon_1 + a_2\epsilon_2 + \ldots + a_n\epsilon_n$, and this will, for the systems under consideration, be equal to E. Thus we must suppose $a_1, a_2, \ldots a_n$ limited not only by equation (92) but also by the equation

$$a_1\epsilon_1 + a_2\epsilon_2 + \ldots + a_n\epsilon_n = E \quad\ldots\ldots\ldots\ldots(96).$$

53. We proceed to find for what distribution K_a is least, when the values of the a's are subject to the restrictions of equations (92) and (96). By variation of equations (95), (92) and (96) we obtain

$$N\delta K_a = \sum_{s=1}^{s=n} \left(\log \frac{na_s}{N} + 1\right) \delta a_s \qquad (97),$$

$$0 = \sum_{s=1}^{s=n} \delta a_s \qquad (98),$$

$$0 = \sum_{s=1}^{s=n} \epsilon_s \delta a_s \qquad (99).$$

Thus the stationary values of K are given by equations of the type

$$\log \frac{na_s}{N} + 1 + \lambda + \mu\epsilon_s = 0 \qquad (100),$$

where λ, μ are multiplying constants.

The density of points in the three-dimensional u, v, w space may, as in § 13, be denoted by τ or Nf. Thus we have

$$a_s = \tau\omega = Nf\omega = Nf\,du\,dv\,dw,$$

so that

$$\frac{na_s}{N} = nf\omega = f\Omega',$$

where $\Omega' = n\omega$, the total volume $8l^3$ of the u, v, w space.

Hence, changing the constants, equation (100) assumes the form

$$f = Ae^{-2h\epsilon_s},$$

or, in terms of the coordinates u, v, w,

$$f(u, v, w) = Ae^{-hm(u^2+v^2+w^2)} \qquad (101),$$

which agrees, as of course it ought, with the law of distribution already obtained by the method of collisions in Chapter II.

54. It could now be shewn, exactly as in the former case, that throughout all but an infinitesimal fraction of the whole of that part of the generalised space in which the energy of the corresponding system is E, the law of distribution of velocities is that given by equation (101). It does not seem necessary to reproduce the details of this proof; the mathematician will be able to construct them for himself, while the physicist will probably not wish to be detained over them.

As in the case of the positional coordinates the law of distribution of velocities could have been predicted from considerations of probability. For selecting a point at random in the region corresponding to systems of energy E is equivalent to assigning velocity coordinates u, v, w to the molecules at random, subject only to the condition that their squares shall be distributed about a certain mean-value. It is therefore natural to find that the distribution of velocities should be in accordance with the law of trial and error.

55. The value of K_a is (cf. equation (95))

$$K = \sum_{s=1}^{s=n} \left(\frac{a_s}{N}\right) \log\left(\frac{na_s}{N}\right),$$

in which $\frac{na_s}{N} = f\Omega'$, $\frac{a_s}{N} = f\omega = fdudvdw$, and $\sum_{s=1}^{s=n}\left(\frac{a_s}{N}\right) = 1$. Using these relations, we find

$$K = \log \Omega' + \iiint f \log f du dv dw,$$

where the integral is taken through the volume Ω' in the u, v, w space, and since $f = 0$ outside this volume, the integral may equally be thought of as extending through the whole space.

It now appears that except for an additive constant, K is identical with the H of the method of the last chapter (cf. equation (13)). The theorem proved there, that H tended continually to decrease, is now seen to mean that K tends to decrease and therefore θ_a tends to increase. In other words, in its passage to the final state, a gas tends always to pass from the less probable to the more probable state, or, we may say, from the abnormal to the normal, where the most normal states are regarded as being those which occur most frequently in the $6N$-dimensional space.

56. The law of distribution of velocities expressed by equation (101) is a special case of the general law found for the "steady state" in the last chapter. We are limited to this special case because we have, at the outset, supposed our containing vessel to be fixed in space. If, on the contrary, the vessel is moving in space with a velocity of components u_0, v_0, w_0 the analysis of this chapter can be made to apply by supposing all coordinates referred to moving axes, moving with a velocity of components u_0, v_0, w_0. In this case equation (101) expresses the law of distribution of velocities relative to these moving axes. The law of distribution of absolute velocities in space is therefore

$$f = Ae^{-hm[(u-u_0)^2+(v-v_0)^2+(w-w_0)^2]} \quad \text{..................(102)},$$

which is the general law for the steady state given by equation (25).

This law, it will have already been noticed, gives $f = 0$ when u, v or w is infinite. There is therefore the *à posteriori* objection to the analysis by which it has been obtained, that if we divide all possible velocities into "cells" in the manner of § 51, the number of molecules in some of these cells cannot legitimately be treated as very great. The difficulty is best met by taking a definite velocity V such that those molecules of which the velocities do not satisfy the inequalities

$$u < V, \quad v < V, \quad w < V$$

form an infinitesimal fraction of the whole. If the velocities which satisfy these inequalities can be partitioned into cells in the manner of § 51, so as to satisfy the condition that the number in each cell is very great, then there is no further difficulty, and equation (102) will give the law of distribution of velocities which are less than V. The law now has no meaning for velocities greater than V. It is obvious, for instance, that the law expressed by equation (102) does not impose any upper limit whatever on the possible values of u, v and w for a single molecule, whereas in point of fact such a limit is definitely imposed by the energy equation.

Molecules of finite size.

57. It is obvious that it is in no way material to the analysis of §§ 51—53, from which the law of distribution of velocities is found, whether the regions mentioned in § 36 are excluded from the generalised space or not. For the exclusion of these regions affects the velocity coordinates equally throughout. Thus the law of distribution of velocities is the same whether the spheres are of finite size or are infinitesimal. It remains the same right up to the extreme limiting case in which the spheres are packed so tightly in the containing vessel that they cannot move.

The Normal State.

58. In the last chapter it was found, with the help of the unwarranted assumption of molecular chaos (§ 15), that the law of distribution expressed by equation (102) represented a "steady state" for the gas. In the present chapter it has been shewn, without making any use of this assumption, that, except for an infinitesimal probability of error, a system selected at random from the generalised space will be in the state specified by equation (102). It will be convenient to refer to this state as the "Normal State" (cf. below, § 87). And when a result is certain except for an infinitesimal probability of error, it will be convenient to speak of the result as "infinitely probable."

If, therefore, a system is selected at random, it is infinitely probable that it will be in the "Normal State." Suppose that a system is selected at random, and then allowed to move under its natural motion for a time t, what do we know now as to its probable state after a time t? The answer is provided by the theorem proved in § 40. The motion of all possible systems is represented in our generalised space by the motion of the supposed fluid. Instead of selecting a system at random and allowing it to move for time t, we may allow the whole fluid in the generalised space to move for a time t, and select a system at random after the motion has proceeded for a time t. The theorem of § 40 proved the motion of the fluid in the generalised space to be a "steady motion." Hence selecting a system after time t is the same thing as selecting a system at time 0, and it is infinitely probable that the system thus selected will be in a normal state.

59. This completes our information about the motion of the gas. At any instant it is infinitely probable that it is in a normal state. In the course of the motion departures from the normal state will occur, but it is infinitely probable that these will only occupy an infinitesimal fraction of the time occupied by the whole motion.

There is in theory a *possibility* of a gas continuing throughout its motion in a state different from the normal. Suppose for instance that the containing vessel is cubical, and that the molecules are started so that all move perpendicular to one edge along a system of parallel lines, no two of which are at a less distance than the diameter of a molecule. Then it is obvious that the molecules will not leave the lines on which they start, and will not change their velocities. In this case any law of velocities $f(u, 0, 0)$ will be permanent, where u is the velocity in the direction of the parallel lines.

The analysis of this chapter breaks down, it will be seen, because the supposition made in § 42 is no longer true, that there are no functions of the coordinates in the generalised space except the energy which remain constant throughout the motion. For obviously we have $u_a^2 = \text{constant}$, $u_b^2 = \text{constant}$, etc., and $v_a = w_a = 0$, $v_b = w_b = 0$, etc.

Our results shew that it is infinitely probable that a system selected at random will not be of this special type. The connection between the trajectories of such systems and the "periodic orbits" of abstract dynamics is interesting, but cannot be discussed here. We shall return to the discussion of cases in which there are constants other than the energy in Chapter V.

Historical Note.

60. The law of distribution of velocities which has been found both in this chapter and the preceding one was discovered by Maxwell, and is generally associated with his name. It first appears in the paper already referred to (§ 9), communicated to the British Association in 1859. The original proof is now universally admitted to be unsatisfactory, but is of interest from its historical importance. Except for a slight change of notation, the form in which it was given is as follows*.

"Let N be the whole number of particles. Let u, v, w be the components of the velocity of each particle in three rectangular directions, and let the number of particles for which u lies between u and $u + du$ be $Nf(u)\,du$, where $f(u)$ is a function of u to be determined.

"The number of particles for which v lies between v and $v + dv$ will be $Nf(v)\,dv$, and the number for which w lies between w and $w + dw$ will be $Nf(w)\,dw$, where f always stands for the same function.

* J. C. Maxwell, *Collected Works*, I. p. 380.

"Now the existence of the velocity u does not in any way affect that of the velocities v or w, since these are all at right angles to each other and independent, so that the number of particles whose velocity lies between u and $u+du$, and also between v and $v+dv$ and also between w and $w+dw$ is

$$Nf(u)f(v)f(w)\,du\,dv\,dw.$$

If we suppose the N particles to start from the origin at the same instant, then this will be the number in the element of volume $du\,dv\,dw$ after unit of time, and the number referred to unit of volume will be

$$Nf(u)f(v)f(w).$$

"But the directions of the coordinates are perfectly arbitrary, and therefore this number must depend on the distance from the origin alone, that is

$$f(u)f(v)f(w)=\phi(u^2+v^2+w^2).$$

Solving this functional equation, we find

$$f(u)=Ce^{Au^2},\quad \phi(u^2+v^2+w^2)=C^3e^{A(u^2+v^2+w^2)}."$$

This proof must be admitted to be unsatisfactory, because it *assumes* the three velocity components to be independent. The velocities do not, however, enter independently into the dynamical equations of collisions between molecules, so that until the contrary has been proved, we should expect to find correlation between these velocities.

On account of this defect, Maxwell attempted a second proof*, which after emendations by Boltzmann† and Lorentz‡ assumes the form given in Chapter II. It is however very doubtful whether this proof can claim any superiority on grounds of logical consistency or completeness over Maxwell's original proof. The later proof finds it necessary to assume that there is no correlation between the velocity and space coordinates, while the earlier proof merely assumed that there was no correlation between the separate velocity components *inter se*. In each case the dynamical conditions equally suggest correlation until the contrary has been proved, and it would be hard to give reasons why one assumption of no correlation is more justifiable than the other. It should be mentioned that Burbury§ was always of opinion that the later proof of Maxwell was not only logically unsound, but led to an inaccurate result. He maintained that correlation actually takes place, except in the limiting case of an infinitely rare gas. This view, however, is not borne out by the analysis of the present and of the succeeding chapter. (Cf. § 69, *infra*.)

* *Collected Works*, II. p. 43.

† *Wiener Sitzungsber.*, LVIII. p. 517 (1868), LXVI. p. 275 (1872), XCV. p. 153 (1887), *Vorlesungen über Gastheorie*, I. p. 15.

‡ *Wiener Sitzungsber.*, XCV. p. 115 (1887).

§ S. H. Burbury, *The Kinetic Theory of Gases*, Cambridge, 1899.

A second class of proof of the law is represented by the proof which has been given in this chapter. In this class of proof the aim is to deduce a law from general dynamical considerations. As important examples of this class of proof may be mentioned a proof due to Kirchhoff, given in his lectures*, and one due to Meyer and Pirogoff, given in Meyer's *Kinetic Theory of Gases*†. Both these proofs are found on analysis to depend upon a use of the calculus of probabilities which cannot be justified. The proof given in this chapter was first given by the present writer in 1903‡.

* Kirchhoff, *Vorlesungen über die Theorie der Wärme*, p. 142.

† Meyer, *Kinetic Theory of Gases*, Eng. Trans. by Baynes, p. 370.

‡ *Phil. Mag.* v. p. 597.

CHAPTER IV

THE LAW OF DISTRIBUTION OF VELOCITIES (*continued*)

COMPARISON BETWEEN THE METHODS OF THE TWO PRECEDING CHAPTERS.

61. THE problem of the present chapter will be to consider the relation between the methods of procedure adopted in Chapters II and III.

The discussion of Chapter II was based upon certain questions of probability, and an answer to these questions was made possible and was obtained by the help of the assumption of molecular chaos enunciated in § 15.

The discussion of Chapter III also rested, although in a different sense, upon the theory of probability. The generalised space filled with fluid supplied a basis for the calculation of probabilities, and as the motion of the fluid was proved to be steady-motion, it followed that this basis was independent of the time. For the present, we continue to take this generalised space as the basis of probability calculations. The question "What is the probability that a system satisfies condition p?" will be taken to mean: "For what proportion of the generalised space is condition p satisfied?" The further question: "Given that a system satisfies condition p, what is the probability that it also satisfies condition q?" will be interpreted to mean: "A point is selected at random from all those parts of the generalised space in which condition p is satisfied: what is the probability that at this point condition q also is satisfied?" And if v_p is the total volume of that part of the space in which condition p is satisfied, and v_{pq} that of the space in which condition q also is satisfied, the value of the probability required is of course the ratio

$$\frac{v_{pq}}{v_p} \quad\text{.................................(103).}$$

If v is the whole volume of this generalised space (or any part of it to which we choose to limit our consideration) the chance that condition q shall be satisfied independently of condition p is

$$\frac{v_q}{v} \quad\text{.................................(104).}$$

Hence the condition that the probabilities of p and q being satisfied may be regarded as "independent" is that expressions (103) and (104) shall be equal, or, written symmetrically, that

$$\frac{v_{pq}}{v} = \frac{v_p}{v}\frac{v_q}{v} \quad \text{.............................(105).}$$

Analysis of the Assumption of Molecular Chaos.

62. The assumption of molecular chaos was tantamount to an assumption that two probabilities might be regarded as independent. Equation (105) accordingly enables us to test whether this assumption is legitimate or not relatively to our present basis of probability—namely, the generalised space filled with homogeneous fluid.

To do this, let us define condition p as the condition that a molecule of class A (defined in § 13) shall be found in the element $dxdydz$ of the gas—in other words, that one of the N molecules shall have coordinates lying between the limits

$$\left.\begin{array}{l} x \text{ and } x + dx, \;\; y \text{ and } y + dy, \;\; z \text{ and } z + dz \\ u \text{ and } u + du, \;\; v \text{ and } v + dv, \;\; w \text{ and } w + dw \end{array}\right\} \text{............(106).}$$

For certain systems this condition is satisfied by molecule A, and these systems are represented in the generalised space by that region for which x_a lies between x and $x + dx$, and for which similar conditions are satisfied by y_a, z_a, u_a, v_a, w_a. This region supplies to v_p a contribution of amount

$$\iiint \ldots dx_a dx_b dx_c \ldots du_a du_b du_c \ldots \quad \text{...............(107),}$$

where the integration extends over all values of the variables which are not excluded by § 36, except in the case of $x_a, y_a, z_a, u_a, v_a, w_a$, for which the limits are those given by (106). The integral may be written in the form

$$dxdydzdudvdw \iint \ldots dx_b dx_c \ldots \int_{-\infty}^{+\infty} du_b \int_{-\infty}^{+\infty} du_c \ldots \quad \text{......(108).}$$

For other systems, condition p is satisfied by molecule B, and these systems again supply a contribution of amount equal to the above. Each of the N molecules contributes in this way to v_p an amount equal to that given by expression (108), so that the total value of v_p is

$$v_p = N dxdydzdudvdw \iint \ldots dx_b dx_c \ldots \int_{-\infty}^{+\infty} du_b \int_{-\infty}^{+\infty} du_c \ldots \quad \text{...(109).}$$

The value of v, the volume of the whole space, is given by (107), if the integrals are taken through all values of all the variables except those values excluded by § 36. This integral may of course be written in the form

$$v = \iiint \ldots dx_a dx_b dx_c \ldots \int_{-\infty}^{+\infty} du_a \int_{-\infty}^{+\infty} du_b \ldots \quad \text{......(110),}$$

and from equations (109) and (110) we now find

$$\frac{v_p}{v} = N \frac{dx\,dy\,dz \iint \dots dx_b\,dx_c \dots}{\iiint dx_a\,dx_b\,dx_c \dots} \frac{du\,dv\,dw}{\int_{-\infty}^{+\infty} du \int_{-\infty}^{+\infty} dv \int_{-\infty}^{+\infty} dw} \quad \text{......(111).}$$

This vanishes through the last factor, for the obvious reason that when the molecules are equally likely to have all velocities, the probability is infinitely against a single molecule belonging to any specified class.

63. Let us now suppose that the velocities of the individual molecules are given, and let us calculate the probability in this case that condition p is satisfied. Let us suppose that the velocities of molecule A are known to lie within the limits

$$u_a \text{ and } u_a + \delta u_a,\ v_a \text{ and } v_a + \delta v_a,\ w_a \text{ and } w_a + \delta w_a \quad \text{.........(112),}$$

and that we have similar knowledge for the other molecules. The value of v, the whole space representing systems for which the molecules have the given velocities, is given by equation (110) if the integration is from u_a to $u_a + \delta u_a$ for u_a instead of from $-\infty$ to $+\infty$, and similarly for the other velocities.

Thus we have as the new value of v,

$$v = \delta u_a \delta v_a \dots \iiint \dots dx_a\,dx_b\,dx_c \dots \quad \text{..............(113).}$$

As before, the systems for which condition p is satisfied by molecule A are represented by those parts of the space v for which x_a lies between x and $x + dx$, and similar conditions are satisfied by y_a, z_a, u_a, v_a, w_a. We shall suppose, as we legitimately may, that the δu_a, δv_a, δw_a of the limits (112) are infinitesimal in comparison with $du\,dv\,dw$. Then provided that the range for u_a given by (112) lies within the range u and $u + du$, and that similar conditions are satisfied by v_a, w_a, the contribution to v_p corresponding to molecule A is given by the right-hand of (113) provided the integration with respect to x_a extends only from x to $x + dx$, and similarly for y_a, z_a. Thus if the velocities of molecule A lie within the specified ranges, there is a contribution from molecule A to v_p of amount

$$dx\,dy\,dz\,\delta u_a \delta v_a \dots \iiiint \dots dx_b\,dx_c \dots dy_b\,dy_c \dots \quad \text{.........(114).}$$

If the velocities given by the limits (112) do not lie within this range $du\,dv\,dw$ the contribution is *nil*. The number of molecules of which the velocities satisfy the condition of lying within this range—in other words, the number of molecules capable of taking the rôle of molecule A in expression (114)—may be taken to be

$$Nf(u, v, w)\,du\,dv\,dw \quad \text{........................(115).}$$

The product of expressions (114) and (115) gives v_p. From this and equation (113) we get

$$\frac{v_p}{v} = Nf(u, v, w)\, du\,dv\,dw\,dx\,dy\,dz\, \frac{\iiiint \ldots dx_b\,dx_c \ldots dy_b\,dy_c \ldots}{\iiiiiint \ldots dx_a\,dx_b\,dx_c \ldots dy_a\,dy_b\,dy_c} \quad \ldots(116).$$

64. Let us define condition q as the condition that there shall be a molecule having its coordinates within the limits x' and $x' + dx'$, etc. The volume v_{pq} for which conditions p and q are both satisfied will consist of contributions from different pairs of molecules. In expression (116) we suppose molecule A to satisfy condition p. If molecule B satisfies condition q the corresponding contribution to v_{pq}/v can be obtained from the right-hand of (116) by limiting the integration in the numerator to the range from x' to $x' + dx'$ as regards x_b, and to similar ranges as regards y_b, z_b. The number of molecules capable of taking the rôle of B is

$$Nf(u', v', w')\, du'\,dv'\,dw'.$$

Hence we obtain as the value of v_{pq}/v

$$\frac{v_{pq}}{v} = N^2 f(u, v, w)\, f(u', v', w')\, du\,dv\,dw\,du'\,dv'\,dw'\,dx\,dy\,dz\,dx'\,dy'\,dz'$$

$$\times \frac{\iint \ldots dx_c \ldots dy_c \ldots}{\iiiiiint \ldots dx_a\,dx_b\,dx_c \ldots dy_a\,dy_b\,dy_c} \quad \ldots\ldots(117).$$

The integration extends throughout all the values of the variables except such as are excluded by the conditions of § 36. In applying these conditions to the numerator, we must replace x_a, y_a, z_a by x, y, z and x_b, y_b, z_b by x', y', z'. We therefore find, as we ought, that v_{pq} vanishes when the points x, y, z and x', y', z' are at a shorter distance than σ, or when either of them is at a distance from the boundary less than $\frac{1}{2}\sigma$. We also see that v_{pq}/v is not equal to the product of v_p/v and v_q/v, so that the fulfilment of conditions p and q cannot be treated as independent events.

Infinitely Small Molecules.

65. In the special case in which the radii of the molecules are vanishingly small, those parts of the generalised space which are excluded in § 36 may be neglected. In the integrals of equations (116) and (117) the integrations with respect to the variables with different suffixes now become independent. We may for instance write

$$\iiiiiint \ldots dx_a\,dx_b\,dx_c \ldots dy_a\,dy_b\,dy_c \ldots = \iiint dx_a\,dy_a\,dz_a \iiint dx_b\,dy_b\,dz_b \ldots = \Omega^N.$$

The other integrals can be simplified in a similar manner, and we obtain

$$\frac{v_p}{v} = \frac{N}{\Omega} f(u, v, w)\, du\, dv\, dw\, dx\, dy\, dz \quad \text{..............(118)},$$

$$\frac{v_{pq}}{v} = \left(\frac{N}{\Omega}\right)^2 f(u, v, w) f(u', v', w')\, du\, dv\, dw\, du'\, dv'\, dw'\, dx\, dy\, dz\, dx'\, dy'\, dz' \text{...(119)}.$$

Writing ν for N/Ω we see that the right-hand of equation (118) becomes identical with our former expression (2) (p. 17). In other words, in the case in which the density is constant throughout the gas, the probability that condition p shall be satisfied for a system selected at random, is equal to the probability calculated in § 14.

From equations (118) and (119) we have the important result

$$\frac{v_{pq}}{v} = \frac{v_p}{v} \frac{v_q}{v} \text{..............................(120)}.$$

Thus the fulfilment of conditions p and q are now independent events. In other words, we have proved that, relatively to our present basis of probability, *the assumption of molecular chaos enunciated in* § 15 *is justifiable in the case in which the radii of the molecules are vanishingly small.*

66. To justify the way in which this assumption was used in Chapter II we must go somewhat further. In expression (4) we found a value which we supposed to be equal to the number of collisions of a certain class α. The actual value of this expression was, however, equal to

(the number of possible collisions of class α)

× (the probability of each collision happening).

This quantity therefore expresses the probable number of collisions, the number which actually occur averaged over a large number of cases, or the "expectation" of collisions, but does not necessarily express the actual number in any particular case. Looked at from another point of view, however, the quantity expressed

= (the sum of a number of small elements of volume)

× (the density of molecules of class B).........(121).

Now there was nothing in the analysis of Chapter III to compel us to take the "cells" to be continuous in space. We may accordingly regard the first factor in expression (121) as one of these cells. We now see in accordance with the results of Chapter III that expression (121) not only gives the "expectation" of molecules of class B in these elements of volume, but that it is infinitely probable that it gives the actual number, to within an infinitesimal fraction of the whole. It follows that the number of collisions of a given type found in Chapter III not only gives the most probable number of collisions, but that it is infinitely probable that it gives the true number, to within an infinitesimal fraction of itself.

Molecules of finite size.

67. When the molecules are not vanishingly small, let us denote the integral

$$\iiiint \dots dx_b dx_c \dots dy_b dy_c \dots,$$

taken over all values of the variables which are not excluded by § 36, by $I(b, c \dots)$. The value of the integral can only depend on x_a, y_a, z_a, and since we have supposed the molecule A to be at the point x, y, z, we may say that $I(b, c \dots)$ is a function of x, y, z only. Equations (116) and (117) now become

$$\frac{\nu_p}{\nu} = Nf(u, v, w)\, du\, dv\, dw\, dx\, dy\, dz\, \frac{I(b, c, d \dots)}{I(a, b, c, d \dots)} \quad \text{.......................(122)},$$

$$\frac{\nu_{pq}}{\nu} = N^2 f(u, v, w) f(u', v', w')\, du\, dv\, dw\, dx\, dy\, dz\, du'\, dv'\, dw'\, dx'\, dy'\, dz' \frac{I(c, d \dots)}{I(a, b, c, d \dots)} \quad \text{.........(123)}.$$

From equation (122) we see that the density of molecules of class A at x, y, z may no longer be taken to be

$$\nu f(u, v, w)\, du\, dv\, dw$$

but must be taken to be

$$\nu_1 f(u, v, w)\, du\, dv\, dw,$$

where

$$\nu_1 = N \frac{I(b, c \dots)}{I(a, b, c \dots)} \quad \text{.........................(124)},$$

a quantity which reduces to $\frac{N}{\Omega}$, and therefore to ν, when the molecules are infinitely small. In general ν_1 is a function of x, y, z but it is not a function of u, v, w. We may conveniently refer to ν_1 as the "effective molecular density" at the point x, y, z. When we require to specify the point x, y, z at which ν_1 is estimated, we shall replace ν_1 by $\nu_{x,y,z}$.

68. The "expectation" of the number of molecules in the whole vessel is equal to the total number of molecules actually present in the vessel, so that

$$\iiint \nu_{x,y,z}\, dx\, dy\, dz = \nu\Omega \quad \text{......................(125)}.$$

Thus ν is the mean value of $\nu_{x,y,z}$ averaged throughout the vessel. We shall see later the importance of the distinction between $\nu_{x,y,z}$ and ν.

For the present the following point may be noticed. In regions so far inside the containing vessel that the boundary may be regarded as far removed compared with the molecular scale of size, the value of $\nu_{x,y,z}$ must be the same at every point, for there is nothing to produce changes in $\nu_{x,y,z}$ as between one point and another. In a vessel of normal or large size (cf. § 160 below) these conditions hold throughout the interior, except for a thin layer, close to the boundary, of which the total volume will be vanishingly small in comparison with that of the whole vessel. It follows from equation (125) that $\nu_{x,y,z}$ will be indistinguishable from ν throughout the interior of the vessel, but may vary appreciably from ν near to the boundary. The evaluation of $\nu_{x,y,z}$ for points near to the boundary is carried out in detail in a later chapter (cf. § 202).

69. From equations (122) and (123) we obtain

$$\frac{\nu_p}{\nu}\frac{\nu_q}{\nu} = \frac{\nu_{pq}}{\nu}\frac{I(b, c, d \ldots) I(a, c, d \ldots)}{I(c, d \ldots) I(a, b, c, d \ldots)} \quad \text{..............(126).}$$

Hence, given that the molecule A is in position at x, y, z, the probability that a second molecule B has a position at x', y', z' is not that which would be given by the assumption of molecular chaos, but is equal to this value multiplied by

$$\frac{I(c, d \ldots) I(a, b, c, d \ldots)}{I(b, c, d \ldots) I(a, c, d \ldots)} \quad \text{....................(127),}$$

a function which is symmetrical as regards the x, y, z coordinates of A and B, *and which is independent of the velocities of the molecules A and B.*

Analysis of the H-theorem.

70. The assumption of molecular chaos (corrected, if necessary, in accordance with § 69) will therefore give correct results, provided it is interpreted with reference to the basis of probability supplied by our generalised space, and provided it is understood that it gives probable, and not certain, results. If we wish to obtain strictly accurate results, the quantities calculated from it must not be regarded as applying to a single system, but must be supposed to be averaged over all the systems in the generalised space which satisfy certain conditions. For instance, the value of $\frac{\partial f}{\partial t}$ given by equation (12) is merely the value of $\frac{\partial f}{\partial t}$ averaged throughout all those parts of the space for which the system has a given f. So also we must interpret the value of $\frac{dH}{dt}$ given by equation (20) as an average value for $\frac{dH}{dt}$, taken over all systems in our space which have a given value for H.

71. We now come to what is, at first sight a paradox. Let us suppose that f is an even function of u, v, w different from the normal function $Ae^{-hm(u^2+v^2+w^2)}$. Then from Chapter II (§ 23) it follows that dH/dt is negative, dH/dt indicating, as we have just seen, the value of dH/dt averaged over all the systems in our space which have this given f. Now these systems may be divided up into pairs. Corresponding to any system there will be a second system of which the positional coordinates will be the same as those of the first system and of which the velocity coordinates will be the same in magnitude but opposite in sign. Since f is an even function of the velocity coordinates, the value of f will be the same for each of these two systems and both systems are equally to be included in the average of dH/dt. But the motion of the first system is exactly the reverse of that of the second system. It would therefore appear as though the values of dH/dt must be equal and opposite for the two systems, so that the average of dH/dt for these two must be zero. Since the whole of the systems corresponding to a given f fall into pairs of this type, it might be inferred that the average value of dH/dt must be zero.

72. The explanation of the apparent paradox is as follows. From § 55 it appears that dH/dt is the same thing as dK/dt, where K is given by equation (95). The law of distribution of the K's for velocity coordinates is easily shewn to be of the same general form as that for the positional coordinates given in § 47. In particular, it can be seen that, of the systems for which H has a value greater than some value H_1 other than the minimum value for H, all except an infinitesimal fraction have a value for H which only differs infinitesimally from H_1. If, therefore, we select at random a point at which the value of H is H_1, it is infinitely probable that H will decrease as we recede from this point *in either direction* along the trajectory through the point. In other words, it is infinitely probable that the value $H = H_1$ is a maximum value of H for the trajectory through the point.

73. It may, perhaps, still be thought paradoxical that dH/dt is not zero at each of these maxima. The explanation is that the variation of H is not governed by the laws of the differential calculus, since this variation is not, strictly speaking, continuous. The value of H is constant between collisions of the molecules, and changes abruptly at every collision. When the number of molecules in the gas is infinite, the interval between successive collisions will become infinitely small, but in general the variation in H will not be continuous. For obviously the differential coefficients of H vanish between collisions and become infinite at every collision, so that H, regarded as a function of t as we follow any trajectory, will be a function of the well-known type which possesses an infinite number of maxima and minima within a finite range of the variable. We can, however, "smooth out" the curve obtained for H as a function of t and in this way obtain the function H as a continuous function of the time. This is the H contemplated by the analysis of Chapter II. But now we can also see that there is no reason to suppose that dH/dt will vanish when H attains a maximum value, but that on the contrary H will in general change sign abruptly at such a point. It is therefore clear that, averaged over all systems which have a given f, dH/dt will be negative except when f is the law of distribution for the normal state, in which case it is zero. This result is now in agreement with that of Chapter II.

CHAPTER V

GENERAL STATISTICAL MECHANICS AND THERMODYNAMICS

74. Two methods of obtaining the Law of Distribution of molecular velocities have now been given and also a comparison between them. These two methods have been limited to the consideration of molecules which may be treated as elastic spheres, exerting no forces on one another except when in actual collision. There is a more general way of treating the question, which permits of the molecules being dynamical systems of the most general type, capable of any kind of internal motion and exerting upon one another any forces we please. This method will be explained in the present chapter.

With a view to obtaining results which will be required later in the development of the subject, we shall not limit the discussion of the present chapter to the problem referred to above; we shall consider the "Statistical Mechanics" of a perfectly general dynamical system, not in the least limited to consisting of a gas. The special applications of the present chapter will be to a gas, while in later chapters we shall have occasion to apply the results to more general systems.

Degrees of Freedom.

75. The total number of independent quantities which need to be known before the configuration and position of any dynamical or other system can be fully known, is called the number of degrees of freedom of the system. It will be seen that this number depends on the motions, or capabilities of motion, of the parts of the system, and not in any way on the forces which produce the motion; it therefore corresponds to geometrical or kinematical properties of the system, and has nothing to do with the dynamics of the system.

For instance the position of a point free to move in space can be determined when three quantities are known, say x, y, z, the coordinates of the point, so that a point free to move in space has three degrees of freedom. The number of degrees of freedom of a rigid body will be six, for the position of the body can be fixed when six quantities are known, say x, y, z the coordinates of the centre of gravity of the body, and three angles θ, ϕ, ψ to determine the orientation of the body. Similarly the number of degrees of freedom of a pair of compasses will be found to be seven, of a nutcracker eight, of two rigid arms connected by a "universal joint" nine, and so on.

76. Numbers of degrees of freedom are additive in the sense that the number of degrees of freedom of a complex system made up of a number of simpler systems is equal to the sum of the numbers of degrees of freedom of the constituent systems. This becomes obvious on noticing that a knowledge of the configuration of the complex system is exactly equivalent to a knowledge of the configurations of all the constituent systems.

For example, if atoms are regarded as points, each atom will have three degrees of freedom, corresponding say to x, y, z its coordinates. A diatomic molecule must therefore necessarily have six degrees of freedom. These can be counted up in a variety of different ways, but the total must always come to six. For instance we might take the six degrees of freedom to consist of the three degrees of freedom of the centre of gravity to move in space, the two degrees of freedom of the line joining the two atoms to change its direction in space, and the one degree of freedom arising from the possibility of the two atoms changing their distance apart.

In general, if atoms are regarded as points, a molecule composed of n atoms will have $3n$ degrees of freedom, while if atoms are regarded as rigid bodies capable of rotational as well as translational motion, a molecule composed of n atoms will have $6n$ degrees of freedom. If electrons are regarded as points, a cluster of n electrons will have $3n$ degrees of freedom. If molecules are treated as points, a gas consisting of N molecules will be a dynamical system having $3N$ degrees of freedom, while if each molecule has n degrees of freedom, the gas will have nN degrees of freedom.

77. In surgery, a joint which has become so stiff as to be incapable of motion is said to be ankylosed. Thus in ankylosis the human body loses some of its "degrees of freedom." The term has been introduced, with great convenience, into dynamical theory by Poincaré. For instance we have seen that a pair of compasses has seven degrees of freedom; if the joint becomes rusted so that the arms cannot turn, the number of degrees of freedom is reduced to six, as in any other rigid body; one degree of freedom has become ankylosed. So in the diatomic-molecule, regarded as made up of two point atoms, the number of degrees of freedom is six. But if the two atoms are, under any conditions, so closely bound together that their distance apart cannot be changed, the number of degrees of freedom is reduced to five; the sixth degree of freedom is ankylosed.

THE MOTION OF A GENERAL DYNAMICAL SYSTEM.

78. In theory the motion of any dynamical system may be traced out, when the initial configuration of the system and the initial velocities of the different parts of the system are given. Statistically, the motion can be examined by a method similar to that already adopted in Chapter III.

Consider any system having n degrees of freedom. Let $q_1, q_2, \dots q_n$ be coordinates specifying the configuration of the system. Then the rates of increase $\dot{q}_1, \dot{q}_2, \dots \dot{q}_n$ of these quantities will specify the velocities of the different parts of the system, and it will be possible to trace out the motion of the system if we are given the initial values of the $2n$ quantities

$$q_1, q_2, \dots q_n, \dot{q}_1, \dot{q}_2, \dots \dot{q}_n \quad \text{.....................(128).}$$

The total energy of the system will be a function of the above $2n$ quantities. It is convenient to introduce the momenta $p_1, p_2, \dots p_n$, defined by

$$p_s = \frac{\partial E}{\partial \dot{q}_s}, \text{ etc.} \quad \text{............................(129),}$$

these momenta being, of course, functions of the $2n$ coordinates and velocities (128). In all ordinary cases under consideration E will be a function of the second degree in the velocities $\dot{q}_1, \dot{q}_2, \dots \dot{q}_n$, so that the momenta will be linear functions of these velocities.

The $2n$ quantities (128) may now be replaced by the $2n$ quantities

$$q_1, q_2, \dots q_n, p_1, p_2, \dots p_n \quad \text{.....................(130).}$$

When the values of these quantities are known at any instant, the configuration and velocities of the system are known at that instant and may therefore be traced throughout all time.

79. In general we shall be concerned only with the study of self-contained conservative systems in which there is no dissipation of energy. For the moment, however, we shall not limit ourselves to conservative systems; we shall suppose that the motion of the system is subject to dissipation of energy at a rate $2F$, where F is a quadratic function of the velocities.

If the system obeys in its motion the ordinary dynamical laws, the equations of motion of the system are the well-known equations

$$\frac{dp_s}{dt} = -\frac{\partial E}{\partial q_s} - \frac{\partial F}{\partial \dot{q}_s} \quad \text{.........................(131),}$$

$$\frac{dq_s}{dt} = \frac{\partial E}{\partial p_s} \quad \text{..................................(132),}$$

in which E is to be expressed as a function of the q's and p's, while the dissipation-function F is expressed as a function of the q's and $\dot{q}$'s. When there is no dissipation of energy, $F=0$, and the equations reduce to the well-known equations of Hamilton

$$\frac{dp_s}{dt} = -\frac{\partial E}{\partial q_s} \quad \text{..............................(133)},$$

$$\frac{dq_s}{dt} = \frac{\partial E}{\partial p_s} \quad \text{..............................(134)}.$$

Representation in a Generalised Space.

80. In Chapter III (§§ 35, 36), the motion of a system consisting of a number of elastic spheres was represented in a generalised space of the appropriate number of dimensions. It is convenient to imagine the motion of the present more general system to be represented in the same way.

We accordingly imagine a $2n$-dimensional space constructed, for which the $2n$-variables

$$q_1, q_2, \dots q_n, p_1, p_2, \dots p_n \quad \text{....................(135)}$$

are orthogonal coordinates. Then any one point in this space will represent one definite set of values of $q_1, q_2, \dots p_n$, and so will represent one definite "state" of the system, as determined by one set of coordinates of position and velocity. As the dynamical system changes its state, the representative point will describe a continuous curve in the $2n$-dimensional space.

The motion of any representative point in such a curve will be determined by equations (131) and (132). These equations are seen to express the components of velocity $\left(\frac{dp_1}{dt}, \frac{dp_2}{dt}, \dots \frac{dq_n}{dt}\right)$ of the representative point, in terms of the coordinates of the point. Clearly, then, these components of velocity are uniquely determined for any given point, so that there can be only one curve through each point in the space.

Also the components of velocity, as given by equations (131) and (132), are functions of the coordinates only, so that the curves determined by these equations may be thought of as permanently fixed in the generalised space. We can imagine all possible curves of this kind mapped out in the generalised space, and this imaginary $2n$-dimensional chart will enable us to follow out the motion of the dynamical system, starting from any initial state that we please.

All the conceptions of the present section are illustrated in the detailed case already worked out in §§ 35—38.

81. What we wish to study is not the motion of the system starting from any one set of initial conditions; we wish rather to find statistical properties of the system, which shall be true for all motions, no matter what the particular state from which the system starts. It is therefore convenient to imagine the whole of the $2n$-dimensional space filled with moving points, each describing its own curve, as determined by equations (131) and (132). These points will be supposed to be so thickly scattered in the space that they may be regarded as forming a continuous dust or fluid, exactly as before in the special case of § 39.

The number of these representative points per unit volume of the generalised space will be denoted by τ, and this will measure the density of the imaginary dust or fluid.

As the representative points move on their paths, the points forming any small continuous group might conceivably close in upon one another, in which case τ would increase, or they might separate out from one another, in which case τ would decrease. The changes which are to be expected in τ are examined in the analysis which follows.

82. Consider any small rectangular element of volume in the generalised space of content dv equal to $dp_1, dp_2, \ldots dq_n$, extending from p_1 to $p_1 + dp_1$, p_2 to $p_2 + dp_2$, etc. The number of representative points inside this element of space at any instant is τdv.

Points are streaming in or out of this element across each of its faces. Let us consider the flow across the pair of opposite faces, perpendicular to the axis of p_1, for which p_1 has the values p_1 and $p_1 + dp_1$. Let the area of each of these faces be dS, so that $dp_1 dS = dv$.

The points which cross the face p_1 have a component of velocity $\frac{dp_1}{dt}$ given by equation (131) normal to the face, so that the number of points which flow across the face into the element dv in time dt will be

$$\tau \frac{dp_1}{dt} dS dt \quad \ldots\ldots\ldots\ldots\ldots\ldots\ldots\ldots(136),$$

this quantity being of course negative if the flow is outwards. Similarly the number which flow in across the opposite face is

$$-\tau \frac{dp_1}{dt} dS dt \quad \ldots\ldots\ldots\ldots\ldots\ldots\ldots\ldots(137).$$

The two quantities (136) and (137) are not equal and opposite, for the former is evaluated at the face p_1 and the latter at the face $p_1 + dp_1$. Their algebraic sum is

$$-\frac{\partial}{\partial p_1}\left(\tau \frac{dp_1}{dt}\right) dp_1 dS dt,$$

or again

$$-\frac{\partial}{\partial p_1}(\tau \dot{p}_1)\, dv dt \quad \ldots\ldots\ldots\ldots\ldots\ldots\ldots\ldots(138),$$

and this represents the net gain to the number of points τdv in the element dv, which is produced by the flow across the pair of faces perpendicular to the axis of p_1.

There are similar gains to be evaluated by considering the flow over each other pair of faces, and the total gain, being equal to the sum of all these contributions, is

$$-\sum_1^n \left\{\frac{\partial}{\partial p_s}(\tau \dot{p}_s) + \frac{\partial}{\partial q_s}(\tau \dot{q}_s)\right\} dv\,dt \quad \text{.................(139)}.$$

This must be equal to the gain to τdv in time dt, and therefore to $\frac{\partial}{\partial t}(\tau dv)\,dt$, so that

$$\frac{\partial \tau}{\partial t} + \sum_1^n \left\{\frac{\partial}{\partial p_s}(\tau \dot{p}_s) + \frac{\partial}{\partial q_s}(\tau \dot{q}_s)\right\} = 0 \quad \text{.................(140)}.$$

Here $\frac{\partial \tau}{\partial t}$ represents the rate of increase of τ inside a fixed element of volume. Let $\frac{D\tau}{Dt}$ be taken to represent the rate of increase in τ as we follow the group of points in its motion. Then, since $\frac{D\tau}{Dt}$ represents the rate of increase of a function of t, p_1, p_2, ... q_n,

$$\frac{D\tau}{Dt} = \frac{\partial \tau}{\partial t} + \frac{\partial \tau}{\partial p_1}\frac{\partial p_1}{\partial t} + \frac{\partial \tau}{\partial p_2}\frac{\partial p_2}{\partial t} + \ldots + \frac{\partial \tau}{\partial q_n}\frac{\partial q_n}{\partial t}$$

$$= \frac{\partial \tau}{\partial t} + \sum_1^n \left\{\frac{\partial \tau}{\partial p_s}\dot{p}_s + \frac{\partial \tau}{\partial q_s}\dot{q}_s\right\} \quad \text{..........................(141)}.$$

It follows that equation (140) can be put in the form

$$\frac{D\tau}{Dt} + \tau \sum_1^n \left\{\frac{\partial \dot{p}_s}{\partial p_s} + \frac{\partial \dot{q}_s}{\partial q_s}\right\} = 0 \quad \text{....................(142)},$$

this being merely the hydrodynamical equation of continuity in the $2n$-dimensional space.

The values of $\dot{p}_s$, $\dot{q}_s$ are given by equations (131) and (132), so that

$$\frac{\partial \dot{p}_s}{\partial p_s} = -\frac{\partial^2 E}{\partial p_s \partial q_s} - \frac{\partial^2 F}{\partial p_s \partial \dot{q}_s}; \quad \frac{\partial \dot{q}_s}{\partial q_s} = -\frac{\partial^2 E}{\partial p_s \partial q_s};$$

and therefore

$$\frac{\partial \dot{p}_s}{\partial p_s} + \frac{\partial \dot{q}_s}{\partial q_s} = -\frac{\partial^2 F}{\partial p_s \partial \dot{q}_s} \quad \text{..........................(143)}.$$

Thus equation (142) reduces to

$$\frac{D\tau}{Dt} = \tau \sum_1^n \frac{\partial^2 F}{\partial p_s \partial \dot{q}_s} \quad \text{..........................(144)},$$

and this is the required equation, giving the change in the density τ as the cloud of representative points moves on its way.

83. It is readily shewn that $\sum_1^n \frac{\partial^2 F}{\partial p_s \partial \dot{q}_s}$ must always be positive, except in the special case in which $F = 0$. For, since F is supposed expressed as a function of the q's and $\dot{q}$'s, we have

$$\sum_1^n \frac{\partial^2 F}{\partial p_s \partial \dot{q}_s} = \sum_{s=1}^{s=n} \sum_{r=1}^{r=n} \frac{\partial^2 F}{\partial \dot{q}_r \partial \dot{q}_s}\frac{\partial \dot{q}_r}{\partial p_s} = \sum_{s=1}^{s=n} \sum_{r=1}^{r=n} \frac{\partial^2 F}{\partial \dot{q}_r \partial \dot{q}_s}\frac{\partial^2 E}{\partial p_r \partial p_s} \quad \text{......(145)}.$$

Equation (144) shews that the value of $\Sigma \frac{\partial^2 F}{\partial p_s \partial \dot{q}_s}$ at any point in the generalised space must be independent of the special choice of axes at that point. By a linear transformation we can change the p's so that E becomes a sum of squares only, say $E = \frac{1}{2}\Sigma \alpha_s p_s^2$, while F has a general quadratic value, say $F = \frac{1}{2}\Sigma \beta_s \dot{q}_s^2 + \Sigma\Sigma \gamma_{rt} \dot{q}_r \dot{q}_t$. The value given by equation (145) now becomes

$$\sum_1^n \frac{\partial^2 F}{\partial p_s \partial \dot{q}_s} = \Sigma \alpha_s \beta_s \quad \text{.........................(146).}$$

Since E and F must necessarily be positive for all values of the variables, all the α's and β's must necessarily be positive or zero, and therefore each member of equation (146) is positive or zero.

Since all the α's are necessarily positive and not zero, the only way in which $\Sigma \alpha_s \beta_s$ can be equal to zero is by all the β's vanishing, and this in turn, since F cannot be negative for any values of the variables, demands that all the remaining coefficients in F should vanish, so that $F = 0$. This proves the required result.

Non-conservative System.

84. For any system for which F is not equal to zero, equation (144) shews that τ must continually increase; in other words the points which form any cluster in the generalised space will, as the motion progresses, continually crowd closer and closer together.

The dissipation-function F of equation (131) is defined to be half the rate at which energy is lost by dissipation, so that

$$\frac{dE}{dt} = -2F \quad \text{..............................(147),}$$

a result which can also of course be deduced from equations (131) and (132) directly.

Since F is supposed positive, this equation expresses that as the representative points move in the $2n$-dimensional space, they continually pass from higher to lower values of E. Thus the motion of the points is one in which τ always increases while E always decreases; it consists of a movement of concentration upon the points in the generalised space at which E has certain minimum values, these points being defined by the condition $F = 0$. No matter how the representative points are started in the generalised space, they will after a sufficient time all be concentrated at these special points.

This result contains the solution of the problem of finding the ultimate final state of any dynamical system whatever, except only in the one special case in which there is perfect conservation of energy. In this case the condition $F = 0$, which in general determines the final resting places of the representative points, is satisfied throughout the whole of the generalised space, and the result just obtained becomes useless.

Conservative System.

Liouville's Theorem.

85. For a dynamical system which is perfectly free from dissipation of energy, $F = 0$, and equation (144) reduces to

$$\frac{D\tau}{Dt} = 0 \quad \text{.................................(148).}$$

This result, first enunciated by Liouville, shews that when the motion of a dynamical system is governed by the equations of Hamilton ((133) and (134)), the density of any cluster of representative points remains unaltered as the motion progresses. Thus there is no tendency for the representative points to crowd into any special region or regions in the generalised space.

A particular instance of this general theorem has been worked out in detail in § 40.

86 This result shews that the problem of searching for the final state of a system must now be treated in a different manner from that followed, for the case of a non-conservative system, in § 84.

Imagine that a system is found invariably to possess a certain property (*e.g.* maximum entropy) after being left to itself for a sufficient time. This might *à priori* be expected to be for one of two reasons: either that the points in the representative space tend to crowd into those regions of the space for which the property is true, or else that the property is true for the whole of the space. For conservative systems, Liouville's theorem excludes the first possibility; the second reason must therefore be the true one. We are therefore led to search for properties such as are true for the whole of the generalised space.

This set of ideas must however be examined somewhat more in detail.

Normal Properties and the Normal State.

87. Let us fix our attention on a certain property P, which is such that the system under consideration possesses this property in some states but not in all.

Since the system now under consideration is supposed to be a conservative system, the value of E will remain the same throughout the motion: the representative points will move on the surfaces $E = \text{constant}$ in the generalised space. Let us confine our attention to that part of the generalised space which represents systems having energy close to some given value of E, say systems of energy between E and $E+dE$. And let us suppose that of this region, a volume W_1 represents states in which the system possesses the property P, while a volume W_2 represents states in which the system does not possess the property P.

Choosing coordinates of position and momenta for the system at random is the same thing as selecting a point at random from the whole of the generalised space. Choosing coordinates at random subject only to the condition that the energy shall lie between E and $E+dE$ is the same thing as selecting a point at random from the region of the space for which the energy lies between E and $E+dE$. It follows that if a system of energy between E and $E+dE$ has its coordinates assigned to it at random, the probability of its possessing the property P will be

$$\frac{W_1}{W_1+W_2} \qquad \text{.............................(149).}$$

A different problem is that of examining what is the probability that a system initially selected at random subject only to the condition of its energy being between E and $E+dE$ shall have the property P after following out its natural motion for a time t. Let us suppose the thin shell in the generalised space which lies between the surfaces E and $E+dE$ to be filled with a cloud of representative points, so close together that they may be regarded as forming a continuous fluid, and let these points be distributed initially so that the density of this fluid is uniform. Then each of these points has an equal chance of representing the system selected initially at random. Let this cloud of points move for any time t, in accordance with the equations of motion of the system. Then from the conservation of energy, it follows that the points will at the end of the time t still lie between the surfaces E and $E+dE$, and from the equation $\frac{D\tau}{Dt}=0$, it follows that the fluid will still be of uniform density. The number of points which, after time t, represent systems possessing the property P will accordingly be a fraction $\frac{W_1}{W_1+W_2}$ of the whole, and therefore the same fraction measures the probability that the system shall possess the property P after time t.

It follows that if a conservative system is found always to possess the property P after a sufficient time has elapsed, this can only be because the ratio $W_1 : W_2$ is infinite. The following definitions will be convenient:

DEFINITIONS. *A property P which is such that the ratio W_1/W_2 is infinite will be called a normal property of the system.*

A system which possesses all the normal properties of which it is capable will be said to be in the normal state.

So long as a system is thought of as having only a finite number $2n$ of coordinates, it is natural to expect the ratio W_1/W_2 corresponding to any property P to have a finite value, but as soon as $2n$ is made infinite, it is not surprising that W_1/W_2, which will in general be a function of n, should become infinite or zero.

88. We are now in a position to answer the question as to what is the final state to be expected in a conservative system.

The system will be capable of possessing certain properties $P_1, P_2, P_3, \ldots$. These properties will, in general, change with the time, some of them very slowly, some more quickly, some with extreme rapidity. We may suppose that the property P_1 may in general be expected to change in a time comparable with t_1, the property P_2 in a time comparable with t_2, and so on.

After a time t which is very large compared with all of the quantities $t_1, t_2, \ldots$, the system will have had ample time to change all its properties. The influence of the initial conditions will in a sense have disappeared; the representative point in the generalised space will have had time to move away from the special regions in which it may have started, where any normal property does not hold. The system may therefore be expected to possess all the normal properties, and therefore to be in the normal state.

89. A complication can arise from the possibility of the system having properties which are not capable of change at all, or for which the time of change is infinite.

For instance, if a system is perfectly self-contained and subject to no external influence, its angular momentum must of necessity remain always equal to its initial value. Thus the property of the system having an angular momentum lying between certain limits is one which the system cannot acquire with the lapse of time; either the system will possess this property at starting, or will never possess it. An examination of the generalised space will shew that one value of the angular momentum, namely zero, is common to all of the generalised space except certain infinitesimal regions—for this property of having zero angular momentum, the ratio W_1/W_2 in the notation introduced in § 87 is infinite. But unless the system happens to have started with zero angular momentum, no time will be sufficient for this value of the angular momentum to be acquired. Thus the having of zero angular momentum, although a normal property of the system, is not to be regarded as an essential of the normal state; it is not required by the definition of the normal state, for it is not a property of which the system is capable.

On the other hand if the system is capable of varying its angular momentum, then the property of having the normal value for its angular momentum must be regarded as one of the properties of the normal state.

For instance, let the system be a gas enclosed in a fixed closed vessel. Except in very special cases, such as that referred to in § 42, the gas can change its angular momentum, and it is easily seen that the possessing of zero angular momentum is one of the normal properties of the system. Hence in the final state of the system we must expect the angular momentum of the gas to be zero.

90. One property which can never be changed in a conservative system is that of having a certain value for the energy. For this reason, in defining the normal state, we considered only systems having a specified amount of energy, namely energy between E and $E+dE$. In the same way if the system has other quantities or properties which are invariable, account must be taken of this invariability in specifying the normal state. The various complications which may arise in this way are somewhat difficult to discuss in general terms, but are not difficult to treat in particular cases, as the various examples which occur in the present book will shew.

The Normal Partition of Energy.

91. The normal properties which may be considered first are those associated with the partition of energy.

If the $2n$ coordinates of position and velocity (135) are now denoted by $\theta_1, \theta_2, \ldots \theta_{2n}$, the energy E will be of the general form

$$E = f(\theta_1, \theta_2, \ldots \theta_{2n}) \quad \text{.......................(150).}$$

Let us however suppose that the energy E can be divided into separate and distinct parts $E_1, E_2, \ldots$, such that E_1 depends only on one group of coordinates, say $\theta_1, \theta_2, \ldots \theta_s$; E_2 depends only on another group, distinct from the former, $\theta_{s+1}, \theta_{s+2}, \ldots, \theta_{s+t}$, and so on. Also let it be supposed that each of these groups contains a number $(s, t, \ldots)$ of coordinates which is so great that it may be treated as infinite. Then

$$E = E_1 + E_2 + \ldots = f_1(\theta_1, \theta_2, \ldots \theta_s) + f_2(\theta_{s+1}, \theta_{s+2}, \ldots \theta_{s+t}) + \ldots \quad \text{...(151).}$$

Let us define the property P as being possessed by the system when there is a certain partition of energy, namely one in which

$$\left.\begin{array}{l} E_1 \text{ lies within a small range } E_1 \text{ to } E_1 + dE_1, \\ E_2 \quad \text{"} \quad \text{"} \quad \text{"} \quad E_2 \text{ to } E_2 + dE_2, \text{ etc.} \end{array}\right\} \text{......(152).}$$

Then the volume W_1 of the generalised space, within which the property P holds, is given by

$$W_1 = \iiint \ldots \int d\theta_1 d\theta_2 \ldots d\theta_{2n} \quad \text{.....................(153),}$$

where the integration is taken throughout the region defined by the conditions

$$E_1 < f_1(\theta_1, \theta_2, \ldots \theta_s) < E_1 + dE_1 \quad \text{.....................(154),}$$

$$E_2 < f_2(\theta_{s+1}, \theta_{s+2}, \ldots \theta_{s+t}) < E_2 + dE_2 \quad \text{..............(155),}$$

and so on.

The integral W_1 may be written in the form of the product

$$W_1 = (\iint \dots \int d\theta_1 d\theta_2 \dots d\theta_s)(\iint \dots \int d\theta_{s+1} d\theta_{s+2} \dots d\theta_{s+t})(\dots) \quad \dots(156),$$

in which the first integral has to be taken within the limits specified by (154), the second integral within the limits (155), and so on. Clearly the first integral in the product can depend only on E_1 and dE_1, and so must be of the form $F_1(E_1)\,dE_1$. Similarly, the second must be of the form $F_2(E_2)\,dE_2$, and so on. Thus we must have

$$W_1 = F_1(E_1)\,F_2(E_2)\dots dE_1 dE_2 \dots \quad \dots\dots\dots\dots(157).$$

It would now be possible to attempt to evaluate the ratio W_1/W_2 (cf. § 87), but it will be easiest to attack first the simpler problem of finding for what values of $E_1, E_2, \dots$ the ratio W_1/W_2 has its maximum value. In other words, we shall search for the most probable partition of energy without at first attempting to prove that it is a normal partition.

92. Since $W_1 + W_2$ represents a constant space for all partitions of energy, namely the total space for which E lies within fixed limits E and $E + dE$, it follows that W_1/W_2 will be a maximum when W_1 is a maximum. If $E_1, E_2, \dots$ refer to that particular partition of energy which makes W_1 a maximum, then $\delta \log W_1$ will vanish when $E_1, E_2, \dots$ are subjected to slight variations $\delta E_1, \delta E_2, \dots$ provided these variations are subject to the condition

$$\delta E_1 + \delta E_2 + \dots = 0 \dots\dots\dots\dots\dots\dots(158),$$

expressing that the total energy remains unaltered. Thus we must have

$$\frac{d \log F_1(E_1)}{dE_1}\,\delta E_1 + \frac{d \log F_2(E_2)}{dE_2}\,\delta E_2 + \dots = 0 \quad \dots\dots\dots(159),$$

for all values of $\delta E_1, \delta E_2, \dots$ which satisfy relation (158).

Replacing δE_1 by $-\delta E_2 - \delta E_3 - \dots$, equation (159) becomes

$$\left\{\frac{d \log F_2(E_2)}{dE_2} - \frac{d \log F_1(E_1)}{dE_1}\right\} \delta E_2 + \left\{\frac{d \log F_3(E_3)}{dE_3} - \frac{d \log F_1(E_1)}{dE_1}\right\} \delta E_3 + \dots = 0 \dots (160),$$

and, since this must now be true for all values of $\delta E_2, \delta E_3, \dots$, we must have

$$\frac{d \log F_1(E_1)}{dE_1} = \frac{d \log F_2(E_2)}{dE_2} = \dots \quad \dots\dots\dots\dots\dots(161).$$

The solution of these equations, together with

$$E = E_1 + E_2 + \dots \quad \dots\dots\dots\dots\dots\dots\dots(162),$$

will give the most probable partition of energy for a system of assigned total energy E.

For the moment we shall not attempt to prove that this partition of energy represents a normal state: we shall assume this provisionally, and prove it later for the special cases in which it is of importance.

Thermodynamics, Entropy and Temperature.

93. Let us put

$$P = \log [F_1(E_1) . F_2(E_2) . \ldots] \quad \text{.....................(163).}$$

Suppose that a quantity dQ of energy is added to the system from outside. Since the system is supposed conservative, the effect of this must be to increase the total energy of the system from E to $E + dE$, where $dE = dQ$, and in doing so, it may be supposed to increase $E_1, E_2, \ldots$ to $E_1 + dE_1, E_2 + dE_2, \ldots$, where

$$dE_1 + dE_2 + \ldots = dQ \quad \text{........................(164).}$$

We shall suppose that before and after, and also during, the addition of heat, the partition of energy is always the most probable, so that equations (161) are true at every instant.

The change produced in P is, from equation (163), given by

$$dP = \frac{d \log F_1(E_1)}{dE_1} dE_1 + \frac{d \log F_2(E_2)}{dE_2} dE_2 + \ldots \quad \text{......(165).}$$

From equation (161), the coefficients of $dE_1, dE_2, \ldots$ in this equation have all the same value. Call this k, then

$$dP = k(dE_1 + dE_2 + \ldots) = kdQ \quad \text{..................(166).}$$

This shews that kdQ is a perfect differential, or in other words that k is an integrating multiplier of the differential dQ.

94. From general thermodynamical theory, another integrating multiplier of the differential dQ is known, namely $\frac{1}{T}$, where T is the temperature measured on the thermodynamical scale. If ϕ is the entropy we know that

$$d\phi = \frac{dQ}{T} \quad \text{..............................(167);}$$

this is in fact the simplest expression of the Second Law of Thermodynamics.

The circumstance that both k and $1/T$ are found to be integrating multipliers of the differential dQ does not of course justify us in identifying k with $1/T$. It does enable us, however, to establish a simple relation between them.

95. The energy E of the system will in general depend on a number of variables $\xi, \eta, \zeta, \ldots$ specifying the physical state of the system or of its different parts. Thus the amount of heat dQ which must be added to produce any specified change must be of the form

$$dQ = Ld\xi + Md\eta + Nd\zeta + \ldots \quad \text{..................(168),}$$

(cf. equation (446) below for a specific instance) and we shall have

$$dP = kLd\xi + kMd\eta + kNd\zeta + \ldots \quad \text{..................(169).}$$

Hence we must have

$$kL = \frac{\partial P}{\partial \xi}, \quad kM = \frac{\partial P}{\partial \eta}, \text{ etc. } \quad \text{.....................(170).}$$

Similarly, from relation (167) we obtain

$$\frac{L}{T}=\frac{\partial\phi}{\partial\xi}, \quad \frac{M}{T}=\frac{\partial\phi}{\partial\eta}, \text{ etc.} \quad \text{.....................(171).}$$

It follows that

$$\frac{\dfrac{\partial P}{\partial\xi}}{\dfrac{\partial\phi}{\partial\xi}}=\frac{\dfrac{\partial P}{\partial\eta}}{\dfrac{\partial\phi}{\partial\eta}}=\frac{\dfrac{\partial P}{\partial\zeta}}{\dfrac{\partial\phi}{\partial\zeta}}=\dots \quad \text{.........................(172),}$$

the value of each fraction being kT. From this it follows that there must be a functional relation between P and ϕ, say $P=f(\phi)$. On substituting this value of P into equation (172), this equation becomes

$$k=\frac{1}{T}f'(\phi) \quad \text{..............................(173),}$$

where $f'(\phi)$ stands for $\dfrac{\partial f(\phi)}{\partial\phi}$, expressing the relation between the two integrating factors k and $1/T$.

96. The quantity k is, however, the value of each of the fractions in equation (161), $\dfrac{d\log F_1(E_1)}{dE_1}$, etc. Thus equation (173) gives the value of E_1 in terms of $\dfrac{1}{T}f'(\phi)$, and so on for $E_2, E_3, \dots$. The value of E_1 would obviously be changed by a change in temperature, but it could not be altered by a change in which the temperature T remained unaltered while the entropy was changed by an alteration in some parts of the system not involving that of energy E_1. Thus E_1 cannot be changed by changes in ϕ, and therefore k as given by equation (173) cannot be changed by changes in ϕ. It follows that $f'(\phi)$ must be a constant.

Let this constant be denoted by $\dfrac{1}{R}$, then $k=\dfrac{1}{RT}$.

97. Equations (161) now become

$$\frac{d\log F_1(E_1)}{dE_1}=\frac{d\log F_2(E_2)}{dE_2}=\dots=\frac{1}{RT} \quad \text{...........(174),}$$

while the relation $P=f(\phi)$ becomes

$$P=\frac{1}{R}\phi+\text{a constant} \quad \text{........................(175).}$$

Thus the entropy is given by

$$\phi=RP+\text{a constant} \quad \text{........................(176),}$$

or again, on comparing equations (163) and (157),

$$\phi=R\log W_1+\text{a constant} \quad \text{.....................(177).}$$

98. This last result throws a flood of light on the meaning of the analysis of the last few sections. It shews that the partition of energy which is most likely—*i.e.* for which W_1 is a maximum—is exactly that which makes the entropy a maximum. If we like to assume, as a general physical principle, that every system tends to a final state in which the entropy is a maximum, then this state must be that for which W_1 is a maximum, and must therefore be given by equations (174). If this assumption is made, it follows at once that the configuration for which W_1 is a maximum is also one for which W_1/W_2 is infinite, and therefore is the normal state as defined in § 87. But this assumption need hardly be made, for, as we shall see (§ 103), a direct proof can be given in all cases which are of physical importance.

Equipartition of Energy.

99. Equations (174) give $E_1, E_2, \ldots$ completely in terms of the temperature, but they can only be solved in the special cases in which the functions $F_1(E_1)$, etc. can be evaluated, and of these the only case which is of any physical importance is that in which E_1 is a homogeneous quadratic function of the coordinates involved. This covers the case of E_1 being kinetic energy, or the potential energy of small displacements from a position of equilibrium, or the energy of any type of isochronous vibration.

100. Let the coordinates which enter in E_1 be supposed, as in § 91, to be s in number, and let E_1 be given by

$$E_1 = f_1(\theta_1, \theta_2, \ldots \theta_s) = c_{11}\theta_1^2 + c_{22}\theta_2^2 + \ldots + 2c_{12}\theta_1\theta_2 + \ldots \quad \ldots(178).$$

Then $F_1(E_1)\,dE_1$ is by definition the value of the integral

$$\iint \ldots \int d\theta_1\, d\theta_2 \ldots d\theta_s \quad \ldots\ldots\ldots\ldots\ldots\ldots\ldots(179),$$

taken over all values of $\theta_1, \theta_2, \ldots \theta_s$ for which $f_1(\theta_1, \theta_2, \ldots \theta_s)$ lies between E_1 and $E_1 + dE_1$.

By a linear transformation of the old coordinates, new coordinates $\phi_1, \phi_2, \ldots \phi_s$ can be obtained such that the value of E_1 in these coordinates is

$$E_1 = \alpha(\phi_1^2 + \phi_2^2 + \ldots + \phi_s^2) \quad \ldots\ldots\ldots\ldots\ldots\ldots(180),$$

and if μ is the modulus of this transformation, namely $\dfrac{\partial(\theta_1, \theta_2, \ldots \theta_s)}{\partial(\phi_1, \phi_2, \ldots \phi_s)}$, the value of the integral (179) will be

$$\mu \iint \ldots \int d\phi_1\, d\phi_2 \ldots d\phi_s \quad \ldots\ldots\ldots\ldots\ldots\ldots(181).$$

The value of this integral, taken for all values of the variables for which $\alpha(\phi_1^2 + \phi_2^2 + \ldots + \phi_s^2)$ is less than E_1, is equal to μ times the volume of a sphere of radius $\sqrt{\left(\dfrac{E_1}{\alpha}\right)}$ in a space of s-dimensions. It is therefore

$$\mu \frac{\pi^{\frac{1}{2}s}}{\Gamma(\frac{1}{2}s + 1)} \left(\frac{E_1}{\alpha}\right)^{\frac{1}{2}s} \quad \ldots\ldots\ldots\ldots\ldots\ldots\ldots(182).$$

By differentiation with respect to E_1, the value of the integral, taken for all values of $\phi_1, \phi_2, \dots \phi_s$ for which $\alpha(\phi_1^2 + \phi_2^2 + \dots + \phi_s^2)$ lies between E_1 and $E_1 + dE_1$, is

$$\mu \frac{\pi^{\frac{1}{2}s}}{\Gamma(\frac{1}{2}s)} \alpha^{-\frac{1}{2}s} E_1^{\frac{1}{2}s-1} dE_1 \quad \dots\dots\dots\dots\dots\dots(183),$$

and this is precisely the quantity which has been called $F_1(E_1)\, dE_1$. Thus we have as the required value of $F_1(E_1)$,

$$F_1(E_1) = \mu \frac{\pi^{\frac{1}{2}s}}{\Gamma(\frac{1}{2}s)} \alpha^{-\frac{1}{2}s} E_1^{\frac{1}{2}s-1} \quad \dots\dots\dots\dots\dots(184).$$

The most probable partition of energy is given by equations (174), so that in this most probable partition, E_1 is given by

$$\frac{d \log F_1(E_1)}{dE_1} = \frac{1}{RT} \quad \dots\dots\dots\dots\dots\dots\dots(185),$$

leading at once, on using the value of $F_1(E_1)$ just found, to

$$E_1 = (\tfrac{1}{2}s - 1)\, RT.$$

Since we have already supposed s to be a very great number, the difference between $\frac{1}{2}s - 1$ and $\frac{1}{2}s$ may be neglected. Also for all the parts of the energy, say $E_2, E_3, \dots$, for which the energy function is quadratic, the most probable values of the energy may be evaluated in the same way, and so we find that the most probable partition of energy is given, as regards those parts of the energy for which the energy-function is quadratic, by the equations

$$E_1 = \tfrac{1}{2}sRT, \quad E_2 = \tfrac{1}{2}tRT, \text{ etc.} \quad \dots\dots\dots\dots\dots(186),$$

where $s, t, \dots$ are the number of coordinates concerned in the quadratic functions $E_1, E_2, \dots$.

101. It is now possible to prove that, as far at least as these parts of the energy are concerned, the partition of energy expressed by equations (186) is not only the most likely partition, but also expresses a "normal" partition in the sense of § 87; that is to say, this partition is infinitely more probable than any other.

102. Let $E_1, E_2, \dots$ specify the most probable partition of energy as given by equations (186), and let $E_1 + \epsilon_1, E_2 + \epsilon_2, \dots$ specify any other partition of energy corresponding to the same total energy. Then we must have

$$\epsilon_1 + \epsilon_2 + \dots = 0 \quad \dots\dots\dots\dots\dots\dots\dots\dots(187).$$

The general value of W_1 is seen from equations (157) and (163) to be

$$W_1 = e^P\, dE_1 dE_2 \dots \quad \dots\dots\dots\dots\dots\dots\dots(188),$$

while the whole value of $W_1 + W_2$ will be

$$W_1 + W_2 = \iint \dots e^P\, d\epsilon_1 d\epsilon_2 \dots \quad \dots\dots\dots\dots\dots(189).$$

For the partition of energy $E_1 + \epsilon_1, E_2 + \epsilon_2, \ldots$

$$P = \Sigma \log F_1 (E_1 + \epsilon_1)$$
$$= \Sigma \log \left(1 + \frac{\epsilon_1}{\frac{1}{2} s RT}\right)^{\frac{1}{2}s} + P_0,$$

where P_0 is a sum of constant terms not involving $\epsilon_1, \epsilon_2, \ldots$, and so is the value of P for the partition of energy $E_1, E_2, \ldots$. Expanding the logarithm, we obtain

$$P - P_0 = \Sigma \frac{\epsilon_1}{RT} \left\{1 - \frac{1}{2} \frac{\epsilon_1}{\frac{1}{2} s RT} + \frac{1}{3} \left(\frac{\epsilon_1}{\frac{1}{2} s RT}\right)^2 - \ldots\right\}.$$

and in virtue of relation (187), the first term in this sum vanishes, leaving

$$P - P_0 = -\Sigma \frac{\epsilon_1^2}{s (RT)^2} + \frac{4}{3} \Sigma \frac{\epsilon_1^3}{s^2 (RT)^3} - \ldots \quad \text{.............(190).}$$

It has already been seen that the only stationary value of P is given by $\epsilon_1 = \epsilon_2 = \ldots = 0$. This makes $P = P_0$, and an inspection of the right-hand of equation (190) shews that this value is a true maximum. It follows that the right-hand of equation (190) is negative for all values of the ϵ's.

As we recede from the value $\epsilon_1 = \epsilon_2 = \ldots = 0$, it is clear that $P - P_0$ becomes finite as soon as ϵ_1 becomes comparable with $\sqrt{s} . RT$, ϵ_2 with $\sqrt{t} . RT$, and so on. For such values of $\epsilon_1, \epsilon_2, \ldots$ the first term of (190) is infinitely greater than any of the succeeding terms, and the value of $P - P_0$ reduces to

$$P - P_0 = -\Sigma \frac{\epsilon_1^2}{s (RT)^2} \quad \text{........................(191).}$$

For values of $\epsilon_1, \epsilon_2, \ldots$ greater than these $P - P_0$ becomes equal to $-\infty$, so that e^{-P} vanishes in comparison with e^{-P_0}.

It follows that the whole value of the integral (189) comes from a small range of values surrounding the values $\epsilon_1 = \epsilon_2 = \ldots = 0$; *i.e.* the values of $E_1, E_2, \ldots$ given by equations (186). Thus the integral (189) reduces to the right-hand member of equation (188), the small range dE_1 being comparable with $\sqrt{s} . RT$, the small range dE_2 being comparable with $\sqrt{t} . RT$, and so on. These small ranges are of course small in comparison with the whole values of $E_1, E_2, \ldots$; thus dE_1 is comparable with $\frac{E_1}{\sqrt{s}}$, dE_2 with $\frac{E_2}{\sqrt{t}}$, and so on.

With such values for the small ranges $dE_1, dE_2, \ldots$, the value of $W_1 + W_2$ given by equation (189) becomes identical with the value of W_1 given by equation (188). Thus we have W_1/W_2 infinite, shewing that the partition of energy now under consideration is a normal property of the system.

We have accordingly shewn that those parts of a system, say $E_1, E_2, \ldots$, in which the energy is of quadratic type, will necessarily tend to the partition of energy specified by equations (186). These equations express the Theorem of Equipartition of Energy:

The energy to be expected for any part of the total energy which can be expressed as a sum of squares is at the rate of $\frac{1}{2}RT$ for every squared term in this part of the energy.

103. The proof that the remaining parts of the system, if any, in which the energy is not of this type, will necessarily tend to the partition of energy given by equations (174) is more difficult. In place of equation (191), we have

$$P - P_0 = \Sigma \tfrac{1}{2}\epsilon_1^2 \frac{\partial^2 \log F_1(E_1)}{\partial E_1^2},$$

and the sign of the terms on the right must necessarily be a matter of uncertainty so long as the form of the energy-function remains unspecified. It is, however, clear that the arrangement of the loci $E_1 =$ cons., $E_2 =$ cons., etc., in the generalised space must, in every case, be of the same general type as that in the simple case just considered, from which we may infer that $P - P_0$ must, in the more general case also, be of negative sign. It again follows that W_1/W_2 must be infinite, so that the most probable partition of energy, as expressed by equations (174), is now seen to be a normal property of the system.

We accordingly see that every system must pass to a final state in which W_1, and therefore also the Entropy, is a maximum. In this way we obtain an analytical proof of the second law of thermodynamics, which may now be regarded as being on a mathematical, instead of on a purely empirical basis.

Law of Distribution of Coordinates.

104. Not only will there be a normal partition of energy, but there will also be a normal way for the separate coordinates to be arranged so as to give this particular energy. This has already been found in Chapter III for the simple case of a gas composed of molecules which behave like hard elastic spheres. A similar method may be applied to the more general problem.

Let us suppose that part of the dynamical system under consideration consists of N similar units, which we may think of as molecules for definiteness, each unit possessing p degrees of freedom, and therefore having its state specified by $2p$ quantities $\phi_1, \phi_2, \ldots \phi_{2p}$, these being coordinates of position and their corresponding momenta, as in § 78.

Imagine a generalised space of $2p$ dimensions constructed, having

$$\phi_1, \phi_2, \ldots \phi_{2p}$$

as orthogonal coordinates. Then the state of any molecule of the system can be represented by a single point in this space, namely the point whose coordinates are equal to the coordinates $\phi_1, \phi_2, \ldots \phi_{2p}$ specifying the state (*i.e.* velocity and positional coordinates) of the molecule. The states of all the molecules can be represented by a collection of points in this space, one point for each molecule. The problem before us is that of finding the law according to which these points are distributed in the space.

Let τ denote the density of points in this space—the quantity which we are trying to find—so that

$$\tau d\phi_1 d\phi_2 \dots d\phi_{2p} \quad \text{.........................(192)}$$

will be the number of points (or molecules) such that ϕ_1 lies between ϕ_1 and $\phi_1 + d\phi_1$, ϕ_2 between ϕ_2 and $\phi_2 + d\phi_2$, and so on.

Thus τ is a function of $\phi_1, \phi_2, \dots \phi_{2p}$.

105. Following the procedure adopted in §§ 43, 50, let the whole space be divided up into n small rectangular elements of volume, each of equal size ω, and let these be identified by numbers 1, 2, 3, ... Let us fix our attention on a special distribution of points, which is such that the number of points in elements 1, 2, 3, ... are respectively $a_1, a_2, a_3, \dots$. Let any distribution of points giving these particular numbers $a_1, a_2, a_3, \dots$ be spoken of as a distribution of class A. Similarly any distribution of points giving another set of numbers $b_1, b_2, b_3, \dots$ may be spoken of as a distribution of class B, and so on.

Each point in the original generalised space of § 80 will correspond to a complete distribution of points in the space now under consideration. The distribution corresponding to some of these original points will be a distribution of points of class A, corresponding to others it will be a distribution of class B, and so on. We proceed to evaluate the volume, say W_A, of the original generalised space which is such that the points in it represent systems for which the distribution of coordinates is of class A.

This volume is readily seen to be given by

$$W_A = \frac{N!}{a_1! \, a_2! \, a_3! \dots} \omega^N \iiint \dots d\chi_1 d\chi_2 d\chi_3 \dots \quad \text{...........(193).}$$

In this expression the first factor represents, as already in § 43, the number of ways in which it is possible to distribute the N points representing the N different molecules, between the n different elements, subject only to the condition of the final arrangement being of type A. The remaining factor, say λ, given by

$$\lambda = \omega^N \iiint \dots d\chi_1 d\chi_2 d\chi_3 \dots \quad \text{....................(194)}$$

represents the volume of the generalised space which corresponds to each one of these arrangements, $\chi_1, \chi_2, \chi_3, \dots$ being coordinates of parts of the system other than the N molecules under consideration. If we write

$$\theta_A = \frac{N!}{a_1! \, a_2! \, a_3! \dots} \quad \text{.........................(195),}$$

and use a similar notation for a system of class B, etc., then the volumes W_A, W_B, ... are given by

$$W_A = \lambda\theta_A, \quad W_B = \lambda\theta_B, \text{ etc.} \qquad \text{(196)}.$$

Using Stirling's Theorem, as in § 44, we find

$$\log \theta_A = (N + \tfrac{1}{2}) \log N - \tfrac{1}{2}(n-1) \log 2\pi - \sum_{s=1}^{s=n} (a_s + \tfrac{1}{2}) \log a_s \ldots \text{(197)},$$

and if, as in equation (72), we put

$$K_a = \frac{1}{N} \sum_{s=1}^{s=n} (a_s + \tfrac{1}{2}) \log \frac{na_s}{N} \qquad \text{(198)},$$

this gives as the value of θ_A (cf. equation (73))

$$\theta_A = n^{N+\frac{1}{2}n} (2\pi N)^{-\frac{1}{2}(n-1)} e^{-NK_a} \qquad \text{(199)}.$$

Since $W_A = \lambda\theta_A$, etc., it is clear that W_A is proportional to e^{-NK_a}.

The most probable partition of energy is obviously obtained by making W_A a maximum, and therefore K a minimum, for different values of $a_1, a_2, \ldots$.

As in § 45, we find for the variation of K

$$\delta K = \frac{1}{N} \sum_{s=1}^{s=n} \left\{ \log \frac{na_s}{N} + 1 + \frac{1}{2a_s} \right\} \delta a_s \qquad \text{(200)}.$$

The variations $\delta a_1, \delta a_2, \ldots$ are not independent. They are necessarily connected by two relations, and in some cases by more. Of the two relations which are certain, the first expresses that the total number of molecules is equal to the prescribed number N; it is therefore expressed by the equation

$$\sum_{s=1}^{s=n} \delta a_s = 0 \qquad \text{(201)},$$

which is obtained on variation of the equation

$$\sum_{s=1}^{s=n} a_s = N \qquad \text{(202)}.$$

The second relation expresses that the total energy of the N molecules is equal to the allotted amount E_1. Let ϵ_1 denote the energy associated with a molecule represented by a point in cell 1, so that ϵ_1 is a function of $\phi_1, \phi_2, \ldots \phi_{2p}$, the coordinates of the first cell. Let ϵ_2 be the energy associated with a molecule represented by a point in cell 2, and so on. Then the total energy of the N molecules, when the distribution of coordinates is of class A, will clearly be $a_1\epsilon_1 + a_2\epsilon_2 + \ldots$, so that we must have

$$\sum_{s=1}^{s=n} \epsilon_s a_s = E_1 \qquad \text{(203)},$$

and on variation of this we obtain the relation

$$\sum_{s=1}^{s=n} \epsilon_s \delta a_s = 0 \qquad \text{(204)}.$$

If there are other relations they will in general be derived from equations of this same type, expressing that the total of some quantity μ summed over all the molecules will have an assigned value. The integral equation will be of the form

$$\sum_{s=1}^{s=n} \mu_s a_s = M \quad \text{.............................(205)},$$

and the corresponding relation between the quantities δa_1, δa_2, ... will be

$$\sum_{s=1}^{s=n} \mu_s \delta a_s = 0 \quad \text{.............................(206)}.$$

In many problems there will be six equations of this type, the different μ's representing three components of linear momentum, and three components of angular momentum. We may, however, be content to take one relation as typical of all, and shall suppose it to be given by the equations just written down.

Following a well-known procedure, we now multiply equations (201), (204), (206) by undetermined multipliers p, q, r, and add corresponding members of these equations and equation (200). We obtain

$$\delta K = \sum_{s=1}^{s=n} \left\{1 + \frac{1}{2a_s} + \log \frac{na_s}{N} + p + q\epsilon_s + r\mu_s\right\} \delta a_s,$$

and the maximum value of K is now given by the equation

$$1 + \frac{1}{2a_s} + \log \frac{na_s}{N} + p + q\epsilon_s + r\mu_s = 0, \quad (s = 1, 2, \ldots n)$$

Since a_1, a_2, ... are all supposed to be large quantities, the term $\frac{1}{2a_s}$ may be neglected, and we obtain

$$\frac{na_s}{N} = e^{-(1+p)} e^{-(q\epsilon_s + r\mu_s)} \quad \text{........................(207)}.$$

If τ is the quantity defined in expression (192) the value of a_s will be $\tau\omega$, where τ refers to the s-th cell. Equation (207) becomes

$$\tau = \frac{N}{n\omega} e^{-(1+p)} e^{-(q\epsilon + r\mu)},$$

and since equation (207) was true for every cell, this equation will hold for all values of ϕ_1, ϕ_2, ... ϕ_{2p}. Changing the constants, this equation may be rewritten

$$\tau = Ce^{-2h\epsilon} e^{-(r_1\mu_1 + r_2\mu_2 + \ldots)} \quad \text{....................(208)},$$

in which the one typical quantity μ is now replaced by the actual series of quantities μ_1, μ_2, ... Using this value for τ, the law of distribution of coordinates (cf. expression (192)) is seen to be given by

$$Ce^{-2h\epsilon} e^{-(r_1\mu_1 + r_2\mu_2 + \ldots)} d\phi_1 d\phi_2 \ldots d\phi_{2p} \quad \text{...........(209)}.$$

In the particular case in which there are no constant quantities except the energy, this reduces to

$$Ce^{-2h\epsilon}\, d\phi_1 d\phi_2 \dots d\phi_{2p} \quad \text{.......................(210).}$$

106. Suppose that certain of the coordinates, say $\phi_1, \phi_2, \dots \phi_s$, enter into the energy ϵ only through their squares, so that the value of ϵ is of the form

$$\epsilon = \tfrac{1}{2}\beta_1\phi_1^2 + \tfrac{1}{2}\beta_2\phi_2^2 + \dots + \tfrac{1}{2}\beta_s\phi_s^2 + \Phi,$$

where Φ does not involve $\phi_1, \phi_2, \dots \phi_s$, but only $\phi_{s+1}, \phi_{s+2}, \dots \phi_{2p}$. Then the law of distribution may be written in the form

$$C(e^{-h\beta_1\phi_1^2}\, d\phi_1)(e^{-h\beta_2\phi_2^2}\, d\phi_2) \dots (e^{-h\beta_s\phi_s^2}\, d\phi_s)\, e^{-2h\Phi}\, d\phi_{s+1} \dots d\phi_{2p}.$$

This shews that there is no correlation between the distributions of $\phi_1, \phi_2, \dots \phi_s$. The law of distribution of any single coordinate, say ϕ_1, is of the form

$$\sqrt{\frac{h\beta_1}{\pi}}\, e^{-h\beta_1\phi_1^2}\, d\phi_1 \quad \text{.......................(211),}$$

the constant being determined from the condition that the integral, taken from $\phi = -\infty$ to $\phi = +\infty$, shall be equal to unity.

The mean value of the contribution from ϕ_1 to the energy is

$$\overline{\tfrac{1}{2}\beta_1\phi_1^2} = \frac{\int_0^\infty \tfrac{1}{2}\beta_1\phi_1^2\, e^{-h\beta_1\phi_1^2}\, d\phi_1}{\int_0^\infty e^{-h\beta_1\phi_1^2}\, d\phi_1} = \frac{1}{4h},$$

and similarly for the other coordinates. The mean value of

$$\tfrac{1}{2}\beta_1\phi_1^2 + \tfrac{1}{2}\beta_2\phi_2^2 + \dots + \tfrac{1}{2}\beta_s\phi_s^2$$

is accordingly $\frac{s}{4h}$, and since this has also been seen to be equal (cf. equations (186)) to $\tfrac{1}{2}sRT$, we must have

$$2hRT = 1,$$

expressing the constant h in terms of the temperature.

Of the coordinates $\phi_1, \phi_2, \dots \phi_s$, which enter the energy only through their squares, three may always be taken to be the velocities u, v, w. The results obtained in this section may accordingly be expressed by the equations

$$\overline{\tfrac{1}{2}mu^2} = \overline{\tfrac{1}{2}mv^2} = \overline{\tfrac{1}{2}mw^2} = \overline{\tfrac{1}{2}\beta_1\phi_1^2} = \overline{\tfrac{1}{2}\beta_2\phi_2^2} = \dots = \frac{1}{4h} = \tfrac{1}{2}RT \quad \text{...(212),}$$

expressing the theorem of equipartition of energy in a new form.

EXAMPLES OF DISTRIBUTION OF COORDINATES.

Gas with infinitesimal hard spherical molecules.

107. As a first instance of the use of the formulae just obtained, we may apply them to the case of the gas already considered in Chapter II, in which the molecules are infinitely small hard elastic spheres, the external physical conditions being the same at every point of the space occupied by the gas. Such molecules have only three degrees of freedom, representing their freedom to move in space, so that the $2p$ coordinates $\phi_1, \phi_2, \dots \phi_{2p}$ reduce to the six coordinates u, v, w, x, y, z. The number of molecules which at any instant will be in collision will be infinitesimally small, so that the potential energy of the gas, arising out of the elastic forces at collisions, will always be infinitesimal, and so may be neglected in comparison with the kinetic energy. The total energy of the gas may accordingly be supposed given by

$$E = \tfrac{1}{2} m \Sigma (u^2 + v^2 + w^2) \dots\dots\dots\dots (213),$$

so that ϵ in formula (209) may be put equal to $\frac{1}{2}m(u^2 + v^2 + w^2)$. We may further take μ_1, μ_2, μ_3 the three components of linear momentum to be identical with mu, mv, mw, and μ_4, μ_5, μ_6 the three components of angular momentum to be given by

$$\mu_4 = m(yw - zv) \text{ etc.}$$

Thus the law of distribution is found to be

$$Ce^{-hm(u^2+v^2+w^2)} e^{-m(r_1u+r_2v+r_3w)+\Sigma r_4(yw-zv)} \, du\,dv\,dw\,dx\,dy\,dz \quad \dots(214).$$

The quantities $C, h, r_1, r_2, r_3, r_4, r_5, r_6$, as yet undetermined, can be evaluated from a knowledge of the total number of molecules, the total energy and the total momenta, by the method already used in § 26.

108. *No mass-rotation.* In the commonest case in which the gas has no motion of rotation as a whole, each of the three components of angular momentum of the whole gas must vanish, and this is easily seen to lead tc the conditions $r_4 = r_5 = r_6 = 0$. The law of distribution is now seen to be

$$Ce^{-hm(u^2+v^2+w^2)} e^{-m(r_1u+r_2v+r_3w)} \, du\,dv\,dw\,dx\,dy\,dz \quad \dots\dots\dots(215).$$

This can also be expressed in the form

$$De^{-hm[(u-u_0)^2+(v-v_0)^2+(w-w_0)^2]} \, du\,dv\,dw\,dx\,dy\,dz \quad \dots\dots\dots(216),$$

in which D, u_0, v_0, w_0 are new constants, and it is easily found that the components of the total momentum of the N molecules are Nmu_0, Nmv_0, Nmw_0, so that u_0, v_0, w_0 are the components of the velocity of mass-motion of the gas, as in § 26. The circumstance that x, y, z now occur only through the differentials $dx\,dy\,dz$, shews that the law of distribution of velocities is the same at every point of the gas. On replacing $D\Omega$ by A, where Ω is the whole volume of gas, the law of distribution of velocities alone is found to be

$$Ae^{-hm[(u-u_0)^2+(v-v_0)^2+(w-w_0)^2]}\,du\,dv\,dw \qquad (217),$$

agreeing exactly with formulae (25) and (102), and incidentally identifying the h of this chapter with the h previously used.

109. *No mass-motion.* When there is no mass-motion u_0, v_0, w_0 vanish, and the law of distribution (215) reduces to

$$Ce^{-hm(u^2+v^2+w^2)}\,du\,dv\,dw\,dx\,dy\,dz \qquad (218),$$

agreeing of course with (101), on putting $C\Omega = A$.

110. *External field of force.* Let the molecules be acted on by an external field of force, such that a molecule at x, y, z has potential energy χ, χ being a function of x, y, z but not of u, v, w. Then, in addition to the kinetic energy expressed by equation (213), the molecules have potential energy $\Sigma\chi$, and the total energy is

$$E = \Sigma\left\{\tfrac{1}{2}m(u^2+v^2+w^2)+\chi\right\} \qquad (219).$$

The value of ϵ in formula (209) may now be taken to be

$$\tfrac{1}{2}m(u^2+v^2+w^2)+\chi.$$

In the case in which there is neither mass-rotation nor mass-motion, the law of distribution of coordinates becomes

$$Ce^{-hm(u^2+v^2+w^2)}\,e^{-2h\chi}\,du\,dv\,dw\,dx\,dy\,dz \qquad (220).$$

This formula shews that at every point of the gas the law of distribution of velocities is simply

$$Ae^{-hm(u^2+v^2+w^2)}\,du\,dv\,dw \qquad (221),$$

and so is identical with that already found for a gas not influenced by an external field of force. Integrating with respect to u, v, w, formula (220) becomes

$$C\sqrt{\frac{\pi^3}{h^3m^3}}\,e^{-2h\chi}\,dx\,dy\,dz \qquad (222),$$

giving the law of distribution with respect to the x, y, z coordinates. This, however, is simply the law of distribution of density in the gas. If ρ is the density in the element $dx\,dy\,dz$, expression (222) must be the same as $\frac{\rho}{m}\,dx\,dy\,dz$, so that

$$\rho = \rho_0 e^{-2h\chi} \qquad (223),$$

where ρ_0 is a constant given by $\rho_0 = C\sqrt{\frac{\pi^3}{h^3m}}$, and so is the density at points at which $\chi = 0$.

Molecules of finite size.

111. By reasoning exactly similar to that used in § 57, it is clear that the law of distribution of velocities (221) must hold even when the molecules are of finite size. The law of distribution of density (223) will not, however, be valid except for molecules of infinitesimal size.

Molecules of complex structure.

112. If the molecule has possibilities of internal motion and of rotation, its energy ϵ will be of the form

$$\epsilon = \tfrac{1}{2} m (u^2 + v^2 + w^2) + \epsilon_i + \epsilon_r,$$

where ϵ_i, ϵ_r denote respectively the energies of internal motion and rotation. Corresponding to these motions there will be coordinates of position and velocity $\phi_1, \phi_2, \ldots \phi_{2s}$, the energies ϵ_i, ϵ_r being functions of these coordinates. The law of distribution of coordinates, in the special case of no mass-motion or mass-rotation, is seen from formula (210) to be

$$Ae^{-hm(u^2+v^2+w^2)} e^{-2h(\epsilon_i+\epsilon_r)} du\,dv\,dw\,dx\,dy\,dz\,d\phi_1 d\phi_2 \ldots d\phi_{2s} \quad \ldots(224),$$

again shewing that the law of distribution of velocities alone is the same as for infinitely small hard spherical molecules, namely that given by formula (221).

A mixture of molecules of different kinds.

113. Exactly the same method can be applied to a mixture of molecules of different kinds. The method is quite general, but for simplicity of statement we may suppose that there are only two kinds of molecules present. The first kind will be supposed identical with the molecules already considered in § 104, the second kind will be supposed to be of mass m', and to have coordinates of position and velocity

$$\psi_1, \psi_2, \ldots \psi_{2p'} \quad \ldots\ldots\ldots\ldots\ldots\ldots\ldots\ldots(225).$$

We follow the methods of §§ 105, 106; there will be two equations of the type of (201), one for each kind of molecules, but only one equation as before of the type of (204), (206). The laws of distribution are readily seen to be as follows:

For the first kind of molecules (cf. formula (209)):

$$Ce^{-2h\epsilon} e^{-(r_1\mu_1 + r_2\mu_2 + \ldots)} d\phi_1 d\phi_2 \ldots d\phi_{2p} \quad \ldots\ldots\ldots\ldots\ldots\ldots(226).$$

For the second kind of molecules:

$$C'e^{-2h\epsilon'} e^{-(r_1\mu_1' + r_2\mu_2' + \ldots)} d\psi_1 d\psi_2 \ldots d\psi_{2p'} \quad \ldots\ldots\ldots\ldots\ldots(227).$$

Here $\epsilon, \mu_1, \mu_2, \ldots$ refer as before to the energy and momenta of molecules of the first kind, while $\epsilon', \mu_1', \mu_2', \ldots$ are the corresponding quantities for molecules of the second kind. The constants C, C' are different for the two kinds of molecules, but the constants $h, r_1, r_2, \ldots$ are the same for both.

The discussions of §§ 107—112 can be applied equally to a mixture of gases. Two particular results may be noticed as being of special importance.

I. When there is mass-motion but no mass-rotation, the laws of distribution of the u, v, w coordinates will be (cf. §§ 108, 112)

$$\begin{cases} Ae^{-hm[(u-u_0)^2+(v-v_0)^2+(w-w_0)^2]}du\,dv\,dw \quad\ldots\ldots\ldots\ldots\ldots(228),\\ A'e^{-hm'[(u-u_0)^2+(v-v_0)^2+(w-w_0)^2]}du\,dv\,dw \quad\ldots\ldots\ldots\ldots\ldots(229),\end{cases}$$

and these formulae are true for molecules of finite size, and capable of rotation or internal motion, as also for molecules acted on by external fields of force. The constants h, u_0, v_0, w_0 are the same in the two formulae. When there is no mass-motion, the formulae assume the simpler forms

$$\begin{cases} Ae^{-hm(u^2+v^2+w^2)}du\,dv\,dw \quad\ldots\ldots\ldots\ldots\ldots\ldots\ldots\ldots(230),\\ A'e^{-hm'(u^2+v^2+w^2)}du\,dv\,dw \quad\ldots\ldots\ldots\ldots\ldots\ldots\ldots(231).\end{cases}$$

II. When the molecules are supposed infinitely small, and are acted on by an external field of force, the laws of distribution of density in space are (cf. formula (223))

$$\begin{cases} \rho=\rho_0 e^{-2h\chi} \quad\ldots\ldots\ldots\ldots\ldots\ldots\ldots\ldots\ldots\ldots(232),\\ \rho'=\rho_0' e^{-2h\chi'} \quad\ldots\ldots\ldots\ldots\ldots\ldots\ldots\ldots\ldots(233).\end{cases}$$

Here ρ, ρ' are the densities of the two kinds of gas, χ is the potential energy of a molecule of the first kind in the field of force, and χ' is the corresponding quantity for a molecule of the second kind. If the field is such that the potential energies are proportional to the masses, we may put $\chi=mV$, $\chi'=m'V$, and the formulae become

$$\begin{cases} \rho=\rho_0 e^{-2hmV} \quad\ldots\ldots\ldots\ldots\ldots\ldots\ldots\ldots\ldots(234),\\ \rho'=\rho_0' e^{-2hm'V} \quad\ldots\ldots\ldots\ldots\ldots\ldots\ldots\ldots(235).\end{cases}$$

Molecular and Atomic Dissociation and Aggregation.

114. A final and important illustration of the method is found in its application to the case in which the molecules are not of fixed permanent types, but are capable of dissociation and recombination, or of forming molecular aggregates.

For simplicity in exposition, it will be assumed that the units under discussion are molecules, and that the compound structures formed of these units are molecular aggregates. The same analysis will apply, with suitable limitations when the units are atoms and the aggregates are molecules, and also, at any rate so far as mathematical theory is concerned, when the units are ions or electrons and the aggregates are atoms or molecules.

Let the different types of units be distinguished by the suffixes $\alpha, \beta, \gamma, \ldots$, and let the various types of aggregates which can be formed out of them be indicated by the suffixes $\alpha\beta$, $\beta\gamma$, ..., $\alpha\beta\gamma$, ..., etc. The aggregate $\alpha\beta$ is of course formed of the two simple units α and β in combination, and so on.

Some convention or definition is necessary to determine the exact stage at which the separate units are to be considered replaced by the aggregate system, or vice versa. Let two units not influenced by each other have energies E_α, E_β; so that the energy of this pair of units is $E_\alpha + E_\beta$. As the units approach within one another's influence this expression for the energy of the combined system will not in general be adequate; we must add to it certain cross terms depending on the coordinates of *both* units. Denote these by $W_{\alpha\beta}$, then the energy of the system, say $E_{\alpha\beta}$, will be given by

$$E_{\alpha\beta} = E_\alpha + E_\beta + W_{\alpha\beta} \quad \text{........................(236).}$$

We shall say that the units α, β lose their identity as separate units, and that the aggregate comes into existence as soon as $W_{\alpha\beta}$ becomes appreciable, and similarly for the reverse process. Thus aggregation takes place as soon as $W_{\alpha\beta}$ attains to a value which differs appreciably from zero; dissociation takes place as soon as $W_{\alpha\beta}$ falls to a value which does not differ appreciably from zero.

115. We shall suppose the number of units of type α not in combination to be N_α, and similar meanings attach to $N_\beta, \ldots N_{\alpha\beta}, \ldots N_{\alpha\beta\gamma}, \ldots$. The total number of units of type α, whether in combination or not, will be taken to be $\mathfrak{N}_\alpha$, and similar meanings are assigned to $\mathfrak{N}_\beta$, $\mathfrak{N}_\gamma$, etc. As motions and changes take place in the whole system, the quantities $N_\alpha, N_\beta, \ldots N_{\alpha\beta}, \ldots N_{\alpha\beta\gamma}, \ldots$ vary, owing to the occurrence of dissociations and recombinations, but the quantities $\mathfrak{N}_\alpha$, $\mathfrak{N}_\beta$, $\mathfrak{N}_\gamma$ remain permanently the same. By a process of pure counting, we arrive at the equations

$$N_\alpha + N_{\alpha\beta} + N_{\alpha\gamma} + \ldots + 2N_{\alpha\alpha} + N_{\alpha\beta\gamma} + \ldots = \mathfrak{N}_\alpha \quad \text{.........(237),}$$

$$N_\beta + N_{\alpha\beta} + N_{\beta\gamma} + \ldots + 2N_{\beta\beta} + N_{\alpha\beta\gamma} + \ldots = \mathfrak{N}_\beta \quad \text{.........(238).}$$

For the N_α separated systems of type α, we shall suppose the law of distribution of the coordinates $\phi_1, \phi_2, \ldots$ of the system to be denoted by $f_\alpha(\phi_1, \phi_2, \ldots)$, so that the number of these systems having coordinates within a range $d\phi_1, d\phi_2, \ldots$ will be

$$N_\alpha f_\alpha(\phi_1, \phi_2, \ldots)\, d\phi_1 d\phi_2 \ldots \quad \text{..................(239),}$$

and this, reverting to a notation already used in § 13, will also be denoted by

$$\tau_\alpha(\phi_1, \phi_2 \ldots)\, d\phi_1 d\phi_2 \ldots \quad \text{.....................(240),}$$

so that $\tau_\alpha = N_\alpha f_\alpha$. Similar meanings will be attached to the symbols $\tau_\beta, \tau_\gamma, \ldots \tau_{\alpha\beta}$, etc. We have, from the meaning of the law of distribution,

$$\iint \ldots f_\alpha(\phi_1, \phi_2, \ldots)\, d\phi_1 d\phi_2 \ldots = 1,$$

so that
$$N_\alpha = \iint \ldots \tau_\alpha(\phi_1, \phi_2, \ldots)\, d\phi_1 d\phi_2 \ldots \quad \ldots\ldots\ldots\ldots\ldots(241).$$

To save printing we shall denote the right-hand member of this equation by $\int \tau_\alpha$, so that our equation becomes

$$N_\alpha = \int \tau_\alpha$$

and equations (237), (238), etc., become

$$\mathfrak{N}_\alpha = \int \tau_\alpha + \int \tau_{\alpha\beta} + \int \tau_{\alpha\gamma} + \ldots + 2\int \tau_{\alpha\alpha} + \int \tau_{\alpha\beta\gamma} + \ldots \quad \ldots\ldots(242),$$

$$\mathfrak{N}_\beta = \int \tau_\beta + \int \tau_{\alpha\beta} + \int \tau_{\beta\gamma} + \ldots + 2\int \tau_{\beta\beta} + \int \tau_{\alpha\beta\gamma} + \ldots \quad \ldots\ldots(243).$$

Throughout all changes, the quantities $\mathfrak{N}_\alpha, \mathfrak{N}_\beta, \ldots$ remain unaltered; the energy also remains unaltered, and if this be denoted by $\mathfrak{E}$, its value is given by

$$\mathfrak{E} = \int E_\alpha \tau_\alpha + \int E_\beta \tau_\beta + \ldots + \int E_{\alpha\beta} \tau_{\alpha\beta} + \ldots \quad \ldots\ldots\ldots\ldots\ldots(244).$$

116. Let us agree to adopt the artifice explained in §§ 50, 105, to limit the variation of the coordinates of the various types of molecules to a finite range. Let us divide up the possible range of coordinates for a single molecule of type α into n_α equal "cells," the possible range for a single molecule of type β into n_β equal cells, and so on. The range for a compound molecule of type $\alpha\beta$ will then be $n_\alpha n_\beta$ cells, if for the moment we regard any combination of an α molecule with a β molecule as a compound of the $\alpha\beta$ type. If we only regard these as forming a double molecule when the intermolecular force exceeds a certain amount, then it follows that double molecules can only occur in certain of these $n_\alpha n_\beta$ cells, and not in all —it does not at present matter in how many.

Let us now consider a special class of system—class A—in which there are

$$\left.\begin{array}{llll} \alpha_1, \alpha_2 \ldots & \text{single molecules of type } \alpha & \text{in the respective } n_\alpha & \text{cells,} \\ \beta_1, \beta_2 \ldots & \quad\text{,,} \qquad\qquad\text{,,} \qquad \beta & \quad\text{,,} \qquad\text{,,} \qquad n_\beta & \text{,,} \\ (\alpha\beta)_1, (\alpha\beta)_2 \ldots & \text{double ,,} \qquad\text{,,} \qquad \alpha\beta & \quad\text{,,} \qquad\text{,,} \qquad n_\alpha n_\beta & \text{,,} \end{array}\right\} (245),$$

and so on.

Each arrangement of molecules which form a system of class A will be represented in an element of the generalised space which forms a fraction

$$\frac{1}{n_\alpha^{N_\alpha} n_\beta^{N_\beta} \ldots (n_\alpha^2)^{N_{\alpha\alpha}} (n_\alpha n_\beta)^{N_{\alpha\beta}} \ldots}$$

of the whole. Using equations (237), (238), etc. this becomes

$$\frac{1}{n_\alpha^{\mathfrak{N}_\alpha} n_\beta^{\mathfrak{N}_\beta} \ldots} \quad \ldots\ldots\ldots\ldots\ldots\ldots\ldots\ldots\ldots(246).$$

Now the number of ways of distributing the rôles of the various constituent molecules so that conditions (245) are satisfied is

$$\frac{\mathfrak{N}_a!\,\mathfrak{N}_\beta!\ldots}{\alpha_1!\,\alpha_2!\ldots\beta_1!\,\beta_2!\ldots(\alpha\beta)_1!\,(\alpha\beta)_2!\ldots}\quad\ldots\ldots\ldots\ldots\ldots\ldots(247).$$

Here the factor $\mathfrak{N}_a!$ is the number of ways in which the $\mathfrak{N}_a$ permanent constituent molecules of type α can be permuted *inter se*, $\alpha_1!$ is the number of ways in which the molecules in the first of the n_a cells can be permuted *inter se*, and so on. Expression (247), then, gives the number of elements which represent systems of class A. Multiplying expressions (247) and (246) together, we find that the fraction of the whole generalised space which is occupied by systems of class A is

$$\theta_a = \frac{1}{n_a^{\mathfrak{N}_a}\, n_\beta^{\mathfrak{N}_\beta}\ldots}\,\frac{\mathfrak{N}_a!\,\mathfrak{N}_\beta!\ldots}{\alpha_1!\,\alpha_2!\ldots}\quad\ldots\ldots\ldots\ldots\ldots\ldots(248).$$

117. The value of θ_a just found is a generalised form of that given by expression (68). If we proceed as in § 44, using Stirling's Theorem [equation (69)] in the form

$$\operatorname*{Lt}_{p=\infty} \log p! = \tfrac{1}{2}\log(2\pi e) + (p+\tfrac{1}{2})\log\frac{p}{e},$$

we obtain
$$\log\theta_a = C - \sum_{n_a}(\alpha_s+\tfrac{1}{2})\log\frac{\alpha_s}{e} - \sum_{n_\beta}(\beta_s+\tfrac{1}{2})\log\frac{\beta_s}{e} - \ldots$$
$$- \sum_{n_a n_\beta}((\alpha\beta)_s+\tfrac{1}{2})\log\frac{(\alpha\beta)_s}{e} - \ldots,$$

where C is a constant depending on the constants $\mathfrak{N}_a, \mathfrak{N}_\beta \ldots n_a, n_\beta \ldots$. From this equation it follows that the normal state is obtained by making $\mathfrak{H}$ a minimum, where

$$\mathfrak{H} = \int \tau_a \log\frac{\tau_a}{e} + \int \tau_\beta \log\frac{\tau_\beta}{e} + \ldots + \int \tau_{a\beta}\log\frac{\tau_{a\beta}}{e} + \ldots\quad\ldots\ldots(249).$$

The variation of $\mathfrak{H}$ is subject to the energy equation (244) and to equations of the type (242) expressing the permanency of the separate types of permanent molecules. If we vary equation (249) and add the variation of equation (244) multiplied by an undetermined multiplier λ, and that of the equations of the type (242), (243) ... multiplied by $\mu_a, \mu_\beta \ldots$, we obtain

$$\delta\mathfrak{H} = \int(\log\tau_a + \lambda E_a + \mu_a)\,\delta\tau_a + \ldots$$
$$+ \int(\log\tau_{a\beta} + \lambda E_{a\beta} + \mu_a + \mu_\beta)\,\delta\tau_{a\beta} + \ldots$$
$$+ \int(\log\tau_{aa} + \lambda E_{aa} + 2\mu_a)\,\delta\tau_{aa} + \ldots,$$

and the condition that $\mathfrak{H}$ shall be a minimum is given by the systems of equations

$$\log \tau_\alpha + \lambda E_\alpha + \mu_\alpha = 0, \text{ etc.,}$$
$$\log \tau_{\alpha\beta} + \lambda E_{\alpha\beta} + \mu_\alpha + \mu_\beta = 0, \text{ etc.,}$$
$$\log \tau_{\alpha\alpha} + \lambda E_{\alpha\alpha} + 2\mu_\alpha = 0, \text{ etc.,}$$
$$\text{etc.}$$

Changing the constants λ, μ_α, μ_β ..., and substituting for $E_{\alpha\beta}$, $E_{\alpha\alpha}$ ... these equations lead at once to the equations

$$\left.\begin{aligned} \tau_\alpha &= Ae^{-2hE_\alpha} \\ \tau_\beta &= Be^{-2hE_\beta} \\ \tau_{\alpha\beta} &= ABe^{-2h(E_\alpha + E_\beta + W_{\alpha\beta})} \\ \tau_{\alpha\alpha} &= A^2e^{-2h(E_\alpha + E_\alpha' + W_{\alpha\alpha})} \end{aligned}\right\} \text{......................(250),}$$

$$\text{etc.}$$

In general, for the multiple molecule $\alpha\beta\gamma$ ... we have

$$\tau_{\alpha\beta\gamma...} = ABC \ldots e^{-2h(E_\alpha + E_\beta + E_\gamma + \ldots + W_{\alpha\beta\gamma\ldots})} \text{..............(251).}$$

These formulae not only give the laws of distribution for molecules which are capable of aggregation, but also for ordinary molecules in collision or exerting force upon one another in any way.

118. From the well-known formula in attractions

$$W = \tfrac{1}{2}\iiint \rho V dx dy dz,$$

where ρ, V are density and potential at the point x, y, z, it follows that we can write $W_{\alpha\beta\gamma...}$ in the form

$$W_{\alpha\beta\gamma...} = \tfrac{1}{2}(\chi_\alpha + \chi_\beta + \chi_\gamma + \ldots),$$

where χ_α is the potential of the molecule of type α in the field of intermolecular forces arising from the other molecules, and so on. Hence in equation (251) we may write

$$\tau_{\alpha\beta\gamma...} = \psi_\alpha \psi_\beta \psi_\gamma \ldots \quad \text{...........................(252),}$$

where
$$\psi_\alpha = Ae^{-2hE_\alpha - h\chi_\alpha}, \text{ etc. ....(253).}$$

We may therefore regard the probability of a combination of molecules having any specified coordinates as the product of the probabilities of the constituent molecules having the appropriate coordinates, if we take the probability of a molecule of type α having its coordinates within the usual range $d\xi_1 d\xi_2$ ... to be

$$Ae^{-2hE_\alpha - h\chi_\alpha} d\xi_1 d\xi_2 \ldots \quad \text{........................(254).}$$

Since the quantity χ_α does not involve the velocity coordinates it is clear that the analysis of § 106 can be made to apply to this case, and hence that the result expressed by equation (212) is true, even when intermolecular forces are taken into account. Thus we see that the law of distribution of velocity coordinates is unaltered by the presence of intermolecular forces, and that the law of equipartition of kinetic energy remains valid independently of the existence of such forces.

In particular the law of distribution of velocities of molecules in collision with one another is the same as that of free molecules. It is frequently but erroneously assumed that molecules which have penetrated a certain distance into one another's fields of force will, on the average, have less kinetic energy than corresponding free molecules. Our analysis has shewn that this is not the case.

Of course a single molecule which moves into a position in which its potential energy is χ will, in so doing, lose kinetic energy χ, but what is often overlooked is that the molecules which do this were not originally average molecules; they were selected molecules, being those of which the kinetic energy initially was greater than χ. Initially their kinetic energy was greater than the average: the work done against repulsive forces just uses up this excess of energy. The matter is perhaps understood most clearly by noticing that the motion of a swarm of molecules into a repulsive field of force lessens the density but not the mean kinetic energy (or temperature) of the swarm (cf. § 110).

Before leaving the subject we must notice the similarity between the effects of an intermolecular and an external field of force. If χ_a, instead of being the potential of a molecule of type a in an intermolecular field of force, had been the potential in a permanent external field of force, then the law of distribution of molecules of type a would, by § 110, be exactly the same as that expressed by (254), except that χ_a would have been replaced by $2\chi_a$.

MAXWELL'S TREATMENT OF THE PARTITION OF ENERGY.

119. The doctrine of the equipartition of energy in a system of molecules of varying masses was discovered and enunciated by Waterston* in 1845, in the paper which has already been referred to. He states the doctrine in the following form: "In mixed media, the mean square molecular velocity is inversely proportional to the specific weight of the molecule. This is the law of the equilibrium of vis-viva." Lord Rayleigh, in a footnote, says "This is the first statement of a very important theorem. The demonstration, however,...can hardly be defended." Exactly the same theorem was brought forward independently by Maxwell in 1859, in the British Association

* *Phil. Trans.* CLXXXIII. p. 1.

paper already referred to*. He states the proposition: "Two systems of molecules move in the same vessel; to prove that the mean vis-viva of each particle will become the same in the two systems." The question was again brought into prominence by the publication of a paper by Boltzmann in 1861†. In 1879 Maxwell also published a paper on equipartition in which he regarded the whole question from a somewhat different standpoint‡. In what follows we shall treat the question from Maxwell's point of view, the only difference being that the mathematical analysis can be put much more concisely by the help of the conception of a generalised space.

120. We again consider the dynamical system of n degrees of freedom already specified in § 78. Its configuration is determined by n coordinates

$$q_1, q_2, \dots q_n \quad \text{.............................(255)},$$

and the n corresponding velocities

$$\dot{q}_1, \dot{q}_2, \dots \dot{q}_n \quad \text{.............................(256)}.$$

The kinetic energy L will be a quadratic function of the n velocities (256), and therefore also a quadratic function of the n momenta

$$p_1, p_2, \dots p_n \quad \text{.............................(257)}$$

defined by the equations (cf. equations (129))

$$p_s = \frac{\partial E}{\partial \dot{q}_s}, \text{ etc.} \quad \text{.............................(258)}.$$

It is known to be possible to transform this quadratic function into a sum of squares of the form

$$L = \tfrac{1}{2}(c_1\eta_1^2 + c_2\eta_2^2 + \dots + c_n\eta_n^2) \quad \text{..............(259)}.$$

where the η's are linear functions of the momenta (257), and are moreover such that the modulus of transformation is unity; that is to say

$$\frac{\partial(\eta_1, \eta_2, \dots \eta_n)}{\partial(p_1, p_2, \dots p_n)} = 1 \quad \text{.......................(260)}.$$

The quantities $\eta_1, \eta_2, \dots$ are spoken of as "momentoids." Since L must necessarily be positive for all configurations of the system, it is clear that each of the coefficients $c_1, c_2, \dots c_n$ must necessarily be positive.

* "Illustrations of the Dynamical Theory of Gases," *Phil. Mag.* Jan. and July, 1860. *Collected Works*, I. p. 378.

† "Studien über das Gleichgewicht der lebendigen Kraft zwischen bewegten materiellen Punkten," *Sitzungsber. der K. Akad. Wien*, LVIII.

‡ "On Boltzmann's Theorem on the average distribution of energy in a system of material points," *Camb. Phil. Trans.* XII. *Collected Works*, II. p. 713.

We may accordingly express the energy E of the system in the form

$$E = V + L \qquad (261)$$

$$= V + \tfrac{1}{2}(c_1\eta_1^2 + c_2\eta_2^2 + \dots + c_n\eta_n^2) \qquad (262),$$

where V is the potential energy, and function of the q's only, and $c_1, c_2, \dots c_n$ are also functions of the q's only.

121. We shall now represent all possible configurations and velocities of this system in a space of $2n$-dimensions, having

$$q_1, q_2, \dots q_n, \eta_1, \eta_2, \dots \eta_n \qquad (263)$$

as coordinates. In virtue of relation (260), it is clear that any assemblage of systems will be represented in the present generalised space by fluid of the same density as that by which it was represented in the former generalised space of § 81. If therefore the fluid in the present space is taken to be initially homogeneous, it will remain homogeneous throughout all time*.

The volume of the generalised space for which $q_1, q_2 \dots q_n$ lie within specified ranges $dq_1, dq_2 \dots dq_n$, while $\eta_1, \eta_2 \dots \eta_n$ have all values such that $E < E_0$, is given by

$$dq_1 dq_2 \dots dq_n \iiint \dots d\eta_1, d\eta_2 \dots d\eta_n \qquad (264),$$

where $\eta_1, \eta_2 \dots \eta_n$ have all values subject to

$$c_1\eta_1^2 + c_2\eta_2^2 + \dots + c_n\eta_n^2 < 2(E_0 - V) \qquad (265).$$

The integral is a Dirichlet Integral† of which the value is known to be

$$\frac{\pi^{\frac{1}{2}n}}{\Gamma(1+\frac{1}{2}n)}(c_1 c_2 \dots c_n)^{-\frac{1}{2}}(2E_0 - 2V)^{\frac{1}{2}n} \qquad (266).$$

Differentiating with respect to E_0, we find that the volume representing systems for which $q_1, q_2 \dots q_n$ lie within the same range as before, while E lies between E_0 and $E_0 + dE$, is

$$2dq_1 dq_2 \dots dq_n dE \frac{\pi^{\frac{1}{2}n}}{\Gamma(\frac{1}{2}n)}(c_1 c_2 \dots c_n)^{-\frac{1}{2}}(2E_0 - 2V)^{\frac{1}{2}n-1} \qquad (267).$$

If we introduce a new condition that η_n is to lie between η_n and $\eta_n + d\eta_n$, the alterations necessary to transform expression (267) to suit the new

* This treatment seems to obviate, in a simple manner, a criticism which has often been urged against Maxwell's original proof. Maxwell takes coordinates in which the kinetic energy is already expressed as the sum of squares, and assumes these to form true Lagrangian coordinates. Unfortunately it is not always possible to find coordinates satisfying these conditions. To take the simplest case, the kinetic energy of rotation of a rigid body can be expressed as a sum of squares in many ways, but in no case are the coordinates true Lagrangian coordinates. If, for instance, we write

$$2L = A\omega_1^2 + B\omega_2^2 + C\omega_3^2,$$

we know that $\int \omega_1 dt$, etc., are not true Lagrangian coordinates.

† Williamson, *Integral Calculus*, p. 320.

conditions will consist in writing $n-1$ for n, $2E_0 - 2V - c_n\eta_n^2$ for $2E_0 - 2V$ and introducing the new differential $d\eta_n$. Making these alterations the expression becomes

$$2dq_1 dq_2 \dots dq_n dE d\eta_n \frac{\pi^{\frac{1}{2}(n-1)}}{\Gamma(\frac{1}{2}n-\frac{1}{2})}(c_1 c_2 \dots c_{n-1})^{-\frac{1}{2}}(2E_0 - 2V - c_n\eta_n^2)^{\frac{1}{2}(n-3)} \quad \dots(268).$$

The ratio of this expression to (267), on replacing $\sqrt{\pi}$ by $\Gamma(\frac{1}{2})$, is found to be

$$\frac{\Gamma(\frac{1}{2}n)}{\Gamma(\frac{1}{2}n-\frac{1}{2})\Gamma(\frac{1}{2})} \frac{(2E_0 - 2V - c_n\eta_n^2)^{\frac{1}{2}(n-3)}}{(2E_0 - 2V)^{\frac{1}{2}(n-2)}} c_n^{\frac{1}{2}} d\eta_n \quad \dots\dots(269).$$

This is the fraction for which η_n lies between η_n and $\eta_n + d\eta_n$, of all the systems for which $q_1, q_2 \dots q_n$, E lie within the specified small ranges. Write

$$K_n = \tfrac{1}{2} c_n \eta_n^2$$

so that K_n is the kinetic energy corresponding to the momentoid η_n; then

$$c_n^{\frac{1}{2}} d\eta_n = \frac{dK_n}{\sqrt{2K_n}}$$

and therefore expression (269) becomes

$$\frac{\Gamma(\frac{1}{2}n)}{\Gamma(\frac{1}{2}n-\frac{1}{2})\Gamma(\frac{1}{2})} \frac{(E_0 - V - K_n)^{\frac{1}{2}(n-3)}}{(E_0 - V)^{\frac{1}{2}(n-2)}} \frac{dK_n}{\sqrt{K_n}} \quad \dots\dots\dots(270).$$

122. The mean value of K_n averaged over all the systems for which $q_1, q_2 \dots q_n$ and E lie within the specified ranges, say $\overline{K}_n$, is therefore

$$\overline{K}_n = \frac{\Gamma(\frac{1}{2}n)}{\Gamma(\frac{1}{2}n-\frac{1}{2})\Gamma(\frac{1}{2})} \int_{K_n=0}^{K_n=E_0-V} \frac{(E_0 - V - K_n)^{\frac{1}{2}(n-3)}}{(E_0 - V)^{\frac{1}{2}(n-2)}} K_n^{\frac{1}{2}} dK_n,$$

of which the value, after integration, is found to be

$$\overline{K}_n = \frac{E_0 - V}{n} \quad \dots\dots\dots\dots\dots\dots(271),$$

so that from symmetry

$$\overline{K}_1 = \overline{K}_2 = \dots = \overline{K}_n \quad \dots\dots\dots\dots\dots\dots(272).$$

In words, this result states that, averaged through all those parts of the generalised space in which $q_1, q_2 \dots q_n$ and E have specified values, the energies of the various momentoids are equal. By addition, it follows that, averaged through all parts of the generalised space for which E has a given value, the energies of the various momentoids are equal.

123. Formula (269) expresses the law of distribution of η_n, and formula (270) that of K_n. These formulae assume special forms when n is very large.

When n is very large,

$$\Gamma(\tfrac{1}{2}n)^2 = \Gamma(\tfrac{1}{2}n+\tfrac{1}{2})\Gamma(\tfrac{1}{2}n-\tfrac{1}{2}) = \tfrac{1}{2}(n-1)\{\Gamma(\tfrac{1}{2}n-\tfrac{1}{2})\}^2,$$

so that
$$\frac{\Gamma(\frac{1}{2}n)}{\Gamma(\frac{1}{2}n-\frac{1}{2})\,\Gamma(\frac{1}{2})}=\sqrt{\frac{n}{2\pi}}.$$

Also
$$\frac{(E_0-V-K_n)^{\frac{1}{2}(n-3)}}{(E_0-V)^{\frac{1}{2}(n-2)}}=(E_0-V)^{-\frac{1}{2}}\left(1-\frac{K_n}{E_0-V}\right)^{\frac{1}{2}(n-3)}$$
$$=\frac{e^{-\frac{1}{2}\frac{nK_n}{E_0-V}}}{\sqrt{E_0-V}}.$$

Putting $\frac{E_0-V}{n}=\kappa$, so that κ is the value of each expression of the equality (272), the law of distribution (270) reduces to

$$\frac{1}{\sqrt{2\pi\kappa K_n}}e^{-\frac{1}{2}\frac{K_n}{\kappa}}\,dK_n \quad \text{.........................(273)},$$

which is easily seen to be identical with the law of distribution already obtained (cf. equation (211)).

124. The result reached in § 122 is Maxwell's main result. If we wish it to apply to the motion of dynamical systems, we must suppose an assemblage of systems started with energies intermediate between the narrow limits E_0 and E_0+dE, in such a way that their density in the generalised space is uniform, *i.e.*, so that all values of the coordinates and momenta which are consistent with the energy lying within the specified limits are equally probable. The separate systems have of course no interaction one with another. It then follows that initially and throughout all time the mean energies of the various momentoids are equal.

By addition over all possible values of the energy, we can arrive at the result that for an assemblage of systems having all possible values for the coordinates and momenta, provided only they are started so that the initial density in the generalised space is uniform, the mean energies of the various momentoids are equal.

We can, however, obtain a result more general than this. The motion in the generalised space is confined to the loci $E=$ constant, so that if we take an initial distribution of density τ in the generalised space such that

$$\tau=\phi(E) \quad \text{.............................(274)},$$

where $\phi(E)$ is any function of the energy, then this distribution is a permanent distribution, *i.e.*, equation (274) is satisfied throughout all time. And by addition of the result obtained in § 122, it follows that in this assemblage the mean values of the energies of the various momentoids are equal.

Continuity of path.

125. The weakness of Maxwell's method lies in the fact that it deals only with the average properties of all the systems represented in the generalised space and gives no information at all as to the properties of any single system. Attempts have accordingly been made to extend or modify Maxwell's result so that instead of applying to an average taken over all systems, it shall apply to the motion of one system averaged through a great length of time. It is obvious that this extension of the result cannot legitimately be made without further assumption of some kind. For instance it may be that the path of the single system is entirely confined to a certain definite region of the energy surface on which it is moving, and in this case it would obviously be fallacious to calculate the time-average by integrating over the whole surface. The assumption which is usually made, in order to make the extension to a time-average possible, is that generally known as the assumption of continuity of path. It is "that the system, if left to itself, will, sooner or later, pass through every phase which is consistent with the conservation of energy"*. Lord Rayleigh† points out that "if we take it quite literally, the assumption is of a severely restrictive character; for it asserts that the systems, starting from any phase, will traverse *every* other phase (consistent with the energy condition) *before* returning to the initial phase. As soon as the initial phase is recovered, a cycle is established and no new phases can be reached, however long the motion may continue."

It is, however, pretty clear that the assumption cannot be justified, if taken quite literally. It is known that in connection with every dynamical problem, there are an infinite number of re-entrant paths—the "periodic orbits" of astronomy—so that obviously a system on one of these paths will never reach the phases outside the one particular path, while a system not on one of these paths can never reach the phases represented by points on them.

This objection might be met by arguing that the re-entrant paths only form an infinitesimal fraction of the whole, and that it is quite conceivable that all the phases outside these re-entrant paths form a single path. If this were so, it would be immaterial, for a system on this single path, whether we averaged over the whole energy surface, or only over the path. This defence, at any rate at first sight, does not seem very plausible. It requires us to suppose that the paths are *all* re-entrant, but that one of them is infinitely longer than all the others added together. It must also be noticed that there are dynamical systems in which all the paths are

* Maxwell, *Collected Works*, II. p. 714.

† "The law of partition of Kinetic Energy," *Phil. Mag.* [5] XLIX. p. 111, 1900.

re-entrant and of finite length, as for example occurs in the case instanced by Lord Rayleigh (*l.c. ante*) of a particle describing an orbit about a centre of force, the law of force being μr.

126. An escape might conceivably be made possible by assuming that the system does not continually traverse a single path undisturbed, but that by the agency of external forces it is at times removed from one path to another. If the action of these external agencies is sufficiently fortuitous it may be that it is legitimate to suppose that the system passes through all phases on the energy surface. A warning must, however, be entered as to the nature of the agencies which may be regarded as fortuitous. The essential elements of the question may all be represented by the simple case of a billiard ball moving on a smooth billiard table. Here the impacts of the ball on the cushions are not fortuitous. In fact the cushions may be replaced by a field of repulsive force which becomes infinite at the cushions and vanishes elsewhere, and the motion of the ball is now *undisturbed* motion in this field of force. Again, if the system consists of two billiard balls moving upon the same table, the collisions between them cannot be regarded as fortuitous, because the impulsive forces between them at collisions can be treated as a special case of a continuous system of forces acting between them. Obviously the same consideration covers the case of a gas of the most general kind moving undisturbed by external agencies, in a closed vessel of any kind.

Extension to time-averages.

127. If Maxwell's assumption were true, the extension to the time-averages of a single system would follow at once*. For the assemblage of systems represented in the generalised space will all pass through the same stages in succession, so that the time-average for any single system, when the average is taken over a sufficiently long time, is exactly the same as the time-average averaged over all the systems. But Maxwell's result gives this time-average averaged over all the systems. For, as regards averages taken over all the systems, equations (271) and (272) are true at every instant, and so are true when averaged over a long time. Hence, *if continuity of path may be assumed,* these equations are true for the time-average of a single system averaged over a very long time.

If, however, Maxwell's assumption is untrue—and it must be borne in mind that no single system has yet been discovered in which it is not untrue—there seems to be nothing to be said in justification of deducing the equality of the time-averages from the theorem of § 122. The main point to be noticed is that if the systems are subject to fortuitous disturbances, there is no reason for supposing that a homogeneous distribution of density in the

* Cf. Lord Rayleigh, *Phil. Mag.* [5] XLIX. p. 108.

generalised space (or, more generally, a distribution satisfying equation (274)) will be permanent, or, conversely, that the permanent state will satisfy the condition expressed by equation (274). And if this is not so, the attempted extension to time-averages fails entirely.

128. It may nevertheless be true that for fortuitous disturbances of a special type the distribution expressed by equation (274) remains permanent, and it may also be that the converse is true, and that the only permanent distribution is that represented by equation (274). For instance, if a gas consists of an infinite number of molecules, we can select one single molecule from the rest, and regard the remainder of the molecules as the dynamical system, while the single molecule plays the part of the fortuitous disturbing agency. The disturbances are not fortuitous in the true sense, but since the single molecule collides only with an infinite number of different molecules in turn, it might be legitimate to regard its action as fortuitous. And again the energy of the system is not constant when the single molecule has been removed from consideration, but it might be legitimate to neglect the deviations of energy which are infinitesimal in comparison with the whole.

In the following sections a treatment of equipartition and law of distribution in a gas is given, which is the outcome of the train of thought just sketched out. The investigation, however, is not based on the somewhat doubtful assumptions which have just been referred to; it is based upon the assumption of molecular chaos, of which the legitimacy has been established in Chapter IV of the present book. We shall consider only the case of a gas.

An alternative treatment of the Partition of Energy in a gas.

129. We shall suppose a gas to be composed of a number of exactly similar dynamical systems—the molecules. We suppose that each molecule is surrounded by a sphere of molecular action of diameter σ, these spheres being of such a size that two molecules exert no action upon one another except when their spheres intersect.

When the spheres of two or more molecules intersect, an "encounter" is said to take place, lasting until the spheres again become clear of one another. The individual molecules are now to be the systems under discussion, and the "encounters" are to play the part of the fortuitous agencies which disturb their motion. Each molecule is to have n degrees of freedom, in addition to the three degrees of freedom represented by the motion of its centre of gravity in space, and the possible states of a single molecule are to be represented in a space of $2n+3$ dimensions, of which $2n$ represent the internal coordinates and momenta of the molecule, and the remaining three represent the velocities of the centre of gravity.

Binary Encounters.

130. We shall begin by considering binary encounters only. That is to say, we work on the hypothesis that the event of a sphere of molecular action being intersected by two other spheres simultaneously is so rare that it may be neglected.

We treat this case as follows. As soon as an encounter begins between two molecules their existence as single molecules is supposed to be abruptly terminated, and their representative points are removed from our generalised space of $2n+3$ dimensions. During the progress of the encounter the two molecules together will be supposed to form a new dynamical system—a double molecule. This system will be specified by $4n+9$ independent coordinates, $2n$ for the internal coordinates of each constituent molecule, six for the velocity and position of the centre of gravity of one molecule relatively to that of the other, and three for the velocity of the centre of gravity of the whole system in space. Hence any such system can be represented by a point in a space of $4n+9$ dimensions. We shall not, however, require the whole of this $4n+9$ dimensional space. If x, y, z, x', y', z' are the coordinates of the centres of the two molecules, the condition that an encounter is beginning or ending is

$$(x-x')^2+(y-y')^2+(z-z')^2=\sigma^2 \quad\text{.................(275)},$$

In the $4n+9$ dimensional space this equation will be the equation of a certain "surface" S (of dimensions $4n+8$), and the representative points of all double molecules will be inside S. We shall find it convenient to denote each double molecule by *two* representative points, since the rôles of first and second molecule can be allotted in two different ways.

Let τ_2 be the density of representative points in any small element of volume in this new space, and τ_1 the density in the original space of $2n+3$ dimensions. Then the necessary and sufficient conditions for a steady state are

$$\frac{d\tau_2}{dt}=0 \quad\text{..............................(276)},$$

$$\frac{d\tau_1}{dt}=0 \quad\text{..............................(277)},$$

in the latter of which the change in τ_1 includes that caused by the formation and dissolution of double molecules.

131. To determine the relation between τ_1 and τ_2, we make the assumption of § 15, namely, that the gas is in a state of molecular chaos. Having made this assumption we proceed to calculate the number of encounters of a given kind which occur in an interval dt. If $\xi_1, \xi_2 \ldots \xi_{2n}$ are the internal coordinates of a molecule, the number of molecules per unit volume for which $\xi_1, \xi_2 \ldots \xi_{2n}, u, v, w$ lie within a range

$$d\xi_1 d\xi_2 \ldots d\xi_{2n} du\,dv\,dw \quad \text{.......................(278)},$$

will be

$$\tau_1 d\xi_1 d\xi_2 \ldots d\xi_{2n} du\,dv\,dw.$$

Hence, as in expression (4), the number of collisions in time dt for which the coordinates of the first molecule lie within the range (278), and those of the second within a similar range in which the variables are accented, while the line joining their centres meets a unit sphere in a given element of surface $d\omega$, will be

$$\tau_1 \tau_1' V \sigma^2 \cos\theta \, d\xi_1 d\xi_2 \ldots d\xi_{2n} d\xi_1' d\xi_2' \ldots d\xi'_{2n} du\,dv\,dw\,du'\,dv'\,dw'\,d\omega\,dt \quad \text{...(279)}.$$

This number of collisions must however be equal to the number of double molecules which cross a certain element of the surface S in the $4n+9$ dimensional space in time dt, and this number will be

$$\tau_2 \frac{\partial \epsilon}{\partial t} dS\,dt \quad \text{..............................(280)},$$

where dS is the element of the surface S representing collisions of the type in question, and $\frac{\partial \epsilon}{\partial t}$ is the velocity in this space at the element of surface dS measured inward along the normal. The equation of the surface S being equation (275), we may clearly suppose the normal to be the shortest distance from dS to

$$x - x' = 0, \quad y - y' = 0, \quad z - z' = 0,$$

and therefore write

$$\epsilon^2 = (x - x')^2 + (y - y')^2 + (z - z')^2.$$

Thus

$$\frac{\partial \epsilon}{\partial t} = \frac{x - x'}{\epsilon} \frac{\partial}{\partial t}(x - x') + \ldots = l(u - u') + \ldots = V\cos\theta \quad \text{....(281)},$$

where l, m, n, as in Chapter II (§ 17), are the direction-cosines of the line of centres. The value of dS corresponding to collisions of the type specified will clearly be

$$dS = d\xi_1 \ldots d\xi_{2n} du\,dv\,dw\,d\xi_1' \ldots d\xi'_{2n} du'\,dv'\,dw'\,\sigma^2 d\omega \quad \text{......(282)}.$$

If we substitute the value for $\frac{\partial \epsilon}{\partial t}$ from equation (281) and for dS from equation (282) into expression (280), and equate the value so obtained to expression (279), we obtain, after dividing through by common multipliers,

$$\tau_1 \tau_1' = \tau_2 \quad \text{..............................(283)}.$$

This equation may be regarded as giving the density τ_2 at all points on the surface S in the $4n+8$ dimensional space, in terms of the densities at points in the $2n+3$ dimensional space.

132. Since the systems represented in the $4n+9$ dimensional space are not acted upon by any external forces, we have, as in § 85 (equation (148)),

$$\frac{D\tau_2}{Dt} = 0 \quad\text{.............................(284)},$$

where D/Dt denotes differentiation with respect to the time as we follow the fluid in its motion. We may however write

$$\frac{D\tau_2}{Dt} = \frac{d\tau_2}{dt} + \frac{\partial s}{\partial t}\frac{d\tau_2}{ds} \quad\text{........................(285)},$$

where $d\tau_2/dt$ denotes the rate of increase at a fixed point, $\partial s/\partial t$ is the velocity along a stream line, and $d\tau_2/ds$ is the increase of τ_2 per unit length along the stream line. In virtue of equations (284) and (276), equation (285) reduces to

$$\frac{d\tau_2}{ds} = 0 \quad\text{.............................(286)},$$

so that τ_2 must be constant along every stream line.

Let τ_1, τ_1' be the densities in the $2n+3$ dimensional space, at points occupied by the representative points of the two component molecules at the formation of a double molecule, and let $\bar{\tau}_1$, $\bar{\tau}_1'$ be the densities at the points representative of the same two molecules at the dissolution of the double molecule. Then by equation (283) $\tau_1\tau_1'$ and $\bar{\tau}_1\bar{\tau}_1'$ are the two values of τ_2 at the two ends of a single stream line in the $4n+9$ dimensional space, and, therefore, by equation (286),

$$\tau_1\tau_1' = \bar{\tau}_1\bar{\tau}_1' \quad\text{..........................(287)},$$

the same result, it will be noticed, as is obtained by the H-theorem of Chapter II (cf. equation (21)).

Since the motion is dynamically reversible we may equally well take $\bar{\tau}_1$, $\bar{\tau}_1'$ to be the densities at formation, then τ_1, τ_1' will be the densities at dissolution, and the same result holds.

From this it follows that in equation (277) the decrease in τ_1 caused by the formation of double molecules of any specified kind is exactly counterbalanced by the increase caused by the dissolution of double molecules of the same kind. Hence in equation (277), $d\tau_1/dt$ may be taken to be the change in τ_1 caused solely by the continuous motion of the fluid, and may be treated as τ_2 has been treated, leading to the result that τ_1 must be constant along every stream line.

133. We have found, therefore, that the conditions for steady motion, on the hypothesis of binary encounters, may be expressed as follows:

(α) Throughout the $2n+3$ dimensional space, τ_1 must be constant along every stream line.

(β) Throughout the $4n+9$ dimensional space, τ_2 must be constant along every stream line.

(γ) At every point on the boundary of the $4n+9$ dimensional space we must have

$$\tau_2 = \tau_1 \tau_1'.$$

To these may be added a fourth condition:

(δ) At every point on the boundary of the $2n+3$ dimensional space (*i.e.* at infinity) the flow across the boundary must be *nil*, this condition securing that steadiness is maintained without a supply of new systems from infinity.

These conditions are necessary and sufficient for steady motion.

Ternary and Higher Encounters.

134. By a simple extension of the method already explained, the possibility of encounters of ternary and higher orders may be considered. For instance, to take ternary encounters into account we imagine systems of triple molecules, these being represented in a suitable space, in which the number of dimensions will be $6n+15$, namely $2n+6$ for each constituent molecule, less three for the position of the centre of gravity of the whole system. The density in this space being τ_3, we have as conditions additional to those just given:

(ϵ) Throughout the $6n+15$ dimensional space, τ_3 must be constant along every stream line.

(ζ) At every point on the boundary of the $6n+15$ dimensional space we must have

$$\tau_3 = \tau_2 \tau_1.$$

135. Encounters of higher orders may be similarly treated. If τ_K is used to denote the density in the space of $2Kn+6K-3$ dimensions, in which K-ple molecules are represented, the complete system of conditions for steady motion is

(i) Along every stream line in the $2Kn+6K-3$ dimensional space

$$\tau_K = \text{constant} \quad \text{..............................(288)}.$$

(ii) At every point on the boundary of this space

$$\tau_K = \tau_a \tau_b \quad \text{..............................(289)},$$

in which τ_a, τ_b refer to the two systems of molecules of orders a, b, of which the encounter results in the particular system of order K which is represented at the point in question (we therefore have always $a+b=K$).

If encounters of all orders are to be taken into account these conditions must be satisfied for all values of K from $K=1$ to $K=\infty$. In the case of $K=1$, equation (289) must be interpreted so as to become identical with the condition (δ) of § 133.

It will be noticed that if these conditions are satisfied for all values up to $K=\infty$ no hypothesis need be made as to the smallness of the radius of molecular action in comparison with the free path. The only assumption now made is that the gas is in a state of molecular chaos.

Solution of Equations.

136. As before, let χ be a quantity, a function of the coordinates of a molecule or system of molecules, such that throughout the undisturbed motion of the molecule or system, χ maintains a constant value, and such that when two molecules or systems combine to form a new system, the χ of the new system is equal to the sum of the χ's of the component systems. Speaking loosely we may say that χ is defined as being capable of exchange between molecules at a collision, but is indestructible.

Then a solution of equations (288) and (289) will be seen to be

$$\log \tau_K = \sum_K \chi \ldots (K=1, 2, \ldots \infty) \quad\ldots\ldots\ldots\ldots\ldots\ldots(290),$$

where $\sum_K \chi$ is the value of χ for a K-ple molecule, being by definition equal to the sum of the χ's of the K constituent molecules. If $\chi_1, \chi_2 \ldots$ are all the possible values of χ, the most general solution is

$$\log \tau_K = \sum_K (A_1\chi_1 + A_2\chi_2 + \ldots) \quad\ldots\ldots\ldots\ldots\ldots\ldots(291).$$

As regards the number and meaning of the χ's the question stands as in § 106; and for the reasons there given we may, in the case of a gas which has no mass-motion, reject all except

$$\chi_1 = 1,$$
$$\chi_2 = E, \text{ the energy of a molecule.}$$

Hence the solution becomes

$$\tau_1 = Ae^{-2hE},$$
$$\tau_2 = A^2 e^{-2h(E+E'+W)},$$
$$\ldots\ldots\ldots\ldots\ldots\ldots\ldots\ldots$$
$$\tau_K = A^K e^{-2h(\sum_K E + W)} \quad\ldots\ldots\ldots\ldots\ldots\ldots\ldots\ldots\ldots(292),$$

where E, E' ... are the energies of the separate molecules, and W is the potential energy of the intermolecular forces acting between molecules, this not now being included in E and E'.

This solution agrees with that of § 117, and equipartition follows at once.

Comparison of the two foregoing Methods.

137. We have now obtained the same result as that obtained in Maxwell's treatment of equipartition, but in place of his assumption of continuity of path, we have made the assumption of molecular chaos, or, more accurately, we have assumed that the number of collisions of a given kind is that given by expression (279). It is infinitely probable, but not certain, that this expression will be accurate, so that it is infinitely probable, but not certain, that (292) will be the solution in a steady state. This is exactly the result arrived at before.

138. It is of interest to notice that it could have been predicted *à priori* that it would be necessary to supplement Maxwell's treatment by some assumption of this kind.

This, as we shall now see, follows from the fact that the problem is a "statistical" problem, and not a dynamical problem of the ordinary type. A dynamical problem may, in accordance with accepted usage, be said to be one of statistical mechanics when the data and objects of inquiry are not the actual values of the various coordinates, but the law of distribution of these coordinates. Since the data of a problem in statistical mechanics do not completely specify the dynamical coordinates of the system, we are, in a problem of statistical mechanics, discussing an infinite number of different systems at once, and without differentiation *inter se*. The motion of these systems will naturally diverge in the course of time. It may be that after the motion a single statistical specification can be given which covers all except an infinitesimal fraction of the systems. In this case a solution may be said to have been found to the problem. It cannot be that a solution can be obtained which covers all the systems, the reason for this being that, even after the initial system has been fully specified statistically, there are still an infinite number of undetermined variables; and, by giving suitable values to these, we can obtain any chosen infinite number of relations between the coordinates of the final result, and can therefore cause this final result to disagree with any single statistical specification. It is therefore clear that a statistical problem must always have an element of uncertainty in its final solution, although in virtue of the infinite number of the variables, this uncertainty may be represented by an infinitesimally small probability of error: it may, in the terminology previously used, be "infinitely probable" that the result is true.

We have found that the assumption of molecular chaos, on the other hand, leads to a definite certain result, and not merely to one which is infinitely probable. Incidentally, this circumstance enables us to trace out some of the inner significance of this assumption. We see at once that the assumption rests on the supposition that the systems with which it deals have at every instant a definite statistical specification. It therefore just excludes those systems, an infinitesimal fraction of the whole, which wander away from the statistical specification obeyed by the majority. In other words, it implies that any system under discussion has the statistical specification of the majority, and therefore naturally leads to a certain result instead of leading merely to one of infinite probability.

As regards Maxwell's treatment of equipartition, enough has perhaps been said to shew that in dealing statistically with a gas, we can never arrive at absolute certainty: it is therefore impossible to reach any definite result unless a loophole of escape from absolute certainty has been introduced into the premises on which we work.

CHAPTER VI

PHYSICAL PROPERTIES: TEMPERATURE, PRESSURE, ETC.

TEMPERATURE.

139. THE preceding chapters have contained a fairly complete investigation of the statistical dynamical properties of the systems under consideration. The task before us now is to interpret these properties, which so far have been expressed almost entirely in purely mathematical language, in terms of the physical conceptions of temperature, pressure, etc. We may first summarise and recapitulate the principal results which have been obtained.

In § 93 we had under discussion a system of a very general nature, and it was there shewn that a quantity dQ of energy added to the system from outside had the effect of producing certain specified changes in the system. It was shewn that for all changes of this type, the quantity $k\,dQ$ must be a perfect differential, where k was a purely mathematical quantity, defined in terms of the constants and variables of the system.

From thermodynamical theory, it is known that $\frac{dQ}{T}$ must, under the same conditions, be a perfect differential, where T is the absolute temperature on the thermodynamical scale. This does not entitle us to identify k with $1/T$, but we proved (§ 96) that there must be a relation of the form $k=\frac{1}{RT}$, where R is a universal constant. In this way the physical conception of temperature is introduced and linked up with the purely mathematical conceptions with which we have so far been dealing.

140. The next stage in the physical development is found in the proof (§ 100) of the equations

$$E_1=\tfrac{1}{2}sRT, \quad E_2=\tfrac{1}{2}tRT, \text{ etc. } \qquad (293),$$

where E_1 is a part of the total energy which is represented by s squared terms in the general expression for the energy, and similarly E_2 is the energy represented by t squared terms, and so on. These equations are only true if

$s, t, \ldots$ are very large numbers, but we may legitimately state the result in the form that the average energy of each squared term is $\frac{1}{2}RT$. This result is known as the Equipartition of Energy. It connects up the physical conceptions of energy and temperature, and assigns a physical meaning to the universal gas-constant R.

Consider any N similar and separate units (molecules, atoms, etc.) of the system under discussion. In the total energy of the whole system there will always be $3N$ squared terms of the form

$$\sum_1^N \tfrac{1}{2}m(u^2+v^2+w^2) \qquad \text{.........................(294)},$$

representing the kinetic energy of translation of the $3N$ units. This energy may be identified with the E_1 of equation (293), in which case s must be put equal to $3N$, and we have the equation

$$\sum_1^N \tfrac{1}{2}m(u^2+v^2+w^2) = \tfrac{3}{2}NRT \qquad \text{.................(295)}.$$

The value of expression (294) may, however, be calculated in another way. It was seen in the last chapter that, no matter how complicated the system may be, the law of distribution of the components of velocity u, v, w is always the same as in the simple cases discussed in Chapter II, namely Maxwell's Law,

$$Ae^{-hm(u^2+v^2+w^2)}\,du\,dv\,dw.$$

Hence, for the complex system, the mean value of $u^2+v^2+w^2$ for all the N units will be $\dfrac{3}{2hm}$ as in equation (45), and the value of expression (294) will be equal to this multiplied by $\frac{1}{2}mN$. Thus we have

$$\sum_1^N \tfrac{1}{2}m(u^2+v^2+w^2) = \frac{3}{4h}N \qquad \text{....................(296)},$$

and on comparing this value with that given by equation (295),

$$\frac{1}{2h} = RT \qquad \text{..............................(297)}.$$

Thus the mean value of each squared term in the energy, which has already been seen to be equal to $\frac{1}{2}RT$, is now seen also to be equal to $1/4h$, giving a physical interpretation to the quantity h which has appeared, so far as a mathematical multiplier, in the law of distribution.

141. To sum up, if the mean value of u^2 for a number of units each of mass m is denoted by $\overline{u^2}$, we have

$$\tfrac{1}{2}m\overline{u^2} = \tfrac{1}{2}m\overline{v^2} = \tfrac{1}{2}m\overline{w^2} = \frac{1}{4h} = \tfrac{1}{2}RT \qquad \text{................(298)},$$

and the mean kinetic energy of translation of any single unit is given by

$$\tfrac{1}{2}m\,(\overline{u^2}+\overline{v^2}+\overline{w^2})=\tfrac{1}{2}m\overline{C^2}=\frac{3}{4h}=\tfrac{3}{2}RT \quad \text{...........(299)}.$$

These equations give, perhaps, the simplest interpretation of temperature in the kinetic theory. It is of the utmost importance to notice that, for the kinetic theory, temperature is a statistical conception; it is meaningless to speak of the temperature of a single molecule.

Equalisation of Temperature.

142. The fundamental necessity for a relation between h and T becomes clear on considering the way in which the quantity h was introduced into the laws of distribution. It will be remembered that h first appeared in every case as an undetermined multiplier, multiplying the energy equation of the system. Different values of h must accordingly represent different values of the total energy, which is again the same thing as saying that different h's correspond to different temperatures.

In § 113, we obtained the law of distribution for a mixture of two different kinds of molecules; it was found that the quantity h was necessarily the same for the two kinds of molecules: the physical interpretation of this is now seen to be simply that the temperatures of the two kinds of gas must be the same. If the gases were initially at different temperatures, they would finally reach a normal or final state in which the laws of distribution would be those obtained in § 113; the value of h would be the same for the two substances, and therefore the temperatures would ultimately be the same. Thus the process of attaining to the normal state would be physically accompanied by a process of equalisation of temperature.

The analysis of § 113 does not in any way require that the two sets of molecules should be those of gases actually mixed; they may be molecules (or atoms) of either gases or solids: the two substances may be actually mixed, or in contact, or entirely separate. All that is required to establish the mathematical result is that it shall be possible for energy to flow from the one substance to the other. If this is the case, there is only one energy equation for the whole system, and so only one value of h: thus equalisation of temperature must ultimately ensue. If, on the other hand, energy cannot flow from one substance, or part of the whole system, to the other, there will be more than one energy equation, namely one for each part of the system, and so more than one value of h, and more than one temperature in the final state.

In the next chapter we shall consider the process of equalisation of temperature from the point of view of thermodynamics and entropy.

CALCULATION OF PRESSURE IN AN IDEAL GAS.

Infinitely Small Molecules.

143. There are in the main two ways of introducing the pressure of a gas into our calculations, these being analogous to the two ways of determining the law of distribution. The method which will be considered first is based on general dynamics; the second method, given in § 146, rests on the conception of detailed collisions between the molecules of the gas and the surface on which the pressure is supposed to be exerted. In the first instance we give the simple analysis appropriate to the case in which the molecules are supposed infinitely small and exerting only negligible forces on one another except when close together.

Determination of Pressure by the Method of General Dynamics.

144. For a gas, or any other aggregation of similar units, we obtained in § 110 the equation

$$\rho = \rho_0 e^{-2hmV} \quad \text{.............................(300)},$$

where ρ is the density of the gas at any point, and V is the potential of an external field of force at the same point, so that mV is the potential energy of a molecule of mass m.

The pressure p is connected with the density ρ by the well-known hydrostatic system of equations

$$\frac{\partial p}{\partial x} = -\rho \frac{\partial V}{\partial x}, \text{ etc.} \quad \text{........................(301)}.$$

On substituting for ρ from equation (300), this becomes

$$\frac{\partial p}{\partial x} = -\rho_0 e^{-2hmV} \frac{\partial V}{\partial x} \quad \text{........................(302)},$$

giving on integration

$$p = \frac{1}{2hm} \rho_0 e^{-2hmV}$$

$$= \frac{\rho}{2hm} \quad \text{....................................(303)},$$

no constant of integration being added since p must vanish with ρ.

Using the equations $\frac{\rho}{m} = \nu$ (§ 11) and $\frac{1}{2h} = RT$ (equation (297)), the equation just found may be put in the simpler forms

$$p = \frac{\nu}{2h} = \nu RT \quad \text{........................(304)},$$

giving the pressure in terms of the density and temperature.

145. *Mixture of gases.* For a mixture of gases, the total density ρ required in equation (301) is the sum of the partial densities of the different constituent gases. Thus equation (302) becomes

$$\frac{\partial p}{\partial x} = -(\rho_0 e^{-2hmV} + \rho_0' e^{-2hm'V} + \ldots)\frac{\partial V}{\partial x},$$

giving on integration $$p = \frac{\rho}{2hm} + \frac{\rho'}{2hm'} + \ldots,$$

or, again, as in equation (304),

$$p = \frac{\nu + \nu' + \ldots}{2h} = (\nu + \nu' + \ldots) RT \quad \ldots\ldots\ldots\ldots(305).$$

The physical interpretation of these laws, as well as their extension to the cases in which the simple assumptions underlying them are no longer valid, will be reserved for a later discussion. In the meantime we shall see how the same laws can be derived by a calculation of the pressure exerted on the boundary of a containing vessel by the impacts of molecules colliding with this boundary.

Determination of Pressure by the Method of Collisions.

146. In fig. 4, let dS be an element of the boundary of a vessel enclosing a gas, and for convenience let the direction of the normal to dS be taken for axis of x.

Let there be ν molecules per unit volume of the gas, and let these be supposed divided into classes, so that all the molecules in any one class have approximately the same velocities, both as regards magnitude and direction. Let $\nu_1, \nu_2, \ldots$ be the numbers of molecules in these classes, so that $\nu_1 + \nu_2 + \ldots = \nu$.

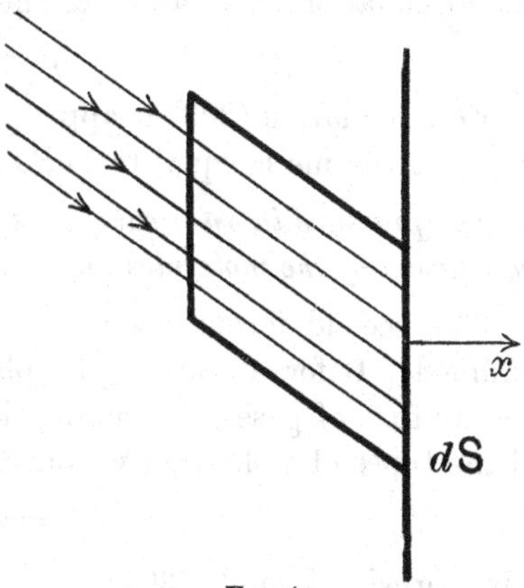

Let u_1, v_1, w_1 denote the components of velocity of molecules of the first class. These molecules may be regarded as forming a shower of molecules of density ν_1 per unit volume, in which every molecule moves with the same velocity.

FIG. 4.

The molecules of this shower which strike dS within an interval of time dt will be those which, at the beginning of the interval, lie within a certain small cylinder inside the vessel (see fig. 4). The cross-section of this cylinder is dS, its height is $u_1 dt$, so that its volume is $u_1 dt dS$. The number of these molecules is accordingly $\nu_1 u_1 dt dS$.

Before impact each of these molecules had momentum normal to the boundary of amount mu_1, so that the momentum normal to the boundary of the whole group of molecules under consideration was $m\nu_1 u_1^2 dSdt$. Summing over all the showers of molecules which strike the area dS in time dt, we find that the total momentum normal to the boundary of these showers was

$$m\,dS\,dt\,\Sigma \nu_1 u_1^2,$$

where the summation extends over all classes of molecules for which u is positive.

This aggregate momentum must, since the gas is in a steady state, be exactly reversed by collision. The change of momentum produced by the pressure pdS on the area dS in time dt will accordingly be

$$p\,dS\,dt = 2m\,dS\,dt\,\Sigma \nu_1 u_1^2 \quad \text{.......................(306)}.$$

The value of $\Sigma \nu_1 u_1^2$ is the sum of the values of u^2 for all the molecules in unit volume for which u is positive, and this is clearly equal to $\frac{1}{2}\nu\overline{u^2}$. Thus equation (306) may be written in the forms

$$p = m\nu\overline{u^2} = \rho\overline{u^2} \quad \text{.............................(307)}.$$

We have seen (§ 141) that

$$m\overline{u^2} = m\overline{v^2} = m\overline{w^2} = \tfrac{1}{3}mC^2 = RT \quad \text{....................(308)},$$

so that equation (307) assumes the forms

$$p = \tfrac{1}{3}\rho C^2 = \nu RT \quad \text{..........................(309)}.$$

From equation (299) it appears that the total kinetic energy of translation in a unit volume is equal to $\frac{3}{2}\nu RT$. Thus, from equation (309),

The pressure in an ideal gas is equal to two-thirds of the kinetic energy of translation of the molecules per unit volume.

The second formula for p is identical with the formula (304) already obtained. Before discussing its physical meaning we may note that if there is a mixture of gases, the summation of equation (306) must be extended to all the types of molecules, so that the final result, instead of equation (309), is

$$p = (\nu + \nu' + \ldots)\,RT \quad \text{.......................(310)},$$

agreeing with formula (305).

If a volume v of homogeneous gas contains N molecules in all, then $(\nu + \nu' + \ldots)\,v = N$, and equations (309) and (310) may be combined in the single equivalent equation

$$pv = NRT \quad \text{............................(311)}.$$

Before leaving this calculation, we may notice that it has not been necessary to make any assumption as to the way in which momenta or velocities are distributed in the various showers after reflection from the boundary.

Knudsen has made the assumption that the directions of motion after impact are distributed without any regard to the directions of motion before impact, and that the number making an angle θ with the normal to the surface is proportional to $\cos\theta d\theta$. From this assumption he deduces certain properties which he finds* to be in good agreement with observations on the flow of gases. The assumption in question has also been tested by R. W. Wood†, who finds that it agrees very closely with experiment. The exact law of reflection, although immaterial for the calculation of normal pressure, is of importance when tangential stresses have to be estimated, as in the flow of gases through tubes.

Physical Laws.

147. It will now be seen that the formulae obtained for the pressure contain within them all the well-known laws of gases.

Avogadro's Law. The value of N is seen, from equation (311), to be equal to pv/RT, a quantity which depends only on the physical state of the gas, and not on the structure of its molecules. Hence we have Avogadro's Law:

Two different gases or mixtures of gases, when at the same temperature and pressure, contain equal numbers of molecules in equal volumes.

The number of molecules in a cubic cm. of gas at standard temperature and pressure has already been taken (§ 8) to be $N_0 = 2{\cdot}705 \times 10^{19}$.

Dalton's Law. Formula (310) shews that the pressure in a mixture of gases is the sum of a number of separate contributions, one from each gas. This is confirmed by Dalton's Law:

The pressure exerted by a mixture of gases is equal to the sum of the pressures exerted separately by the several components of the mixture.

The Laws of Boyle and Charles. Clearly equations (309) and (311) imply the laws of Boyle and Charles:

The pressure of a gas is proportional to its density, so long as the temperature remains unaltered; and is proportional to the temperature, so long as the volume remains unaltered.

148. The various laws which have been predicted by theory, and are found to be confirmed by the experimental laws of the last section, are of course true only within the limits imposed by the assumptions made. The principal of these assumptions has been that the molecules (or other units by which the pressure is exerted) are so small that they may be treated as points in comparison with the scale of intermolecular distances. Thus the laws may best be regarded as ideal laws, the conditions for which can never be absolutely satisfied, but which are satisfied very approximately in a gas

* *Ann. d. Physik*, XLVIII. (1915), p. 1113.

† *Phil. Mag.* XXX. (1915), p. 300; see also *Phil. Mag.* XXXII. (1916), p. 364.

of great rarity. An imaginary gas in which the molecules have dimensions so small in comparison with the other distances involved that they may be regarded as points is spoken of as an ideal gas. Thus the foregoing laws are always true for an ideal gas; for real gases they will be true to within varying degrees of closeness, the accuracy of the approximation depending on the extent to which the gas approaches the state of an ideal gas.

149. As regards the first method of evaluating the pressure (§ 144), the analysis in no way required that the medium should be gaseous, although the resulting laws of Dalton, Boyle and Charles are usually thought of only in relation to gases. Clearly, however, these laws must apply to any substance with a degree of approximation which will depend only on the nearness to the truth of the assumptions just referred to.

In point of fact the laws are found to be true (as they ought to be) for the osmotic pressure of weak solutions. The intermolecular forces between the molecules of the solvent and those of the solute can be allowed for in the value assigned to V in the analysis of § 144, and the forces between pairs of molecules of the solute may be neglected if the solution is sufficiently weak.

In a similar manner, the foregoing conception of pressure may be extended to the pressure exerted by free electrons moving about in the interstices of a conducting solid, and also to the pressure exerted by the "atmosphere" of electrons surrounding a hot solid. Each of these pressures p may be assumed to be given by formula (309), where ν is the number of free electrons per unit volume*.

Numerical Estimate of Velocities.

150. We have seen that the pressure and density in a gas are connected by the relation

$$p = \tfrac{1}{3}\rho C^2 \quad \text{.................................(312)},$$

where (cf. § 30) C is a velocity, equal to 1·086 times $\bar{c}$, the mean velocity of all the molecules, or again is such that C^2 is exactly equal to $\overline{c^2}$, the mean value of c^2 for all the molecules in the gas.

We have also found the relation

$$C^2 = \frac{3RT}{m} \quad \text{.............................(313)},$$

shewing that C is proportional to the square root of the absolute temperature, and, for different gases, varies inversely as the square root of the molecular weight.

151. As soon as corresponding values of p and ρ are known for any gas, we can determine the value of C from equation (312).

For instance, the mass of a litre of oxygen at 0° C. and at the standard

* See Richardson, *The Electron Theory of Matter*, pp. 445, 468 and elsewhere.

pressure of $1{\cdot}01323 \times 10^6$ dynes per square cm. is $1{\cdot}42900$ grammes. Hence for oxygen at 0° C. we have as corresponding values

$$p = 1{\cdot}01323 \times 10^6, \quad \rho = 1{\cdot}42900 \times 10^{-3},$$

and equation (312) now gives us the value of C for oxygen at 0° C.,

$$C = 461{\cdot}2 \text{ metres per sec.}$$

At 0° C. the value of T is 273·1 (see § 8), whence equation (313) gives us the value of R/m for oxygen,

$$\frac{R}{m} = 259{\cdot}6 \times 10^4.$$

From this value of R/m for oxygen we can calculate the value of R/m for any other substance, and equation (313) will then give C and $\bar{c}$ for any temperature we please. In this way the following table has been calculated:

Gas (or other substance)	Molecular Weight (O=16)	$\frac{R}{m}$	C (cms. per sec. at 0° C.)	$\bar{c}$ (cms. per sec. at 0° C.)
Hydrogen...........	2·016	4127×10^4	1839×10^2	1694×10^2
Helium..............	4	2077×10^4	1310×10^2	1207×10^2
Water-vapour	18·016	462×10^4	615×10^2	565×10^2
Neon................	20·15	412×10^4	584×10^2	538×10^2
Carbon-monoxide..	28	297×10^4	493×10^2	454×10^2
Nitrogen	28	297×10^4	493×10^2	454×10^2
Ethylene	28	297×10^4	493×10^2	454×10^2
Nitric oxide........	30	277×10^4	476×10^2	438×10^2
Oxygen..............	32	260×10^4	461×10^2	425×10^2
Argon	39·91	208×10^4	431×10^2	380×10^2
Carbon-dioxide ...	44	189×10^4	393×10^2	362×10^2
Nitrous oxide	44	189×10^4	393×10^2	362×10^2
Krypton	82·9	100×10^4	286×10^2	263×10^2
Xenon	130·2	64×10^4	228×10^2	210×10^2
Mercury vapour ...	200	$41{\cdot}6 \times 10^4$	185×10^2	170×10^2
Air	—	$[287 \times 10^4]$	485×10^2	447×10^2
Free electron	$\frac{1}{1835}$ (H=1)	$1{\cdot}515 \times 10^{11}$	$1{\cdot}114 \times 10^7$	$1{\cdot}026 \times 10^7$

We have seen that for oxygen $R/m = 259{\cdot}6 \times 10^4$, while the value of m is found, as in § 8, to be $52{\cdot}83 \times 10^{-24}$ grammes. Hence, by multiplication,

$$R = 1{\cdot}372 \times 10^{-16} \quad\quad (314).$$

This quantity is a universal constant, depending only on the particular scale of temperature employed. It will be remembered that $\frac{3}{2}R$ is the kinetic energy of translation of any molecule whatever at a temperature of 1° absolute (cf. equation (299)).

This quantity $\frac{3}{2}R$ is sometimes denoted by α, so that αT is the kinetic energy of translation of a molecule (or free atom or electron) at a temperature of T degrees absolute. The value of α is

$$\alpha = \tfrac{3}{2}R = 2{\cdot}058 \times 10^{-16} \quad\text{.......................(315).}$$

Other numerical values which are frequently of service are

$$RT_0 = 3{\cdot}747 \times 10^{-14}, \quad \alpha T_0 = 5{\cdot}620 \times 10^{-14},$$

where $T_0 = 273{\cdot}1^\circ$ (centigrade) and so is the temperature of melting ice (0° C.).

It must be understood that the accuracy of these evaluations of R, α, etc., is no greater than that of our estimation of Avogadro's number for which we have assumed the value $N_0 = 2{\cdot}705 \times 10^{19}$.

152. We have now obtained our first insight, as regards quantitative measurements, into the mechanism of the molecular motions of gases. The order of magnitude of the molecular-velocities could, however, have been predicted without actual detailed calculation.

For instance, if gas is allowed to stream out into a vacuum through a small hole in the containing vessel, the velocity of efflux is nothing else than the velocities of the individual molecules, which would have been simply molecular-velocities inside the vessel had the hole not been present. Thus the mean molecular-velocity must be comparable with the velocity of efflux of the main stream of gas, and this velocity is known to be of the order of magnitude of the velocities tabulated in the last column of the table on the preceding page.

Or again, a disturbance at any point in a gas will produce an effect on the molecules in its immediate neighbourhood. When these molecules collide with those in the next layer of gas, the effect of this disturbance is carried on into that layer, and so on indefinitely. Thus the molecules act as carriers of the effect of any disturbance, so that the disturbance is propagated, on the whole, with a velocity comparable with the mean velocity of motion of the molecules, just as, for instance, news which is carried by relays of messengers, spreads with a velocity comparable with the mean rate of travelling of the messengers. The propagation of a disturbance in the gas is, however, nothing but the passage of a wave of sound, and the velocity of sound is known to be comparable with the values of C given in the table.

Velocity of Sound.

153. It is easy to find an exact formula for the velocity of sound. For, if a is this velocity, we have the well-known formula where γ is the ratio of the two specific heats of the gas in question (cf. § 243 below). On substituting for p its value, $\frac{1}{3}\rho C^2$, this equation becomes

$$a = \sqrt{\tfrac{1}{3}\gamma}\, C.$$

For diatomic gases at ordinary temperatures, $\gamma = 1\frac{2}{5}$, so that for these gases

$$a = \cdot 683 C = \cdot 742 \bar{c} \qquad (316),$$

shewing the actual relation between the velocity of sound and the velocities C and $\bar{c}$.

For instance, the table gives for air at 0° C., $C = 485$ metres per second, whence formula (316) leads to $a = 331 \cdot 3$ metres per second for the velocity of sound in air, which approximates very closely to the true value.

Velocity of Effusion of Gases.

154. It is equally possible to find an exact formula for the rate of effusion of a gas. In fig. 4 (p. 115), imagine that the element dS forms a trap-door, capable of being opened at any instant. When this trap-door is opened, the gas will stream out through the opening dS, and we have the phenomenon of effusion through a small aperture.

We imagine the various molecules inside the vessel divided up into showers of molecules moving with equal velocities, as in § 146. The number of molecules of any specified shower, say the first, which will stream through the aperture dS in time dt, will of course be the same as the number which would have impinged on the element dS of the boundary had the trap-door remained closed. It is therefore equal to $u_1 \nu_1 dt dS$ as in § 146.

Thus the rate of effusion, measured in mass per unit time, is

$$\Sigma u_1 \nu_1 m dS \qquad (317),$$

where the summation extends over all the showers which can meet the element dS from the inside, and therefore, with the convention of § 146, over all classes of molecules for which u_1 is positive. Using Maxwell's law, we may replace ν_1 by

$$\nu \left(\frac{hm}{\pi}\right)^{\frac{1}{2}} e^{-hmu^2} du,$$

and the rate of effusion (317) becomes

$$\nu m \left(\frac{hm}{\pi}\right)^{\frac{1}{2}} \int_0^\infty e^{-hmu^2} u\, du = \frac{1}{2}\frac{\rho}{\sqrt{\pi h m}} = \rho \sqrt{\frac{RT}{2\pi m}} \qquad (318).$$

The first of these formulae shews that the rate of effusion is the same as if the whole gas of density ρ moved out of the aperture with a uniform velocity $\frac{1}{4}\bar{c}$, while the second implies the well-known law that

The rates of efflux of different gases at the same density and temperature vary inversely as the square roots of the molecular weights of the gases.

This law is confirmed in a striking manner by some experiments published by Graham* in 1846. The following table shews some of the rates of efflux found by Graham for various gases coming through fine holes in a perforated brass plate:

Gas	$\sqrt{}$(specific gravity) (air = 1)	Rate of efflux (air = 1)
Hydrogen	0·263	0·276
Marsh gas	0·745	0·753
Ethylene.................	0·985	0·987
Nitrogen.................	0·986	0·986
Air	1·000	1·000
Oxygen	1·051	1·053
Carbon-dioxide	1·237	1·203

The figures will give some idea of the degree of accuracy with which the law is obeyed. It is of interest to note that early investigators used the law as a means of determining the molecular weights of various gases†.

155. *Thermal Effusion.* From formula (318) it appears that the rate of effusion of a gas increases with its temperature, being in fact proportional to the square root of the absolute temperature when the density is kept constant. Thus the rate of efflux of a gas into a vacuum is increased by heating the gas, as is of course obvious from a consideration of the molecular mechanism of efflux.

Formula (318) is strictly applicable only to the case of efflux into a perfect vacuum. If there is a gas on the further side of the orifice, some of the molecules of the issuing gas will collide with the molecules of the external gas and will be driven back, reducing the rate of efflux. If, however, the density of the external gas is small, the number of collisions of this kind will be few, and formula (318) will still give a good approximation to the rate of efflux.

For experimental purposes, instead of using a single orifice or perforation, it is convenient to use the large number of very small orifices provided by the interstices in a plug of porous material—say of earthenware or meerschaum. The phenomenon is then spoken of as "transpiration" rather than "effusion."

Imagine a vessel of gas divided into two parts by a division, part of which consists of a porous plug of the type just described. There will be

* *Phil. Trans.* 136, p. 573.

† Leslie, *Gilb. Annalen*, xxx. (1808), p. 260; Bunsen, *Gasometrische Methoden* (Braunschweig, 1857), p. 127.

transpiration or effusion going on from each side of this plug to the other. If the two chambers into which the vessel is divided are spoken of as A and B, there will be some gas from A crossing through the porous plug into B, and similarly some from B crossing into A. And, if the pressures in the two chambers A and B are each sufficiently low, the rates of transpiration may, as an approximation, be supposed given by formula (318). If the gases in the two chambers are the same in all respects, the two rates of effusion will of course be the same. If, however, one chamber is kept warmer than the other, then the rates of effusion will not be the same, and we have the phenomenon of thermal transpiration.

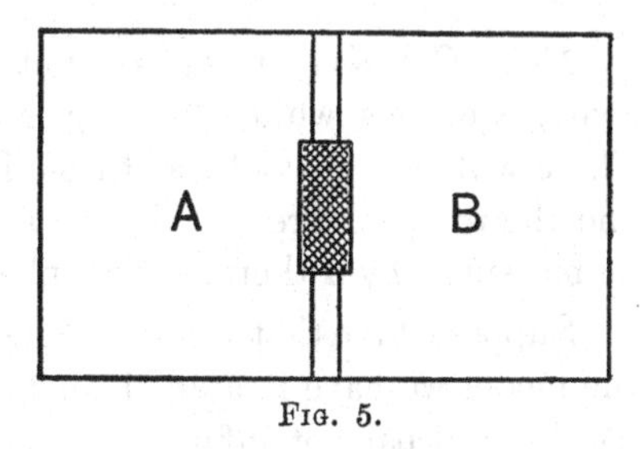

Fig. 5.

Let T_A, T_B be the temperatures of the two chambers, and let the corresponding densities and pressures be ρ_A, ρ_B and p_A, p_B. If the temperature difference is permanently maintained, the flow of gas will go on until a steady state is attained in which the flow from A to B is exactly equal to that from B to A. From formula (318) this state will be reached when

$$\rho_A \sqrt{T_A} = \rho_B \sqrt{T_B} \quad \text{.........................(319).}$$

From the pressure equation, the ratio of the pressures p_A, p_B is given by

$$\frac{p_A}{p_B} = \frac{\rho_A T_A}{\rho_B T_B},$$

and therefore in the steady state, as specified in equation (319),

$$\frac{p_A}{p_B} = \sqrt{\frac{T_A}{T_B}} \quad \text{.............................(320).}$$

Thus if the two chambers are kept unequally heated, a flow of gas will be set up which will continue until a difference of pressure between the two sides is established, such as is expressed by equation (320).

This phenomenon has been investigated in a series of experiments by Osborne Reynolds*. The two chambers were kept at steady temperatures of 8° C. and 100° C. When a steady state was attained, the pressures were measured, and it was found that, in cases in which the pressure was sufficiently low, equation (319) was satisfied with very considerable accuracy. For higher pressures this equation failed, as was to be expected.

156. Suppose that the chambers A and B in fig. 5, in addition to being connected by the porous plug, are also connected by an external pipe, of which the effect is to equalise the pressures in A and B. Then a steady

* *Phil. Trans.* 170, II. (1879), p. 727.

state cannot be attained so long as the temperatures are kept permanently at different temperatures T_A, T_B, and there occurs a steady flow of gas through the cycle formed by the chambers A, B and the pipe, a flow which is suggestive of and analogous to that of a thermoelectric current.

157. *Cohesion of Gases.* Let us suppose that the chamber B in fig. 5 contains no gas, while chamber A is filled with gas kept at temperature T_A. There will be a flow of gas through the plug or orifice into the chamber B, and the temperature of this gas as it arrives in the chamber B, say T_B, could be measured by a thermometer placed in B.

Suppose first that the molecules of the gas had corresponded exactly to the model we have imagined for them, and that they had been hard spheres, like billiard balls of infinitesimal size, exerting no force on each other except when actually in collision. Assume, as can easily be arranged, that no conduction of heat takes place between the effluent molecules and the walls of the orifice (or material of the plug) during their passage through it. Then the molecules would retain their velocities during their passage through the plug and, temperature being measured by the mean squares of these velocities, the temperature T_B would be equal to the temperature T_A.

Suppose next that the molecules of the gas had been held together by strong forces of cohesion, so that each molecule was attracted by the other molecules of the gas, or at least by those in its immediate proximity. Then each molecule, while passing through the plug, would be under an attraction towards the molecules in the chamber B, and this attraction would reduce its velocity, so that the average velocity of molecules arriving in B would be less than the average velocity of molecules in A.

It is accordingly clear that an examination of the temperature of a gas after transpiration or effusion will give important information as to the existence or non-existence of forces of cohesion in a gas. Experiments to test this question were devised and conducted by Gay-Lussac and Joule, and afterwards a more delicate and crucial set of experiments was devised by Lord Kelvin, and carried out by himself and Joule*. The earlier experiments had failed to detect any temperature change in the gas, shewing that the forces of cohesion in a gas were at least very small. The more elaborate experiments of Joule and Kelvin established definitely the existence of a slight temperature change, thus proving the existence of forces of cohesion in gases. In an experiment in which air passed by transpiration from a pressure of about four atmospheres to a pressure of one atmosphere, the change of temperature observed was a fall of 0·9° C. In general it was found that for air and many of the more permanent gases the cooling, although appreciable,

* The original papers will be found in the *Phil. Trans. of the Royal Society of London* (143, p. 357, 144, p. 321, 150, p. 325 and 152, p. 579). See also Lord Kelvin's *Collected Works*, I. p. 333.

was very slight; for carbon-dioxide, however, there was a much larger cooling, while for hydrogen there was observed a very slight heating.

For an ideal or perfect gas there would, as we have seen, be no change in temperature. Thus as regards forces of cohesion, air and the permanent gases may be said to be nearly "perfect," while carbon-dioxide is very far from "perfect," as is generally the case with this particular gas in all the properties with which the Kinetic Theory is concerned. The anomalous behaviour of hydrogen led Regnault to describe this gas as "plus que parfait."

CALCULATION OF PRESSURE IN AN "IMPERFECT" GAS.

158. It is now clear that a real gas will differ from the ideal or "perfect" gas which was under consideration in §§ 143—146 in at least two respects. The molecules which in the ideal gas were treated as spherical points must have size and shape, and the forces of cohesion which were supposed to be non-existent in the ideal gas will not be altogether negligible in the real gas. Hence it comes about that equations (307) to (311), which gave the pressure accurately in an ideal gas, will only give approximations when applied to a real gas. We must accordingly examine in what way these equations need to be corrected, so as to be made applicable to a real gas.

159. The best known correction of this type is that given by Van der Waals, in his essay *On the Continuity of the Liquid and Gaseous States**. We shall first give an explanation of the corrections introduced by Van der Waals, which, it will be found, lead to an equation expressing the deviations from Boyle's Law to a first approximation only, and we shall afterwards attempt a more general calculation of the pressure, which will not be restricted to small deviations from Boyle's Law.

Van der Waals' Equation.

160. According to Van der Waals, equation (311),

$$pv = RNT \qquad (321),$$

must be corrected in two ways. The first correction is a correction to be applied to the term v to represent the finite size of the molecules, and the second is a correction to be applied to the term p to represent the influence upon the pressure of the forces of cohesion in the gas.

* The original edition (1873) is in Dutch, published by Sigthoff, Leyden. There is a German translation by Roth (1881, Barth, Leipzig) and this has been translated into English by Threlfall and Adair (1890, *Physical Memoirs* published under the direction of the Physical Society, Taylor and Francis, London).

Reference ought also to be made to the very complete and masterly treatment of the subject by Kamerlingh Onnes and Keesom in the *Encyclopädie der Mathematischen Wissenschaften* (v. 10, pp. 615—945). This is reprinted as Vol. XI. of the *Communications from the Physical Laboratory of Leiden.*

The argument of Van der Waals as to the first correction is as follows*. In the volume v, let there be N molecules supposed still to be spherical, but now of finite size, each being of diameter σ, and let us imagine the centre of each surrounded by a sphere of radius σ, and therefore of volume $\frac{4}{3}\pi\sigma^3$. In considering possible positions for the centre of any molecule A, we know that it cannot lie within any of the $N-1$ spheres surrounding the $N-1$ other molecules, so that the space available for the centre of A must not be taken to be v but $v - \frac{4}{3}\pi\sigma^3(N-1)$.

This expression requires correction on account of the possibility of two or more of the $N-1$ spheres overlapping, but this correction will be of a higher order of small quantities than that already made, and may therefore be neglected. The expression also requires correction owing to the impossibility of the centre of a sphere being within a distance $\frac{1}{2}\sigma$ of the boundary. This correction requires us further to reduce v to the extent of the volume of a layer of thickness $\frac{1}{2}\sigma$ taken round the boundary of the containing vessel, but clearly this correction may be neglected if σ vanishes in comparison with the dimensions of the vessel. This condition is, of course, entirely different from the condition that the sum of the volumes of the molecules shall be small compared with the volume of the vessel. The former condition is satisfied if $\sigma v^{-\frac{1}{3}}$ may be neglected, the latter is satisfied if $N\sigma^3/v$ can be neglected. Using the figures given in § 8, and taking the case of a gas at atmospheric pressure in a vessel of 1 litre capacity, we find

$$\sigma v^{-\frac{1}{3}} = 2 \times 10^{-9}, \quad N\sigma^3/v = 2{\cdot}2 \times 10^{-4}.$$

It is therefore rational to neglect the one correction, while taking the other into account.

Hence in any element dv which is not within a distance $\frac{1}{2}\sigma$ of the boundary, or included in any one of the spheres surrounding each molecule, the probability of finding the centre of a molecule is

$$\frac{N\,dv}{v - \frac{4}{3}\pi\sigma^3(N-1)} \qquad \text{.........................(322).}$$

If, however, the element is selected at random we must consider what is the probability that the conditions postulated as to its not lying inside a sphere, or within a distance $\frac{1}{2}\sigma$ of the boundary, shall be satisfied.

The particular element of volume which is ultimately of importance for the calculation of the pressure is one of which the distance from the boundary is just greater than $\frac{1}{2}\sigma$. The second condition, therefore, is satisfied as a

* As regards method of presentation, I have followed Boltzmann (*Gastheorie*, II. p. 7) more closely than the original work of Van der Waals. The use of the calculus of probabilities which is made in this argument is probably open to criticism, but we shall subsequently arrive at exactly the same result by a method which does not rely on the calculus of probabilities for its justification (see § 201 below).

matter of course. To calculate the probability of the other condition being satisfied, namely that the element dv shall not lie inside any one of the $N-1$ spheres of radius σ, we notice that if it does lie in any one of these spheres, then the centre of the sphere, being at a distance not less than $\frac{1}{2}\sigma$ from the boundary, must be at least as far away from the boundary as the element dv. In other words, if the sphere in question is divided into two hemispheres by a plane parallel to the boundary, the element dv can only lie in that hemisphere which is the nearer of the two to the boundary.

Hence the probability that dv, selected at random, shall lie inside any particular sphere is $\frac{2}{3}\pi\sigma^3/v$, so that the probability that it shall *not* lie in any of the $N-1$ spheres in question is, as far as the first order of small quantities,

$$1 - \frac{\frac{2}{3}\pi\sigma^3 (N-1)}{v}.$$

The probability that a molecule shall be found in the small element dv of which the distance from the boundary is just greater than $\frac{1}{2}\sigma$, is measured by the product of this expression and expression (322), and this is

$$\frac{N dv}{v} \frac{v - \frac{2}{3}\pi\sigma^3 (N-1)}{v - \frac{4}{3}\pi\sigma^3 (N-1)} \quad \text{.......................(323).}$$

As far as the first order of small quantities this expression is the same as

$$\frac{N dv}{v-b},$$

where

$$b = \frac{2}{3} N \pi \sigma^3 \quad \text{..............................(324),}$$

in which the distinction between $N-1$ and N is now ignored.

The effect of allowing for the finite size of the molecules in the calculation of the pressure is therefore the same as that of reducing the volume from v to $v-b$, and to allow for this we replace equation (321) by

$$p(v-b) = RNT \quad \text{.........................(325).}$$

The value of b, it is of interest to notice, is four times the aggregate sum of the volumes of the molecules in the gas.

161. The principle underlying Van der Waals' second correction is as follows. It is supposed that when two molecules are near to one another, although not in contact, the forces between the molecules, although small, are not negligible. In other words, we suppose that a molecule in the gas is subjected to forces of cohesion acting between it and all the neighbouring molecules. The resultant of these forces varies continually both in direction and magnitude with the position of the molecules. When the molecule is sufficiently far removed from the surface, all directions are equally likely for this resultant, and hence the aggregate force, averaged over a sufficient length of time, will be *nil*. When, however, the molecule is at or near the surface this is no longer true. Let the force from each adjacent molecule be resolved into components tangential and normal to the boundary. Then all directions in the tangent plane are equally likely for the tangential components, but the normal component is in the majority of cases directed inwards. Averaged over a sufficient length of time the resultant force will be a normal force always directed inwards.

We may suppose the radii of curvature of the surface to be so large compared with molecular dimensions that the surface may at every point be regarded as plane. In this case the conditions will be the same at every point of the surface, and the normal force will depend only on the density of the gas and the distance from the boundary of the point at which this force is estimated.

It follows that the *average* effect of the forces of cohesion can be represented by a permanent field of force acting upon each molecule at and near the surface. It is this field of force which may be regarded as giving rise to the phenomena of capillarity and surface-tension in liquids. This field of force can be regarded as exerting an inward pressure, say p_1 per unit area, upon the outermost layer of molecules of the gas. Clearly this pressure must be supposed proportional jointly to the number of molecules per unit area in this layer, and to the intensity of the normal component of force. Each of these two factors is directly proportional to the density of the gas, so that p_1 will be proportional to the square of the density. Thus we may suppose that

$$p_1 = c\rho^2,$$

where c is a constant depending only on the nature of the gas. The molecules are now deflected upon reaching the boundary, not by impact alone, but as the total result of their impact with the boundary and of the action of the supposed field of force. In other words their change of momentum may be supposed to be produced by a total pressure $p + p_1$ or $p + c\rho^2$, instead of by the simple pressure p_1.

Hence equation (325) must be further amended by writing it in the form

$$(p + c\rho^2)(v - b) = RNT \quad \text{.....................(326)},$$

or again, replacing ρ by Nm/v, and putting $cN^2m^2 = a$,

$$\left(p + \frac{a}{v^2}\right)(v - b) = RNT \quad \text{.....................(327)}.$$

This is Van der Waals' equation connecting p, v and T. It will be noticed that a and b are constants for the same mass of gas, but depend on the amount of gas as well as on its nature, a being proportional to the square and b to the first power of the amount of gas.

162. One factor which is overlooked in the argument by which this equation is obtained, is that when cohesion forces exist, some molecules which would have reached the boundary had there been no cohesion forces, may never reach the boundary at all, being deflected by the cohesion forces before their paths meet the boundary. Actually, then, these molecules exert no pressure on the boundary, whereas Van der Waals' argument supposed them to exert a negative pressure. As a consequence, equation (327) admits of negative values for p, whereas an examination of the physical conditions shews that p is necessarily positive.

This objection is of no weight so long as it is clearly recognised that equation (327) is true only to the first order, as regards deviations from Boyle's Law.

Calculation of the Pressure from the Virial of Clausius.

163. Clausius attempted, in a way entirely different from that followed by Van der Waals, to calculate the relation between pressure, volume and temperature in an imperfect gas*.

If X, Y, Z are the components of force acting upon any molecule in a gas, its motion will be governed by the equations

$$m\frac{d^2x}{dt^2} = X, \text{ etc. } \qquad \qquad (328).$$

With the help of these equations, we find for the kinetic energy of the molecule

$$\begin{aligned} \tfrac{1}{2}m\left[\left(\frac{dx}{dt}\right)^2 + \left(\frac{dy}{dt}\right)^2 + \left(\frac{dz}{dt}\right)^2\right] &= \tfrac{1}{4}m\frac{d^2}{dt^2}(x^2+y^2+z^2) - \tfrac{1}{2}m\left(x\frac{d^2x}{dt^2} + y\frac{d^2y}{dt^2} + z\frac{d^2z}{dt^2}\right) \\ &= \tfrac{1}{4}m\frac{d^2}{dt^2}(x^2+y^2+z^2) - \tfrac{1}{2}(xX + yY + zZ). \end{aligned}$$

Hence the total kinetic energy of translation in the gas is

$$\tfrac{1}{2}\Sigma mc^2 = \tfrac{1}{4}\Sigma m\frac{d^2}{dt^2}(x^2+y^2+z^2) - \tfrac{1}{2}\Sigma(xX + yY + zZ),$$

where Σ denotes summation over all the molecules of the gas.

Averaged over all instants of time from $t=0$ to $t=\tau$, this equation becomes

$$\begin{aligned} \frac{1}{\tau}\int_{t=0}^{t=\tau}\tfrac{1}{2}\Sigma mc^2\,dt = \frac{1}{4\tau}\Sigma m\left[\frac{d}{dt}(x^2+y^2+z^2)\right]_{t=0}^{t=\tau} \\ - \frac{1}{\tau}\int_{t=0}^{t=\tau}\tfrac{1}{2}\Sigma(xX + yY + zZ)\,dt \qquad (329). \end{aligned}$$

* *Phil. Mag.* August, 1870.

In the steady motion of a gas, the quantities

$$\tfrac{1}{2}\Sigma mc^2 dt, \quad \Sigma m \frac{d}{dt}(x^2 + y^2 + z^2) \text{ and } \tfrac{1}{2}\Sigma (xX + yY + zZ)$$

are approximately constant throughout the motion. Hence as we increase τ indefinitely in equation (329), the first and last terms will remain approximately constant, while the middle term tends to vanish. Taking τ sufficiently large, the equation reduces to

$$\tfrac{1}{2}\Sigma mc^2 = -\tfrac{1}{2}\Sigma (xX + yY + zZ) \quad\quad (330),$$

in which both sides, which are in any case constant except for the slight departures from the steady state which occur in the motion of the gas, are averaged over a time sufficient for them to be regarded as sensibly constant. The mean value of $-\tfrac{1}{2}\Sigma (xX + yY + zZ)$ was termed by Clausius the *virial* of the system. We have therefore shewn that when a gas moves, undisturbed from its steady state, the kinetic energy of its motion is equal to its virial.

164. The virial depends solely on the forces acting upon the molecules, and not upon the motion of the molecules. In the case of a gas these forces consist of the pressure exerted upon the gas by the walls of the containing vessel, and the forces exerted by the molecules upon one another.

If dS is an element of the surface of the containing vessel, and l, m, n the direction cosines of its outward normal, the pressure of the element dS exerts upon the gas a force of which the components are $-lpdS$, $-mpdS$, $-npdS$, so that the value of that part of ΣxX which is contributed by the pressure will be $\iint -lpxdS$. The present treatment compels us to *assume* that the pressure is the same at all points of the containing vessel. On making this assumption the quantity just obtained may be written $-p\iint lxdS$. Hence the contribution of the pressure to the virial is, in all,

$$\tfrac{1}{2}p \iint (lx + my + nz)\, dS,$$

which, by Green's Theorem,

$$= \tfrac{1}{2}p \iiint \left(\frac{\partial x}{\partial x} + \frac{\partial y}{\partial y} + \frac{\partial z}{\partial z}\right) dx\,dy\,dz = \tfrac{3}{2}pv,$$

where v is the volume of the vessel.

We suppose that the force between two molecules at distance r is a repulsive force $\phi(r)$, a function of the distance r only. If the centres of the molecules are at x, y, z, x', y', z', and if X, Y, Z, X', Y', Z' are the components of the forces acting on them, then

$$X = \phi(r)\frac{x - x'}{r}, \quad X' = \phi(r)\frac{x' - x}{r}.$$

The contribution to ΣxX made by the force between these two particles is

$$xX + x'X' = \frac{\phi(r)}{r}(x - x')^2.$$

The contribution to $\Sigma(xX + yY + zZ)$ is therefore

$$\frac{\phi(r)}{r}\{(x - x')^2 + (y - y')^2 + (z - z')^2\} = r\phi(r),$$

so that the part of the virial which arises from intermolecular forces is

$$-\tfrac{1}{2}\Sigma\Sigma r\phi(r),$$

where the summation extends over all pairs of molecules.

Equation (330) may now be replaced by

$$\tfrac{1}{2}\Sigma mc^2 = \tfrac{3}{2}pv - \tfrac{1}{2}\Sigma\Sigma r\phi(r),$$

so that the pressure is given by

$$pv = \tfrac{1}{3}\Sigma mc^2 + \tfrac{1}{3}\Sigma\Sigma r\phi(r) \quad \text{.....................(331).}$$

165. Clerk Maxwell* makes an important observation on the subject of this equation. By obliterating one or other of the terms on the right-hand side, we notice that a pressure may be produced either wholly by molecular motion or wholly by intermolecular force. The latter is a hypothesis on which attempts have been made to account for the pressure in a gas†. If this were the true account, then Boyle's Law that pv is constant could be satisfied only by making $\Sigma\Sigma r\phi(r)$ constant, and therefore by taking $\phi(r) = 1/r$. In other words, two molecules would have to repel one another with a force proportional to the inverse distance. This is, however, an impossible law for a gas; it would make the action of the distant parts of the mass preponderate over that of the contiguous parts, and would not give a pressure which, for a given volume and temperature, would be constant as we passed from one vessel to another, or even from one part to another of the surface of the same vessel. We therefore conclude that the pressure of a gas cannot be explained by assuming repulsive forces between the molecules; it must arise, at any rate in part, from the *motion* of the molecules.

166. Returning to the general problem, it appears that if we could calculate the term $\Sigma\Sigma r\phi(r)$ in equation (331) for any law of intermolecular force, we should have a complete knowledge of the corrections to be applied to Boyle's Law. Unfortunately this is hardly possible even in the simplest cases.

167. Since there are N molecules in the gas, the total number of possible pairs will be $\tfrac{1}{2}N(N-1)$. Let A, B be any two molecules forming such a pair.

* "The Dynamical Evidence of the Molecular Constitution of Bodies," *Collected Works*, II. p. 422.

† Cf. Newton's *Principia*, II. Prop. 23, and a note by Maxwell in Cavendish's *Electrical Researches* (Note 6, Art. 97).

If the molecules were simply points exerting no forces on one another, the chance of A and B being at a distance between r and $r + dr$ from one another would be

$$\frac{4\pi r^2 dr}{v} \quad \text{.................................(332)},$$

the numerator being the volume of a shell of thickness dr surrounding molecule A, and the denominator v being the whole space possible for the centre of B. In forming this expression, we disregard the possibility of the centre of A being within a distance r of the boundary of the containing vessel, as we legitimately may if r is sufficiently small. Thus the number of pairs of molecules having their centres within a distance r of one another would, in this case, be

$$\frac{2\pi N^2}{v} r^2 dr \quad \text{.............................(333)},$$

since it is obviously legitimate to neglect the difference between $N-1$ and N.

When, however, molecules at a distance r repel one another with a force $\phi(r)$, it will be seen, upon examination of the results of § 117, that expression (331) must be modified in two ways. The probability of finding two molecules at a distance r apart is, by § 117, less than the probability of finding the same two molecules at a distance ∞ apart (∞ here denoting any distance great enough for the molecules to be out of range of each other's action) in the ratio

$$e^{-2h\int_r^\infty \phi(r)\,dr} \quad \text{............................(334)}.$$

And again the probability of finding two molecules at a distance ∞ apart is *greater* than it would be if there were no intermolecular forces, because some positions for the molecules, not at great distances apart, are *less* likely, on account of the intermolecular forces, than they would otherwise be. The former consideration requires us to multiply expression (333) by the factor (334); the latter requires us to modify the factor $\frac{N^2}{v}$ in expression (333), which is only accurate if all positions of the molecules are equally probable. Let us for the moment suppose that this latter modification can be represented by replacing $\frac{N^2}{v}$ by $(1+\beta)\frac{N^2}{v}$.

Making these corrections to expression (333), the number of pairs of molecules at a distance r apart will be

$$\frac{2\pi N^2(1+\beta)}{v} r^2 e^{-2h\int_r^\infty \phi(r)\,dr}\,dr,$$

and on multiplying by $r\phi(r)$ and integrating from 0 to ∞, we obtain

$$\Sigma\Sigma r\phi(r) = \frac{2\pi N^2(1+\beta)}{v}\int_0^\infty r^3\phi(r)\,e^{-2h\int_r^\infty \phi(r)\,dr}\,dr \quad \text{......(335)}.$$

168. In the simple case in which the molecules approximate closely to elastic spheres of diameter σ, the value of $\phi(r)$ will be very small when r is much greater than σ, and the value of the exponential in equation (335) will be very small when r is much less than σ. Thus the right-hand member of equation (335) derives all its value from values of r which are very near to σ. We may accordingly replace the factor r^3 by σ^3 and take it outside the integral. This makes the integration possible, and equation (335) reduces to

$$\Sigma\Sigma r\phi(r) = \frac{2\pi N^2(1+\beta)}{v}\frac{\sigma^3}{2h} \quad\text{.....................(336)},$$

or, if we again introduce b, defined by $b = \frac{2}{3}N\pi\sigma^3$ (cf. equation (324)), this becomes

$$\Sigma\Sigma r\phi(r) = \frac{3Nb}{2hv}(1+\beta) \quad\text{.....................(337)}.$$

169. If σ is small, b/v will be a small quantity of the first order, so that in an approximation which is carried only to the first order, we may be content to neglect β. Equation (331) then becomes

$$pv = \tfrac{1}{3}\Sigma mc^2 + \frac{Nb}{2hv},$$

or, introducing the temperature,

$$pv = RNT\left(1 + \frac{b}{v}\right) \quad\text{.....................(338)},$$

which agrees with equation (325) as far as first powers of b.

170. By evaluating the factor β, and also taking account of the possibility of more than two of the spheres of force surrounding the molecules intersecting one another, Boltzmann* has carried the approximation to second order terms. This corrected equation is found to be

$$pv = RNT\left\{1 + \frac{b}{v} + \frac{5}{8}\left(\frac{b}{v}\right)^2 + \ldots\right\} \quad\text{.................(339)}.$$

171. Keesom, using an alternative method which will be explained later (§ 201), has calculated the coefficients in the equation on the supposition that the molecules are rigid ellipsoids†, and also on the suppositions that they are of spherical structure, charged electrically with electric doublets‡ and with electric quadruplets§. The object of the investigation was to examine which assumption as to molecular structure gave results in best agreement with experiment, but it does not seem possible to arrive at any very definite conclusions.

* *Vorlesungen über Gastheorie*, II. § 51.

† *Communications from the Physical Laboratory of Leiden*, Supp. 24 *a* (1912), p. 15.

‡ *l.c.* Supp. 24 *b* (1912), p. 32.

§ *l.c.* Supp. 39 (1916), p. 1.

172. A more general case in which the right-hand member of equation (337) admits of integration is that in which the repulsive force $\phi(r)$ varies as some inverse power of the distance, say $\phi(r) = \mu r^{-s}$. We then have

$$\int_r^\infty \phi(r)\, dr = \frac{1}{s-1}\frac{\mu}{r^{s-1}},$$

$$\Sigma\Sigma r\phi(r) = \frac{2\pi N^2(1+\beta)}{v}\int_0^\infty \frac{\mu}{r^{s-3}} e^{-\left(\frac{2h}{s-1}\frac{\mu}{r^{s-1}}\right)} dr$$

$$= \frac{2\pi N^2(1+\beta)}{2hv}\left(\frac{2h\mu}{s-1}\right)^{\frac{3}{s-1}}\Gamma\left(1-\frac{3}{s-1}\right).$$

It appears that this equation can be made to agree with (336) if we regard the molecules as having a diameter σ given by

$$\sigma^3 = \left(\frac{2h\mu}{s-1}\right)^{\frac{3}{s-1}}\Gamma\left(1-\frac{3}{s-1}\right) \quad \text{.................(340)},$$

so that σ must be regarded as depending on the temperature; on replacing $2h$ by $\frac{1}{RT}$, it appears that σ is proportional to $T^{-\frac{1}{s-1}}$.

We may still introduce a quantity b defined by $b = \frac{2}{3}N\pi\sigma^3$. If b_0 is the value of this quantity at 0° C. ($T = 273{\cdot}1$), the general value of b will be

$$b = b_0\left(\frac{T}{273{\cdot}1}\right)^{-\frac{3}{s-1}} \quad \text{.....................(341)},$$

and equation (338) will be true, with this value for b.

173. If forces of cohesion of the kind specified in § 161 are also supposed to act, these forces will have a contribution to make to the virial. To the first order of small quantities, we may, in calculating $\Sigma\Sigma r\phi(r)$, ignore the effect of the forces of cohesion on the distribution of density of the gas. The value of $\Sigma\Sigma r\phi(r)$ is therefore obviously proportional simply to ρ^2 per unit volume of the gas. Allowing for this addition to the virial, equation (338) becomes

$$pv = RNT\left(1+\frac{b}{v}\right) - c\rho^2 v,$$

where c is the same as the c of § 161, and is independent of the temperature. Or again this last equation may be written

$$\left(p+\frac{a}{v^2}\right)v = RNT\left(1+\frac{b}{v}\right),$$

agreeing with Van der Waals' equation (327) as far as the first order of small quantities.

Physical Interpretation of the Equations.

174. The equation of Van der Waals undoubtedly provides the most convenient basis for discussing the behaviour of a gas over those ranges of pressure, density and temperature within which the equation may be regarded as approximately true—that is to say, ranges within which the deviation from Boyle's Law is small. We consider now some of the physical properties of a gas, derived from the equation of Van der Waals. We shall first examine the rates at which the pressure and volume change when the gas is heated.

Changes of Constant Volume.

175. Imagine the volume of a gas satisfying Van der Waals' equation (327) to be kept constant, and let us suppose the temperature first to be T_0 and afterwards T_1.

If p_0, p_1 are the corresponding pressures, we have

$$\left(p_0 + \frac{a}{v^2}\right)(v-b) = RNT_0 \quad \text{....................(342)},$$

$$\left(p_1 + \frac{a}{v^2}\right)(v-b) = RNT_1 \quad \text{....................(343)}.$$

By subtraction we get

$$(p_1 - p_0)(v-b) = RN(T_1 - T_0) \quad \text{.................(344)},$$

and on elimination of $v-b$ from this and equation (342),

$$\frac{p_1 - p_0}{T_1 - T_0}\frac{T_0}{p_0} = 1 + \frac{a}{p_0 v^2} \quad \text{........................(345)}.$$

If T_0, p_0 refer to a fixed temperature, the increase in T is proportional to that in p. Hence a gas kept at constant volume may be regarded as a thermometer giving readings on the thermodynamical scale of temperature, the reading being proportional to p.

The relation between p_1 and p_0 can be put in the form

$$p_1 = p_0\{1 + \kappa_p(T_1 - T_0)\} \quad \text{....................(346)},$$

where κ_p is what is commonly called the "pressure-coefficient" of the gas in question for the range of temperature T_0 to T_1. Using relation (346), equation (345) reduces to

$$\kappa_p = \left(1 + \frac{a}{p_0 v^2}\right)\frac{1}{T_0} \quad \text{..........................(347)}.$$

Thus κ_p depends on the density, but not on the temperature, so that

for a given density of gas, the pressure-coefficient is independent of the temperature.

This law naturally is only true within the limits in which Van der Waals' equation is true. It was shewn to be very approximately true under ordinary conditions by Regnault*, who found that gas thermometers filled with different gases gave identical readings over a large range of temperature. More recent and more exact experiments have shewn, as might be expected, that the law is by no means absolutely exact or of universal validity. Full tables of values of κ_p will be found in the *Recueil de constantes physiques*†. As a specimen may be given the following values, obtained by Chappuis in 1903‡:

VALUES OF κ_p.

Temperature	For nitrogen ($p_0=1001{\cdot}9$ mm. at 0° C.)	For CO_2 ($p_0=998{\cdot}5$ mm. at 0° C.)
0° to 20°	$\kappa_p=\cdot0036754$	$\kappa_p=\cdot0037335$
0° to 40°	·0036752	·0037299
0° to 100°	·0036744	·0037262

Callendar§ gives the following values for the pressure-coefficients (0° to 100° at initial pressure 1000 mm.) of three of the more permanent gases:

Air ·00367425,
Nitrogen ·00367466,
Hydrogen ·00366254,

while for neon and argon, Leduc|| has found the values:

Neon $T_0 = 5{\cdot}47°$ C. to $T_1 = 29{\cdot}07°$ C., $\kappa_p = \cdot003664$,
Argon $T_0 = 11{\cdot}95°$ C. to $T_1 = 31{\cdot}87°$ C., $\kappa_p = \cdot003669$.

For a perfect gas the value would of course be $\frac{1}{273{\cdot}10}$ or ·0036617. It will be noticed that the pressure-coefficient of hydrogen approximates very nearly to that of a perfect gas, shewing that the value of a is extremely small for hydrogen. For this reason the *Comité internationale des poids et mesures* decided on the constant volume hydrogen thermometer as standard thermometer. The small value which is known to exist for a is recognised in the stipulation of the committee that the volume at which the gas is used is to be such that there is a pressure of 1000 mm. at 0° C.

* *Mém. de l'Acad.* XXI. p. 180.

† pp. 234—240. The pressure-coefficient κ_p from range θ to θ' is there denoted by $\beta^{\theta'}_{\theta}$.

‡ *l.c.* p. 234.

§ *Phil. Mag.* v. p. 92.

|| *Comptes Rendus*, 164 (1917), p. 1003.

Changes at Constant Pressure.

176. Consider next a gas in which the pressure is kept at the constant value p, while the volume of the gas is changed by heating from v_0 to v_1. The two equations analogous to (342) and (343) of § 175 are

$$\left(p + \frac{a}{v_0^2}\right)(v_0 - b) = RNT_0 \quad \text{.....................(348)},$$

$$\left(p + \frac{a}{v_1^2}\right)(v_1 - b) = RNT_1 \quad \text{.....................(349)}.$$

Neglecting the product ab which is small and of the second order, we obtain, on subtraction,

$$\left(p - \frac{a}{v_0 v_1}\right)(v_1 - v_0) = RN(T_1 - T_0) \quad \text{..............(350)}.$$

We can introduce a "volume-coefficient" κ_v for the range of temperature from T_0 to T_1 such that

$$v_1 = v_0\{1 + \kappa_v(T_1 - T_0)\},$$

so that κ_v is given by

$$\kappa_v = \frac{v_1 - v_0}{v_0(T_1 - T_0)} \quad \text{..........................(351)}.$$

Equation (350) then becomes

$$\left(p - \frac{a}{v_0 v_1}\right) v_0 \kappa_v = RN,$$

giving

$$\kappa_v = \frac{RN}{pv_0 - a/v_1}.$$

On eliminating RN between this and equation (348), we obtain

$$\kappa_v = \left\{1 + \frac{a}{pv_0}\left(\frac{1}{v_0} + \frac{1}{v_1}\right) - \frac{b}{v_0}\right\} \frac{1}{T_0} \quad \text{..................(352)}.$$

This is more complicated than the formula for the pressure-coefficient (347) in that it depends both on the volume and the pressure.

The following table, similar to that given on the opposite page, will shew some values for κ_v.

VALUES OF κ_v.

Temperature	For nitrogen (p=1001·9 mm.)	For CO_2 (p=998·5 mm.)	For CO_2 (p=517·9 mm.)
0° to 20°	κ_v=·0036770	κ_v=·0037603	κ_v=·0037128
0° to 40°	·0036750	·0037536	·0037100
0° to 100°	·0036732	·0037410	·0037073

Further values will be found in the *Recueil de constantes physiques*, from which the above are taken.

Evaluation of a and b.

177. From an experimental evaluation of the "pressure-coefficient" κ_p given by equation (347), the quantity a can be obtained at once, and when a is known, the value of b can be obtained from the value of the volume-coefficient.

For instance, using Callendar's value for κ_p for air, we have (equation (347)), with $T_0 = 273 \cdot 10$,

$$\kappa_p = \left(1 + \frac{a}{p_0 v^2}\right)\frac{1}{T_0} = \cdot 00367425,$$

while
$$\frac{1}{T_0} = \cdot 0036617.$$

The value of κ_p refers to a pressure of 1000 mm. of mercury, or 1·3158 atmospheres. At this pressure, therefore,

$$\frac{a}{v^2} = p_0 T_0 \times \cdot 0000125 = \cdot 00453 \text{ atmospheres pressure.}$$

Thus for air at 1·3158 atmospheres pressure at the boundary, the forces of cohesion result in an apparent diminution of pressure of ·00453 atmospheres, or about one-threehundredth of the whole, so that the pressure in the interior of the gas is 1·3203 atmospheres. This will give an idea of the magnitude of the forces of cohesion.

Let us now suppose that we are dealing with a special mass of gas, say one for which the volume is unity at a pressure of 1 atmosphere. At a pressure of 1000 mm. of mercury the value of v is 7599, and this leads to the value

$$a = 2649 \cdot 5 \text{ in C.G.S. units} = \cdot 00260 \text{ atmospheres,}$$

for this particular mass of gas. When a has been determined in this way, we can determine b from the observed values of κ_v.

178. The determination of b is of special interest, because from it we can calculate directly the value of σ, the diameter of the molecule or of its sphere of molecular action.

The values for b which Van der Waals deduced, by the method just explained, from the discussion of a great number of experiments by Regnault were as follows:

Air	·0026,
Carbon-dioxide	·0030,
Hydrogen	·00069.

These values refer to a mass of gas which occupies unit volume at a pressure of 1000 mm. of mercury.

A more recent method of determining b depends on the measurement of the Joule-Thomson effect. From calculations by Rose-Innes*, Callendar† deduces the following values for b:

Air 1·62, Nitrogen 2·03, Hydrogen 10·73.

These values are in cubic centimetres referred to unit mass of gas. The coresponding values referred to a cubic centimetre of gas at normal pressure are found to be

Air ·00209,
Nitrogen ·00255,
Hydrogen ·00096.

For helium, Kamerlingh Onnes‡ has determined the value for b:

Helium ·000432.

Values of Molecular radius $\frac{1}{2}\sigma$.

179. The value of b is, as in equation (324), equal to $\frac{2}{3}N\pi\sigma^3$, and since the values of b have been determined for a cubic cm. of gas at normal pressure, we may take $N = 2{\cdot}705 \times 10^{19}$, and so determine σ immediately.

The values of $\frac{1}{2}\sigma$ deduced from the best values of b are as follows:

Gas	Value of b (1 c.c. of gas)	Observer	Value of $\frac{1}{2}\sigma$
Hydrogen	·00096	Rose-Innes	$1{\cdot}27 \times 10^{-8}$
Helium	·000432	Kamerlingh Onnes	$0{\cdot}99 \times 10^{-8}$
Nitrogen...........	·00255	Rose-Innes	$1{\cdot}78 \times 10^{-8}$
Air	·00209	Rose-Innes	$1{\cdot}66 \times 10^{-8}$
Carbon-dioxide ...	·00228	Van der Waals	$1{\cdot}71 \times 10^{-8}$

Isothermals.

180. One of the most instructive ways of representing the relation between the pressure, volume and temperature of a gas, is by drawing "isothermals" or graphs shewing the relation between pressure and volume when the temperature is kept constant. There will of course be one isothermal corresponding to every possible temperature, and if all the isothermals are imagined drawn on a diagram in which the ordinates and abscissae represent pressure and volume respectively, we shall have a complete representation of the relation in question.

* *Phil. Mag.* II. p. 130.

† *Phil. Mag.* V. p. 48, or *Proc. Phys. Soc.* XVIII. p. 282.

‡ *Communications from the Physical Laboratory of Leiden*, 102 *a*, p. 8.

Isothermals of an Ideal Gas.

181. For an ideal or perfect gas the relation between pressure, volume and temperature is expressed by the equation

$$pv = RNT \quad\text{..............................(353).}$$

To represent this relation by means of isothermals, we take p and v as rectangular axes and draw the various curves obtained by assigning different constant values to T in equation (353). The curves all have equations of the form $pv =$ const., and so are a system of rectangular hyperbolas, lying as in fig. 6. These are the isothermals of an ideal gas.

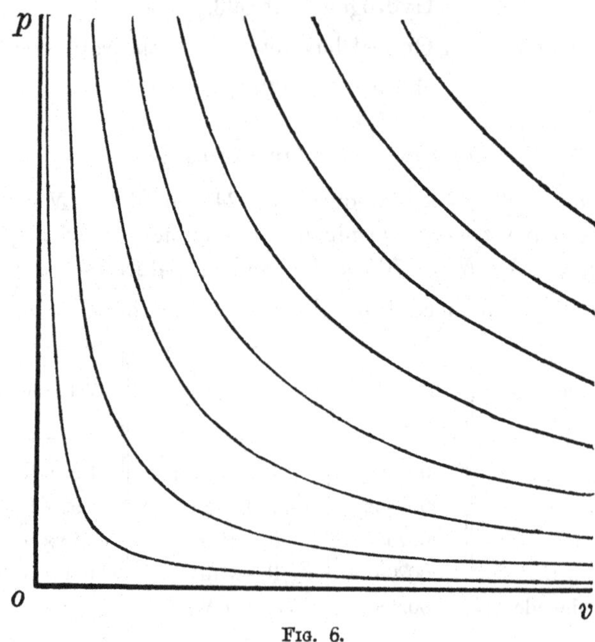

Fig. 6.

Isothermals of a Real Gas.

182. Let us consider next what isothermals will correspond to Van der Waals' equation

$$\left(p + \frac{a}{v^2}\right)(v - b) = RNT \quad\text{....................(354),}$$

bearing in mind, however, that this equation must be expected to give the true relation between p, v and T only within a range in which the deviations from the ideal relation (353) are small.

It will be noticed that if the system of curves shewn in fig. 6 are pushed bodily through a distance b parallel to the axis of v, they will give the system of isothermals represented by the equation

$$p(v - b) = RNT \quad\text{..........................(355),}$$

and on further drawing down every ordinate through a distance a/v^2 parallel to the axis of p, we shall obtain the system of isothermals represented by equation (354).

These isothermals are shewn in fig. 7, in which the thick line AB is the curve $p = -a/v^2$, while the line BCD is $v = b$.

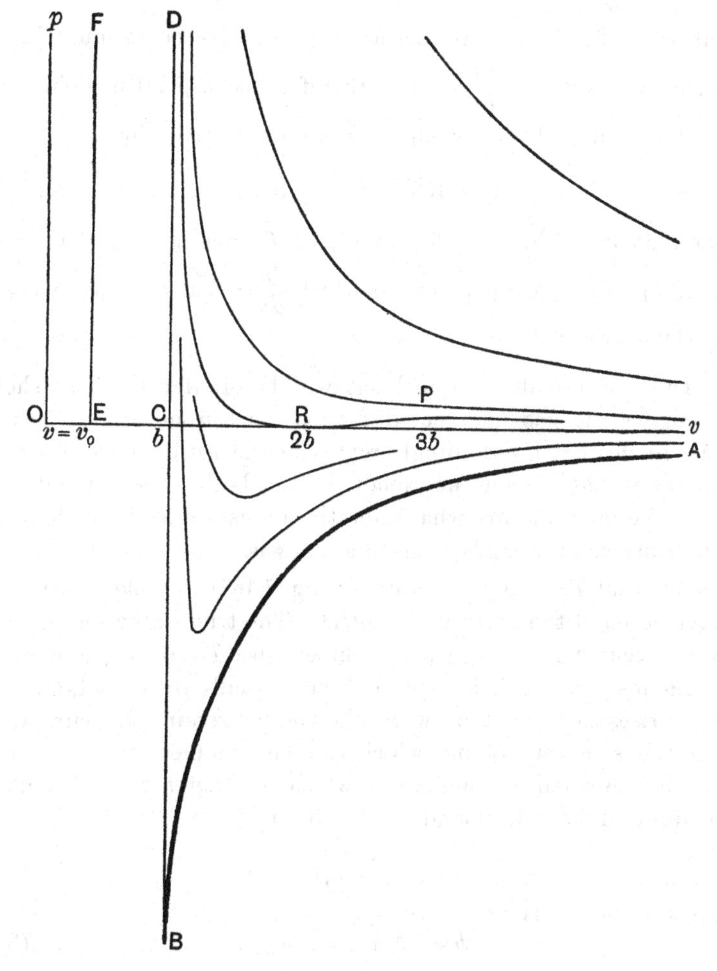

Fig. 7.

From equation (354) it is readily found that the points at which the isothermals are parallel to the axis of v—*i.e.* the points on the isothermals at which $\frac{dp}{dv} = 0$—all lie on the curve

$$p = \frac{a(v - 2b)}{v^3} \quad \text{..............................(356).}$$

This curve cuts the axis of v at $v = 2b$, so that there is an isothermal which touches the axis of v at $v = 2b$; this is found to be the isothermal

$$RNT = \tfrac{1}{4}a/b.$$

In equation (356), the maximum value possible for p is found to be given by $v = 3b$, $p = \frac{a}{27b^2}$, so that at the point of which these are the coordinates (the point P in fig. 7) the isothermal through the point must have two coincident points at which $\frac{dp}{dv} = 0$, and therefore has a point of inflexion with a horizontal tangent. This particular isothermal is given by

$$RNT = \frac{8}{27}\frac{a}{b} \quad\text{.............................(357).}$$

It is clear that all isothermals for values of T greater than that given by equation (357) can have no points at which $\frac{dp}{dv} = 0$, and so are everywhere convex to the axis of v.

183. The isothermals of a real gas will be similar to those shewn in fig. 7 so long as the gas does not differ too much from an ideal gas. The isothermals in fig. 7 will accordingly represent the isothermals of a real gas with accuracy in the regions far removed from both axes, but not near to these axes. We must inquire what alterations must be made in these curves in order to represent the isothermals of a real gas.

The isothermal $T = 0$ is represented in fig. 7 by the broken line made up of the curve AB and the vertical line BCD. The true isothermal is however known with accuracy. As a gas at temperature $T = 0$ is compressed, the pressure remains zero until the molecules are actually in contact, after which the pressure rises to any extent, while the volume retains the same value v_0, this being the smallest volume which can be occupied by the molecules. Now v_0, being the smallest volume into which N spheres each of diameter σ can be compressed, is easily found to be given by

$$v_0 = N\frac{\sigma^3}{\sqrt{2}},$$

while from equation (324),

$$b = \tfrac{2}{3}N\pi\sigma^3 = 2{\cdot}96 v_0 \quad\text{.......................(358).}$$

Thus the isothermal $T = 0$, instead of being the curve $ABCD$ in fig. 7, must consist of the two lines vE, EF. If we imagine the curves in fig. 7 so distorted that the point B is made to coincide with the point E, and the curve $ABCD$ with the lines vEF, we shall obtain an idea of the run of the isothermals of a real gas. The curves may be imagined to lie somewhat as in fig. 8, in which both the vertical and horizontal scales have been largely increased over those employed in fig. 7, but the vertical scale has been increased much more than the horizontal.

184. In this figure the isothermal having a point of inflexion with a horizontal tangent is represented by the line P_1PP_2, the point of inflexion being P. This isothermal corresponds to the maximum value of T for which it is possible for $\frac{dp}{dv}$ to vanish.

For any isothermal corresponding to a smaller value of T, say $SQRN$ in fig. 8, there will be two points Q, R at which dp/dv vanishes, and there will accordingly be a range QXR over which dp/dv will be negative. Any point, say X, within this range, will represent a state such that a decrease in volume, keeping the temperature constant, is accompanied by a *decrease* in pressure. The state represented by the point X is accordingly a collapsible or unstable state, any slight decrease in volume producing at once a tendency to a further decrease in the form of an unbalanced external pressure. All points inside the curve RPQ (the locus of points at which $dp/dv = 0$) will represent unstable states.

On the isothermal through X there must clearly be two other points Y, Z which represent states having the same temperature and pressure as X. At each of these two points dp/dv is positive, so that the two states in question are both stable, and so ought both to be known to observation. The point Z obviously represents the gaseous state; *the point Y corresponding to lesser volume is believed to represent the liquid state.*

With this interpretation it is at once clear that if the gas is kept at a temperature above that of the isothermal P_1PP_2, no amount of compression can force the substance into the liquid state. Thus the temperature of the isothermal P_1PP_2 must be the "critical temperature" of the substance.

So long as the temperature is kept above the critical temperature, no pressure, however great, can liquefy the substance.

Continuity of the liquid and gaseous states.

185. It is usual to speak of a gas, when below the critical temperature, as a vapour. We therefore see that the line PP_2 in fig. 8 is the line of demarcation between the gaseous and vapour states, and that PP_1 is the line of demarcation between the gaseous and liquid states. We must now examine the demarcation between the liquid and vapour states, which is at present represented by the unstable region in which dp/dv is positive. If U is any point in this region it is clear from physical considerations that there must be some *stable* state in which the pressure and volume are those of the point U. What is this state?

Through U draw a line parallel to the axis of v. Let this cut any isothermal in the points X, Y, Z, the two latter representing stable states—liquid and vapour respectively. These states have the same pressure, so that a quantity of vapour in the state Z can rest in equilibrium with a quantity of liquid in state Y. By choosing these quantities in a suitable ratio, the volume of the whole will be that represented by the point U. Here, then, we have an interpretation of the physical meaning of the point U. As the vapour is compressed at the temperature of the isothermal $SZQXRY$, the substance remains a vapour until the point Z is reached. At this point condensation sets in, and as the condensation proceeds the representative point moves along the straight line $ZXUY$ until, by the time the point Y

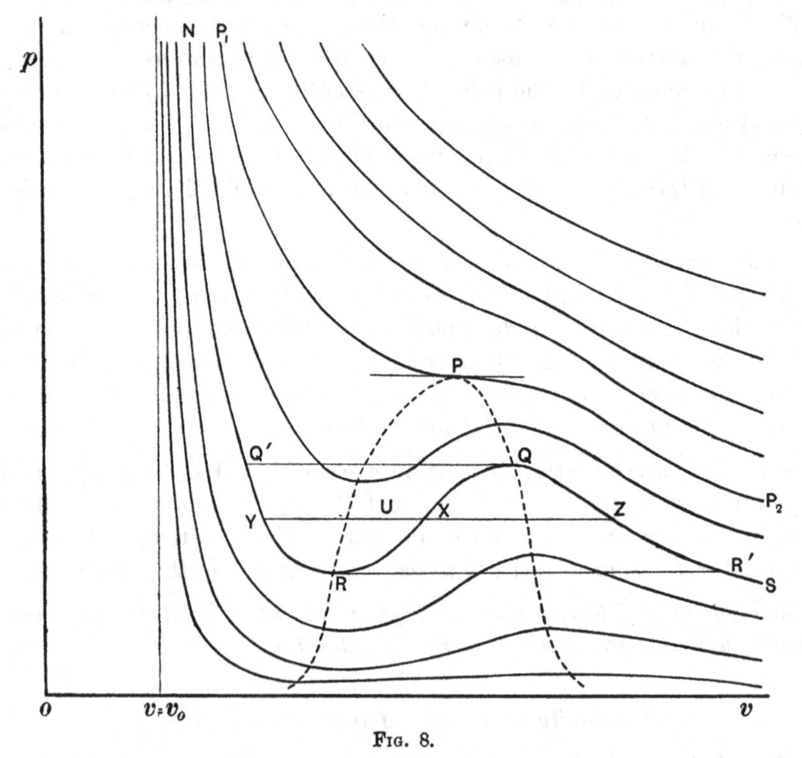

FIG. 8.

is reached, the whole of the matter is in the liquid state. After this the substance, wholly in the liquid state, moves through the series of changes represented by the path $YQ'N$.

It will be seen that there is an element of arbitrariness in this, for instead of describing the path $SZUYN$ the substance might equally well be supposed to describe the path $SR'RYN$, keeping at the same temperature throughout; or any other path composed of two stable branches of an isothermal joined by a line of constant pressure. In other words there is no unique relation between the pressure and temperature of evaporation or condensation. This is however in accordance with the known properties of matter, the range ZQ in fig. 8 representing super-cooled vapour, and the range YR representing super-heated liquid.

186. When, however, there are no complications arising from surface-tensions, particles of dust, or other extraneous agencies, there must be a definite boiling point corresponding to each pressure, so that the path of the substance from one state to another, given the same external conditions, must be quite definite. So far we have not arrived at any such definiteness.

Maxwell* and Clausius† have both attempted to obtain definite paths for a substance changing at a constant temperature. The conclusion they arrive at is that the line $SZXYN$ in fig. 8 will represent the actual isothermal path from S to N, if the line ZXY is so chosen that the areas ZQX, XRY are equal. The argument by which this conclusion is justified is as follows. Imagine the substance starting from Z, and caused to pass through the cycle

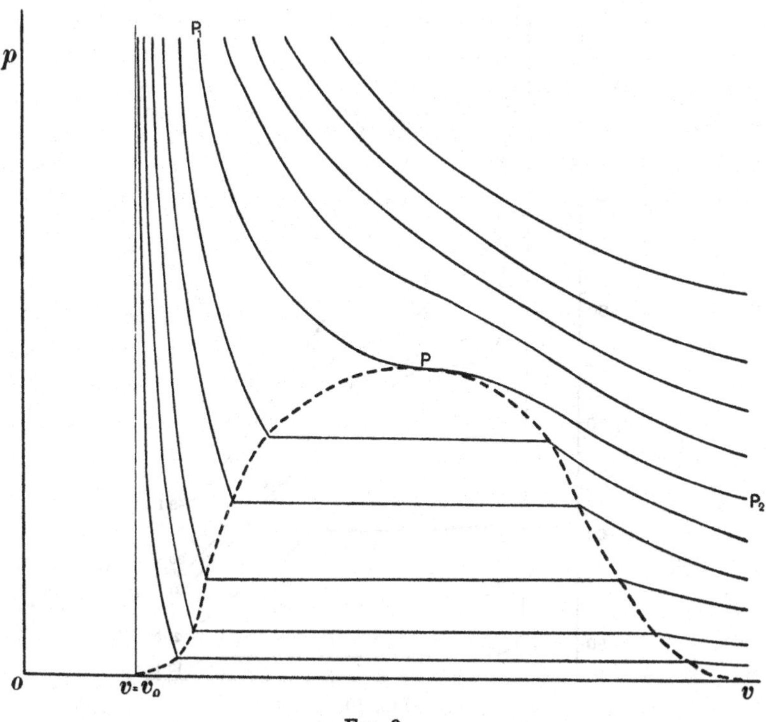

FIG. 9.

of changes represented in fig. 8 by the path $ZQXRYXZ$, the first part of the path $ZQXRY$ being along the curved isothermal, and the second part YXZ along the straight line. Since this is a closed cycle of changes, it follows from the second law of thermodynamics that $\int \frac{dQ}{T} = 0$, where dQ is the total heat supplied to the substance in any small part of its path in fig. 8, and

* *Nature*, II. 1875; *Collected Works*, II. p. 425.

† *Wied. Ann.* IX. p. 337 (1880).

the integral is taken round the whole closed path representing the cycle. Since the temperature is constant throughout the motion, this equation becomes $\int dQ = 0$, so that the integral work done on the gas throughout the cycle is *nil* This work is, however, equal to $\int pdv$ and therefore to the area, measured algebraically, of the curve in fig. 8 which represents the cycle. Hence this area must vanish, which is the result already stated. Objections have been urged against this argument, but to discuss the question any further would carry us too far into the domain of thermodynamics.

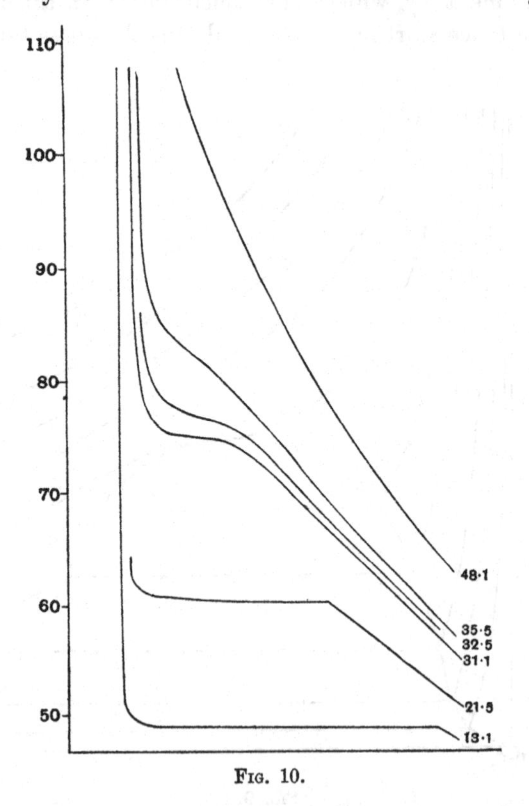

FIG. 10.

187. The figure which is obtained from fig. 8, upon replacing the curved parts of isothermals such as $ZQXRY$ by the straight line ZXY, is represented in fig. 9. This figure ought accordingly to represent the main features of the observed systems of isothermals of actual substances.

188. *Comparison with Experiment.* In fig. 10 the curves are the isothermals of carbon-dioxide as found in the classical experiments of Andrews*. The figures on the left-hand denote pressure measured in atmospheres, the

* *Phil. Trans.* 159, p. 575 (1869) and 167, p. 421 (1876).

isothermals only being shewn for pressures above 47 atmospheres. The figures on the right-hand denote the temperatures centigrade of the corresponding isothermals.

The isothermal corresponding to the temperature 31·1° is of great interest, as being very near to the critical isothermal, the value of the critical temperature being given by Andrews* as 30·92°. On this isothermal, as on all those above it, the substance remains gaseous, no matter how great the pressure.

On the next lower isothermal, corresponding to temperature 21·5° C., we notice a horizontal range at a pressure of about 60 atmospheres. As the representative point moves over this range, boiling or condensation is taking place. Thus at a pressure of about 60 atmospheres the boiling point of carbon-dioxide is about 21·5° C. The ratio of volumes in the liquid and vapour states is equal to the ratio of the two values of v at the extremities of the horizontal range—a ratio of about one to three.

The lowest isothermal of all corresponds to a temperature of 13·1° C. Here the inequality between the volumes of the liquid and the gas is greater than before. In fact an examination of the general theoretical diagram given in fig. 9 shews that as the temperature decreases the inequality must become more and more marked, so that in all substances the distinction between the liquid and gaseous states must become continually more pronounced as we recede from the critical temperature.

The Critical Point.

189. The critical point, as has been seen, is the point at which the isothermal passing through it has a point of inflexion with a horizontal tangent. It is therefore determined by the equations

$$\frac{dp}{dv} = 0, \quad \frac{d^2p}{dv^2} = 0 \qquad (359).$$

If we suppose the pressure determined by Van der Waals' equation

$$p + \frac{a}{v^2} = \frac{RNT}{v - b} \qquad (360),$$

then these equations become

$$\frac{2a}{v^3} = \frac{RNT}{(v-b)^2},$$

$$\frac{6a}{v^4} = \frac{2RNT}{(v-b)^3}.$$

* Keesom (1903) gives 30·98°, Amagat's experiments lead to the value 31·35°.

Solving these we obtain for the values of the critical volume, temperature and pressure, v_c, T_c and p_c,

$$v_c = 3b \quad\text{.............................(361)},$$

$$RNT_c = \frac{8a}{27b} \quad\text{.............................(362)},$$

$$p_c = \frac{a}{27b^2} \quad\text{.............................(363)},$$

these equations of course giving the coordinates of the point P in fig. 7 (p. 141).

By combination of these equations we find as the value of pv at the critical point

$$p_c v_c = \tfrac{3}{8} RNT_c \quad\text{..........................(364)},$$

shewing that at this point the deviation from an ideal gas is represented by a factor $\frac{3}{8}$.

190. It is now clear that the critical point is not within the region within which Van der Waals' equation can be regarded as a good approximation, and consequently equations (361) to (363) must not be expected to determine the critical point with any accuracy. In point of fact it is found* that for most gases the critical volume v_c is nearer to $2b$ than to $3b$, while the value of RNT_c is generally about $3{\cdot}7 p_c v_c$ instead of being equal to $2{\cdot}66 p_c v_c$, as predicted by equation (364). These figures, however, indicate that equations (361) to (363) may determine the critical point to within an error of 20 or 30 per cent., and conversely the equations may be used to determine the values of a, b with the same degree of accuracy when the critical point is known. The following table gives an example of values of a and b calculated in this way:

Substance	T_c obs.	p_c obs.	a calculated	b calculated	b (p. 139)
Hydrogen	$-239{\cdot}9°$ C.	12·80 atm.	·00056	·00137	·00096
Helium	$-267{\cdot}8°$ C.	2·3 ,,	·0000615	·000995	·000432
Nitrogen............	$-147{\cdot}1°$ C.	33·49 ,,	·00259	·00165	·00255
Air	$-140{\cdot}7°$ C.	37·2 ,,	·00257	·00156	·00209
Carbon-dioxide ...	$31{\cdot}1°$ C.	73 ,,	·00717	·00191	·00228

Reduced Equation of State.

191. Let us introduce quantities $\mathfrak{t}$, $\mathfrak{p}$ and $\mathfrak{v}$ defined by

$$\mathfrak{t} = \frac{T}{T_c}, \quad \mathfrak{p} = \frac{p}{p_c}, \quad \mathfrak{v} = \frac{v}{v_c} \quad\text{....................(365)},$$

* See below, § 208; also the *Recueil de Constantes Physiques*, table 83, p. 243.

so that $\mathfrak{t}$ denotes the ratio of the temperature of any substance to its critical temperature, and so on. The quantities $\mathfrak{t}$, $\mathfrak{p}$, $\mathfrak{v}$ are called the reduced temperature, pressure and volume respectively.

If we suppose Van der Waals' equation to hold, we have (cf. equations (361) to (363))

$$p = \frac{a}{27b^2}\mathfrak{p}, \quad v = 3b\mathfrak{v}, \quad RNT = \frac{8a}{27b}\mathfrak{t} \quad \text{.............(366)},$$

and Van der Waals' equation reduces to

$$\left(\mathfrak{p} + \frac{3}{\mathfrak{v}^2}\right)\left(\mathfrak{v} - \frac{1}{3}\right) = \frac{8}{3}\mathfrak{t} \quad \text{........................(367)}.$$

It will be noticed that this equation is the same for all gases, for the quantities a, b, which vary from one gas to another, have entirely disappeared. An equation, such as that of Van der Waals, which aims at expressing the relation between pressure, volume and temperature in a gas, is called an equation of state, or sometimes a characteristic equation or gas-equation. Equation (367) may be called the "reduced" equation of state of Van der Waals, and is the same for all gases.

Corresponding States.

192. Assuming for the moment that Van der Waals' equation might be regarded as absolutely true for all gases, it appears from equation (367) that when any two of the quantities $\mathfrak{t}$, $\mathfrak{p}$, $\mathfrak{v}$ are given, the third also is given, and is the same for all gases. In other words, there is a relation of the form

$$\mathfrak{p} = f(\mathfrak{t}, \mathfrak{v}) \quad \text{............................(368)},$$

in which the coefficients in f are independent of the nature of the gas.

193. Again, suppose that the equation of state of a gas is more general than that of Van der Waals, but depends on only *two* quantities which determine the particular structure of the gas in question—say for instance the same two as in the equation of Van der Waals, representing the size of the molecules and the cohesion-factor. If a and b are these two constants, it is clear that the equation of state can be expressed in the form

$$p = f\left(RT, \frac{v}{N}, a, b\right) \quad \text{........................(369)}.$$

In this we can replace the constants by new ones, and a consideration of physical dimensions will shew that it must be possible to do this in such a way that the equation assumes the form

$$\frac{p}{p_0} = f\left(\frac{v}{v_0}, \frac{T}{T_0}\right) \quad \text{..........................(370)},$$

in which p_0, v_0, T_0 are constants—*i.e.* functions of a, b, N and R—whose physical dimensions are those of a pressure, a volume and a temperature respectively.

The critical point will be given by

$$\frac{dp}{dv}=0,\quad \frac{d^2p}{dv^2}=0,$$

and the solutions of these equations must be of the form

$$v_c=c_1v_0,\quad T_c=c_2T_0.$$

From equation (370), it now follows that there must also be the equation $p_c=c_3p_0$, in which c_1, c_2, c_3 are pure numerical constants. If now the substitution (365) is made, it appears that equation (370) can be put in the form

$$\mathfrak{p}=f(\mathfrak{t},\mathfrak{v})\quad\quad(371),$$

in which the coefficients in f are independent of the nature of the gas. This is the result which has been already found true for the special equation of Van der Waals (cf. equation (368)).

Assuming that the gas-equation can be expressed in the form (371), two gases which have the same values of $\mathfrak{t}$, $\mathfrak{p}$ and $\mathfrak{v}$ are said to be in "corresponding" states. Clearly for two gases to be in corresponding states it is sufficient for any two of the three quantities $\mathfrak{t}$, $\mathfrak{p}$ and $\mathfrak{v}$ to be the same for both.

194. *The Law of Corresponding States.* It is sometimes asserted as a natural law, that when two of the quantities $\mathfrak{t}$, $\mathfrak{p}$ and $\mathfrak{v}$ are the same for two gases, then the third quantity will also be the same, and this supposed law is called the "Law of Corresponding States." The condition for the truth of the law is, as has been seen, that the reduced equation of state can be put in the form of equation (371), and this in turn demands that the nature of the gas shall be specified by only two physical constants, as for instance the a and b of Van der Waals. Evidence as to the extent to which the law is true will appear later. There is of course no question that the law is true as a first approximation, because Van der Waals' equation is true as a first approximation.

Obviously the law of corresponding states asserts that by contraction or expansion of the scales on which p and v are measured, the isothermals of all gases can be made exactly the same. As Raveau* has pointed out, this statement can be put in the alternative form that if the law of corresponding states is true graphs in which $\log p$, $\log \rho$ and $\log T$ are plotted against one another must be the same for all gases.

A large amount of work has been undertaken with a view to testing the truth of the law of corresponding states. Some of the results obtained will be discussed later (§§ 196—199, below).

* *Journ. de Phys.* VI. (1897), p. 432.

OTHER EQUATIONS OF STATE.

The Empirical Equation of State of Kamerlingh Onnes.

195. Following Kamerlingh Onnes, let us introduce a quantity K defined for any gas by

$$K = \frac{RNT_c}{p_c v_c} \quad \text{..........................(372)}.$$

From a consideration of physical dimensions, it is seen that K must be a pure number. According to Van der Waals' equation, K is equal to 2·66, although the value of K for actual gases is about 3·7 (cf. § 208, below). Let us further put

$$\mathfrak{v}_K = \frac{\mathfrak{v}}{K} \quad \text{..........................(373)}.$$

With this notation, Van der Waals' equation (367) reduces to

$$\left(\mathfrak{p} + \frac{27}{64\mathfrak{v}_K^2}\right)\left(\mathfrak{v}_K - \frac{1}{8}\right) = \mathfrak{t} \quad \text{..................(374)},$$

which can also be written in the form

$$\mathfrak{p}\mathfrak{v}_K = \frac{\mathfrak{t}}{1 - \dfrac{1}{8\mathfrak{v}_K}} - \frac{27}{64\mathfrak{v}_K}$$

$$= \mathfrak{t}\left\{1 + \frac{1}{\mathfrak{v}_K}\left(\frac{1}{8} - \frac{27}{64}\frac{1}{\mathfrak{t}}\right) + \frac{1}{64\mathfrak{v}_K^2} + \frac{1}{512\mathfrak{v}_K^3} + \dots\right\} \text{......(375)}.$$

It has been found by innumerable observers that an equation of this type is not adequate to represent the various states of a gas, and so Kamerlingh Onnes has assumed the more general empirical form

$$\mathfrak{p}\mathfrak{v}_K = \mathfrak{t}\left\{1 + \frac{\mathfrak{B}}{\mathfrak{v}_K} + \frac{\mathfrak{C}}{\mathfrak{v}_K^2} + \frac{\mathfrak{D}}{\mathfrak{v}_K^4} + \frac{\mathfrak{E}}{\mathfrak{v}_K^6} + \frac{\mathfrak{F}}{\mathfrak{v}_K^8}\right\} \quad \text{.........(376)},$$

where $\mathfrak{B}$, $\mathfrak{C}$, $\mathfrak{D}$ are themselves series of the form

$$\mathfrak{B} = \mathfrak{b}_1 + \frac{\mathfrak{b}_2}{\mathfrak{t}} + \frac{\mathfrak{b}_3}{\mathfrak{t}^2} + \frac{\mathfrak{b}_4}{\mathfrak{t}^4} + \frac{\mathfrak{b}_5}{\mathfrak{t}^6} \text{...................(377)}.$$

This expansion contains no fewer than 25 adjustable coefficients. If Van der Waals' Law were true, all of these ought to vanish except $\mathfrak{b}_1$, $\mathfrak{c}_1$, $\mathfrak{d}_1$, $\mathfrak{e}_1$, $\mathfrak{f}_1$, and $\mathfrak{b}_2$, of which the values ought to be given by the following scheme:

	1	2
$10^3\,\mathfrak{b}$	125·000	−416·666
$10^4\,\mathfrak{c}$	156·25	0·0
$10^5\,\mathfrak{d}$	195·3125	0·0
$10^7\,\mathfrak{e}$	?	0·0
$10^9\,\mathfrak{f}$	?	0·0

in which the values for $\mathfrak{e}_1$, $\mathfrak{f}_1$ have not been inserted, since these values cannot be determined by comparison with (375) owing to the omission of terms in $\mathfrak{v}^{-4}$, $\mathfrak{v}^{-7}$, etc. in (376).

If Van der Waals' Law were not true, but the general law of corresponding states were true, the coefficients in the expansions (373), (374) ought to be the same for all gases.

196. Kamerlingh Onnes* finds that a general equation of state of type (371) can be obtained which expresses with fair accuracy the observations of Amagat on hydrogen, oxygen, nitrogen and $C_4H_{10}O$, also of Ramsay and Young on $C_4H_{10}O$ and of Young on isopentane (C_5H_{12}). The coefficients in this equation are found to be those given in the following table:

	1	2	3	4	5
$10^3\,\mathfrak{b}$	117·796	−228·038	−172·891	−72·765	−3·172
$10^4\,\mathfrak{c}$	135·580	−135·788	295·908	160·949	51·109
$10^5\,\mathfrak{d}$	66·023	−19·968	−137·157	55·851	−27·122
$10^7\,\mathfrak{e}$	−179·991	648·583	−490·683	97·940	4·582
$10^9\,\mathfrak{f}$	142·348	−547·249	508·536	−127·736	12·210

The circumstance that the same equation of state is valid for all these six gases shews of course the wide applicability of the law of corresponding states. On the other hand the coefficients of the equation of state of argon have been determined by Kamerlingh Onnes and Crommalin † and are found to differ substantially from those in the table just given. In a later discussion‡ the same authors express the opinion that the divergence of argon from the standard equation may be ascribed to the absence in argon of the usual compressibility of the molecules of normal substances; in other words, the argon molecule approximates more closely to an ideal elastic sphere than do the molecules of the substances mentioned above.

197. A comparison of the coefficients in the table just given and those in the table of § 195 will at once shew that Van der Waals' equation fails to a very great extent to represent the true equation of state: in fact, as has been already seen, its accuracy is limited to the range of states in which the gas behaves nearly like an ideal gas, or in other words to large values of $\mathfrak{t}$ and $\mathfrak{v}$.

* *Encyc. d. Math. Wissenschaften*, v. 10, p. 729, or *Communications from the Physical Laboratory of Leiden*, XI. Supplement 23, p. 115.

† *Communications from the Physical Laboratory of Leiden*, 118 *b* (1910), p. 24.

‡ *l.c.* 121 *b* (1911), p. 25.

Kamerlingh Onnes has suggested* that Van der Waals' Law must be expected to agree better with observation when applied to a gas in which the molecular conditions conform more closely to those contemplated by Van der Waals. He has accordingly examined the isothermals of helium, and has found that they can be represented very fairly between 100° C. and − 217° C. by the equation

$$\mathfrak{p}\mathfrak{v} = NRT + \frac{NRTb - a}{\mathfrak{v}} + \frac{5}{8}\frac{NRTb^2}{\mathfrak{v}^2},$$

which is simply Van der Waals' equation adjusted by the inclusion of Boltzmann's second-order correction. It must however be noticed that the value of T_c for helium is very low, about 5·3° abs., so that the range of temperatures studied by Prof. Kamerlingh Onnes is about from $\mathfrak{t} = 10$ to $\mathfrak{t} = 70$, and for these high values of $\mathfrak{t}$ it is inevitable that Van der Waals' equation should in any case give a good approximation.

The Equation of State of Clausius.

198. Various attempts have been made to improve Van der Waals' equation by the introduction of a few more adjustable constants, which can be so chosen as to make the equation agree more closely with experiment.

The best known of these equations is that of Clausius, namely

$$\left(p + \frac{a'}{T(v+c)^2}\right)(v - b) = RNT \quad\quad (378).$$

If in this we put $c = 0$, the equation becomes similar to that of Van der Waals, except that the a of Van der Waals' equation is replaced by a'/T; in other words, instead of a being constant, it is supposed to vary inversely as the temperature. For some gases the equation of Clausius, reduced in this way, is found to fit the observations better than the equation of Van der Waals.

If, however, c is not put equal to zero in (378) but is treated as an adjustable constant and selected to fit the observations, there is found to be no tendency for c to vanish. The following table shews the values of a', b and c which are found by Sarrau† to give the best approximation to the observations of Amagat:

Gas	a'	b	c	c/b
Nitrogen	0·4464	·001359	·000263	0·19
Oxygen	0·5475	·000890	·000686	0·75
Ethylene	2·688	·000967	·001919	1·98
Carbon-dioxide	2·092	·000866	·000949	1·10

* *Communications from the Physical Laboratory of Leiden*, 102 *a* (1907).

† *Comptes Rendus*, 114 (1882), pp. 639, 718, 845.

199. *Law of Corresponding States.* We have already seen that, if Van der Waals' equation were true, the law of corresponding states would follow as a necessary consequence, because in Van der Waals' Law there are only two constants, a and b, which respectively provide the scales on which the pressure and volume can be measured in reduced coordinates.

On the other hand in the equation of Clausius (equation (378)) there are three separate constants, a', b and c, and of these two, namely b and c, provide different scales on which the volume can be measured: these two scales only become identical if b and c stand in a constant ratio to one another. Thus if the law of corresponding states were true, the ratio c/b would be the same for all gases. The last column of the above table shews, however, that there is no approximation to constancy in the values of c/b.

200. Clausius originally devised formula (378) in an effort to fit a formula to the observations of Andrews on carbon-dioxide. It was found that this formula could be made to agree well with observations on carbon-dioxide at high densities, but that at low densities the formula of Van der Waals, as might be expected, fitted the observations better*. It was next found that equation (378), although partially successful in the case of carbon-dioxide, was not equally successful with other gases, and Clausius suggested the more general form

$$\left\{p + \left(\frac{a''}{T^{n-1}} - a'''T\right)\frac{1}{(v+c)^2}\right\}(v-b) = RNT,$$

which contains five adjustable constants. For carbon-dioxide, it is found that $n = 2$ and $a''' = 0$, so that the equation reduces to (378), but for other gases it is not found that n and a''' approximate to these values. For instance Clausius finds† that for ether $n = 1{\cdot}192$, for water-vapour $n = 1{\cdot}24$, while to agree with observations on alcohol, n itself must be regarded as a function of the temperature and pressure, having values which vary from 1·087 at 0° C. to 0·184 at 240° C.

It is obvious, then, that there is no finality in any of these formulae, and it is possible to go on extending them indefinitely without arriving at a fully satisfactory formula, as might indeed be anticipated from the circumstance that they are purely empirical, and not founded on any satisfactory theoretical basis.

General Calculation of Pressure.

201. We may now attempt to examine what type of formula is predicted by the Kinetic Theory for the general relation between pressure, volume and temperature, although it will be found that the formula obtained is of so complicated a nature that it is not possible to progress very far.

* See a diagram by Berthelot, *Arch. Néerl.* v. (1900), p. 420, reproduced in the *Recueil de Constantes Physiques* (p. 246).

† Cf. Preston, *Theory of Heat*, p. 511.

We shall modify the calculation of § 146, so as to apply when no assumption is made as to the size or structure of the molecules. In this calculation, we estimated the number of molecules having velocities within given limits $du\,dv\,dw$, which impinged on the element dS in time dt, and this was shewn to be equal to the number of such molecules which at the beginning of the interval dt were to be found within a certain element of volume $udSdt$ (cf. fig. 11).

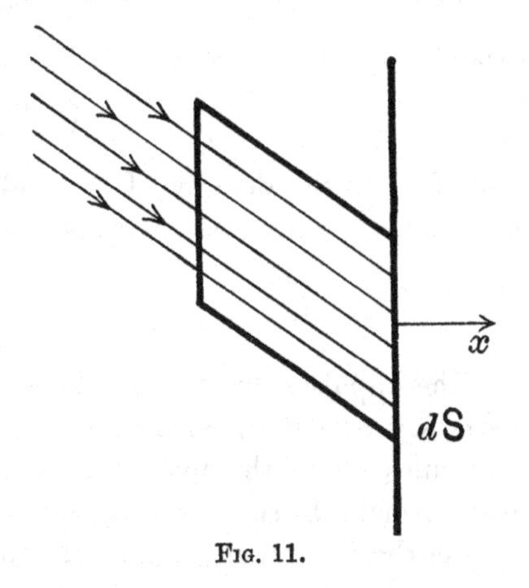

FIG. 11.

The molecules of the more general type now to be considered must be supposed to have a number of coordinates $\xi_1, \xi_2, \ldots$ specifying their orientation and internal state in addition to their coordinates of position and velocity of the centre of gravity u, v, w, x, y, z. All the molecules in the vessel, whether chemically similar or not, may be divided into classes $\alpha, \beta, \gamma, \ldots$ such that all the molecules in any one class are all chemically the same, have their coordinates $\xi_1, \xi_2, \ldots$ lying within a specified range $d\xi_1, d\xi_2, \ldots$, and have their components of velocity u, v, w lying within a specified range $du\,dv\,dw$. Thus all the molecules in any one class form a shower of parallel moving molecules, which throughout their motion remain all similar to one another as regards dynamical specification, except in so far as they are affected by collisions.

For any particular class of molecule, say α, let q denote the perpendicular distance, at the beginning of the interval dt, from the centre of gravity of the molecule on to that tangent plane to the surface of the molecule which is perpendicular to the axis of x and so is parallel to dS. The corresponding perpendicular distance at the end of the interval dt will be

$$q + \frac{dq}{dt}\,dt.$$

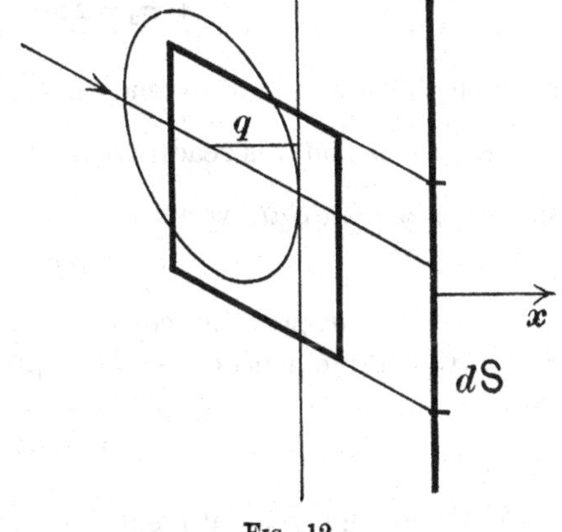

FIG. 12.

The molecules of class α which will impinge on dS during an interval dt are those of which the centres of gravity at the beginning of this interval lie within a certain cylindrical volume, as in fig. 12. The cross-section of this area is dS and it is bounded by planes at distances

q and $q + udt + \frac{dq}{dt} dt$ from dS. The volume of this cylindrical element is accordingly

$$\left(u + \frac{dq}{dt}\right) dS dt,$$

and if ν_a is the density of molecules of class α in this particular element of volume, the number of impacts on dS within the interval dt will be

$$\nu_a \left(u + \frac{dq}{dt}\right) dS dt \quad \text{.........................(379)}.$$

The impulses exerted on the area dS in time dt can be divided into two parts, say ϖ_1 and ϖ_2, such that ϖ_1 is the sum of the impulses exerted by the molecules up to the instant at which their centre of gravity is reduced to rest normally to the boundary (*i.e.* the instant at which $u = 0$), and ϖ_2 is the sum of the remaining parts of the impulses.

Each of the impacts enumerated in expression (379) brings a contribution mu to ϖ_1, so that

$$\varpi_1 = \Sigma m\nu_a \left(u + \frac{dq}{dt}\right) u dS dt \quad \text{.....................(380)},$$

where the summation is over all classes of molecules $\alpha, \beta, \gamma, \ldots$ for which u is positive.

The value of ϖ_2 will of course be exactly equal to that of ϖ_1. It may be supposed to be given by an expression exactly similar to (380) except that the summation is over all classes of molecules for which u is negative.

By addition of the values of ϖ_1 and ϖ_2, we find

$$\varpi_1 + \varpi_2 = \Sigma m\nu_a \left(u + \frac{dq}{dt}\right) u dS dt \quad \text{.................(381)},$$

in which the summation extends over all classes of molecules. On carrying out the summation it is readily seen that the terms in $\frac{dq}{dt}$ must vanish. Thus, since $\varpi_1 + \varpi_2 = p dS dt$, we have

$$p = \Sigma m\nu_a u^2 \quad \text{...........................(382)},$$

analogous to our previous equation (306). The summation covers all values of u, so that the equation may be expressed in the equivalent forms

$$p = (\Sigma\nu_a) RT = \frac{\Sigma\nu_a}{2h} \quad \text{......................(383)}.$$

Although this general result has proved to be expressible in a very simple form, it must be remembered that the meaning of the symbol ν_a is one of great complexity. Consequently, it is only in very simple cases that the evaluation of p can be carried out to the full.

202. We may consider the form assumed by the problem, when the molecules are supposed to be all spherical, and of a definite diameter σ which may not be treated as very small. In this case q becomes identical with $\frac{1}{2}\sigma$, so that $\frac{dq}{dt}$ disappears. The classes $\alpha, \beta, \gamma, \ldots$ of molecules are now, as before in § 146, differentiated only by the different values for the velocity components u, v, w, and $\Sigma\nu_a$ may be replaced by ν_b, where ν_b is the density of centres of molecules in an element of volume at a distance just greater than $\frac{1}{2}\sigma$ from the boundary.

In the notation of § 67 (equation (124)), the "effective molecular density" ν_1 at any point was found to be given by the equation

$$\nu_1 = N\frac{I(b, c, d\ldots)}{I(a, b, c, d\ldots)} \quad\quad (384),$$

expressing the value of ν_1 at the point x_a, y_a, z_a. Here $I(a, b, c, d \ldots)$ is the element of volume integrated throughout the whole of the generalised space except those parts excluded by § 36.

The element of volume integrated throughout the whole of the generalised space is, as in § 65, equal to Ω^N, where Ω is the volume of the vessel in which the gas is supposed to be contained. The excluded parts are of two types.

We have in the first place to exclude the region given by equations (60) in which $\phi(x_a, y_a, z_a) < \frac{1}{2}\sigma$. This exclusion can be represented fully by limiting the size of our vessel, and supposing it to have a layer of thickness $\frac{1}{2}\sigma$ removed from the interior. Let the remaining volume be Ω', then the corrected value of the integral $I(a, b, c \ldots)$ will obviously be Ω'^N.

Secondly we must exclude from the generalised space regions of the type given by

$$(x_a - x_b)^2 + (y_a - y_b)^2 + (z_a - z_b)^2 < \sigma^2 \quad\quad (385).$$

Let us consider the contribution to the whole integral Ω'^N which must be removed on account of this particular exclusion. The whole integral may be taken to represent all possible ways of distributing the centres of the molecules $A, B, \ldots$ throughout a volume Ω', each position being equally likely for each molecule. In this case the contribution which satisfies condition (385) represents all arrangements for which the centres of A and B lie within a distance σ of one another.

In fig. 13 let the shaded part represent the layer of thickness $\frac{1}{2}\sigma$ close to the boundary of the vessel, and let the spheres represent possible positions for the molecule B of diameter σ.

If B is in a position in which its centre is at a distance greater than σ from the boundary of the volume Ω', the proportion of configurations represented in the generalised space in which condition (385) is satisfied—or, what is the same thing, in which the centre of A lies within a sphere of radius σ

surrounding B—is equal to the ratio of the volume of this sphere to the whole volume which is available for the centre of A, and is therefore equal to $\frac{4}{3}\pi\sigma^3/\Omega'$. If, however, B is in a position in which its centre is at a distance less than σ from the boundary of the volume Ω', including being in a position such as I. in fig. 13 in which its centre is at a distance greater than $\frac{1}{2}\sigma$ from the boundary, and also being in a position such as II. in fig. 13, in which its centre is at a distance less than $\frac{1}{2}\sigma$ from the boundary, then the proportion in question is less than that just found, since it is only possible for A to lie in that portion of a sphere of radius σ surrounding B which lies inside the volume Ω'. Ultimately, when B is in position III., the proportion is only one-half of the above quantity, namely, $\frac{2}{3}\pi\sigma^3/\Omega'$, for it is now only possible

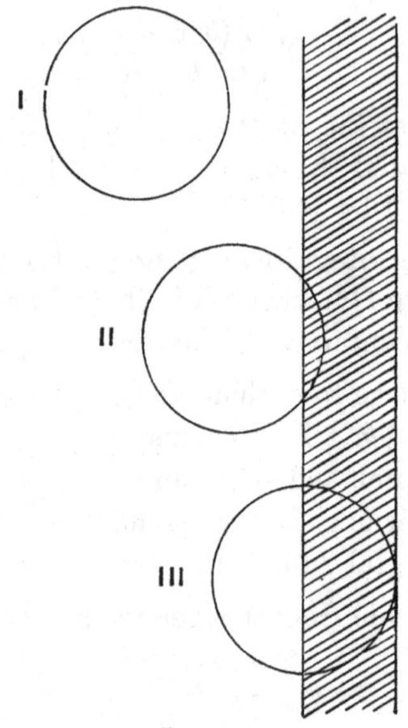

FIG. 13.

for A to lie inside a single hemisphere of the sphere about B. On averaging over all positions of B it is obviously legitimate to disregard these exceptional cases, and we arrive at the result that the proportion of cases in which condition (385) is satisfied is $\frac{4}{3}\pi\sigma^3/\Omega'$. Thus if condition (385) defined the only region to be excluded, the whole integral Ω'^N would have to be reduced by $\frac{4}{3}\pi\sigma^3/\Omega'^{N-1}$.

There are, however, $\frac{1}{2}N(N-1)$ such conditions, corresponding to all possible pairs of molecules. The total reduction is therefore $\frac{1}{2}N(N-1)$ times the foregoing amount. From this must be subtracted a quantity representing regions in which two of the conditions of the type of (385) are satisfied at once. This again must be corrected on account of the possibility of more than two of these conditions being satisfied at once, and so on, indefinitely.

In this way we obtain an expansion in descending powers of Ω', of which the first two terms have been shewn to be

$$I(a, b, c, d \ldots) = \Omega'^N - \tfrac{2}{3}\pi\sigma^3 N(N-1)\,\Omega'^{N-1} + \ldots \qquad \ldots\ldots(386).$$

The integral $I(b, c, d \ldots)$ can be evaluated in a similar way except that in this case the molecule A is supposed already to be in collision with the boundary, and therefore in position III. in fig. 13. Hence the available space for the centres of the molecules $B, C, \ldots$ is $\Omega' - \frac{2}{3}\pi\sigma^3$ instead of the former Ω', and the number of molecules is $N-1$ instead of N. Making the necessary alterations in equation (386) we obtain an expansion in the form

$$I(b, c, d \ldots) = (\Omega' - \tfrac{2}{3}\pi\sigma^3)^{N-1} - \tfrac{2}{3}\pi\sigma^3 (N-1)(N-2)(\Omega' - \tfrac{2}{3}\pi\sigma^3)^{N-2} \qquad \ldots\ldots(387).$$

Equation (384) accordingly leads to the value

$$\nu_b = N\frac{(\Omega' - \frac{2}{3}\pi\sigma^3)^{N-1} - \frac{2}{3}\pi\sigma^3(N-1)(N-2)(\Omega' - \frac{2}{3}\pi\sigma^3)^{N-2} + \ldots}{\Omega'^N - \frac{2}{3}\pi\sigma^3 N(N-1)\,\Omega'^{N-1} + \ldots}$$

$$= \frac{N}{\Omega'}\left(1 + \tfrac{2}{3}\pi\sigma^3\frac{N-1}{\Omega'} + \ldots\right),$$

on expanding as far as the first two terms; and since, in the limit, we may put

$$\frac{N-1}{\Omega'} = \frac{N}{\Omega'} = \frac{N}{\Omega} = \nu,$$

we obtain ν_b in the form

$$\nu_b = \nu + \tfrac{2}{3}\pi\sigma^3\nu^2 + \ldots \qquad \ldots\ldots\ldots\ldots\ldots\ldots(388).$$

Using this value for ν_b in equation (383) we obtain the equation

$$p = \frac{\nu_b}{2h} = \left(\nu + \frac{2\pi\sigma^3}{3}\nu^2 + \ldots\right)RT \qquad \ldots\ldots\ldots\ldots(389),$$

which agrees with equation (325) as far as the first two terms.

This method can of course be used to obtain the expansion (388) for ν_b to as many terms as we please, but the calculations prove to be laborious, and the results are of little value, since they involve the supposition that the molecules are spherical. This supposition may perhaps lead to fairly accurate results when we are concerned only with first-order terms, but can hardly be expected to lead to accurate results as regards terms of second and higher orders.

203. We have next to examine how to allow for the existence of forces of cohesion in the pressure-equation. The physical effect of forces of cohesion has already been explained (§ 161) in deducing the equation of Van der Waals, so that we may start at once with the conception of a permanent field of force, at and near to the boundary of the gas. The effect of this field of force upon the gas as a whole will be to alter the density of the gas at points near the boundary.

Let χ be the amount of work which must be done on a molecule to drag it from a point in the interior of the gas to the boundary, in opposition to the attractive forces of cohesion, then the density at the boundary, say ρ_0, must by equation (232) be given by

$$\rho_0 = \rho e^{-2h\chi} \quad \text{.....(390)},$$

where ρ is the density inside the gas, and so is equal to the mean density of the gas as a whole. If the effect of the forces of cohesion is small, the density of the gas will be approximately the same everywhere, and χ will be of the form $\rho\psi$, where ψ is independent of the density.

Obviously the pressure in the gas depends only on the density of gas in the layer nearest to the boundary, and so is the same for the gas we have now under consideration as it would be for a gas which was unaffected by forces of cohesion, and was of uniform density ρ_0 throughout. Hence the pressure is given by

$$p = \nu_b(\rho_0) RT \quad \text{.....(391)},$$

where $\nu_b(\rho_0)$ denotes the effective density ν_b for a gas of density ρ_0, the value of ρ_0 being given by equation (390).

204. Equation (391) expresses the general relation between pressure, volume and temperature. Using the value of ν_b provided by equation (388), it becomes

$$p = RT(\nu e^{-2h\chi} + \tfrac{2}{3}\pi\sigma^3\nu^2 e^{-4h\chi} + \ldots) \quad \text{.....(392)}.$$

The bracket on the right may be regarded as the product of $e^{-2h\chi}$ and an expansion of powers $\sigma^3 e^{-2h\chi}$. If we introduce a new quantity σ_c defined by

$$\sigma_c = \sigma e^{-\frac{2}{3}h\chi} \quad \text{.....(393)},$$

the pressure will be given by

$$p = RT\nu_b(\sigma_c) e^{-2h\chi} \quad \text{.....(394)},$$

where

$$\nu_b(\sigma_c) = \nu + \frac{2\pi\sigma_c^3}{3}\nu^2 + \ldots \quad \text{.....(395)},$$

and so is the same function of σ_c as is ν_b of σ.

Small Deviations from Boyle's Law.

205. When the deviations from Boyle's Law are small, χ may be replaced by $\rho\psi$, and σ_c identified with σ. Thus in equation (394), $e^{-2h\chi}$ may be replaced by $1 - 2h\rho\psi$ or $1 - \frac{\rho\psi}{RT}$, and $\nu_b(\sigma_c)$ becomes identical with ν_b or $\nu(1 + \frac{2}{3}\pi\sigma^3\nu)$, which again, in the notation of § 160, becomes $\nu\left(1 + \frac{b}{v}\right)$. Thus equation (394) becomes

$$p = RT\nu\left(1 + \frac{b}{v}\right)\left(1 - \frac{\rho\psi}{RT}\right),$$

which, as far as the first order of small quantities, may equally be written in the form

$$p=\frac{RNT}{v-b}-\frac{a}{v^2} \quad \text{.........................(396)},$$

where $a=N^2m\psi$. This is identical with Van der Waals' equation.

The Equation of State of Dieterici.

206. Making use of the relation just obtained between ψ and a, it appears that, for small deviations from Boyle's Law, equation (394) becomes identical with

$$p=\frac{RNT}{v-b}e^{-\frac{a}{RNTv}} \quad \text{.........................(397)},$$

an equation of state first suggested by Dieterici *.

Theoretically, this equation has exactly the same range of validity as Van der Waals' Law, being true to the first order of small quantities only. In a former section we examined what errors were involved in assuming Van der Waals' Law to be true beyond the first order of small quantities: it was found that at the critical point Van der Waals' Law predicted a ratio of b/v equal to 3, whereas the true value is about 2, while for the ratio K defined by equation (372), Van der Waals' Law predicted a value $2\frac{2}{3}$, whereas the true value is about 3·7. It is worth examining whether equation (397) agrees better or worse with experiment than Van der Waals' Law in ranges where the deviations are no longer small.

If equation (397) were true throughout the whole range of pressure, volume and temperature, the equations giving the critical point, namely

$$\frac{dp}{dv}=0, \quad \frac{d^2p}{dv^2}=0,$$

would be

$$\frac{1}{(v-b)}-\frac{a}{RNTv^2}=0 \quad \text{.......................(398)},$$

$$\frac{1}{(v-b)^2}-\frac{2a}{RNTv^3}=0 \quad \text{.......................(399)}.$$

From these we obtain at once, at the critical point,

$$v_c=2b \quad \text{...............................(400)},$$

and

$$\frac{a}{RNT}=4b.$$

Thus

$$p_c=\frac{RNT_c}{b}e^{-2},$$

* *Wied. Ann.* LXV. (1898), p. 826 and LXIX. (1899), p. 685.

so that the ratio K is given by

$$K = \frac{RNT_c}{p_c v_c} = \tfrac{1}{2}e^2 = 3{\cdot}695 \quad \text{.....................(401).}$$

It is at once apparent that the values of v_c and K given by equations (400) and (401) agree very much better with experiment than the values $3b$ and 2·66 given by Van der Waals' equation. Thus, although equation (397) has the same validity as Van der Waals' equation within the range in which these equations are applicable, yet it follows the true laws much more closely than Van der Waals' equation when we pass outside this range.

On substituting the values of T_c, p_c and v_c, the "reduced" form of equation (397) is found to be

$$\mathfrak{p}\,(2\mathfrak{v} - 1) = \mathfrak{t} e^{2\left(1 - \frac{1}{\mathfrak{v}}\right)} \quad \text{.......................(402).}$$

207. *Comparison with experiment.* If equation (397) were accurately true, we should have

$$e^{\frac{a}{RNTv}} = \frac{RNT}{p\,(v-b)},$$

whence, on taking logarithms of both sides, multiplying by v and neglecting $(b/v)^2$,

$$\frac{a}{RNT} = v \log\left(\frac{RNT}{pv}\right) + b.$$

Thus, to the first order of small quantities, $v \log\left(\frac{RNT}{pv}\right)$ ought to be constant along any isothermal, and from what has been seen in § 205, it follows that the same result could be obtained from Van der Waals' equation.

The following table gives corresponding values of p and v observed at the critical temperature of isopentane, by Young*. This temperature according to Young is 187·8° C., the critical volume being 4·266 cubic centimetres for 1 gramme of the gas, and the value of K being 3·739. The values of p in the third column are calculated by Dieterici† from equation (402), and those of $v \log \frac{RNT}{pv}$ in the fourth column from Young's observations.

It appears that $v \log \frac{RNT}{pv}$ is approximately constant for all pressures less than about 12 atmospheres, shewing the range within which a first approximation holds. There is, however, very tolerably good agreement between the observed and calculated values of p far beyond this range, indicating considerable accuracy in the equation of state.

* S. Young, *Proc. Roy. Soc.* xv. p. 126 and xvi. p. 11.

† *Annalen der Physik*, v. p. 58, 1901.

Critical Isothermal of Isopentane.

Critical temperature = 187·8° C. Critical volume = 4·266 c.c. per gramme.

Volume v, per gramme	Pressure p, mm. of mercury	Pressure p (calc.)	$v \log \frac{RNT}{pv}$
2·4	49080	42730	1·271
2·5	40560	35810	1·486
2·6	34980	32090	1·669
2·8	28940	28390	1·938
3·0	26460	26780	2·103
3·2	25490	26000	2·205
3·6	25050	25420	2·326
4·0	25020	25320	2·402
4·3	25010	25300	2·447
4·6	25000	25300	2·483
5	24990	25240	2·520
6	24840	24880	2·564
7	24400	—	2·577
8	23710	23400	2·582
9	22930	—	2·576
10	22040	21590	2·575
12	20300	19850	2·568
15	17980	17540	2·548
20	14840	14560	2·564
30	10950	10770	2·526
40	8570	8508	2·624
50	7068	7025	2·625
60	6001	5978	2·652
80	4614	4604	2·680
90	4132	4127	2·637
100	3750	3740	2·680

Values of Constants at the Critical Point.

208. The following table gives some values observed for the quantity K by reliable observers:

Gas	K	Observer
Oxygen	3·419	Kamerlingh Onnes, Dorsman and Holst*
Nitrogen	3·421	,,
Argon	3·424	,,
Xenon	3·605	Patterson†
Marsh gas	3·67	Berthelot‡
Carbon-dioxide	3·61	,,
,, ,,	3·61	Young§
$C\,Cl_4$	3·68	,,
$C_6\,H_6$	3·71	,,
$Sn\,Cl_4$	3·75	,,
$C_6\,H_5\,Br$	3·78	,,
$C_6\,H_5\,Cl$	3·81	,,

* *Communications from the Physical Laboratory of Leiden*, 145 (1914).

† Patterson, Cripps and Whytlaw-Gray, *Proc. Roy. Soc.* 86 A. (1912), p. 579.

‡ *Bull. de la Soc. Franç. de Phys.* 167 (1901), p. 4.

§ *Phil. Mag.* 1892, p. 503, 1894, p. 1, and L. (1900), p. 303.

It may be added that Young is of opinion that for all substances which can attain the critical state without chemical change, K is nearly constant, and has a value of about 3·7.

It is more difficult to give values of the ratio of v_c to b, for b is difficult to determine, and for many substances, different experimenters give widely different values for the critical constants. The following values are, however, fairly reliable:

Hydrogen*	$v_c = \cdot 00269$,	$b = \cdot 00096$,	$v_c/b = 2\cdot 80$,
Nitrogen*	$v_c = \cdot 00382$,	$b = \cdot 00255$,	$v_c/b = 1\cdot 50$,
Air†	$v_c = \cdot 00392$,	$b = \cdot 00209$,	$v_c/b = 1\cdot 89$,
Oxygen‡	$v_c = \cdot 00332$,	$b = \cdot 00228$,	$v_c/b = 1\cdot 46$,
Carbon-dioxide§	$v_c = \cdot 00424$,	$b = \cdot 00228$,	$v_c/b = 1\cdot 86$,
Argon‖	$v_c = \cdot 00328$,	$b = \cdot 00232$,	$v_c/b = 1\cdot 41$.

Thus what evidence there is suggests that the values $K = 3\cdot 7$ and $v_c/b = 2\cdot 0$ are fairly close to the truth¶. These are the values predicted by Dieterici's equation of state. We may notice that the variations in the values of K and of v_c/b for different gases give an indication of the degree to which the law of corresponding states is inaccurate. If this law had been exactly true the values of these quantities would have been the same for all gases.

* The critical volumes are deduced from Dewar's values of the critical density (from the rectilinear diameter).

† Obtained by combining Witkowski's isothermals for air at low temperatures with Olszewski's values for the critical temperature and pressure (see p. 148).

‡ The value of v_c is that given by Mathias and Kamerlingh Onnes (Leyden Comm. No. 117). The value of b is deduced from the value of $\frac{1}{2}\sigma$ given by the coefficient of viscosity (cf. § 386 below).

§ Obtained from Amagat's experiments, giving a critical density 0·464. This value of v_c agrees closely with that resulting from Keesom's experimental study of the critical isothermal of CO_2.

‖ The value of v_c is that given by Kamerlingh Onnes (Leyden Comm. No. 120 *a*). The value of b is deduced in the same way as for oxygen.

¶ Berthelot states as a general rule that the critical volume is about equal to four times the co-volume v_0 given by our equation (358), which, however, he also takes to be the same as b (see *Recueil de Constantes Physiques*, p. 244). The relation $v_c = 4v_0$ gives $v_c/b = 1\cdot 35$. Many investigators have taken Berthelot's statement to mean that $v_c/b = 4$.

CHAPTER VII

PHYSICAL PROPERTIES (*continued*)

MASS MOTION, THERMODYNAMICS, CALORIMETRY AND DISSOCIATION

The equations of Mass Motion.

209. In the last chapter some insight was obtained into the way in which pressure is exercised by a collection of moving molecules. We shall now attempt to carry further the study of the physical properties of a medium consisting of moving molecules, especially with reference to the similarities between the behaviour of such a medium and a continuous fluid.

We shall no longer consider only problems in which the gas is in a steady state, so that the law of distribution is Maxwell's law at every point. We shall consider problems in which the law of distribution of velocities is the general law $f(u, v, w)$, and it will no longer be assumed that the law of distribution is the same at every point: we shall suppose the law of distribution to depend on x, y, z, as well as on u, v, w. When it is required to indicate this we shall write the law of distribution in the form

$$f(u, v, w, x, y, z).$$

The number of molecules having their centres within the element of volume $dxdydz$ surrounding the point x, y, z, and having their velocities within a range $dudvdw$ surrounding the values u, v, w, will be

$$\nu f(u, v, w, x\ y, z)\, dudvdwdxdydz \quad \text{...............(403)}.$$

In this expression ν will in general be a function of x, y, z, and will be evaluated at the same point as f. It is often convenient to think of the product νf as a single variable.

Hydrodynamical Equation of Continuity.

210. Consider the small element of volume $dxdydz$ inside the gas, having its centre at ξ, η, ζ and bounded by the six planes parallel to the coordinate planes, of which the equations are

$$x = \xi \pm \tfrac{1}{2}dx, \quad y = \eta \pm \tfrac{1}{2}dy, \quad z = \zeta \pm \tfrac{1}{2}dz \quad \text{...........(404)}.$$

The number of molecules of class A (defined on p. 16) which cross the plane $x = \xi - \frac{1}{2}dx$ into this element of volume in time dt will be

$$\iint \nu f(u, v, w, \xi - \tfrac{1}{2}dx, y, z)\, dy\, dz\, du\, dv\, dw\, u\, dt,$$

where the integration is with respect to y, z and is between the limits $y = \eta \pm \frac{1}{2}dy$, $z = \zeta \pm \frac{1}{2}dz$. As far as squares of small quantities this may be written in the form

$$\nu f(u, v, w, \xi - \tfrac{1}{2}dx, \eta, \zeta)\, dy\, dz\, du\, dv\, dw\, u\, dt \qquad \text{..........(405).}$$

Similarly the number of molecules of class A which cross the plane $x = \xi + \frac{1}{2}dx$ out of the element of volume is given by expression (405) if $\xi - \frac{1}{2}dx$ is replaced by $\xi + \frac{1}{2}dx$. On subtraction, the resulting loss to the element, of molecules of class A, caused by motion through the two faces perpendicular to the axis of x, is found to be

$$\frac{\partial}{\partial x}[\nu f(u, v, w)]\, dx\, dy\, dz\, du\, dv\, dw\, u\, dt,$$

in which the differential coefficient is evaluated at ξ, η, ζ. The net loss of molecules of class A caused by motion through all the faces is therefore

$$\left(u\frac{\partial}{\partial x} + v\frac{\partial}{\partial y} + w\frac{\partial}{\partial z}\right)[\nu f(u, v, w)]\, du\, dv\, dw\, dx\, dy\, dz\, dt \qquad \text{......(406).}$$

By integration over all values of u, v and w, we obtain the total number of molecules which are lost to the element $dx\,dy\,dz$ in time dt. If we write

$$\iiint u f(u, v, w)\, du\, dv\, dw = u_0, \text{ etc.} \qquad \text{..................(407),}$$

so that u_0, v_0, w_0 are the components of mass velocity of the gas at x, y, z, this number is seen to be

$$\left(\frac{\partial}{\partial x}(\nu u_0) + \frac{\partial}{\partial y}(\nu v_0) + \frac{\partial}{\partial z}(\nu w_0)\right) dx\, dy\, dz\, dt \qquad \text{...........(408).}$$

Since, however, the number of molecules in the element at time t is $\nu\, dx\, dy\, dz$, and at time $t + dt$ is $\left(\nu + \frac{d\nu}{dt}dt\right) dx\, dy\, dz$, the net loss must be

$$-\frac{d\nu}{dt}\, dt\, dx\, dy\, dz.$$

Equating this to expression (408), we obtain

$$\frac{d\nu}{dt} + \frac{\partial}{\partial x}(\nu u_0) + \frac{\partial}{\partial y}(\nu v_0) + \frac{\partial}{\partial z}(\nu w_0) = 0 \qquad \text{..............(409).}$$

This is the hydrodynamical equation of continuity, expressing the permanence of the molecules of the gas—in other words, the conservation of mass.

Hydrodynamical Equations of Motion.

211. The loss of each molecule of class A to the element means a loss of momentum parallel to the axis of x equal to mu. The total loss of momentum parallel to the axis of x arising from this cause in time dt is, by expression (406),

$$m\,dx\,dy\,dz\,dt \iiint \left(u^2 \frac{\partial}{\partial x} + uv \frac{\partial}{\partial y} + uw \frac{\partial}{\partial z} \right) [\nu f(u, v, w)]\, du\,dv\,dw \quad \ldots (410).$$

Let us write

$$\iiint u^2 f(u, v, w)\, du\,dv\,dw = \overline{u^2} \quad \ldots\ldots\ldots\ldots\ldots\ldots (411),$$

$$\iiint uv f(u, v, w)\, du\,dv\,dw = \overline{uv} \quad \ldots\ldots\ldots\ldots\ldots\ldots (412),$$

etc.,

so that $\overline{u^2}$, $\overline{uv}$, etc. are the mean values of u^2, uv, etc. at the point x, y, z. Then expression (410) becomes

$$m\,dx\,dy\,dz\,dt \left(\frac{\partial}{\partial x} (\nu \overline{u^2}) + \frac{\partial}{\partial y} (\nu \overline{uv}) + \frac{\partial}{\partial z} (\nu \overline{uw}) \right) \ldots\ldots\ldots\ldots (413).$$

With this must be compounded a gain of momentum caused by the action of forces on the molecules. The gain of momentum parallel to the axis of x accruing to any single molecule in time dt is $X\,dt$, where X is the component of force parallel to the axis of x, including of course intermolecular forces and forces at collision, acting upon the molecule in question. Combining the sum of these gains with the loss given by expression (413), we find for the net increase of momentum inside the element $dx\,dy\,dz$ in time dt,

$$\left[\Sigma X - m\,dx\,dy\,dz \left(\frac{\partial}{\partial x} (\nu \overline{u^2}) + \frac{\partial}{\partial y} (\nu \overline{uv}) + \frac{\partial}{\partial z} (\nu \overline{uw}) \right) \right] dt \ldots\ldots (414),$$

where Σ denotes summation over all the molecules which were inside the element $dx\,dy\,dz$ at the beginning of the interval dt.

The total x-momentum inside the element $dx\,dy\,dz$ at time t is, however, $m\nu u_0\,dx\,dy\,dz$. Hence the gain in time dt may be expressed as

$$\frac{d}{dt} (\nu u_0)\, m\,dx\,dy\,dz\,dt,$$

and on equating this to expression (414), we obtain

$$\left[\frac{d}{dt} (\nu u_0) + \frac{\partial}{\partial x} (\nu \overline{u^2}) + \frac{\partial}{\partial y} (\nu \overline{uv}) + \frac{\partial}{\partial z} (\nu \overline{uw}) \right] m\,dx\,dy\,dz = \Sigma X \quad \ldots (415).$$

These and the similar equations in y, z are the hydrodynamical equations of motion of the gas, expressing that momentum is conserved, except in so far as this momentum is changed by the action of external forces.

212. If, as in § 26, we write

$$u = u_0 + \mathsf{U}, \text{ etc.}$$

we have $\overline{u} = u_0$, and $\overline{\mathsf{U}} = 0$. Thus $\overline{uv} = u_0 v_0 + \overline{\mathsf{UV}}$, and similarly for $\overline{u^2}$ and $\overline{uw}$. Hence we have

$$\frac{\partial}{\partial x}(\nu\overline{u^2}) + \frac{\partial}{\partial y}(\nu\overline{uv}) + \frac{\partial}{\partial z}(\nu\overline{uw})$$

$$= \frac{\partial}{\partial x}(\nu u_0^2) + \frac{\partial}{\partial y}(\nu u_0 v_0) + \frac{\partial}{\partial z}(\nu u_0 w_0) + \frac{\partial}{\partial x}(\nu\overline{\mathsf{U}^2}) + \frac{\partial}{\partial y}(\nu\overline{\mathsf{UV}}) + \frac{\partial}{\partial z}(\nu\overline{\mathsf{UW}}) \quad \text{...(416).}$$

Again, using the equation of continuity (409),

$$\frac{d}{dt}(\nu u_0) = \nu\frac{du_0}{dt} + u_0\frac{d\nu}{dt}$$

$$= \nu\frac{du_0}{dt} - u_0\left(\frac{\partial}{\partial x}(\nu u_0) + \frac{\partial}{\partial y}(\nu v_0) + \frac{\partial}{\partial z}(\nu w_0)\right),$$

so that on adding corresponding sides of this equation and (416), we obtain

$$\frac{d}{dt}(\nu u_0) + \frac{\partial}{\partial x}(\nu\overline{u^2}) + \frac{\partial}{\partial y}(\nu\overline{uv}) + \frac{\partial}{\partial z}(\nu\overline{uw})$$

$$= \nu\left(\frac{d}{dt} + u_0\frac{\partial}{\partial x} + v_0\frac{\partial}{\partial y} + w_0\frac{\partial}{\partial z}\right)u_0 + \frac{\partial}{\partial x}(\nu\overline{\mathsf{U}^2}) + \frac{\partial}{\partial y}(\nu\overline{\mathsf{UV}}) + \frac{\partial}{\partial z}(\nu\overline{\mathsf{UW}}).$$

The left-hand member of the equation is, however, identical with the bracket in equation (415), so that this equation now becomes

$$\nu\left(\frac{d}{dt} + u_0\frac{\partial}{\partial x} + v_0\frac{\partial}{\partial y} + w_0\frac{\partial}{\partial z}\right)u_0 m\,dx\,dy\,dz$$

$$= \Sigma X - \left[\frac{\partial}{\partial x}(\nu\overline{\mathsf{U}^2}) + \frac{\partial}{\partial y}(\nu\overline{\mathsf{UV}}) + \frac{\partial}{\partial z}(\nu\overline{\mathsf{UW}})\right] m\,dx\,dy\,dz \quad \text{...(417).}$$

213. We now proceed to examine more closely the system of forces which act upon those molecules of which the centres are inside the element $dx\,dy\,dz$—the system of forces which we have so far been content to denote by ΣX, ΣY, ΣZ.

These forces will be supposed to arise partly from the action of an external field of force upon the molecules of the gas, and partly from the actions of the molecules upon one another.

If there is a field of external force of components Ξ, H, Z per unit mass the contribution to ΣX will be

$$\Xi m\nu\,dx\,dy\,dz \quad \text{.............................(418).}$$

The remaining contribution of ΣX arises from the intermolecular forces. As regards the forces between a pair of molecules, both of which are inside the element $dx\,dy\,dz$, we see that, since action and reaction are equal and opposite, the total contribution to ΣX will be *nil*. We are left with forces between pairs of molecules such that one is inside and the other outside the element $dxdydz$, that is to say, intermolecular forces which act across the boundary of this element. The range of intermolecular forces may be treated as small even compared with the dimensions of the element $dxdydz$, so that the forces arise from pairs of molecules which lie close to the boundary, but on opposite sides.

Let the sum of the components of all the forces of this kind per unit area across a plane perpendicular to the axis of x be denoted by ϖ_{xx}, ϖ_{xy}, ϖ_{xz}, and let us adopt a similar notation as regards pressures across planes perpendicular to the axes of y and z. Then the contribution to ΣX which comes from forces acting across the planes $x = \xi \pm \frac{1}{2} dx$ is

$$(\varpi_{xx})_{x=\xi-\frac{1}{2}dx}\, dy\, dz - (\varpi_{xx})_{x=\xi+\frac{1}{2}dx}\, dy\, dz = -\frac{\partial \varpi_{xx}}{\partial x}\, dx\, dy\, dz.$$

On adding similar contributions from the other planes, we find that the x-component of all the molecular forces which act on the element $dxdydz$ can be put in the form

$$-\left(\frac{\partial \varpi_{xx}}{\partial x} + \frac{\partial \varpi_{yx}}{\partial y} + \frac{\partial \varpi_{zx}}{\partial z}\right) dx\, dy\, dz.$$

Combining this contribution to ΣX with that already found (expression (418)), we find, as the whole value of ΣX,

$$\Sigma X = \left[m\nu \Xi - \left(\frac{\partial \varpi_{xx}}{\partial x} + \frac{\partial \varpi_{yx}}{\partial y} + \frac{\partial \varpi_{zx}}{\partial z}\right)\right] dx\, dy\, dz,$$

and there are, of course, similar equations giving the values of ΣY and ΣZ.

Hence equation (417), on dividing throughout by $dxdydz$ and replacing $m\nu$ by ρ, becomes

$$\rho\left(\frac{d}{dt} + u_0 \frac{\partial}{\partial x} + v_0 \frac{\partial}{\partial y} + w_0 \frac{\partial}{\partial z}\right) u_0$$

$$= \rho \Xi - \frac{\partial}{\partial x}(\varpi_{xx} + \rho \overline{\mathrm{U}^2}) - \frac{\partial}{\partial y}(\varpi_{yx} + \rho \overline{\mathrm{UV}}) - \frac{\partial}{\partial z}(\varpi_{zx} + \rho \overline{\mathrm{UW}}) \quad \text{......(419)},$$

and this gives another form, alternate to equation (417), of the equations of motion of the gas.

Evaluation of the Stresses in a Gas.

214. Let us write

$$\left.\begin{aligned} P_{xx} &= \varpi_{xx} + \rho \overline{\mathrm{U}^2}, \\ P_{yx} &= \varpi_{yx} + \rho \overline{\mathrm{UV}}, \quad \text{etc.} \end{aligned}\right\} \quad \text{......................(420)},$$

then equation (419) may be written

$$\rho\left(\frac{d}{dt} + u_0 \frac{\partial}{\partial x} + v_0 \frac{\partial}{\partial y} + w_0 \frac{\partial}{\partial z}\right) u_0 = \rho \Xi - \frac{\partial P_{xx}}{\partial x} - \frac{\partial P_{yx}}{\partial y} - \frac{\partial P_{zx}}{\partial z} \quad \text{...(421)}.$$

This equation is identical in form with the hydrodynamical equation of a fluid of density ρ moving with a velocity u_0, v_0, w_0, the body forces being Ξ, H, Z per unit mass, and the pressures being the system

$$P_{xx},\ P_{yx},\ P_{zx},\ P_{xy},\ \ldots\ldots \text{ etc.}$$

We may, therefore, speak of this system of pressures as the *total* pressure at the point x, y, z. As is clear from equation (420), the pressures arise partly from the intermolecular forces and partly from the molecular agitation in the gas.

215. If we multiply equation (409) throughout by m, it becomes

$$\frac{d\rho}{dt} + \frac{\partial}{\partial x}(\rho u_0) + \frac{\partial}{\partial y}(\rho v_0) + \frac{\partial}{\partial z}(\rho w_0) = 0 \quad \ldots\ldots\ldots\ldots\ldots(422),$$

which is formally the same as the hydrodynamical equation of continuity, and the hydrodynamical analogy is now complete.

We have, therefore, seen that we may regard a gas as a continuous fluid, of which the motion is subject to hydrodynamical equations of motion and continuity of the usual type. This is the justification, in discussing the mass-motion of gases—the propagation of sound, for instance,—for treating the gas as if it was a continuous fluid of the kind contemplated in hydrodynamics.

Molecules of Finite Size.

216. In order to separate the difficulties as much as possible, and so simplify the treatment of the subject, it has been found convenient to defer the difficulties introduced by the finite size of the molecules. The finite size can, as has already been explained, be supposed to have been allowed for in the field of intermolecular force, but in order to get results which admit of easy interpretation it is best to suppose the molecules to be of finite size, in addition to possessing fields of intermolecular force.

If $\nu_{x,y,z}$ is the "effective molecular density" at x, y, z as defined in § 67, then the expectation of the number of molecules of which the centres lie within any element of volume must be taken to be

$$\iiint \nu_{x,y,z}\, dx\,dy\,dz,$$

where the integration extends throughout the element of volume in question. This, however, as has been seen in § 68, must be equal to

$$\nu \iiint dx\,dy\,dz,$$

provided the integral is taken through a volume large enough to contain a great number of molecules, and this is identical with the value assumed when the molecules were supposed infinitesimal.

Similarly the number of molecules crossing an element of surface dS will contain as a factor

$$\iint \nu_{x,y,z}\, dS \quad \text{..............................(423)},$$

but if this element of surface is great compared with the size of a molecule, while at the same time ν is approximately constant while we pass in any direction over a distance comparable with the size of dS, then the factor may also be written

$$\nu \iint dS \quad \text{................................(424)},$$

which, again, was the value assumed for infinitesimal molecules.

A complication occurs in the neighbourhood of the boundary, because here, as appeared in the last chapter, the value of ν may vary perceptibly over a distance comparable with the molecular diameter. If, then, dS is an element of surface at a distance $\frac{1}{2}\sigma$ from the boundary and parallel to the boundary, expression (423) must be replaced, not by expression (424), but by

$$\nu_b \iint dS \quad \text{................................(425)},$$

where ν_b is the same as the ν_b of § 202.

It will now be clear that all the analysis of the present chapter will hold, even after allowing for the finite size of the molecules, if we take ν to be the density everywhere except at the boundary, provided that, in considering an element of surface at distance $\frac{1}{2}\sigma$ from the boundary, we replace ν by ν_b.

In calculating the pressure, it will be remembered, we assumed the element of volume to be so great that intermolecular forces could not act across it. At the same time we assume it to be so small that the density is approximately constant throughout. These assumptions become incompatible at the boundary, because the density, as is clear from § 161, varies over a range across which intermolecular forces can and do act. In this case, however, we can abandon the supposition that the density will be approximately constant throughout. All that is required is a knowledge of the mean density in the element of volume, and this, on account of the thinness of the layer near the surface in which the mean density differs perceptibly from ν, may still be taken to be ν.

Gas in Equilibrium.

217. Equations (420) provide general formulae for the pressures of a gas in motion. We shall now examine the form assumed by these equations when the gas is in equilibrium, and shall find that it is possible to obtain alternative proofs of a number of the theorems which have already been proved about the equilibrium state of a gas. We may take the law of distribution of velocities to be Maxwell's Law at every point, so that

$$\rho \overline{\mathsf{U}^2} = \rho \overline{u^2} = \nu \overline{m u^2} = \frac{\nu}{2h},$$

by equations (298), and

$$\overline{\mathsf{UV}} = \overline{\mathsf{VW}} = \overline{\mathsf{WU}} = 0.$$

Thus of the system of pressures specified by equations (420), that part which arises from molecular agitation reduces to a simple hydrostatical pressure of amount $\frac{\nu}{2h}$. Clearly the ϖ system of pressures, arising from the intermolecular forces, also reduces to a simple hydrostatical pressure ϖ, and we therefore have

$$P_{xx} = P_{yy} = P_{zz} = \varpi + \frac{\nu}{2h},$$

$$P_{xy} = P_{xz} = P_{yz} = 0.$$

The total pressure at the point x, y, z is therefore a simple hydrostatical pressure of amount P given by

$$P = \varpi + \frac{\nu}{2h} \quad \text{..............................(426)},$$

an equation which may be compared with the virial equation (331).

218. The equations of equilibrium become

$$\rho \Xi = \frac{\partial P}{\partial x}, \text{ etc.} \quad \text{..............................(427)}.$$

The conditions for equilibrium to be possible are, therefore, three of the type

$$\frac{\partial}{\partial y}(\rho \Xi) = \frac{\partial}{\partial x}(\rho \mathrm{H}).$$

In general, the result of eliminating the unknown variable ρ from these equations is

$$\Xi \left(\frac{\partial \mathrm{H}}{\partial z} - \frac{\partial \mathrm{Z}}{\partial y}\right) + \mathrm{H} \left(\frac{\partial \mathrm{Z}}{\partial x} - \frac{\partial \Xi}{\partial z}\right) + \mathrm{Z} \left(\frac{\partial \Xi}{\partial y} - \frac{\partial \mathrm{H}}{\partial x}\right) = 0 \quad \text{......(428)}.$$

If the gas is at the same temperature throughout, the intermolecular pressure ϖ will depend solely on ρ, so that we can regard P as a function of ρ only, and equation (427) can be written in the form

$$\Xi = \frac{\partial}{\partial x} \int \frac{dP}{\rho} \quad \text{..............................(429)}.$$

There are of course two similar equations for H, Z, and the system of three equations taken together simply expresses that the forces must have a potential. This condition being satisfied, that expressed by equation (428) is satisfied also, but the two conditions are not identical. It is only in the case of a gas at a uniform temperature throughout that the one can be deduced from the other.

219. If the temperature is supposed uniform, and if the forces Ξ, H, Z are derived from a potential V, it is clear from equations (429) that this potential must be

$$V = -\int \frac{dP}{\rho} \quad \text{.............................(430)},$$

or, substituting for P from equation (426),

$$V = -\int \frac{d\varpi}{\rho} - \frac{1}{2hm}\int \frac{d\rho}{\rho}$$

$$= -\int \frac{d\varpi}{\rho} - \frac{1}{2hm} \log \rho + \text{a constant},$$

leading to the equation for ρ,

$$\rho = \rho_0 e^{-2hm\left(V + \int \frac{d\varpi}{\rho}\right)} \quad \text{.......................(431)}.$$

Clearly equation (223) obtained in Chapter V was a special case of the equation just found, for χ in that equation has the same meaning as the present mV, and the assumption of infinitesimal hard spherical molecules requires us to take $\varpi = 0$.

220. When no external forces act, the equations of equilibrium (427) shew that P must be constant throughout the gas. Indeed, it is obvious that when a gas is in equilibrium under no forces the total pressure P must be the same throughout. And at the boundary the total pressure must become identical with the pressure p at the boundary calculated in the last chapter. This, then, is the constant value of P; it is what is commonly called the pressure of the gas.

Equation (426) shews that at every point in the gas

$$p = \varpi + \frac{\nu}{2h} \quad \text{.............................(432)}.$$

If we calculate p at a distance just greater than $\frac{1}{2}\sigma$ from the boundary, we may put $\varpi = 0$, for the number of molecules whose centres lie between this layer and the boundary is infinitesimal. In doing this we must, as explained in § 216, take ν_b to be the value of ν, and equation (432) becomes

$$p = \frac{\nu_b}{2h} \quad \text{..................................(433)}.$$

As in the treatment of Van der Waals' cohesion (§ 203), we can denote the density inside the gas by ρ, and that at the boundary by ρ_0. Equation (433) may now be expressed in the more complete form

$$p = \frac{\nu_b(\rho_0)}{2h} \quad \text{.............................(434)},$$

agreeing with the value of the pressure calculated in the last chapter (equation (391)).

From equations (432) and (433) we obtain

$$\varpi = \frac{\nu_b(\rho_0) - \nu}{2h} \quad \text{.........................(435)},$$

giving a general expression for the intermolecular pressure ϖ.

221. Let us write $$\varpi = \varpi_c + \varpi_i,$$

where ϖ_c is the part of ϖ produced by collisions—*i.e.* encounters in which the molecules are so close that they cannot approach any nearer,—and ϖ_i is the part of ϖ produced by the outstanding intermolecular forces between pairs of molecules not in contact. Obviously ϖ_i will be negative when these outstanding intermolecular forces are attractive, as with forces of cohesion.

We have already calculated a value for ϖ; we can now calculate ϖ_i and ϖ_c separately.

Imagine a rigid plane surface of area unity set up and held at rest at some point inside the gas. This operation will not affect the distribution of density inside the gas, for the field of intermolecular force will exist on both sides of the area. The pressure produced by molecular impacts on either side of the area will of course be the same, say p_i. Treating this surface as we have already (§ 220) treated the surface which forms the true boundary of the gas, we obtain equations giving the constant p in the forms

$$p = \varpi_i + \frac{\nu_b(\rho)}{2h} \quad \text{.........................(436)},$$

by considering an area at a distance just greater than $\frac{1}{2}\sigma$ from the fixed area; and

$$p = \varpi_i + p_i,$$

by considering an area at a distance just less than $\frac{1}{2}\sigma$ from the fixed area.

Incidentally we may notice that by comparing the two we obtain for the internal pressure p_i the value

$$p_i = \frac{\nu_b(\rho)}{2h} = R\nu_b(\rho)\,T \quad \text{.......................(437)}.$$

Returning to the value found for p in equation (436), and replacing p by its known value $p = \frac{\nu_b(\rho_0)}{2h}$, we obtain

$$\varpi_i = \frac{\nu_b(\rho_0) - \nu_b(\rho)}{2h} \quad \text{.......................(438)}.$$

We also have found that

$$\varpi = \varpi_i + \varpi_c = \frac{\nu_b(\rho_0) - \nu}{2h},$$

whence, by subtraction,

$$\varpi_c = \frac{\nu_b(\rho) - \nu}{2h}.$$

These equations give ϖ_i and ϖ_c separately.

222. To obtain some insight into the meaning of these pressures, let us examine what they become when the deviations from Boyle's Law are small.

If we put $p = \frac{\nu_b(\rho_0)}{2h}$ in Van der Waals' equation, we get

$$\left(\frac{\nu_b(\rho_0)}{2h} + \frac{a}{v^2}\right)(v - b) = RNT,$$

or again, for the pressure on an area inside the gas, noticing that the cohesion term must be supposed to vanish in this case,

$$\frac{\nu_b(\rho)}{2h}(v - b) = RNT \quad \text{.......................(439).}$$

From these equations we obtain at once

$$\frac{\nu_b(\rho_0)}{2h} + \frac{a}{v^2} = \frac{\nu_b(\rho)}{2h},$$

so that

$$\varpi_i = \frac{\nu_b(\rho_0) - \nu_b(\rho)}{2h} = -\frac{a}{v^2} \quad \text{....................(440),}$$

and ϖ_i becomes identical except for sign with the cohesion tension assumed in arriving at Van der Waals' equation, as we should expect.

We have also

$$\varpi_c = \frac{\nu_b(\rho) - \nu}{2h} = \frac{RNT}{v - b} - \frac{RNT}{v} = \frac{b}{v}\left(\frac{RNT}{v - b}\right) = \frac{b}{v}\left(p + \frac{a}{v^2}\right) \quad \text{...(441),}$$

so that ϖ_c turns out to be proportional to the b of Van der Waals' equation, being a fraction $\frac{b}{v}$ of the total internal pressure $\left(p + \frac{a}{v^2}\right)$.

Thus the two pressures ϖ_i and ϖ_c are the two physical agencies which necessitate the constants a and b in Van der Waals' Law. In fact equation (432), namely

$$p = \varpi + \frac{\nu}{2h},$$

can be written in the form

$$p - \varpi_i = \varpi_c + \frac{RNT}{v},$$

or, again

$$(p - \varpi_i)\left(v - \frac{\varpi_c v}{p - \varpi_i}\right) = RNT \quad \text{.................(442),}$$

which is of Van der Waals' form, and immediately gives the values of a, b by comparison.

It could by no means have been predicted without consideration that the pressure ϖ_c would have a finite value. This pressure originates in the impacts between pairs of molecules in collision. The number of pairs in collision at any instant is infinitesimal, while the pressure between each pair in infinite. It is the product of this infinitesimal number and infinite pressure that produces the finite pressure ϖ_c.

THERMODYNAMICS.

Equation of Energy.

223. The total energy of the gas may be divided into internal and external. The former consists of the kinetic energy of molecular motion and the potential energy of intermolecular forces. The external energy is measured by the work which could be performed by the gas in expanding to a state of infinite rarity starting from its present configuration, this work of course being performed by the pressure of the gas. The external energy will be denoted by $\mathfrak{W}$, the internal energy by $\mathfrak{E}$. We may write

$$\mathfrak{E} = N\bar{E} + \Phi \qquad (443),$$

where $\bar{E}$ is the mean energy of the N molecules, and Φ is the potential energy of the intermolecular forces.

224. Let us suppose that a quantity dQ of energy, either in the form of heat or otherwise, is absorbed by the gas from some external source, so that as a consequence the pressure, volume and temperature of the gas change.

Let us suppose that after this absorption of heat the gas again assumes a steady state, and that in the new state the values of $\mathfrak{W}$ and $\mathfrak{E}$ have become changed to $\mathfrak{W} + d\mathfrak{W}$ and $\mathfrak{E} + d\mathfrak{E}$. The energy equation—obtained by equating the total energies before and after the change—is

$$dQ = d\mathfrak{E} + d\mathfrak{W} = Nd\bar{E} + d\mathfrak{W} + d\Phi \qquad (444).$$

If we suppose the volume of the gas increased, by a small element dS of the surface moving a small distance dn parallel to itself normally outwards, then obviously the work done by the gas is that of moving a force pdS through a distance dn, and is therefore $pdSdn$. The resulting increase in volume (dv) is of course the sum of the contributions $dSdn$ from all the elements of surface which are moved, so that the value of $d\mathfrak{W}$ is

$$d\mathfrak{W} = pdv \qquad (445).$$

Thus equation (444) becomes

$$dQ = Nd\bar{E} + pdv + d\Phi \qquad (446).$$

225. We know on experimental grounds (§ 157) that $d\Phi$ is very small. If we take $d\Phi = 0$, equation (446) assumes the form

$$dQ = Nd\bar{E} + pdv \qquad (447),$$

or, replacing p by its value $R\nu_b T$,

$$dQ = Nd\bar{E} + R\nu_b Tdv \qquad (448).$$

226. If we do not neglect the term $d\Phi$, the calculation becomes considerably more difficult, for the intermolecular pressure ϖ will do work when the gas expands.

It is at once obvious that the work done by the part ϖ_c of this pressure is *nil*, for the number of collisions in being at any instant is infinitesimal, and at most only a finite amount of work can be done by each of these.

To evaluate the work done by the pressure ϖ_i we may suppose that in the process of expansion any element $dxdydz$ of the gas expands to a volume $(1+\epsilon)\,dxdydz$. The resulting contribution to $d\Phi$ from this element is $-\varpi_i\epsilon dxdydz$, the negative sign indicating that work is done by the pressure on the gas, and not as before by the gas against the pressure. The total value of $d\Phi$ is the sum of all these contributions, and is accordingly

$$\begin{aligned} d\Phi &= -\iiint \varpi_i \epsilon\, dxdydz \\ &= -\varpi_i \iiint \epsilon\, dxdydz \\ &= -\varpi_i dv \quad \text{.....................(449)} \\ &= -\frac{\nu_b(\rho_0) - \nu_b(\rho)}{2h}\, dv \quad \text{.....................(450),} \end{aligned}$$

by equation (438).

227. The energy equation (446) now assumes the form

$$\begin{aligned} dQ &= Nd\bar{E} + pdv + d\Phi \\ &= Nd\bar{E} + \left[\frac{\nu_b(\rho_0)}{2h} - \frac{\nu_b(\rho_0) - \nu_b(\rho)}{2h}\right] dv \\ &= Nd\bar{E} + \frac{\nu_b(\rho)}{2h}\, dv \\ &= Nd\bar{E} + R\nu_b(\rho)\, Tdv \quad \text{.....................(451).} \end{aligned}$$

This general equation of energy is now seen to be exactly the same in form as equation (448) which was obtained by the neglect of the forces of cohesion, but the value of $\nu_b(\rho)$ must now be calculated on the supposition of a uniform density ρ equal to the mean density of the gas. The value of $\nu_b(\rho)$ is obviously a function only of the constants and of the volume or density of the gas (cf. for instance equations (388) or (439) in which $\nu_b(\rho)$ is calculated in special cases).

The First Law of Thermodynamics.

228. The law which is commonly called the first law of thermodynamics is that contained in equation (444) and simply expresses that heat is energy, which is capable of transformation into other forms of energy, such, for instance, as the kinetic energy of motion of material masses. This law is, however, included in the hypothesis upon which the kinetic theory is based, so that, for the kinetic theory, the first law of thermodynamics is reduced to a truism. For the special problem of a gas such as we have had under consideration, the first law of thermodynamics will be expressed by equation (451).

The Second Law of Thermodynamics.

229. If equation (451) is divided throughout by T, it becomes

$$\frac{dQ}{T} = N\frac{d\bar{E}}{T} + R\nu_b(\rho)\,dv \quad\text{.....................(452).}$$

In this equation, the gas is supposed to remain of the same constitution throughout its change of state. Thus $\bar{E}$ is a function of T only, and $\nu_b(\rho)$ of v only, so that the two terms $\frac{d\bar{E}}{T}$ and $\nu_b(\rho)\,dv$ are each complete differentials. It follows that the right-hand member of (452) is a complete differential, say $d\phi$, so that

$$\frac{dQ}{T} = d\phi \quad\text{...............................(453).}$$

The quantity ϕ introduced in this way is, however, nothing but the entropy, already introduced in § 94. The circumstance that the gas obeys an equation of the form of (453) merely shews that the gas is subject to the second law of thermodynamics. The value of the entropy ϕ is now seen to be given by

$$\phi = N\int\frac{d\bar{E}}{T} + R\int\nu_b(\rho)\,dv \quad\text{.....................(454),}$$

and it appears that the entropy is the sum of two terms, one depending only on the motion and internal energies of the molecules, and the other only on their positions.

230. In § 94, considering the more general case of a gas not necessarily in its normal state, we found for ϕ the general value

$$\phi = R\log W + \text{a constant} \quad\text{.....................(455).}$$

In this equation W is the whole volume of that part of the generalised space of Chapter V which represents systems having a given arrangement for the velocities and positions of the molecules.

The whole volume of this generalised space, say Ω, may be regarded as the product of two spaces Ω_V and Ω_P, Ω_V being a space in which all velocity coordinates are represented, and Ω_P a space in which all positional coordinates are represented.

Let $\frac{W_V}{\Omega_V}$ be the fraction of the first space in which the velocities have any specified arrangement, and let $\frac{W_P}{\Omega_P}$ be the fraction of the second space in which the positional coordinates have their required arrangement. Then the fraction of the whole space in which both velocities and positional coordinates have the assigned arrangement will be given by

$$\frac{W}{\Omega} = \left(\frac{W_V}{\Omega_V}\right)\left(\frac{W_P}{\Omega_P}\right),$$

so that $W = W_V W_P$, and equation (455) may be replaced by

$$\phi = R \log W_V + R \log W_P + \text{a constant} \quad \text{............(456)},$$

shewing that the entropy is the sum of two parts, one depending on the velocities only, and one on the positional coordinates only, say $\phi = \phi_V + \phi_P$.

The maximum value of ϕ leads to the normal state, and so is given by equation (454). It accordingly appears that the two terms on the right-hand side of equation (454) represent the maximum values of ϕ_V and ϕ_P respectively.

231. Let us examine this question somewhat more in detail, confining ourselves for simplicity to the particular case in which the molecules are the elastic spheres of Chapter III.

We have seen that the value of ϕ_V is

$$\phi_V = R \log W_V + \text{a constant} \quad \text{.................(457)},$$

where (cf. § 51)

$$W = \theta_a \times (\text{a constant}),$$

and

$$\log \theta_a = \text{a constant} - \sum_{s=1}^{s=n} a_s \log a_s.$$

Thus the value of ϕ_V given by equation (457) reduces to

$$\phi_V = \text{a constant} - R \sum_{s=1}^{s=n} a_s \log a_s \quad \text{..................(458)}.$$

We may now replace a_s by $f(u, v, w)\, du\, dv\, dw$, and replace the summation over the n possible ranges into which all possible velocities are supposed to be divided by an integration over all values of u, v, w. Equation (458) becomes

$$\phi_V = \text{a constant} - R \iiint f \log f \, du\, dv\, dw,$$

or, if H is defined, as in § 22, by the equation

$$H = \iiint f \log f \, du\, dv\, dw,$$

we have

$$\phi_V = \text{a constant} - RH \quad \text{........................(459)}.$$

Thus finding the minimum value of H in § 23 was in effect the same thing as finding the maximum value of ϕ_V and led to the steady or normal state accordingly.

In the same way, the equation for ϕ_P, namely

$$\phi_P = R \log W_P + \text{a constant},$$

can be shewn to lead, with the help of the analysis of § 43, to

$$\phi_P = \text{a constant} - R \sum_{s=1}^{s=n} a_s \log a_s$$

$$= \text{a constant} - R \iiint \nu \log \nu \, dx dy dz \quad \text{............(460)},$$

and again the steady state found in § 49 is simply that for which ϕ_P is a maximum.

Thus the method employed in Chapter III to find the steady state was simply that of finding the state which made the entropy a maximum (cf. § 98).

The Physical Meaning of Entropy.

232. The importance of the entropy function ϕ, looked at from a purely physical aspect, can be seen as follows. Let two systems, distinguished by suffixes 1 and 2, have initially entropies ϕ_1 and ϕ_2, and let a quantity dQ of heat pass from the first to the second. The loss to ϕ_1 is $\frac{dQ}{T_1}$, while the gain to ϕ_2 is $\frac{dQ}{T_2}$. Thus the total change of entropy is

$$d(\phi_1 + \phi_2) = dQ\left(\frac{1}{T_2} - \frac{1}{T_1}\right) \quad \text{....................(461)}.$$

In a process in which the entropy increases, $dQ\left(\frac{1}{T_2} - \frac{1}{T_1}\right)$ must be positive, and therefore heat must pass from the hotter body to the colder; there is a process of equalisation of temperature. In a process in which the entropy decreases, heat must pass from the colder body to the hotter, so that the hot body gets hotter, and the cold body gets still colder. In natural processes, the entropy increases, and the physical process is one of equalisation of temperature.

Apparent Irreversibility.

233. We appear at this stage to have arrived again at an irreversible phenomenon similar to that already encountered in connection with the function H in Chapter IV. In each case the equations of motion from which the phenomenon has been deduced were strictly reversible, and yet these equations seem to lead to an irreversible phenomenon. To put the matter more concretely: the machine of the universe, assuming its motion to be governed by the canonical equations of motion, is just as capable of running in one direction as in the reverse direction. If it can pass from a state A to a state B, the equations of motion shew that it can also pass from a state B to a state A. If the passage from A to B involves an increase of entropy, the passage from B to A must involve a decrease of entropy: in this latter motion heat will pass from the colder bodies to the hotter.

We have seen that the entropy ϕ is closely allied with the earlier function H, and the explanation of the apparent irreversibility of ϕ is the same as that already given for H. We saw in Chapter IV that a certain minimum value H_0 was possible for H; if the initial value of H were different from H_0, it was infinitely probable that H would decrease, but on the other hand, it was infinitely probable—assuming the basis of probability supplied by the generalised space—that the initial value of H would be equal to H_0, in which case, as H could not decrease further, the "expectation" was of a slight increase. Thus the large probability of a small increase in H just balanced the small probability of a finite decrease in H, and on the whole the "expectation" of change in H was *nil*.

A precisely similar explanation holds with reference to ϕ. An increase in ϕ presupposes an initial difference of temperature between the two component systems; and these initial conditions, looked at from the point of view of abstract dynamics, and judged with reference to the basis of probability supplied by a generalised space, are infinitely improbable. With reference to the same basis of probability, it is infinitely probable that the initial conditions will be those of equilibrium of temperature, in which case the only change possible in ϕ is a decrease. Or, physically, the only possible alteration in the state of the system is the production of inequalities of temperature. The production of such an inequality, although improbable when the motion is confined to a short time, is not impossible, and indeed becomes infinitely probable when the motion is continued for a sufficient time. Thus the increase of entropy, even granted the infinitely improbable (from the dynamical point of view) initial conditions which make such an increase possible, is only a probability and not a certainty; and when the entropy starts initially at its maximum, it is infinitely probable that, granted sufficient time, the entropy will decrease.

234. When applied to concrete instances, these results seem at first sight somewhat startling. To borrow an illustration from Lord Kelvin, if we have a bar of iron initially at uniform temperature, and subject neither to external disturbance nor to loss of energy, it is infinitely probable that, given sufficient time, the temperature of one half will at some time differ by a finite amount from that of the other half. Or again, if we place a vessel full of water over a fire, it is only probable, and not certain, that the water will boil instead of freezing. And moreover, if we attempt to boil the water a sufficient number of times, it is infinitely probable that the water will, on some occasions, freeze instead of boil. The freezing of the water, in this case, does not in any way imply a contravention of the laws of nature: the occurrence is merely what is commonly described as a "coincidence," exactly similar in kind to that which has taken place when the dealer in a game of whist finds that he has all the trumps in his hand.

The analogy of the distribution of a pack of cards will help us to see further into the problem presented by the entropy of a gas. In dealing cards, it is just as likely that the dealer will have the thirteen trumps as that he will have any other thirteen cards that we like to specify. The occurrence of a hand composed of thirteen trumps might, however, be justly regarded as a "coincidence," whereas the occurrence of any specified hand in which the cards were more thoroughly mixed, could not reasonably be so regarded. The explanation is that there are comparatively few ways in which a hand which is all trumps can be dealt, but a great number in which a mixed hand can be dealt.

A similar remark applies to the result of putting cold water over a hot fire. There are comparatively few ways in which the fire can get hotter, and the water colder, but a great many ways in which the fire can impart heat to the water—a proposition which becomes obvious on looking at it from the dynamical point of view of the generalised space. Speaking loosely, it is just as likely that the water will freeze as that it will boil in any specified way. There are, however, so many ways in which the water can boil, all these ways being indistinguishable to us, that we can say that it is practically certain that the water will boil.

The increase of entropy, then, simply means the passage from a more easily distinguishable state to a less easily distinguishable state, or, in terms of the generalised space, from a less probable to a more probable configuration.

235. A reference to equation (456) shews that the entropy consists of two parts, the former depending on the energy of the molecules of the gas, and the latter on their positions. So far we have considered only variations in the first term, resulting from inequalities in the temperature of the gas. Similar remarks could, however, be made about the variations of the second term, these denoting inequalities in the density of the gas. A single illustration, suggested by Willard Gibbs*, will, perhaps, make clear what is meant.

If we put red and blue ink together in a vessel, and stir them up, common experience tells us that, if we assume the inks initially to differ in nothing more than colour, the result of stirring is a uniform violet ink. Here we have the passage from a more easily distinguishable to a less easily distinguishable arrangement of coloured inks. If, however, we start by stirring

* *Elementary Principles of Statistical Mechanics*, p. 144.

a uniform violet ink composed of a mixture of red and blue inks, then it is possible, although not probable, that the effect of the stirring will be to separate the inks of different colour, so that one half of the vessel is occupied solely by red, and the other solely by blue ink. And from the dynamical standpoint it is no less probable that this should occur, than that we should be able to start stirring inks which were separated initially as regards colour.

236. With reference to this subject, some well-known remarks of Maxwell* are of interest. He says: "One of the best established facts in thermodynamics is that it is impossible in a system enclosed in an envelope which permits neither change of volume nor passage of heat, and in which both the temperature and the pressure are everywhere the same, to produce any inequality of temperature or of pressure without the expenditure of work. This is the second law of thermodynamics, and it is undoubtedly true so long as we can deal with bodies only in mass and have no power of perceiving or handling the separate molecules of which they are made up. But if we conceive a being whose faculties are so sharpened that he can follow every molecule in its course, such a being, whose attributes are still as essentially finite as our own, would be able to do what is at present impossible to us. For we have seen that the molecules in a vessel full of air at uniform temperature are moving with velocities by no means uniform though the mean velocity of any great number of them, arbitrarily selected, is almost exactly uniform. Now let us suppose that such a vessel is divided into two portions A and B, by a division in which there is a small hole, and that a being, who can see the individual molecules, opens and closes this hole, so as to allow only the swifter molecules to pass from A to B, and only the slower ones to pass from B to A. He will thus, without expenditure of work, raise the temperature of B and lower that of A, in contradiction to the second law of thermodynamics."

Thus Maxwell's sorting demon could effect in a very short time what would probably take a very long time to come about if left to the play of chance. There would, however, be nothing contrary to natural laws in the one case any more than in the other.

Calorimetry.

Specific Heats of a Perfect Gas.

237. We now turn to an investigation of the specific heats of a gas, and shall begin by considering the simplest case, namely that of a perfect gas in which the relation between pressure, volume, and temperature is

$$p = \frac{RNT}{v} \quad \text{.............................(462).}$$

* *Theory of Heat*, p. 328.

The equation of energy in this case becomes

$$dQ = Nd\bar{E} + pdv$$

$$= Nd\bar{E} + RNT\frac{dv}{v} \quad \text{......................(463)},$$

and this is the general equation of calorimetry for a perfect gas.

238. *Specific heat at constant volume.* Let us first suppose that a quantity dQ of energy in the form of heat is absorbed by the gas, while the volume of the gas is maintained constant. In this case all the heat goes towards raising the temperature of the gas, equation (463) assuming the form

$$dQ = Nd\bar{E} \quad \text{..............................(464)}.$$

Let C_v be the specific heat of the gas at constant volume, *i.e.*, the amount of heat required to raise the temperature of a unit mass of gas by one degree, then the amount of heat required to raise the mass Nm of gas through a temperature difference dT will be C_vNmdT. Thus if J is the mechanical equivalent of heat,

$$dQ = JC_vNmdT,$$

and equation (464) becomes

$$C_v = \frac{1}{Jm}\frac{d\bar{E}}{dT} \quad \text{..............................(465)}.$$

239. *Specific heat at constant pressure.* Let us next suppose that the absorption of heat takes place at constant pressure. In this case both the volume and temperature will change, but from equation (462) they must change in such a manner that

$$\frac{T}{v} = \text{constant.}$$

If we differentiate this equation logarithmically, we obtain

$$\frac{dT}{T} = \frac{dv}{v},$$

as the relation between dT and dv when the pressure is maintained constant, and using this relation, equation (463) becomes

$$dQ = Nd\bar{E} + RNdT \quad \text{......................(466)}.$$

The value of dQ is now JC_pNmdT, where C_p is the specific heat at constant pressure. Hence equation (466) leads to the relation

$$C_p = \frac{1}{Jm}\frac{d\bar{E}}{dT} + \frac{R}{Jm} \quad \text{..........................(467)}.$$

240. From equations (465) and (467) we obtain by subtraction

$$C_p - C_v = \frac{R}{Jm} \quad\quad (468).$$

Since m is proportional to the molecular weight of the particular gas we are discussing, this equation expresses Carnot's Law:

The difference of the two specific heats of a gas is inversely proportional to the molecular weight of the gas.

This law can be expressed in a different form. The specific heats referred to unit volume instead of to unit mass are of course $C_p\rho$, $C_v\rho$, and equation (468) may be written

$$C_p\rho - C_v\rho = \frac{R\rho}{Jm} = \frac{p}{JT} \quad\quad (469),$$

the last transformation depending on equation (462). Hence:

At a given temperature and pressure the difference of the two specific heats per unit volume is the same for all gases.

241. It is found by experiment that, at any rate for a large number of gases, C_p and C_v are approximately independent of the temperature over a large range of temperatures and pressures. This, as is shewn by a reference to the formulae (465) and (467), must mean that $\frac{d\bar{E}}{dT}$ is a constant, and therefore that the mean energy of a molecule of the gas stands in a constant ratio to the translational energy. Let us denote this ratio by $(1+\beta)$, so that β is the ratio of internal to translational energy. Then

$$\bar{E} = (1+\beta)\tfrac{1}{2}m\overline{C^2}$$

$$= (1+\beta)\tfrac{3}{2}RT \quad\quad (470),$$

so that

$$\frac{d\bar{E}}{dT} = \tfrac{3}{2}R(1+\beta) \quad\quad (471).$$

Substituting this value for $\frac{d\bar{E}}{dT}$ in equations (465) and (467), we obtain

$$C_v = \tfrac{3}{2}(1+\beta)\frac{R}{Jm} \quad\quad (472),$$

$$C_p = [1 + \tfrac{3}{2}(1+\beta)]\frac{R}{Jm} \quad\quad (473).$$

If we denote the ratio C_p/C_v by γ, we obtain by division

$$\gamma \equiv \frac{C_p}{C_v} = \frac{1+\tfrac{3}{2}(1+\beta)}{\tfrac{3}{2}(1+\beta)}$$

$$= 1 + \frac{2}{3(1+\beta)} \quad\quad (474).$$

Adiabatic Motion.

242. Suppose the pressure, volume and temperature to change in such a way that no heat enters or leaves the gas. Then since $dQ = 0$, we have from equation (463)

$$N d\bar{E} + RNT \frac{dv}{v} = 0,$$

or, on substituting for $\bar{E}$ from equation (470), and dividing by RNT,

$$\tfrac{3}{2}(1+\beta)\frac{dT}{T} + \frac{dv}{v} = 0.$$

Hence upon integration

$$Tv^{\frac{2}{3(1+\beta)}} = \text{constant} \quad \text{......................(475)},$$

or again, since T is proportional to pv,

$$pv^{1+\frac{2}{3(1+\beta)}} = \text{constant},$$

or

$$pv^{\gamma} = \text{constant} \text{......................(476)}.$$

This is the general relation between pressure and volume in a motion of the gas in which no heat enters or leaves the gas—a type of motion which is known as "adiabatic."

Since β cannot be negative, we see from equation (475) that in adiabatic motion an increase in v is accompanied by a decrease in T, and vice versa—a gas necessarily cools on expanding, and is heated on being compressed.

243. Since the conduction of heat in gases is a very slow process, it results that in many physical phenomena the rates of change are so rapid that the temperature of the gas has not time to equalise itself. Frequently we may suppose that the process is so rapid that conduction of heat plays no part at all, so that the change of each element of the gas may be treated as adiabatic. A well-known instance of this is provided by the motions of the currents of air in the lower strata of the atmosphere.

A second instance, of importance for our present problem, is provided by the propagation of sound in a gas. The different elements of volume in a gas undergo expansion and contraction as the waves of sound pass over them—these expansions and contractions are readily seen to be too rapid for the conduction of heat to be of any importance, and so the changes in each element of gas obey the adiabatic law.

The velocity of sound, say a, is given by the well-known formula*

$$a = \sqrt{\left(\frac{dp}{d\rho}\right)},$$

* See for instance Lord Rayleigh's *Sound*, II. § 246.

and if we suppose p and ρ to be connected by the adiabatic law $p = c\rho^\gamma$, this gives as the value of $\frac{dp}{d\rho}$,

$$\frac{dp}{d\rho} = \gamma c \rho^{\gamma-1} = \gamma \frac{p}{\rho},$$

so that

$$a = \sqrt{\gamma \frac{p}{\rho}} \qquad (477).$$

Since observations on the wave-length of sound are effected with comparative ease, this equation provides a ready means of determining the value of γ. It ought, however, to be particularly noticed that the equation is only true for a gas which may be treated as perfect.

General Calculation of Specific Heats.

244. If we attempt to repeat the investigation of the specific heats of a gas without making the simplifying assumption that the gas is a perfect gas, the relation between pressure, volume and temperature must be taken to be the general relation found in equation (391),

$$p = R\nu_b(\rho_0)\, T \qquad (478),$$

and the equation of energy will be the general equation (451),

$$dQ = N d\bar{E} + R\nu_b(\rho)\, T dv \qquad (479).$$

For a change at constant volume, we obtain, as in § 238,

$$N d\bar{E} = dQ = JC_v N m\, dT,$$

so that, as before,

$$C_v = \frac{1}{Jm} \frac{d\bar{E}}{dT} \qquad (480).$$

For a change at constant pressure, the value of Q in equation (479) must be put equal to $JC_p N m\, dT$, giving

$$C_p = \frac{1}{Jm} \left\{ \frac{d\bar{E}}{dT} + \frac{R}{N} \nu_b(\rho)\, T \left(\frac{dv}{dT} \right)_{p \text{ cons.}} \right\} \qquad (481).$$

Using equation (478),

$$\left(\frac{dv}{dT} \right)_{p \text{ cons.}} = - \frac{\frac{\partial p}{\partial T}}{\frac{\partial p}{\partial v}} = - \frac{\frac{d}{dT}[\nu_b(\rho_0)\, T]}{T \frac{d}{dv}[\nu_b(\rho_0)]},$$

and equation (481) becomes

$$C_p = \frac{1}{Jm} \left\{ \frac{d\bar{E}}{dT} - \frac{R}{N} \frac{\nu_b(\rho) \frac{d}{dT}[\nu_b(\rho_0)\, T]}{\frac{d}{dv}[\nu_b(\rho_0)]} \right\} \qquad (482).$$

This equation is too complicated for any further discussion to be profitable in the general case. We may notice, however, that if we neglect the forces of cohesion, $\nu_b(\rho_0)$ and $\nu_b(\rho)$ become identical, and independent of the temperature, so that equation (482) reduces to

$$C_p = \frac{1}{Jm}\left\{\frac{d\bar{E}}{dT} + \frac{R}{N\dfrac{d}{dv}\left(\dfrac{1}{\nu_b}\right)}\right\} \quad \text{......................(483).}$$

If ν_b is further assumed to have the value assigned to it by Van der Waals,

$$\frac{d}{dv}\left(\frac{1}{\nu_b}\right) = \frac{d}{dv}\left(\frac{v-b}{N}\right) = \frac{1}{N},$$

and equation (483) becomes

$$C_p = \frac{1}{Jm}\frac{d\bar{E}}{dT} + \frac{R}{Jm} \quad \text{........................(484),}$$

which agrees exactly with the equation previously obtained for the simpler case of an ideal gas.

Dependence of Specific Heats on Molecular Structure.

245. The quantities $\frac{d\bar{E}}{dT}$ and β (the two being connected by relation (471)) can only be evaluated when the internal structure of the molecule is known. We have not sufficient knowledge of this internal structure to evaluate these quantities directly, but their values can to some extent be determined from a comparison of the specific heat formulae and the experimentally determined values of the specific heats, and the values obtained in this way provide a basis for the discussion of the structure of molecules.

246. As an example of this procedure, we may examine the case of air which for the moment, as frequently in the kinetic theory, may be thought of as consisting of similar molecules.

For γ, the ratio of the specific heats, under a pressure of 1 atmosphere, the following values have been obtained by Koch and others*:

Values of γ for air at 760 mm. pressure:

$\theta = -79{\cdot}3°$ C.,	$\gamma = 1{\cdot}405$,
$\theta = 0°$ C.,	$\gamma = 1{\cdot}404$.
$\theta = 100°$ C.,	$\gamma = 1{\cdot}403$,
$\theta = 500°$ C.,	$\gamma = 1{\cdot}399$,
$\theta = 900°$ C.,	$\gamma = 1{\cdot}39$.

These numbers shew that at this pressure γ is almost independent of the temperature, and approximately equal to $1\frac{2}{5}$. Since the ratio of the specific heats, and also, by equation (468), their difference, is independent of the temperature, it follows that the specific heats themselves must be independent, so that β must have a definite value. Equation (474), namely

$$\gamma = 1 + \frac{2}{3(1+\beta)} \quad \text{..........................(485),}$$

shews that this value is $\beta = \frac{2}{3}$.

* Kaye and Laby's *Physical Constants* (1911), or *Recueil de Constantes Physiques* (1913).

At higher pressures the value of γ by no means remains constant, as is shewn by the following observations by Koch*.

VALUES OF γ FOR AIR AT HIGH PRESSURES.

	$\theta = -79\cdot3°$ C.	$\theta = 0°$ C.
$p =$ 1 atmos.	$\gamma = 1\cdot405$	$\gamma = 1\cdot404$
25 ,,	1·57	1·47
100 ,,	2·21	1·66
200 ,,	2·33	1·85

247. Generally speaking, it is found that for monatomic gases, and for the more permanent diatomic gases, there exists a range of the kind we have been considering, within which the specific heats remain approximately constant. Frequently in the case of more complex gases no such range appears to exist.

248. The table on p. 190 gives values of C_p and of γ for a number of the more common gases of both types. The last column contains values of β calculated from formula (485), but it will be remembered that these values have no obvious physical meaning except within the range in which the specific heats remain approximately steady.

From the figures in this table, confirmed by a large mass of other experimental evidence, it appears that the value of β is approximately equal to zero ($\gamma = 1\frac{2}{3}$) for the monatomic gases, mercury, krypton, argon and helium. It is nearly equal to $\frac{2}{3}$ ($\gamma = 1\frac{2}{5}$) throughout the steady range for a number of diatomic gases, hydrogen, nitrogen, oxygen, carbon-monoxide and others. When we pass to temperatures below the steady range, β is found to decrease with great rapidity.

249. The energy of a molecule will consist always of three squared terms representing the kinetic energy of motion, to which may be added any number of other terms representing energy of rotation, of internal vibration, etc.

Let us consider a molecule having, in its energy, n squared terms in addition to the three representing its kinetic energy of motion, so that

$$E = \tfrac{1}{2}m(u^2 + v^2 + w^2) + \tfrac{1}{2}\alpha_1\phi_1^2 + \tfrac{1}{2}\alpha_2\phi_2^2 + \ldots + \tfrac{1}{2}\alpha_n\phi_n^2 \quad\ldots\ldots(486).$$

The mean value of each squared term is, as in § 140, equal to $\frac{1}{2}RT$, so that

$$\bar{E} = \tfrac{1}{2}RT(3 + n) \quad\ldots\ldots(487),$$

* *Ann. d. Phys.* XXVI. (1908), p. 551.

VALUES OF C_p AND γ.

Gas	Observer	Value of C_p	Value of γ	Value of β
Mercury vapour	Kundt and Warburg	—	1·666	·000
Krypton	Rayleigh and Ramsey	—	1·666	·000
Argon	Niemeyer	—	1·667	·000
Helium at 18° C.	Scheel and Heuse *	1·260	1·654	·02 §
„ − 180° C.	„ „	1·245	1·667	·00 §
Hydrogen at 16° C. ...	„ „	3·403	1·407	·637
„ − 76° C. ...	„ „	3·157	1·453	·470
„ − 181° C. ...	„ „	2·644	1·597	·124
Nitrogen at 20° C. ...	„ „	0·2492	1·400	·666
„ − 181° C. ...	„ „	0·2556	1·468	·424
Oxygen at 20° C.	„ „	0·218	1·399	·670
„ − 76° C.	„ „	0·214	1·416	·603
„ − 181° C.	„ „	0·228	1·447	·493
Air at 100° C.	Swann †	0·24301	—	—
„ 20° C.	„	0·24173	—	—
„ 20° C.	Scheel and Heuse *	0·2406	1·401	·663
„ − 76° C.	„ „	0·2430	1·401	·663
„ − 181° C.	„ „	0·2496	1·450	·481
Carbon-monoxide at 18° C.	„ „	0·2502	1·398	·675
„ „ − 188° C.	„ „	0·2587	1·472	·412
Carbon-dioxide at 0° C.	Holborn and Henning ‡	0·2010	—	1·30
„ „ 20° C.	Swann †	0·20202	—	1·31
„ „ 50° C.	Holborn and Henning ‡	0·2095	—	1·43
„ „ 100° C.	Swann †	0·22141	—	1·61

and comparing this with the mean value assumed for $\bar{E}$ in equation (470), namely

$$\bar{E} = \tfrac{3}{2} RT(1 + \beta) \qquad (488),$$

we at once see that $n = 3\beta$.

This immediately gives a simple explanation of the tendency for the various steady values of β to cluster round the values $\beta = 0, \tfrac{2}{3}$, etc., for it appears that these are exactly the values which correspond to integral values of n. In other words, these values are just such as would be expected on the hypothesis that the molecular energy is of the form (486).

* *Ann. d. Physik*, XXXVII., p. 79 and XL., p. 473.

† *Phil. Trans.* 210 A. p. 199.

‡ *Sitzungsber d. k. Akad. d. Wissen.* 1905, p. 175.

§ Calculated from $\frac{R}{m} = 2077 \times 10^4$.

For a gas of which the molecular energy is of this form, formulae (472) and (474) become

$$C_v = \tfrac{1}{2}(3+n)\frac{R}{Jm} \qquad \text{.........................(489)},$$

$$\gamma = 1 + \frac{2}{3+n} \qquad \text{............................(490)}.$$

Monatomic Gases.

250. There are four gases in the table, namely mercury, krypton, argon and helium, for which, very approximately, $\gamma = 1\frac{2}{3}$, $\beta = 0$, and $n = 3$. For these gases, then, there is no molecular energy except that of translation. This seems to indicate that the molecules of the gas must behave at collision like hard spherical bodies. If they did not do so, an appreciable fraction of the molecular energy of translation would be transformed into rotational or vibratory energy at each collision. The four gases have the peculiarity that they are all monatomic; the molecule is identical with the atom.

Although the atoms of these substances behave like hard spherical bodies at collision, there is abundant evidence that they have a highly complicated internal structure. The helium atom for instance is believed, on almost incontrovertible evidence, to be made up of three parts—a positive nucleus, which is identical with the α-ray particle of radioactivity, and two negative electrons. The helium atom made up in this way, must, as a matter of geometry, have six degrees of freedom in addition to its three degrees of freedom of motion in space.

The explanation of why, under the circumstances, the specific heats are accurately given by taking $n = 0$ in formulae (489) and (490) presented for many years a problem of the utmost gravity. It is now generally accepted that no satisfactory explanation can be given in terms of the classical system of dynamics. In recent years a new system of dynamics has arisen, commonly called the quantum-dynamics. It will perhaps suffice to remark here that the quantum-dynamics has provided an explanation not only of the behaviour of the monatomic molecule but of many other problems of specific heats in addition. We shall not further discuss the quantum-dynamics here as it forms the subject of a separate chapter (chap. XVII).

251. After $n = 0$, the next value theoretically possible would be $n = 1$, giving $\gamma = 1\frac{1}{2}$. In the table opposite there is no gas for which $n = 1$, ($\beta = \frac{1}{3}$), and there is no gas known for which β and γ have these values, even approximately. This, however, upon closer examination, might be regarded as additional confirmation of the truth of the kinetic theory, since no molecular system could have its energy of the form (486) with $n = 1$. Either a rotation or a vibration would necessarily introduce at least two squared terms into the energy. Thus a molecule for which $n = 1$ is an impossibility, and the value $\gamma = 1\frac{1}{2}$ cannot be expected to occur for any type of gas.

Diatomic Gases.

252. After $n=0$, then, the next value theoretically possible will be $n=2$, giving $\gamma=1\frac{2}{5}$. From the table it appears that n and γ have very approximately these values for air, hydrogen, nitrogen, oxygen, and carbon-monoxide. The molecules of these gases are all diatomic and the same values of n and γ are found to hold for most, but not all, of the diatomic gases. The two terms

$$\tfrac{1}{2}\alpha_1\phi_1^2+\tfrac{1}{2}\alpha_2\phi_2^2$$

in the energy as given by formula (486) may either represent a vibration of the two atoms relative to one another along the line ioining them, or a possibility of a rotation. We must notice that in general the energy of rotation of a rigid body will be represented by three squared terms in formula (486), but the energy of rotation of a body symmetrical about an axis may be represented by only two, no rotation about the axis of symmetry being set up by collisions.

Whatever is the true physical origin of these two terms, the table on p. 190 shews that their energy falls off rapidly as the temperature falls, particularly in the case of hydrogen. Eucken and others* have found experimentally that as the absolute zero of temperature is approached, the molecules of diatomic gases tend to lose all energy except that of translation, and so behave like the molecules of monatomic gases, with the values $\beta=0$ and $\gamma=1\frac{2}{3}$. We shall return to a discussion of this remarkable fact later (Chap. XVII).

More complex Gases.

253. It is difficult to discover any law or regularity in the values of n and γ for more complex gases. Various attempts have been made to connect the values of n and γ with the number of atoms in the molecule. Naumann†, for instance, suggested that n is identical with the number of atoms in the molecule, while J. J. Thomson‡ suggested that in the special case of a symmetrically arranged molecule, $n+3$ might be found to be proportional to the number of atoms in the molecule.

The experiments of Capstick§ have, however, shewn quite conclusively that no general law can be expected to relate γ with the number of atoms, independently of the nature of the atoms. For instance, he finds the following values for the methane derivatives:

		γ	$n+$
Methane	CH_4	1·313	6·4
Methyl Chloride	CH_3Cl	1·279	7·2
Methylene Chloride	CH_2Cl_2	1·219	9·0
Chloroform	$CHCl_3$	1·154	13·0
Carbon tetrachloride	CCl_4	1·130	15·4

* Eucken, *Sitzungsber. Berlin Akad. d. Wissensch.* VI. (1912) p. 141. Scheel and Heuse, *Ann. d. Physik*, XL. (1913) p. 473. Schreiner, *Zeits. phys. chem.* CXII. (1924) p. 1.

† *Annalen der Chemie*, CXLIII. (1867) p. 284. ‡ *Watts' Dictionary of Chemistry*, I. p. 89.

§ *Phil. Trans.* CLXXXVI. (1895) p. 564; CLXXXV. (1894) p. 1.

and somewhat similar values for n can be deduced from Regnault's determinations of C_p for this series. Thus the introduction of the series of chlorine atoms increases n very perceptibly at every step.

A similar result was obtained by Strecker*, who found that hydrochloric, hydrobromic, hydriodic acids all have approximately the same values as hydrogen, namely

$$\gamma = 1{\cdot}4, \qquad n + 3 = 5,$$

while for chlorine, bromine and iodine, the values are approximately

Chlorine	$\gamma = 1{\cdot}333$	$n + 3 = 6$,
Bromine, Iodine	$\gamma = 1{\cdot}293$	$n + 3 = 6{\cdot}8$.

Similarly for the iodides of bromine and chlorine,

Bromine iodide	$\gamma = 1{\cdot}33$	$n + 3 = 6$,
Chlorine iodide	$\gamma = 1{\cdot}317$	$n + 3 = 6{\cdot}3$.

From these figures it appears that one halogen can be put in the place of hydrogen without producing any difference in the values of γ and n, but that the substitution of the second halogen atom causes a marked increase in n. Capstick† finds a similar phenomenon in the case of the paraffin derivatives. In general the second chlorine atom introduced into the molecule causes a large change, although the first may or may not do so.

Molecular aggregation.

254. The discussion of the physical properties of gases given in this and the preceding chapter has been based upon the supposition that a gas can be regarded as a collection of separate dynamical systems, namely molecules, each of which retains its identity through all time. As a close to this discussion, we may examine what changes are to be expected if the supposition is regarded as an approximation to the truth only, and not as being wholly true. We shall first consider what complications are introduced by the possibilities of molecular aggregation, leaving the discussion of the converse process of dissociation until later.

255. We have already seen that there must be a small attractive force between those molecules in a gas which are sufficiently near to one another, or, more precisely, that the potential energy of the total intermolecular forces in a gas is negative.

This result, it is worth noticing, is intelligible without assuming that there is any definitely attractive force inherent in a single molecule. In § 117 we obtained as the laws of distribution for those molecules which were free

* *Wied. Ann.* XIII. p. 20 and XVII. p. 85.

† *l.c.* ante.

from intermolecular force and for those molecules which were under the influence of intermolecular force, equations of the forms

$$\left.\begin{aligned}\tau_a &= Ae^{-2hE_a}\\ \tau_{aa} &= A^2e^{-2h(E_a+E'_a+W_{aa})}\\ &\text{etc.}\end{aligned}\right\} \quad\text{.....................(491)},$$

it being sufficient for our present purpose to consider a gas in which only one kind of molecule is present, *i.e.* a gas which is chemically pure. In the above equations W_{aa} is the potential of the intermolecular forces between the two molecules. If we denote the potential of the intermolecular forces between three molecules by W_{aaa}, and so on, we obtain as the total intermolecular potential energy of the gas,

$$\Phi = A^2\iint\ldots W_{aa}e^{-2h(E_a+E'_a+W_{aa})}\,d\xi_1 d\xi_2\ldots$$

$$+A^3\iint\ldots W_{aaa}e^{-2h(E_a+E'_a+E''_a+W_{aaa})}\,d\xi_1 d\xi_2\ldots+\text{etc. ......(492)},$$

where the integrations extend over all configurations in which the intermolecular forces are appreciable. Now if, when the configuration of two molecules is selected at random from all possible configurations, W_{aa} is as likely to be positive as negative, then the whole of the first integral can be expressed as a sum of terms of the form

$$W_{aa}e^{-2hW_{aa}} - W_{aa}e^{2hW_{aa}},$$

this term being obtained by combining two configurations in which the values of W_{aa} are equal in magnitude but opposite in sign. This expression is, however, negative for all values of W_{aa}. The second integral can be similarly treated, so that we arrive at the final result that Φ is negative.

Degree of Aggregation.

256. We may now simplify the problem by regarding molecules as point-centres of force, acting on one another with a force depending only on their distance apart. The chance of finding a free molecule of class A inside an element of volume $dxdydz$ is now, by equation (491),

$$Ae^{-hmc^2}\,dudvdwdxdydz \quad\text{.....................(493)},$$

while the chance of finding two molecules of classes A and B in adjacent elements $dxdydz$ and $dx'dy'dz'$ is

$$A^2e^{-hm(c^2+c'^2)-2h\Psi}\,dudvdwdxdydzdu'dv'dw'dx'dy'dz'.$$

If we take the element $dx'dy'dz'$ to be a spherical shell of radii r and $r+dr$ surrounding the centre of the first molecule, this last expression becomes

$$A^2e^{-hm(c^2+c'^2)-2h\Psi}\,dudvdwdu'dv'dw'4\pi r^2drdxdydz,$$

Ψ being a function of r. If, as in § 33 (p. 35), we use the transformations

$$\mathbf{u} = \tfrac{1}{2}(u + u') \text{ etc.}, \quad \alpha = u' - u, \text{ etc.},$$

and write

$$\mathbf{u}^2 + \mathbf{v}^2 + \mathbf{w}^2 = \mathbf{c}^2, \quad \alpha^2 + \beta^2 + \gamma^2 = V^2,$$

we can transform the foregoing expression into

$$A^2 e^{-2hm\mathbf{c}^2} d\mathbf{u}\, d\mathbf{v}\, d\mathbf{w}\, dx\, dy\, dz\, e^{-\frac{1}{2}hmV^2 - 2h\Psi} d\alpha\, d\beta\, d\gamma\, 4\pi r^2 dr \quad \text{......(494)}.$$

The first factor after the A^2 expresses the law of distribution of translational velocities for a double molecule. It is exactly the same as if each double molecule were a permanent structure of mass $2m$. The remaining factors express the distribution of those coordinates which may be regarded as internal to the double molecule.

257. Throughout the motion of a double molecule, so long as it is undisturbed by collisions, $\mathbf{c}^2$ will remain constant, so that from the energy equation it follows that $\frac{1}{2}mV^2 + 2\Psi$ remains constant. The possible orbits which the component molecules can describe about their common centre of gravity fall into two classes, according as they pass to infinity or not. Analytically these two classes are differentiated by the sign of $\frac{1}{2}mV^2 + 2\Psi$. Double molecules for which $\frac{1}{2}mV^2 + 2\Psi$ is positive consist of two molecules which have approached one another from outside each other's sphere of action, and which after passing once within a certain minimum distance of each other, will again recede out of each other's sphere of influence. On the other hand, double molecules for which $\frac{1}{2}mV^2 + 2\Psi$ is negative consist of two molecules describing orbits about one another, these orbits being entirely within the two spheres of action, and this motion continues except in so far as it is interrupted by collisions with other molecules. It is clear that double molecules of the first kind are simply pairs of molecules in collision. In discussing molecular aggregation we must confine our attention to double molecules of the second kind, *i.e.* those for which $\frac{1}{2}mV^2 + 2\Psi$ is negative. It is to be noticed that double molecules of this kind cannot be produced solely by the meeting of two single molecules. It is necessary that while the single molecules are in collision something should happen to change the motion—in fact to change the sign of $\frac{1}{2}mV^2 + 2\Psi$. This might be effected by collision with a third molecule, or possibly if $\frac{1}{2}mV^2 + 2\Psi$ were small at the beginning of an encounter, sufficient energy might be dissipated by radiation for $\frac{1}{2}mV^2 + 2\Psi$ to become negative before the termination of the encounter. We may leave the consideration of this second possibility on one side for the present, with the remark that if this were the primary cause of aggregation, we should no longer be able to use the equations with which we have been working, since they rest upon the assumption of conservation of energy.

258. Integrating expression (493) over all values of u, v and w, we find for ν_1, the molecular density of uncombined molecules,

$$\nu_1 = A\left(\frac{\pi}{hm}\right)^{\frac{3}{2}} \quad\text{.............................(495).}$$

Similarly, if ν_2 is the molecular density of double molecules, we obtain by integration of expression (494),

$$\nu_2 = A^2\left(\frac{\pi}{2hm}\right)^{\frac{3}{2}} \iiiint e^{-h[\frac{1}{4}mV^2+2\Psi]}\, d\alpha\, d\beta\, d\gamma\, 4\pi r^2 dr$$

$$= A^2\left(\frac{\pi}{2hm}\right)^{\frac{3}{2}} \iint e^{-h[\frac{1}{4}mV^2+2\Psi]}\, 16\pi^2 V^2 r^2 dV dr \quad\text{...........(496),}$$

in which the integration extends over all values of V and r for which $\frac{1}{2}mV^2 + 2\Psi$ is negative.

The total number of constituent molecules per unit volume is

$$\nu = \nu_1 + 2\nu_2 + 3\nu_3 + \ldots$$

$$= \nu_1\left(1 + \frac{A}{\sqrt{2}}\iint e^{-h[\frac{1}{4}mV^2+2\Psi]}\, 16\pi^2 V^2 r^2 dV dr + A^2(\ldots) + \ldots\right) \quad\text{...(497),}$$

so that if q denotes the fraction of the whole mass which is free,

$$q = \frac{\nu_1}{\nu} = \frac{1}{1 + \frac{A}{\sqrt{2}}\iint e^{-h[\frac{1}{4}mV^2+2\Psi]}\, 16\pi^2 V^2 r^2 dV dr + \ldots} \quad\text{........(498).}$$

Eliminating A from equations (495), (496), etc., we obtain a series of relations of the form

$$\left.\begin{aligned} \nu_2 &= \nu_1^2 \phi(T) \\ \nu_3 &= \nu_1^3 \psi(T), \text{ etc.} \end{aligned}\right\} \quad\text{..........................(499),}$$

where ϕ, ψ, ... are functions of the temperature only. Equations of this form are the basis of practically every theory of aggregation and dissociation*.

To study the variation of aggregation with temperature a knowledge of the exact form of the functions $\phi(T)$, $\psi(T)$, etc. is necessary, but we can examine the dependence of aggregation on density without this knowledge.

Dependence of aggregation on Density.

259. For a number of substances, it is probable that no greater degree of aggregation need be considered than that implied in the formation of double molecules. For such substances

$$\nu = \nu_1 + 2\nu_2.$$

* Compare, for instance, Boltzmann's Theory, *Wied. Ann.* XXXII. p. 39, or *Vorlesungen über Gastheorie*, II. § 63; Natanson's Theory, *Wied. Ann.* XXXVIII. p. 288, or Winkelmann's *Handbuch d. Physik*, III. p. 725, or the theory of J. J. Thomson, *Phil. Mag.* [5] XVIII. (1884), p. 233. These theories are based on widely different physical assumptions, but all lead to equations of the same general form as (499). The difference of the physical assumptions made shews itself in the different forms for the functions $\phi(T)$, etc.

Neglecting the Van der Waals' corrections, the pressure is given by (cf. § 145)

$$p = RT(\nu_1 + \nu_2) = \tfrac{1}{2}RT(\nu + \nu_1) = \tfrac{1}{2}R\nu T(1+q) \qquad (500),$$

where q is introduced from equation (498). Thus it appears that q, the fraction of the whole mass which is free, can be readily obtained from readings of pressure and temperature.

The following table gives the values of $1-q$ calculated in this way from the observations of Natanson* on the density of peroxide of nitrogen:

AGGREGATION OF NO_2.

Temp.	Value of $1-q=\frac{2\nu_2}{\nu}$			
	p=115 mm.	p=250 mm.	p=580 mm.	p=760 mm.
$\theta=-12{\cdot}6°$	0·919	—	—	—
$\theta=0°$	0·837	0·901	—	—
$\theta=21°$	—	—	0·824	—
$\theta=49{\cdot}7°$	0·253	0·370	0·550	—
$\theta=73{\cdot}7°$	0·084	0·149	0·263	—
$\theta=99{\cdot}8°$	0·031	0·050	0·093	0·117
$\theta=151{\cdot}4°$	in	appre	ciab	le

In this case the single molecule is NO_2, the double molecule is N_2O_4, and more complex structures are supposed not to occur. The value of $1-q$ is $2\nu_2/\nu$, and so measures the proportion by mass which occurs in the form N_2O_4.

Equation (499) now predicts that the ratio of ν_2 to ν_1^2 ought to be the same for all readings at the same temperature. We have from this equation

$$\frac{1}{q} = \frac{\nu}{\nu_1} = \frac{\nu_1 + 2\nu_2}{\nu_1} = 1 + 2\nu_1\phi(T),$$

so that

$$\frac{1}{q}\left(\frac{1}{q} - 1\right) = \left(\frac{\nu}{\nu_1}\right)[2\nu_1\phi(T)] = 2\nu\phi(T),$$

or

$$1 - q = 2q^2\nu\phi(T).$$

Combining this with equation (500) we obtain

$$1 - q^2 = 4pq^2\frac{\phi(T)}{RT},$$

shewing that the ratio $\dfrac{1-q^2}{pq^2}$ ought to be the same for all readings at the same temperature.

* *Recueil de Constantes Physiques*, p. 168.

The following table, calculated from the observations in the table of p. 197, will shew to what extent this prediction of theory is borne out by experiment:

AGGREGATION OF NO_2.

Temp.	Value of $\frac{1-q^2}{pq^2} = \frac{4\phi(T)}{RT}$			
	$p=115$ mm.	$p=250$ mm.	$p=580$ mm.	$p=760$ mm.
$\theta=49{\cdot}7°$	·689	·608	·680	—
$\theta=73{\cdot}7°$	·167	·152	·145	—
$\theta=99{\cdot}8°$	·056	·043	·037	·037

Dependence of aggregation on Temperature.

260. It has been seen that

$$\frac{1}{q^2} = 1 + \frac{4p\phi(T)}{RT},$$

where, from equations (495) and (496),

$$\phi(T) = \frac{\nu_2}{\nu_1^2} = \left(\frac{hm}{2\pi}\right)^{\frac{3}{2}} \iint e^{-h[\frac{1}{2}mV^2+2\Psi]} 16\pi^2 V^2 r^2 dV dr \quad \text{......(501).}$$

The exact relation of the degree of aggregation to the temperature accordingly depends on the evaluation of the function $\phi(T)$, and so involves a difficult problem.

In general terms we can easily see that at high temperatures (h very small) the value of $\phi(T)$ will be insignificant, while after it becomes appreciable, it must be expected to increase rapidly with falling temperature.

Our knowledge of the structure of matter is not sufficient to enable us to evaluate $\phi(T)$, as given by equation (501), with precision. Progress can only be made by the introduction of simple hypotheses as to the interaction of molecules, which may prove to lead to results near to the truth.

Boltzmann* assumes that potential energy exists between two molecules only when the centre of the second lies within a small and clearly defined region which is of course fixed relative to the first, and that when the second molecule has its centre within this "sensitive region," the potential energy has always the same value Ψ. This region does not necessarily consist of a spherical shell, but if ω denotes its total volume, equation (501) may be written in the form

$$\phi(T) = \left(\frac{hm}{2\pi}\right)^{\frac{3}{2}} \omega e^{-2h\Psi} \int e^{-\frac{1}{2}hmV^2} 4\pi V^2 dV,$$

* *Vorlesungen über Gastheorie*, II. chap. VI.

where ω replaces the integral $4\pi\int r^2 dr$, which has represented the extent of the "sensitive region" in our analysis. If we replace $\frac{1}{2}hmV^2$ by x^2, this equation may be expressed in the form

$$\phi(T) = \frac{4\omega}{\sqrt{\pi}} e^{-2h\Psi} \int_0^{\xi} e^{-x^2} x^2 dx \qquad (502).$$

The upper limit of integration is determined by the condition that $\frac{1}{2}mV^2 + 2\Psi$ shall vanish, and is therefore given by $\xi^2 = -2h\Psi$, the value of Ψ being necessarily negative. If we put $-\Psi = R\beta$, so that β is positive the value of ξ^2 is $2hR\beta$ or β/T.

For some substances Ψ may be so large that a good approximation can be obtained by taking the integral in equation (502) between the limits $x = 0$ to $x = \infty$ (cf. § 267, below). In this case the integration is readily effected, and we find

$$\phi(T) = \omega e^{-2h\Psi} = \omega e^{-\frac{\Psi}{RT}} = \omega e^{\frac{\beta}{T}},$$

in which Ψ is negative. The degree of dissociation is then given by

$$\frac{1}{q^2} = 1 + \frac{4p\omega}{RT} e^{\frac{\beta}{T}} \qquad (503),$$

which is Boltzmann's formula for molecular aggregation and dissociation. Numerical values, obtained by the comparison of this formula with experiment, are given by Boltzmann*.

Willard Gibbs† also has treated the subject by a method which, although at first appearing very different from that of Boltzmann, will be found, as Boltzmann remarks‡, to rest ultimately upon exactly the same physical basis, and so leads necessarily to essentially the same equations.

The following table given by Willard Gibbs contains the densities of peroxide of nitrogen observed at various temperatures by Deville and Troost§, the pressure being one atmosphere throughout, and also the values calculated from equation (503).

For other substances Ψ may be so small that a good approximation can be obtained by expanding in powers of ξ. By repeated integration by parts, we obtain the expansion

$$\int_0^{\xi} e^{-x^2} x^2 dx = \frac{1}{2}\xi e^{-\xi^2}\left\{\frac{2\xi^2}{1.3} + \frac{(2\xi^2)^2}{1.3.5} + \frac{(2\xi^2)^3}{1.3.5.7} + \dots\right\},$$

* *l.c.* § 66.

† *Trans. Connecticut Acad.* III. p. 108 (1875) and p. 343 (1877); *Silliman Journal*, XVIII. (1879), p. 277. Also *Coll. Works*, I. pp. 55 and 372.

‡ *Vorlesungen über Gastheorie*, II. p. 211.

§ *Comptes Rendus*, LXIV. (1867), p. 237.

which is convergent for all values of ξ. Using this equation, (502) becomes

$$\phi(T) = \frac{4\omega}{\sqrt{\pi}}\left(\frac{\beta}{T}\right)^{\frac{3}{2}}\left\{\frac{1}{3} + \frac{2}{3.5}\frac{\beta}{T} + \frac{2^2}{3.5.7}\left(\frac{\beta}{T}\right)^2 + \dots\right\}.$$

AGGREGATION OF NO_2.

Temperature	Density (observed	Density (calc.)	Temperature	Density (observed)	Density (calc.)
183·2	1·57	1·592	80·6	1·80	1·801
154·0	1·58	1·597	70·0	1·92	1·920
135·0	1·60	1·607	60·2	2·08	2·067
121·5	1·62	1·622	49·6	2·27	2·256
111·3	1·65	1·641	39·8	2·46	2·443
100·1	1·68	1·676	35·4	2·53	2·524
90·0	1·72	1·728	26·7	2·65	2·676

Another possible expansion, also convergent for all values of ξ, is

$$\int_0^\xi e^{-x^2}x^2\,dx = \tfrac{1}{3}\xi^3 - \frac{1}{1.5}\xi^5 + \frac{1}{1.2.7}\xi^7 - \frac{1}{1.2.3.9}\xi^9 + \dots,$$

leading to

$$\phi(T) = \frac{4\omega}{\sqrt{\pi}}\left(\frac{\beta}{T}\right)^{\frac{3}{2}}\left\{\frac{1}{3} - \frac{1}{1.5}\frac{\beta}{T} + \frac{1}{1.2.7}\left(\frac{\beta}{T}\right)^2 - \dots\right\}e^{\frac{\beta}{T}}.$$

Jäger* gives a table, taken from Neumann†, shewing that the dissociation of hyponitric acid between temperatures of 27° C. and 135° C. can be well represented by assuming that $\phi(T)$ is proportional to

$$\frac{1}{T^2}e^{\frac{\beta}{T}}.$$

A theory is given leading to this form for $\phi(T)$, but it is not consistent with the dynamical principles given in the present book.

Continuity of Liquid and Gaseous States.

261. At very high temperatures, the series (497) reduces to its first term, so that $q = 1$, and there are no molecules in permanent combination.

At lower temperatures h is greater, so that not only is A greater, but the exponential $e^{-h[\frac{1}{2}mV^2+2\Psi]}$, in which it will be remembered that the index is always positive, is also greater. The relative importance of the later terms of the series (497) is therefore greater. Finally, we reach values

* Winkelmann's *Handbuch der Physik*, Vol. III. (Wärme). Article, "Die Kinetische Theorie der Gase," p. 731.

† Neumann, *Thermochemie*, p. 177.

of the temperature for which h has so great a value that the series (497) becomes divergent. At this point the molecules tend, according to our analysis, to form into clusters, each containing an infinitely great number of molecules, or, ultimately, into one big cluster absorbing all the molecules. By the time this stage is reached the analysis has ceased to apply, as the assumption that the molecular clusters are small, made in § 117, is now invalidated. It is, however, easy to give a physical interpretation of the point now reached: obviously it is the point at which liquefaction begins, and the collection of molecular clusters is a saturated vapour.

Regarded as a series in terms of A, the series (497) is a power series in ascending powers of A. Thus for a given value of h, say h_0, there is a single value of A, say A_0, such that the series is convergent for all values of A less than A_0 and is divergent for all values of A greater than A_0. In other words, corresponding to a given temperature, there is a definite density at which the substance liquefies. This of course is the vapour-density corresponding to this temperature. Clearly as h increases, A decreases, and conversely, so that an increase of pressure is accompanied by a rise in the boiling-point of the substance.

Since A depends on ν, the relation between corresponding values h_0, A_0 which has just been obtained may be expressed in the form

$$f(\nu, T) = 0 \quad \text{.............................(504)},$$

expressing the relation between ν and T at the boiling-point of a liquid.

The Critical Point.

262. It has already been noticed that ν_1 and ν become identical for very small values of h, so that the series (497) cannot become divergent. Thus for very high values of T equation (501) can have no root corresponding to a physically possible state. Let T_c be the lowest value of T for which a root of equation (501) is possible, then T_c will be a temperature above which liquefaction cannot possibly set in, no matter how great the density of the gas; in other words, T_c is the critical temperature.

From ordinary algebraic theory, it appears that there must be two coincident values of ν given by equation (501) to correspond to the critical temperature T_c, agreeing with what is already known as to the slope of the isothermals at the critical point.

Pressure, Density and Temperature.

263. It will now be clear that when a gas or vapour is at a temperature which is only slightly greater than its boiling-point at the pressure in question, it cannot be regarded as consisting of single molecules, but must be supposed to consist partly of single molecules and partly of clusters of two, three or more molecules. If m is the mass of a single molecule, and if $\nu_1, \nu_2, \nu_3, \ldots$ have the same meaning as before, the density is given by

$$\rho = m(\nu_1 + 2\nu_2 + 3\nu_3 + \ldots).$$

In calculating the pressure, we must treat each type of cluster as a separate kind of gas, exerting its own partial pressure. We accordingly obtain for the pressure, as in § 145,

$$p = \frac{1}{2h}(\nu_1 + \nu_2 + \nu_3 + \ldots) = \frac{\rho RT}{m}\left(\frac{\nu_1 + \nu_2 + \nu_3 + \ldots}{\nu_1 + 2\nu_2 + 3\nu_3 + \ldots}\right).$$

From a comparison of this equation with equation (305), remembering that $\nu_1, \nu_2, \ldots$ are functions of T and ρ, it is clear that neither Boyle's Law, Charles' Law nor Avogadro's Law will be satisfied with any accuracy.

264. The observed deviations from the laws obeyed by a perfect gas must of course be attributed partly to aggregation, as has just been explained, and partly to the causes which have already been discussed in Chapter VI. The two sets of causes are not, however, altogether independent; so that it is not sufficient to consider the effects separately, and then add. The state of the question is, perhaps, best regarded as follows.

The effect of the forces of cohesion is too complex for an exact mathematical treatment to be possible. We have therefore, in Chapter VI and the present chapter, examined their effect with the help of two separate simplifying assumptions. In Chapter VI, following Van der Waals, we regarded the gas as a single molecular cluster containing an infinite number of molecules; and in replacing the whole system of the forces of cohesion by a permanent *average* force, we virtually neglected the effect of any formations of small clusters inside the large cluster. In the present chapter, on the other hand, we have been concerned solely with the formation of small clusters, and have disregarded the large cluster altogether. As a consequence of the omission of the former treatment to take account of the formation of small clusters, this treatment led to the erroneous result (equation (436)) that the internal pressure is exactly proportional to the temperature, whereas as a consequence of the omission of the present treatment to consider the clustering of the gas as a whole, we are led in the present chapter to the erroneous conclusion that the internal pressure is identical with the boundary pressure. The situation may then be summed up by saying that the treatment of Chapter VI considers only the tendency to *mass-clustering*, while that of the present chapter considers only the tendency to *molecular-clustering*.

So long as the deviations from the behaviour of a perfect gas are small, the two tendencies may be considered separately, and the total deviation regarded as the sum of the two deviations caused by these tendencies separately. On the other hand, as we approach the critical point the phenomena of mass-clustering and molecular-clustering merge into one another and ultimately become identical at the critical point. The two effects are no longer additive, for each has become identical with the whole effect.

It must be borne in mind that we have only found an exact mathematical treatment of either effect to be possible by making the assumption that the effect itself is small. In other words, so far as our results apply, the effects are additive. It may be noticed that the deviations from the laws of a perfect gas, which were discussed in Chapter VI, fell off proportionally to $\frac{1}{T}$ and $\frac{1}{T^2}$, whereas the deviations discussed in the present chapter fall off much more rapidly as the temperature increases.

Calorimetry.

265. It is clear that the formulae which have been obtained for the specific heats may be greatly affected by the possibilities of molecular aggregation. For in raising the temperature of the gas work is done not only in increasing the energy of the various molecules, but also in separating a number of molecules from one another's attractions. This latter work will involve an addition to the values of C_p and C_v such as was not contemplated in the earlier analysis of §§ 237—241. We should therefore expect the values of C_p and C_v to be in excess of the values obtained from our earlier formulae, throughout all regions of pressure and temperature in which molecular aggregation can come into play. For instance, the specific heats of nitrogen peroxide have been studied by Berthelot and Ogier*, who give the following values for C_p:

From	27° to 67°,	$C_p = 1{\cdot}62$,
„	27° to 100°,	1·46,
„	27° to 150°,	1·115,
„	27° to 200°,	0·85,
„	27° to 300°,	0·64.

The excess in the values of C_p at the low temperatures may be reasonably attributed to the work required to separate molecules of N_2O_4 into pairs of molecules of NO_2.

As a further illustration of a somewhat different nature we may take the case of steam. Wet steam is steam in which large molecular clusters occur, dry steam is steam in which the molecules are all separate, and our quantity q measures what engineers speak of as the dryness of wet steam. For the value of γ for wet (saturated) steam, Rankine and Zeuner give respectively

* *Bull. Soc. Chimie*, [2], XXXVII. (1882), p. 434; *Comptes Rendus*, XCII. (1882), p. 916; *Ann. d. Chimie et Physique*, [5], XXX. (1883), p. 382; *Recueil de Constantes Physiques*, p. 108.

the values 1·0625, 1·0646. For dry steam ("steam gas") the recognised value is 1·30. If we used the formula

$$\gamma = 1 + \frac{2}{n+3},$$

for the calculation of n, we should come to the conclusion that $n+3$ had the value 32 for wet steam, and 6·6 for dry steam.

The large value of n in the former case is fully in keeping with the existence of large clusters of molecules, so large that each has about 32 degrees of freedom.

DISSOCIATION.

266. So far as the mathematical analysis goes, there is nothing in the preceding treatment to prevent it being applied to dissociation. The former molecules must be replaced by atoms, and the former clusters of molecules by single molecules.

Let us consider a gas in which the complete molecules are each composed of two atoms, of types α, β respectively. As in equations (491) the laws of distribution of dissociated atoms and complete molecules are

$$\tau_\alpha = Ae^{-2hE_\alpha},$$
$$\tau_\beta = Be^{-2hE_\beta},$$
$$\tau_{\alpha\beta} = ABe^{-2h(E_\alpha + E_\beta + \Psi)},$$

where Ψ is the potential energy of the two atoms forming the molecule. The analysis will be simplified, and the theory sufficiently illustrated, by regarding the atoms as point centres of force, of masses which will be supposed to be m_1, m_2 respectively. Thus we obtain as the laws of distribution of dissociated atoms

$$\left.\begin{aligned} &Ae^{-hm_1c^2}du\,dv\,dw \\ &Be^{-hm_2c^2}du\,dv\,dw \end{aligned}\right\} \quad \text{.........................(505)},$$

and as the law of distribution of complete molecules

$$ABe^{-h(m_1+m_2)c^2}du\,dv\,dw\;e^{-h\frac{m_1m_2}{m_1+m_2}V^2 - 2h\Psi}d\alpha\,d\beta\,d\gamma\,4\pi r^2dr \quad \text{...(506)},$$

the law being arrived at in the same way as the law (494), except that the scheme of transformation of velocities must be taken to be

$$\mathbf{u} = \frac{m_1u + m_2u'}{m_1 + m_2}, \quad \alpha = u' - u, \text{ etc.},$$

this being a generalisation of the transformation previously used (cf. § 338, below).

267. Although the mathematical analysis is similar to that of aggregation there is an important difference in the physical conditions. The law of distribution (506) is limited to values of the variable such that

$$\frac{m_1m_2}{m_1+m_2}V^2 + 2\Psi$$

is negative; as soon as this quantity becomes positive the molecule splits up into its component atoms. Now in the case of molecular aggregation, the attraction between complete molecules is not great, so that Ψ is a *small* negative quantity, and the range of values for V is correspondingly small. In the case of chemical dissociation Ψ is a *large* negative quantity, and the range for V is practically unlimited.

Thus in the theory of § 260, the former of the two evaluations of $\phi(T)$, namely that which assumes Ψ' to be very large, ought in general to be suited to problems of true dissociation, and the latter to problems of aggregation.

An estimate of the value of Ψ can be formed by considering the amount of heat evolved when chemical combination takes place. For instance when 2 grammes of hydrogen combine with 16 grammes of oxygen to form 18 grammes of water the amount of heat developed according to Thomsen's determination, is 68,376 units,—sufficient to raise the temperature of the whole mass of water by 3,600° C. The value of V necessary for dissociation to occur is therefore comparable with the mean value of V at 3,600° C., and these high values of V will be very rare in a gas at ordinary temperatures. The exclusion from the law of distribution (506) of high values of V will therefore have but little effect either on the law of distribution or on the energy represented by the internal degrees of freedom, and we may, without serious error, regard the law of distribution as holding for all values of V.

In such a case, it appears that the molecule may be treated exactly as an ordinary diatomic molecule, supposed incapable of dissociation, but possessing six degrees of freedom, three translational degrees represented by the differentials $du\,dv\,dw$, and three internal degrees represented by the differentials $d\alpha\,d\beta\,d\gamma$.

Since there are six degrees of freedom, the value of γ, even if we neglect potential energy, will be as low as $1\frac{1}{3}$, and will be even less if potential energy be taken into account. We have, however, seen that for diatomic molecules γ is fairly uniformly equal to $1\frac{2}{5}$, and this shews that the ordinary diatomic molecule must not be treated as consisting of two atoms describing orbits in the way we have imagined.

We are here brought back to the difficulties which have already been encountered in § 250 in connection with the specific heats of gases. The solution of these difficulties is not provided by the old classical dynamics but by the new quantum dynamics. We accordingly leave the question at this stage, to return to it in Chap. XVII, which deals especially with the quantum dynamics.

CHAPTER VIII

PHENOMENA OF A GAS NOT IN A STEADY STATE

268. In Chapters VI and VII we discussed the physical properties of gases in which the molecular motion at every point was symmetrical with respect to every direction in space. We now approach a much more complex class of problems for which this property is not true. If we refer to the expression obtained for the law of distribution of velocities at any point of a gas in the normal state, namely,

$$\nu f = \nu \left(\frac{hm}{\pi}\right)^{\frac{3}{2}} e^{-hm[(u-u_0)^2+(v-v_0)^2+(w-w_0)^2+2\chi]} \quad \text{............(507)},$$

we notice that there are five independent constants u_0, v_0, w_0, h and ν. The constancy of u_0, v_0, w_0 indicates that the mass motion of the gas is the same throughout the gas: if this mass motion varies from point to point in the gas, the layers of gas move relatively to one another, and we have the problem of determining the viscosity of the gas. Similarly the constancy of h indicates the equality of temperature throughout the gas: if this varies from point to point we have the problem of conduction of heat. Finally the constancy of ν indicates the mass-equilibrium of the gas: if this equilibrium does not exist we have the problem of diffusion. These three problems of viscosity, conduction, and diffusion have now to be approached.

The treatment would be easier than it actually proves to be, if expression (507) could be assumed to give the law of distribution at every point, subject only to the circumstance of u_0, v_0, w_0, h and ν varying from point to point. Unfortunately the analysis now to be given will shew that this assumption would not be a legitimate one.

General equation satisfied by f.

269. As in § 209, let the number of molecules whose centres at any instant t lie within an element of volume $dxdydz$, while the velocity components lie within a range $dudvdw$, be denoted by

$$\nu f(u, v, w, x, y, z, t)\, dudvdwdxdydz \quad \text{..............(508)}.$$

If the law of distribution f is known at the instant t, it will clearly be possible to follow out the motion of each group of molecules, and so obtain the law of distribution at the next instant $t+dt$, and similarly at every subsequent instant. Thus the law of distribution (508) is determined for all time when its value is given at any one instant.

It follows that the function νf defined above must satisfy a characteristic equation, of such a form that $\frac{d}{dt}(\nu f)$ is given as depending on νf. And, for a steady state, νf must satisfy an equation which is derived from the previous equation by equating $\frac{d}{dt}(\nu f)$ to zero. Thus in a problem of steady motion, we may not legitimately choose the values of νf so as to satisfy the physical conditions in the simplest way: the only values for νf which are eligible are those which satisfy the characteristic equation. We proceed to investigate the form of this equation, following a method given by Boltzmann*.

270. Let the molecules be supposed to move in a permanent field of force, such that a molecule at x, y, z is acted on by a force (X, Y, Z) per unit mass. Thus the equations of motion of a molecule, apart from collisions, are

$$\frac{du}{dt}=X, \quad \frac{dv}{dt}=Y, \quad \frac{dw}{dt}=Z \quad \text{......................(509).}$$

The number of molecules which at any instant t have velocity components u, v, w within a small range $du\,dv\,dw$, and coordinates x, y, z within a small range $dx\,dy\,dz$ is given by formula (508).

Let these molecules pursue their natural motion for a time dt. At the end of this interval, if no collisions have taken place in the meantime, the u, v, w components of velocity of each molecule will have increased respectively by amounts $X\,dt$, $Y\,dt$, $Z\,dt$, while the coordinates x, y, z will have increased respectively by amounts $u\,dt$, $v\,dt$, $w\,dt$. Thus after the interval dt, the original molecules will have velocities lying within a small range $du\,dv\,dw$ surrounding the values $u+X\,dt$, $v+Y\,dt$, $w+Z\,dt$, and coordinates lying within a small element $dx\,dy\,dz$ surrounding the point $x+u\,dt$, $y+v\,dt$, $z+w\,dt$. Moreover, by tracing the motion backwards, it appears that the molecules which we have had under consideration are the only ones which, at the instant $t+dt$, can have values of x, y, z, u, v, w lying within this range.

The number of molecules having values of x, y, z, u, v, w lying within this range at time $t+dt$ is however

$$\nu f(u+X\,dt,\ v+Y\,dt,\ w+Z\,dt,\ x+u\,dt,\ y+v\,dt,\ z+w\,dt)\,du\,dv\,dw\,dx\,dy\,dz \quad \text{......(510).}$$

Hence if no collisions occur, this expression must be exactly equal to expression (508).

* *Vorlesungen über Gastheorie*, I. chapters II and III.

Expanding expression (510) as far as first powers of dt, and equating to expression (508), we obtain the relation

$$\frac{\partial}{\partial t}(\nu f) = -\left[X\frac{\partial}{\partial u} + Y\frac{\partial}{\partial v} + Z\frac{\partial}{\partial w} + u\frac{\partial}{\partial x} + v\frac{\partial}{\partial y} + w\frac{\partial}{\partial z}\right](\nu f) \quad \text{...(511)},$$

expressing the rate at which νf changes on account of the motion of the molecules, and the forces acting on them.

When collisions occur, these produce an additional change in νf which can be evaluated as in § 21 (cf. equation (12)). It was there found that when the molecules were elastic spheres of diameter σ, this change was expressed by the equation

$$\frac{\partial}{\partial t}(\nu f) = \iiiint\!\!\int \nu^2 (\bar{f}\bar{f}' - ff') V\sigma^2 \cos\theta \, du' dv' dw' d\omega \quad \text{......(512)}.$$

In the more general case in which the molecules may be supposed to have any structure we please, let the contribution to $\frac{\partial}{\partial t}(\nu f)$ produced by collisions be denoted by

$$\left[\frac{\partial}{\partial t}(\nu f)\right]_{\text{coll.}}.$$

On combining the two causes of change in (νf), we arrive at the general equation

$$\frac{\partial}{\partial t}(\nu f) = -\left[X\frac{\partial}{\partial u} + Y\frac{\partial}{\partial v} + Z\frac{\partial}{\partial w} + u\frac{\partial}{\partial x} + v\frac{\partial}{\partial y} + w\frac{\partial}{\partial z}\right](\nu f) + \left[\frac{\partial}{\partial t}(\nu f)\right]_{\text{coll.}} \quad \text{......(513)}.$$

This equation must, under all circumstances, be satisfied by νf. When the gas is in a steady state the right-hand member must of course vanish.

271. No progress can be made with the development or solution of this equation until the term $\left[\frac{\partial}{\partial t}(\nu f)\right]_{\text{coll.}}$ has been evaluated, and this unfortunately can only be effected to a very limited extent.

Let us consider the form assumed by the problem when the molecules are regarded as point centres of force, attracting or repelling with a force which depends only on their distance apart.

We fix our attention on an encounter between two molecules, the velocities before the encounter begins being u, v, w and u', v', w'. The relative velocity before encounter will be V, given by

$$V^2 = (u'-u)^2 + (v'-v)^2 + (w'-w)^2 \quad \text{..............(514)}.$$

In fig. 14 let O represent the centre of the first molecule moving in some direction QO with a velocity u, v, w, and let MNP represent the path described *relatively to O* by the second molecule before the encounter begins, the relative velocity before encounter being V. When the second molecule comes to within such a distance of O that the action between the two molecules becomes appreciable, it will be deflected from its original rectilinear path MNP, and will describe a curved orbit such as MNS, this orbit being of course in the plane $MNPO$.

Let ROP be a plane through O perpendicular to MN, and let MN meet this plane in a point P. Let the polar coordinates of P in the plane ROP be p, ϵ, the point O being taken as origin, so that $OP = p$, and any line RO in this plane being chosen for initial line. Clearly p is the perpendicular from the first molecule O on to MN, the relative path of the second molecule before encounter.

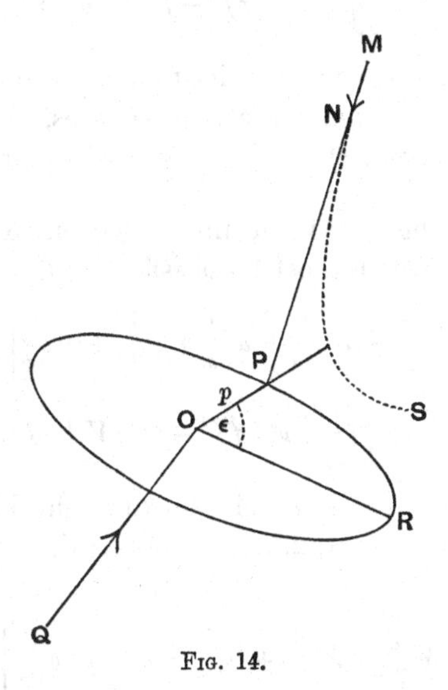

Fig. 14.

Let us examine what is the frequency of collisions such that the second molecule has a velocity u', v', w' whose components lie within a small specified range $du'dv'dw'$, while its path before the encounter is such that p, ϵ lie within a small range dp, $d\epsilon$. For all such collisions the line MP must meet the plane ROP within a small element of area $p\,dp\,d\epsilon$. The number of such collisions to be expected within an interval dt will therefore be equal to the number of molecules which at a certain instant lie within a small volume $p\,dp\,d\epsilon\,V\,dt$, and have velocities within the specified range $du'dv'dw'$. This number is

$$\nu f(u', v', w')\, du'dv'dw' p\,dp\,d\epsilon\,V\,dt \qquad (515).$$

The number of molecules per unit volume having velocities between u and $u + du$, v and $v + dv$, w and $w + dw$ is

$$\nu f(u, v, w)\, du\,dv\,dw,$$

so that the total number per unit volume of collisions of the kind we now have under consideration is

$$\nu^2 f(u, v, w) f(u', v', w')\, du\, dv\, dw\, du'\, dv'\, dw'\, p\, dp\, d\epsilon\, V\, dt \;\ldots\ldots(516).$$

The class of collisions now under consideration is similar to that which we called class α in § 17, and expression (516) is obviously a generalisation of our former expression (4).

The argument can proceed as in § 17, expression (516) replacing expression (4). Formula (9) is still true, being now a consequence of the general theorem of Liouville (§ 85), and we obtain, just as in formula (11),

$$\left[\frac{\partial}{\partial t}(\nu f)\right]_{\text{coll.}} = \iiiint\!\!\int \nu^2 (\bar{f}\bar{f}' - ff')\, V\, du'\, dv'\, dw'\, p\, dp\, d\epsilon \;\ldots\ldots(517).$$

This, then, is the required generalisation of equation (512). It clearly reduces to this latter equation for elastic spheres, the factor $\sigma^2 \cos\theta\, d\omega$ of equation (512) being exactly the factor $p\, dp\, d\epsilon$ of equation (517).

272. On substituting this value into equation (513), we obtain as the characteristic equation which must be satisfied by f,

$$\frac{\partial}{\partial t}(\nu f) = -\left[X\frac{\partial}{\partial u} + Y\frac{\partial}{\partial v} + Z\frac{\partial}{\partial w} + u\frac{\partial}{\partial x} + v\frac{\partial}{\partial y} + w\frac{\partial}{\partial z}\right](\nu f)$$
$$+ \iiiint\!\!\int \nu^2 (\bar{f}\bar{f}' - ff')\, V\, du'\, dv'\, dw'\, p\, dp\, d\epsilon \;\ldots(518).$$

For a mixture of gases, in which the different kinds of molecules are distinguished by the suffixes 1, 2, ..., we obtain in a similar way a series of equations such as

$$\frac{\partial}{\partial t}(\nu_1 f_1) = -\left[X\frac{\partial}{\partial u} + Y\frac{\partial}{\partial v} + Z\frac{\partial}{\partial w} + u\frac{\partial}{\partial x} + v\frac{\partial}{\partial y} + w\frac{\partial}{\partial z}\right](\nu_1 f_1)$$
$$+ \Sigma \iiiint\!\!\int \nu_1 \nu_2 (\bar{f}_1\bar{f}_2' - f_1 f_2')\, V\, du'\, dv'\, dw'\, p\, dp\, d\epsilon \;\ldots(519).$$

273. It at once appears that putting

$$f = A e^{-hm[(u-u_0)^2 + (v-v_0)^2 + (w-w_0)^2]} \;\ldots\ldots\ldots\ldots\ldots\ldots(520)$$

makes the right-hand member of equations (518) and (519) vanish, and so provides a solution when $X = Y = Z = 0$, and νf is independent of x, y, z. This is the solution already found in § 25.

On substituting

$$f = A e^{-hm(c^2 + 2\chi)} \;\ldots\ldots\ldots\ldots\ldots\ldots\ldots\ldots(521)$$

into equation (518), we obtain

$$Xu + Yv + Zw + u\frac{\partial\chi}{\partial x} + v\frac{\partial\chi}{\partial y} + w\frac{\partial\chi}{\partial z} = 0 \;\ldots\ldots\ldots\ldots(522),$$

which is satisfied if

$$X = -\frac{\partial \chi}{\partial x}, \quad Y = -\frac{\partial \chi}{\partial y}, \quad Z = -\frac{\partial \chi}{\partial z}.$$

Thus (521) is a solution when χ is the potential of the forces acting on the molecule, the result obtained in § 110.

274. If, however, u_0, v_0, w_0, h and ν vary from point to point, formulae (520) and (521) do not provide a solution, for on substituting them into equation (518) we find that the right-hand member vanishes, while the left does not.

To search for a solution appropriate to this case, assume

$$f = f_0 [1 + \Phi(x, y, z, u, v, w)] \quad \text{.................(523)},$$

where Φ is a small quantity of the first order, and

$$f_0 = Ae^{-hm[(u-u_0)^2+(v-v_0)^2+(w-w_0)^2]} \quad \text{.................(524)}.$$

Since $f = f_0$ is a solution when u_0, v_0, w_0, h and ν do not vary from point to point, it follows as a matter of necessity that equation (523) must provide a solution when these quantities vary to the first order of small quantities.

275. The integrand of equation (519) contains a term $\nu_1 \nu_2 f_1 f_2'$ of which the value, by equation (523), is

$$\nu_1 \nu_2 f_1 f_2' = \nu_1 \nu_2 f_{01} f_{02}' (1 + \Phi_1 + \Phi_2') \quad \text{..............(525)}.$$

Here f_{01} denotes the value of f_0 for a molecule of the first kind, and so on, and the product $\Phi_1 \Phi_2'$ is omitted as being of the second order of small quantities. Similarly

$$\nu_1 \nu_2 \bar{f}_1 \bar{f}_2' = \nu_1 \nu_2 \bar{f}_{01} \bar{f}_{02}' (1 + \bar{\Phi}_1 + \bar{\Phi}_2') \quad \text{..............(526)},$$

and, from the conservation of energy and momenta at an encounter,

$$\bar{f}_{01} \bar{f}_{02}' = f_{01} f_{02}' \quad \text{..............................(527)},$$

so that

$$\nu_1 \nu_2 (\bar{f}_1 \bar{f}_2' - f_1 f_2') = \nu_1 \nu_2 f_{01} f_{02}' (\bar{\Phi}_1 + \bar{\Phi}_2' - \Phi_1 - \Phi_2').$$

On substituting solution (523) into equation (519), f may be replaced by f_0 everywhere except in the integrals; for if we retained terms in Φ in the remaining parts of the equation, we should be including terms of the second order of small quantities. Equation (519) accordingly reduces to

$$\left(\frac{\partial}{\partial t} + X\frac{\partial}{\partial u} + Y\frac{\partial}{\partial v} + Z\frac{\partial}{\partial w} + u\frac{\partial}{\partial x} + v\frac{\partial}{\partial y} + w\frac{\partial}{\partial z}\right)(\nu_1 f_{01})$$

$$= \nu_1 f_{01} \Sigma \iiiiint \nu_2 f_{02}' (\bar{\Phi}_1 + \bar{\Phi}_2' - \Phi_1 - \Phi_2') V du' dv' dw' p dp d\epsilon \quad \text{...(528)},$$

an equation in which every term is of the first order of small quantities.

On dividing out by $\nu_1 f_{01}$ and replacing f_{01} by its value from equation (565), this equation becomes

$$\left(\frac{\partial}{\partial t} + X\frac{\partial}{\partial u} + Y\frac{\partial}{\partial v} + Z\frac{\partial}{\partial w} + u\frac{\partial}{\partial x} + v\frac{\partial}{\partial y} + w\frac{\partial}{\partial z}\right)$$
$$\times \{\log \nu A - hm\,[(u-u_0)^2 + (v-v_0)^2 + (w-w_0)^2]\}$$
$$= \Sigma \iiiiint \nu_2 f_{02}' (\bar{\Phi}_1 + \bar{\Phi}_2' - \Phi_1 - \Phi_2')\, V du' dv' dw' p dp d\epsilon \;\ldots(529).$$

276. There is a certain indeterminateness about the proposed solution (523), in that changes in u_0, v_0, w_0 or in νA or h are not separate from changes in Φ: thus changes in f_0 may be absorbed in Φ, or vice versa. For instance the total momentum parallel to the axis of x of unit volume of the gas is

$$\iiint mu\nu f du dv dw = \iiint mu\nu f_0 du dv dw + \iiint mu\nu f_0 \Phi du dv dw$$
$$= m\nu u_0 + \iiint mu\nu f_0 \Phi du dv dw \;\ldots\ldots\ldots\ldots\ldots(530),$$

and an increase in this can equally be represented by increasing u_0 or by changing Φ.

We can make the solution (523) perfectly definite if we agree that u_0, v_0, w_0, h and νA are to have the same physical interpretation in the general solution (523) as they have in the steady-state solution. If this is agreed on, the components of the velocity of mass-motion must be u_0, v_0, w_0. Thus the momentum parallel to the axis of x must be $m\nu u_0$, and so, from equation (530), we must have

$$\iiint \nu f_0 u \Phi du dv dw = 0 \;\ldots\ldots\ldots\ldots\ldots\ldots\ldots(531),$$

together with two similar equations in v and w.

Similarly we may agree that we must have

$$A = \left(\frac{hm}{\pi}\right)^{\frac{3}{2}} \;\ldots\ldots\ldots\ldots\ldots\ldots\ldots\ldots\ldots(532),$$

and the condition for this is found to be

$$\iiint \nu f_0 \Phi du dv dw = 0 \;\ldots\ldots\ldots\ldots\ldots\ldots\ldots(533).$$

Finally we may agree to make

$$\mathbf{c}^2 = \frac{3}{2hm} \;\ldots\ldots\ldots\ldots\ldots\ldots\ldots\ldots\ldots\ldots(534),$$

as in equation (45) for the steady state, $\mathbf{c}^2$ being the mean value of c^2 for all the molecules in any small element of volume: the condition for this will be

$$\iiint \nu f_0 (u^2 + v^2 + w^2)\, \Phi du dv dw = 0 \;\ldots\ldots\ldots\ldots(535).$$

277. If Φ is restricted in this way, the equation of continuity for the gas, neglecting small quantities of the second order (cf. equation (409)), becomes

$$\frac{\partial}{\partial t}\log \nu = -\left(\frac{\partial u_0}{\partial x}+\frac{\partial v_0}{\partial y}+\frac{\partial w_0}{\partial z}\right) \quad \text{..................(536).}$$

Using the values for $\log \nu$ and $\log A$ provided by equations (536) and (532), and simplifying by omitting certain terms which are small quantities of the second order, equation (529) reduces to

$$(1+\tfrac{2}{3}hmc^2)\frac{\partial}{\partial t}\log\nu + \frac{\partial}{\partial t}\{\tfrac{3}{2}\log h - hm[(u-u_0)^2+(v-v_0)^2+(w-w_0)^2]\}$$

$$-2hm(uX+vY+wZ)+\left(u\frac{\partial}{\partial x}+v\frac{\partial}{\partial y}+w\frac{\partial}{\partial z}\right)\log\nu$$

$$-\left(mc^2-\frac{3}{2h}\right)\left(u\frac{\partial h}{\partial x}+v\frac{\partial h}{\partial y}+w\frac{\partial h}{\partial z}\right)$$

$$-2hm\left[(u^2-\tfrac{1}{3}c^2)\frac{\partial u_0}{\partial x}+(v^2-\tfrac{1}{3}c^2)\frac{\partial v_0}{\partial y}+(w^2-\tfrac{1}{3}c^2)\frac{\partial w_0}{\partial z}+uv\left(\frac{\partial v_0}{\partial x}+\frac{\partial u_0}{\partial y}\right)+\dots\right]$$

$$=\Sigma\iiiint\!\!\int \nu_2 f_{02}'(\overline{\Phi}_1+\overline{\Phi}_2'-\Phi_1-\Phi_2')\,V\,du'\,dv'\,dw'\,p\,dp\,d\epsilon \quad \text{...(537).}$$

278. It will be remembered that this equation is only accurate when Φ satisfies five relations, expressed by equations (531), (533) and (535). The solutions in Φ will however be additive, since the equations are linear; five solutions which contribute nothing to either side are

$$\Phi = 1,\ mu,\ mv,\ mw,\ mc^2 \quad \text{....................(538),}$$

so that to any solution for Φ which satisfies equation (537) may be added terms of the form

$$\Phi = B + Cmu + Dmv + Emw + Fmc^2 \quad \text{..............(539),}$$

and the constants B, C, D, E, F may be adjusted so as to satisfy the five necessary conditions.

Law of Force μr^{-s}.

279. Further progress with equation (537) can only be made by assuming definite laws for the interaction between molecules at collisions. We shall therefore suppose that the molecules are centres of force repelling according to the law μr^{-s}*.

If two molecules of masses m_1, m_2 at a distance r apart exert a repulsive force

$$m_1 m_2 \frac{K}{r^s} \quad \text{............................(540),}$$

* The method of §§ 279—284 is that of Maxwell, *Collected Works*, II. p. 36.

then their potential Ω at this distance is

$$\Omega = \int_{\infty}^{r} m_1 m_2 \frac{K}{r^s} dr = m_1 m_2 \frac{K}{(s-1) r^{s-1}} \quad \text{..............(541).}$$

Let the coordinates of the two molecules be denoted by x_1, y_1, z_1 and x_2, y_2, z_2. Let them be acted on by their mutual repulsive force, and also by a force of components X, Y, Z per unit mass, which may be supposed not to vary over distances comparable with the distance r between the two molecules in an encounter.

The equations of motion of the two molecules are

$$m_1 \ddot{x}_1 = \frac{\partial \Omega}{\partial x_1} + m_1 X, \text{ etc.} \quad \text{........................(542),}$$

$$m_2 \ddot{x}_2 = \frac{\partial \Omega}{\partial x_2} + m_2 X, \text{ etc.} \quad \text{........................(543),}$$

from which we obtain

$$m_1 m_2 (\ddot{x}_1 - \ddot{x}_2) = m_2 \frac{\partial \Omega}{\partial x_1} - m_1 \frac{\partial \Omega}{\partial x_2} \quad \text{.................(544),}$$

and two similar equations.

Let x, y, z be the coordinates of the first molecule relative to the second, so that $x = x_1 - x_2$, etc. Then Ω is a function of x, y, z and equation (544) reduces to

$$m_1 m_2 \ddot{x} = (m_1 + m_2) \frac{\partial \Omega}{\partial x} \quad \text{......................(545).}$$

Thus the motion of the first molecule relative to the second is that of a particle of unit mass about a fixed centre of force, the potential energy at distance r being

$$\frac{m_1 + m_2}{m_1 m_2} \Omega = \frac{(m_1 + m_2) K}{(s-1) r^{s-1}} \quad \text{.....................(546).}$$

280. To investigate this orbit, we change from the coordinates x, y, z to polar coordinates r, θ in the plane of the orbit.

We have the two usual integrals of momentum and energy,

$$r^2 \dot{\theta} = h \quad \text{..............................(547),}$$

$$\tfrac{1}{2} (\dot{r}^2 + r^2 \dot{\theta}^2) = C - \frac{(m_1 + m_2) K}{(s-1) r^{s-1}} \quad \text{.................(548).}$$

Eliminating the time, the differential equation of the orbit is

$$\frac{1}{2} \frac{h^2}{r^4} \left\{ \left(\frac{\partial r}{\partial \theta} \right)^2 + r^2 \right\} = C - \frac{(m_1 + m_2) K}{(s-1) r^{s-1}},$$

and this has the integral

$$\theta = \int_{\infty}^{r} \frac{dr}{\sqrt{\left(\frac{2C}{h^2} r^4 - r^2 - \frac{2(m_1+m_2)K}{(s-1)h^2} r^{5-s}\right)}} \text{ ....(549)},$$

in which the direction of the asymptote to the orbit is taken to be the initial line $\theta = 0$.

281. From equations (547) and (548) we have

$$h = pV, \quad C = \tfrac{1}{2}V^2 \text{ ....(550)},$$

where V is the velocity in the orbit at infinity (*i.e.* the relative velocity of the two molecules before the encounter begins) and p is the perpendicular from the centre on to the asymptote described with this velocity. Thus p, V are used in the same sense as in § 271.

Using relations (550), and further writing η for p/r, equation (549) becomes

$$\theta = \int_0^{\eta} \frac{d\eta}{\sqrt{\left\{1 - \eta^2 - \frac{2(m_1+m_2)K}{(s-1)V^2}\left(\frac{\eta}{p}\right)^{s-1}\right\}}}$$

$$= \int_0^{\eta} \frac{d\eta}{\sqrt{\left\{1 - \eta^2 - \frac{2}{s-1}\left(\frac{\eta}{\alpha}\right)^{s-1}\right\}}} \text{ ....(551)},$$

where

$$\alpha = p\left(\frac{V^2}{(m_1+m_2)K}\right)^{\frac{1}{s-1}} \text{ ....(552)}.$$

282. The apses of the orbit are given by $\frac{\partial r}{\partial \theta} = 0$, and therefore by

$$1 - \eta^2 - \frac{2}{s-1}\left(\frac{\eta}{\alpha}\right)^{s-1} = 0.$$

From a simple graphical treatment, or from Sturm's theorem on the roots of algebraic equations, it is clear that this equation can only have one real root for all values of s greater than 1. Call this root η_0, then the angle, say θ_0, between the asymptote and the apsidal distance will be given by equation (551) on taking the upper limit to be η_0. The angle between the asymptotes, say θ', is equal to twice this, and so is given by

$$\theta' = 2\theta_0 = 2\int_0^{\eta_0} \frac{d\eta}{\sqrt{\left\{1 - \eta^2 - \frac{2}{s-1}\left(\frac{\eta}{\alpha}\right)^{s-1}\right\}}} \text{ ....(553)}.$$

After the encounter, the velocities parallel and perpendicular to the initial line are of course $-V\cos\theta'$ and $-V\sin\theta'$.

283. For any value of s, there will naturally be a doubly infinite series of possible orbits corresponding to different values of p and V. Except for a difference of linear scale, however, these may be reduced to a singly infinite system corresponding to the variation of α or $pV^{\frac{2}{s-1}}$. In fig. 15* some members of this singly infinite system are shewn for the law of force μ/r^5.

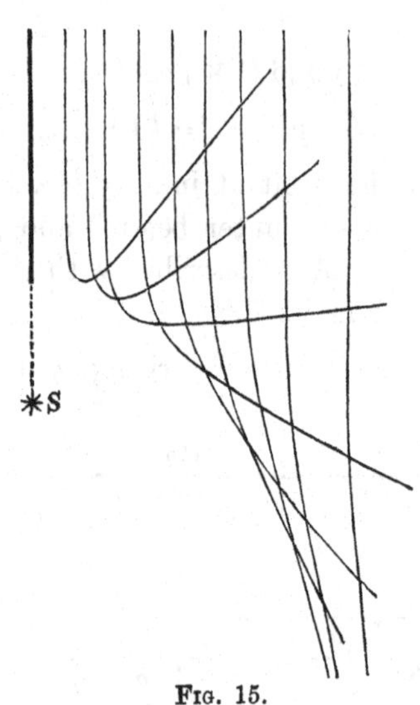

FIG. 15.

284. Let us consider a collision between two molecules, the velocities before collision being u, v, w and u', v', w', so that the relative velocity V is given by

$$V^2 = (u' - u)^2 + (v' - v)^2 + (w' - w)^2 \quad \text{..............(554)}.$$

In fig. 14 of § 271, the line OR from which ϵ was measured was supposed to be an arbitrarily chosen line. For definiteness, let this now be supposed to be the intersection of the plane POR with a plane through O containing the direction of NP and the axis of x, as in fig. 16.

In fig. 17, let OR, OX be the directions of the line OR of fig. 16 and of the axis of x. Let OG be the direction of V, the relative velocity before collision, so that OR, OX, OG all lie in one plane. Let these lines be supposed each of unit length, so that the points GXR lie on a sphere of unit radius about O as centre.

Let OY, OZ be unit lines giving the directions of the axes of y and z, and let OG' give the direction of the relative velocity after the encounter. Then

* This figure is given by Maxwell, *Collected Works*, II. p. 42. I am indebted to the University Press for the use of the original block.

GOG' is the plane of the orbit, which is the plane NPO in fig. 16. Thus the angle RGG' is the ϵ of § 271, while the angle GOG' is θ'.

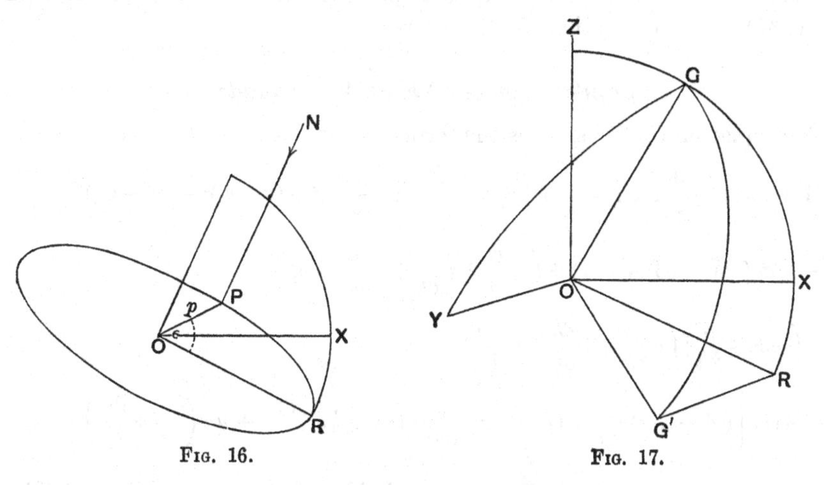

FIG. 16. FIG. 17.

From the spherical triangle $G'GX$,

$$\cos G'X = \cos GX \cos GG' + \sin GX \sin GG' \cos \epsilon,$$

in which we have $\cos G'X = -\frac{\overline{u'} - \overline{u}}{V}$, $\cos GX = \frac{u' - u}{V}$, so that

$$\overline{u} - \overline{u'} = (u' - u) \cos \theta' + \sqrt{V^2 - (u' - u)^2} \sin \theta' \cos \epsilon \text{ ......(555)}.$$

Denoting the angle XGY by ω_2 and XGZ by ω_3, we have, in a similar way,

$$\overline{v} - \overline{v'} = (v' - v) \cos \theta' + \sqrt{V^2 - (v' - v)^2} \sin \theta' \cos (\epsilon - \omega_2) \text{ ...(556)},$$

$$\overline{w} - \overline{w'} = (w' - w) \cos \theta' + \sqrt{V^2 - (w' - w)^2} \sin \theta' \cos (\epsilon - \omega_3) \text{ ...(557)}.$$

We determine ω_2 by noticing that in the triangle GXY, $XY = \frac{1}{2}\pi$ and $XGY = \omega_2$; thus

$$(u' - u)(v' - v) + \sqrt{[V^2 - (u' - u)^2][V^2 - (v' - v)^2]} \cos \omega_2 = 0 \text{ ...(558)},$$

and similarly

$$(u' - u)(w' - w) + \sqrt{[V^2 - (u' - u)^2][V^2 - (w' - w)^2]} \cos \omega_3 = 0 \text{ ...(559)}.$$

In addition to these equations, there are three equations of momentum, such as

$$m_1 \overline{u} + m_2 \overline{u'} = m_1 u + m_2 u' \text{ ....(560)}.$$

Eliminating $\overline{u'}$ from equations (555) and (560), we obtain

$$\overline{u} = u + \frac{m_2}{m_1 + m_2} [2 (u' - u) \cos^2 \tfrac{1}{2}\theta' + \sqrt{V^2 - (u' - u)^2} \sin \theta' \cos \epsilon] \text{ ...(561)},$$

giving $\overline{u}$ in terms of the velocities before collision, and there are of course similar equations giving $\overline{v}$ and $\overline{w}$.

Solutions for Φ.

285. We are now in a position to proceed with the discussion of equation (537). From equation (552) we have

$$p\,dp\,d\epsilon = [(m_1+m_2)K]^{\frac{2}{s-1}} V^{-\frac{4}{s-1}} \alpha\, d\alpha\, d\epsilon,$$

so that equation (537) assumes the form

$$(1+\tfrac{2}{3}hmc^2)\frac{\partial}{\partial t}\log\nu + \frac{\partial}{\partial t}\{\tfrac{3}{2}\log h - hm[(u-u_0)^2+(v-v_0)^2+(w-w_0)^2]\}$$

$$-2hm(uX+vY+wZ) + \left(u\frac{\partial}{\partial x}+v\frac{\partial}{\partial y}+w\frac{\partial}{\partial z}\right)\log\nu$$

$$-\left(mc^2-\frac{3}{2h}\right)\left(u\frac{\partial h}{\partial x}+v\frac{\partial h}{\partial y}+w\frac{\partial h}{\partial z}\right)$$

$$-2hm\left[(u^2-\tfrac{1}{3}c^2)\frac{\partial u_0}{\partial x}+(v^2-\tfrac{1}{3}c^2)\frac{\partial v_0}{\partial y}+(w^2-\tfrac{1}{3}c^2)\frac{\partial w_0}{\partial z}+uv\left(\frac{\partial v_0}{\partial x}+\frac{\partial u_0}{\partial y}\right)+\dots\right]$$

$$=\Sigma\nu_2[(m_1+m_2)K]^{\frac{2}{s-1}} I \quad\quad (562),$$

where I stands for the quintuple integral

$$I=\iiiiint(\overline{\Phi}_1+\overline{\Phi}_2'-\Phi_1-\Phi_2')\,\alpha\, d\alpha\, d\epsilon\, V^{\frac{s-5}{s-1}} f_{02}'\,du'dv'dw' \quad \dots(563).$$

In this integral, it will be remembered that Φ_1 is a function, as yet undetermined, of u, v and w; Φ_2' is the similar function of u', v', w' for a molecule of the second kind, while $\overline{\Phi}_1$, $\overline{\Phi}_2'$ have corresponding meanings in terms of the velocities after collision. Our task is not to evaluate expression (563) for given values of Φ, but to find values of Φ such that after integration expression (563) shall be equal to a certain algebraic function containing terms of degrees 0, 1, 2 and 3 in u, v and w.

286. Consider tentatively a value of Φ which is algebraic and of degree p in u, v, w. It is readily seen from the equations of § 284 that both

$$\overline{\Phi}_1+\overline{\Phi}_2'-\Phi_1-\Phi_2'$$

and

$$\iint(\overline{\Phi}_1+\overline{\Phi}_2'-\Phi_1-\Phi_2')\,\alpha\, d\alpha\, d\epsilon \quad\quad (564)$$

will be of degree p in u, v, w.

To obtain the integral I given by equation (563), the integral (564) must be multiplied by $V^{\frac{s-5}{s-1}}$ and the product averaged over all values of u', v', w', Maxwell's law being assumed to hold. The result is an expression of degree

$$p+\frac{s-5}{s-1} \quad\quad (565)$$

in the velocities.

In order that equation (562) may be satisfied, Φ must consist of terms for which expression (565) has the values 0, 1, 2 and 3. But in general, unless s has very special values, the values for p obtained in this way will not be integral, and the evaluation even of the preliminary integral (564) cannot be effected in finite terms. For this reason the method fails to provide an exact solution in the general case. In particular cases the integration can be effected.

287. The special values of s for which the integration may be possible are those for which $\frac{s-5}{s-1}$ is integral, and are therefore $s=\infty$ and $s=5$.

The value $s=\infty$ corresponds to the case of elastic spheres, but it does not prove to be possible to carry out the integration in finite terms in this case.

The value $s=5$, corresponding to molecules repelling with a force μ/r^5, has been fully treated by Maxwell. Maxwell's method is somewhat different from that we are now considering, and so his treatment is reserved for the next chapter, where it is given in full. For the present the following points may be noticed about the special case of $s=5$.

When $s=5$, the factor $V^{\frac{s-5}{s-1}}$ disappears entirely from expression (563) and the value of I reduces simply to the value of the integral (564) averaged for all velocities of the second molecule. Also when $s=5$, expression (565) reduces simply to p, so that in the correct solution Φ will consist of terms of degrees 0, 1, 2 and 3 in u, v, w*. When a value of this type is assumed for Φ, the integral (564) is easily evaluated from equations such as (561), and the value of I is then easily found, leading directly to the correct value for Φ.

Leaving aside the special case of $s=5$, it has been seen that the preliminary integral (564) cannot be evaluated for values of Φ such as are required to give a solution of the general equation (562). There is however one special case in which the integration can be carried out without limitation to the special value $s=5$. We proceed to consider this case, partly as an illustration of the method, but mainly because the case is itself of great physical interest, and because the result obtained will be required later†.

* It is clear from what has been said that in no case except $s=5$ can Φ consist of terms of degrees 0, 1, 2 and 3 only in u, v and w. Chapman published an interesting paper ("On the Kinetic Theory of a Gas constituted of spherically symmetrical molecules," *Phil. Trans.* A, 211 (1912), p. 433) in which he assumed that *for all values* of s, Φ might be supposed to consist of terms of degrees 0, 1, 2 and 3 in u, v, w. One of the main results of his paper was that certain formulae, obtained by Maxwell for the special case $s=5$, were approximately true for all laws of force. This conclusion could not be regarded as rigorously established since the author virtually limited himself to the case of $s=5$ by the form he assumed for Φ. In a later paper (*Phil. Trans.* 216 A, 1915, p. 279) the same author has examined the error introduced by this assumption, and shews that it is very small. See below, §§ 373, 401, 424.

† The method of §§ 288—295 agrees in its main lines with one first given by Lorentz (*Theory of Electrons*, Note 29).

MIXTURE OF TWO KINDS OF MOLECULES, THE MASS-RATIO BEING VERY GREAT.

288. Suppose that there are only two kinds of molecules (or other units), and that of these the molecules of the second kind are enormously heavier than those of the first. Let us also suppose that the light molecules are few in number compared with the heavy ones, so that the deflections of the paths of the light molecules may be supposed to be caused entirely by encounters with the heavy molecules: we neglect the encounters of the light molecules with one another. To fix the ideas, we may think of the light "molecules" as electrons in a solid, and the heavy "molecules" as the atoms or molecules of the solid, between which the electrons thread their way. The results obtained will subsequently be applied to this particular case, although the analysis is naturally applicable to more general problems.

On account of the assumed inequality of mass, it follows that the velocities of the heavy molecules will be very slight compared with those of the light molecules. We shall accordingly neglect u', v', w' in comparison with u, v, w, and so may think of the heavy molecules as being at rest. The dynamics of collisions are greatly simplified, for we may now regard the heavy molecules as massive obstacles which deflect the light molecules without altering their velocities. The velocity c of the light molecule before collision is equal to the relative velocity V, and also to the velocity $\bar{c}$ after collision.

289. Since u', v', w' are supposed to be so small as to be negligible, Φ_2' and $\overline{\Phi}_2'$ may be neglected, and the preliminary integral (564) is given by

$$I_P = \iint (\overline{\Phi}_1 - \Phi_1)\, \alpha d\alpha d\epsilon \qquad (566),$$

which is a function of u, v, w only. Furthermore V is simply equal to c, and the value of I_P does not depend on u', v', w', so that equation (563) gives as the value of I,

$$I = I_P c^{\frac{s-5}{s-1}},$$

and the right-hand member of equation (562) reduces to the single term

$$\nu_2 (m_2 K)^{\frac{2}{s-1}} I_P c^{\frac{s-5}{s-1}} \qquad (567).$$

The preliminary integral I_P, given by equation (566), is readily evaluated in the special case now under consideration, for equation (562) reduces to

$$\bar{u} = u - 2u\cos^2 \tfrac{1}{2}\theta' + \sqrt{c^2 - u^2}\sin\theta' \cos\epsilon \qquad (568),$$

and there are corresponding equations giving $\bar{v}$ and $\bar{w}$.

290. Consider, for instance, a tentative solution

$$\Phi = u\phi(c) \qquad (569),$$

where $\phi(c)$ is any function of the velocity c. Since c remains unchanged by the encounter, we have $\overline{\Phi} = \bar{u}\phi(c)$, so that

$$\overline{\Phi}_1 - \Phi_1 = -(2u\cos^2\tfrac{1}{2}\theta' - \sqrt{c^2-u^2}\sin\theta'\cos\epsilon)\,\phi(c).$$

Multiplying by $\alpha d\alpha d\epsilon$ and integrating from $\alpha=0$ to $\alpha=\infty$ and from $\epsilon=0$ to $\epsilon=2\pi$, we obtain

$$I_P = -4\pi u\phi(c)\int_0^\infty \cos^2\tfrac{1}{2}\theta'\,\alpha d\alpha \quad\text{..................(570).}$$

The integral in this expression depends only on s (cf. equation (553)). It cannot be evaluated in finite terms, but for a given value of s it can of course be evaluated by quadrature. If we write

$$4\pi\int_0^\infty \cos^2\tfrac{1}{2}\theta'\,\alpha d\alpha = I_1(s) \quad\text{.....................(571),}$$

then it appears that the solution (569) gives to I_P the value $I_P = -I_1(s)\,\Phi$.

By combining solutions of this type, it appears that if ψ_1 is any function of u, v, w and c, which is linear in u, v and w, the solution

$$\Phi = \psi_1 \quad\text{...............................(572)}$$

gives

$$I_P = -I_1(s)\,\psi_1 \quad\text{..........................(573).}$$

291. Consider next a solution

$$\Phi = (u^2 - \tfrac{1}{3}c^2)\,\phi(c) \quad\text{........................(574).}$$

With the help of equation (568), we obtain

$$\overline{\Phi}_1 - \Phi_1 = (\bar{u}^2 - u^2)\,\phi(c)$$

$$= [-u^2 + (c^2-u^2)\cos^2\epsilon]\sin^2\theta' + 2u\sqrt{c^2-u^2}\sin\theta'\cos\theta'\cos\epsilon,$$

so that

$$I_P = \pi(c^2 - 3u^2)\int_0^\infty \sin^2\theta'\,\alpha d\alpha.$$

If we write

$$\pi\int_0^\infty \sin^2\theta'\,\alpha d\alpha = I_2(s) \quad\text{.......................(575),}$$

then the solution (574) is found to give to I_P a value

$$I_P = -3I_2(s)\,\Phi \quad\text{..........................(576).}$$

292. As regards the solution just obtained, imagine the axes of coordinates transformed, so that u becomes replaced by $lu + mv + nw$. Then the solution

$$\Phi = [(lu+mv+nw)^2 - \tfrac{1}{3}c^2]\,\phi(c)$$

$$= [l^2(u^2 - \tfrac{1}{3}c^2) + \ldots + 2lmuv + \ldots]\,\phi(c) \quad\text{............(577)}$$

is seen to give the value

$$I_P = -3I_2(s)\,[l^2(u^2 - \tfrac{1}{3}c^2) + \ldots + 2lmuv + \ldots] \quad\text{........(578)}$$

But we have just seen that the terms in l^2, m^2, n^2 in (577) give exactly the terms in l^2, m^2, n^2 in (578). Hence we may equate coefficients in the remaining terms and find that a solution

$$\Phi = uv\phi(c) \qquad (579)$$

gives

$$I_P = -3I_2(s)\Phi \qquad (580).$$

293. By combination of solutions of the type (574) and (579) it appears that if ψ_2 is a function of u, v, w and c, of degree 2 in u, v and w, and such that the sum of the coefficients of u^2, v^2 and w^2 vanishes, then the solution

$$\Phi = \psi_2 \qquad (581)$$

gives

$$I_P = -3I_2(s)\psi_2 \qquad (582).$$

Clearly ψ_2 regarded as a function of u, v, w must be a spherical harmonic, as also is ψ_1 in § 290.

294. Combining the two solutions ψ_1 and ψ_2, a solution

$$\Phi = \psi_1 + \psi_2 \qquad (583)$$

gives

$$I_P = -[I_1(s)\psi_1 + 3I_2(s)\psi_2],$$

and therefore makes expression (567) equal to

$$-\nu_2(m_2K)^{\frac{2}{s-1}}c^{\frac{s-5}{s-1}}[I_1(s)\psi_1 + 3I_2(s)\psi_2] \qquad (584).$$

Since this is the value of the right-hand member of equation (562), it appears that a solution of the type (583) will provide an adequate solution to equation (562) in the particular case of a steady state. For, in this particular case, the equation to be satisfied becomes

$$\begin{aligned}
&-2hm(uX + vY + wZ) + \left(u\frac{\partial}{\partial x} + v\frac{\partial}{\partial y} + w\frac{\partial}{\partial z}\right)\log\nu \\
&\quad -\left(mc^2 - \frac{3}{2h}\right)\left(u\frac{\partial h}{\partial x} + v\frac{\partial h}{\partial y} + w\frac{\partial h}{\partial z}\right) \\
&\quad -2hm\left[(u^2 - \tfrac{1}{3}c^2)\frac{\partial u_0}{\partial x} + \ldots + uv\left(\frac{\partial v_0}{\partial x} + \frac{\partial u_0}{\partial y}\right) + \ldots\right] \\
&\quad = -\nu_2(m_2K)^{\frac{2}{s-1}}c^{\frac{s-5}{s-1}}[I_1(s)\psi_1 + 3I_2(s)\psi_2] \qquad (585),
\end{aligned}$$

and this is satisfied by taking

$$\psi_1 = \frac{u\left[2hmX - \dfrac{1}{\nu}\dfrac{\partial \nu}{\partial x} + \left(mc^2 - \dfrac{3}{2h}\right)\dfrac{\partial h}{\partial x}\right] + \ldots}{\nu_2(m_2K)^{\frac{2}{s-1}}c^{\frac{s-5}{s-1}}I_1(s)} \qquad (586),$$

$$\psi_2 = \frac{2hm\left[(u^2 - \tfrac{1}{3}c^2)\dfrac{\partial u_0}{\partial x} + \ldots + uv\left(\dfrac{\partial v_0}{\partial x} + \dfrac{\partial u_0}{\partial y}\right) + \ldots\right]}{3\nu_2(m_2K)^{\frac{2}{s-1}}c^{\frac{s-5}{s-1}}I_2(s)} \qquad (587).$$

Thus a solution $\Phi = \psi_1 + \psi_2$ is adequate to satisfy equation (585), but in order also to satisfy the conditions expressed by equations (531), (533) and (535) we may require to take the more general solution of § 278,

$$\Phi = \psi_1 + \psi_2 + B + Cmu + Dmv + Emw + Fmc^2,$$

where B, C, D, E, F are constants, as yet undetermined. In the present problem, C, D, E must be the same for both kinds of gas, and the molecules of the heavier gas are supposed to be at rest. Thus C, D, E must all vanish, leaving the solution

$$\Phi = \psi_1 + \psi_2 + B + Fmc^2 \quad \text{.....................(588).}$$

295. The law of distribution of velocities has been supposed (cf. equation (523)) to be

$$\nu f(u, v, w, x, y, z) = \nu \left(\frac{hm}{\pi}\right)^{\frac{3}{2}} e^{-hm[(u-u_0)^2+(v-v_0)^2+(w-w_0)^2]} \{1 + \Phi\} \text{ ...(589),}$$

and on substituting for Φ the value given by equation (588) we obtain the law appropriate to the special case now under consideration, namely that in which deflections in the paths of molecules of the first kind are produced solely by encounters with very much heavier molecules of a second kind. We may illustrate the nature of the solution by the following examples.

Viscosity.

296. Suppose that the light gas is in a steady state, and at a uniform temperature throughout, but that it has a mass-velocity u_0, v_0, w_0 which varies slightly from point to point. The uniformity of temperature involves also uniformity of density, for otherwise inequalities of pressure would set up further mass-motions. Thus the solution appropriate to this case will be obtained by putting h and ν each constant, and $X = Y = Z = 0$ in the equations of § 294.

It is clear from equation (586) that $\psi_1 = 0$, so that ψ_2 occurs alone in the solution (588). Moreover, B and F may be omitted, since their retention would merely result in infinitesimal changes in h and ν.

Thus an adequate solution is $\Phi = \psi_2$, given by equation (587), and the substitution of this value for Φ in equation (589) will give the true law of distribution.

297. We may now calculate the pressures in the gas from formulae (418) of the last chapter, leaving out the components ϖ_{xx}, etc., which arise from intermolecular forces (cf. § 329 below). We have

$$P_{xx} = \rho \overline{U^2} = m\nu \left(\frac{hm}{\pi}\right)^{\frac{3}{2}} \iiint e^{-hm(U^2+V^2+W^2)} (1 + \psi_2)\, U^2 du\, dv\, dw.$$

Substituting for ψ_2, and writing for brevity

$$\Delta_2 = \nu_2 (m_2 K)^{\frac{2}{s-1}} I_2(s) \qquad \text{.......................(590)},$$

this becomes

$$P_{xx} = \frac{\nu}{2h} - \frac{2hm^2\nu}{\Delta_2}\left(\frac{hm}{\pi}\right)^{\frac{3}{2}} \int_0^\infty e^{-hmc^2} c^{6-\frac{s-5}{s-1}} dc$$

$$\times \frac{4\pi}{15}\left[2\frac{\partial u_0}{\partial x} - \frac{2}{3}\left(\frac{\partial u_0}{\partial x} + \frac{\partial v_0}{\partial y} + \frac{\partial w_0}{\partial z}\right)\right]$$

$$= p - \frac{4m\nu}{45\sqrt{\pi}\Delta_2} \frac{\Gamma\left(\frac{2}{s-1}+3\right)}{(hm)^{\frac{2}{s-1}+\frac{1}{2}}}\left[2\frac{\partial u_0}{\partial x} - \frac{2}{3}\left(\frac{\partial u_0}{\partial x} + \frac{\partial v_0}{\partial y} + \frac{\partial w_0}{\partial z}\right)\right] \text{...(591)},$$

and similarly

$$P_{xy} = \rho\overline{UV} = -\frac{4m\nu}{45\sqrt{\pi}\Delta_2} \frac{\Gamma\left(\frac{2}{s-1}+3\right)}{(hm)^{\frac{2}{s-1}+\frac{1}{2}}}\left[\frac{\partial v_0}{\partial x} + \frac{\partial u_0}{\partial y}\right] \qquad \text{......(592)}.$$

These equations shew the amount of the additional pressures which are superposed on to the hydrostatic pressure p by the mass-motion of the gas.

298. In a viscous fluid, having a coefficient of viscosity κ, the system of pressures at any point is given by the equations*

$$P_{xx} = p - 2\kappa\frac{\partial u_0}{\partial x} + \frac{2}{3}\kappa\left(\frac{\partial u_0}{\partial x} + \frac{\partial v_0}{\partial y} + \frac{\partial w_0}{\partial z}\right) \qquad \text{...........(593)},$$

$$P_{xy} = -\kappa\left(\frac{\partial v_0}{\partial x} + \frac{\partial u_0}{\partial y}\right) \text{..................................(594)},$$

in which u_0, v_0, w_0 are the components of the mass-velocity of the fluid. It accordingly appears that the pressures given by equations (591) and (592) will be exactly accounted for by regarding the gas as a viscous fluid having a coefficient of viscosity κ given by

$$\kappa = \frac{4m\nu}{45\sqrt{\pi}\Delta_2} \frac{\Gamma\left(\frac{2}{s-1}+3\right)}{(hm)^{\frac{2}{s-1}+\frac{1}{2}}} \qquad \text{....................(595)}.$$

Conduction of Heat.

299. Consider next a gas which is in a steady state, and devoid of mass-motion, but which is not at a uniform temperature. For simplicity suppose the gas arranged in parallel strata of equal temperature, so that the temperature is a function of z only.

* Lamb, *Hydrodynamics*, p. 512.

Turning to the general solution of § 294, it appears that in this special case ψ_2 does not occur, while, on taking $X = Y = Z = 0$, the value of ψ_1 is found to be

$$\psi_1 = \frac{u}{\Delta_1} c^{-\frac{s-5}{s-1}} \left[-\frac{1}{\nu}\frac{\partial \nu}{\partial x} + \left(mc^2 - \frac{3}{2h} \right) \frac{dh}{dx} \right] \quad \text{............(596),}$$

where

$$\Delta_1 = \nu_2 (m_2 K)^{\frac{2}{s-1}} I_1(s) \quad \text{.......................(597).}$$

An adequate solution is now seen to be $\Phi = \psi_1$, where, in order that there may be no mass-motion—*i.e.* in order that the process of conduction may not be complicated by the addition of convection—we must have

$$\iiint \nu f_0 u \psi_1 du dv dw = 0 \quad \text{........................(598).}$$

On substituting for ψ_1 in this equation, and carrying out the integrations, we obtain the relation

$$\left(\frac{1}{\nu}\frac{\partial \nu}{\partial h} + \frac{3}{2h}\frac{\partial h}{\partial x} \right) = \frac{1}{h}\frac{\partial h}{\partial x}\left(\frac{2s}{s-1} \right) \quad \text{.................(599).}$$

This is found to give as the value of ψ_1,

$$\psi_1 = \frac{u}{\Delta_1} c^{-\frac{s-5}{s-1}} \left[mc^2 - \frac{1}{h}\left(\frac{2s}{s-1} \right) \right] \frac{\partial h}{\partial x} \quad \text{..............(600).}$$

300. The translational energy of a molecule $\frac{1}{2}mc^2$ represents an amount of heat equal to $\frac{1}{2}mc^2/J$, where J is the mechanical equivalent of heat. Summing over all molecules, we find as the total flow of heat per unit area perpendicular to the axis of x, which arises from the translational energy of the molecules,

$$\frac{m}{2J}\iiint \nu A e^{-hmc^2} c^2 u \psi_1 du dv dw \quad \text{.................(601),}$$

of which the value, after integration over all values of u, v, w, is found to be

$$\frac{m^2 \nu}{3\sqrt{\pi} J} \frac{\Gamma\left(\frac{2}{s-1} + 3 \right)}{(hm)^{\frac{2}{s-1}+\frac{5}{2}}} \frac{\frac{\partial h}{\partial x}}{\Delta_1} \quad \text{........................(602).}$$

301. If ϑ is the coefficient of conduction of heat, the total flow of heat per unit area perpendicular to the axis of x is $-\vartheta \frac{\partial T}{\partial x}$, or, since $T = \frac{1}{2hR}$,

$$\frac{\vartheta}{2h^2 R}\frac{\partial h}{\partial x} \quad \text{..............................(603).}$$

If we assume that the ratio of internal energy to translational energy involved in this flow is β*, as in § 261, then expression (603) must be equal to $(1+\beta)$ times expression (602). Hence we obtain

$$\vartheta = \frac{2}{3\sqrt{\pi}} \frac{R\nu(1+\beta)}{J\Delta_1} \frac{\Gamma\left(\frac{2}{s-1}+3\right)}{(hm)^{\frac{2}{s-1}+\frac{1}{2}}} \quad \text{..............(604).}$$

Using the values of the specific heat at constant volume given by equation (471), and of the coefficient of viscosity given by equation (595), we find that this value of ϑ can be expressed in the form

$$\vartheta = \epsilon\kappa C_v \quad \text{..............................(605),}$$

where ϵ is a pure number, being given by

$$\epsilon = 5\frac{I_2(s)}{I_1(s)} \quad \text{............................(606).}$$

For the special case of $s = 5$, I_1 and I_2 have been evaluated by Maxwell (cf. below, §§ 323, 324). Using these values, equation (606) gives $\epsilon = 2{\cdot}507$.

For the case of $s = \infty$ (elastic spheres), the values of I_1 and I_2 are readily found to be $I_1(\infty) = \pi\sigma^2$ and $I_2(\infty) = \frac{1}{3}\pi\sigma^2$, whence equation (606) is found to give $\epsilon = 1{\cdot}666$.

Conduction of Heat in a Solid.

302. According to Drude's theory of metallic conduction (cf. below, § 404), conduction of heat in a solid takes place through the agency of free electrons, moving about, like the molecules of a gas, through the interstices of the solid, and having their motion checked at intervals by collisions with the atoms or molecules of the solid. The physical conditions assumed by this theory of Drude's are accordingly almost exactly the same as those which we assumed in § 286.

On this theory, then, the coefficient of conduction of heat in a solid will be given by formula (604), with β put equal to zero, to represent that the whole energy of the electron is its energy of translation.

On the same theory, the coefficient of conduction of electricity may easily be found from the analysis just given.

Conduction of Electricity.

303. We suppose the current to flow parallel to the axis of x, and to be produced by an electric force of intensity Ξ. We may then put

$$X = \frac{e}{m}\Xi$$

* This is a very debateable assumption: see § 394 below.

in equation (585); we have h, ν, u_0, v_0 and w_0 independent of x, and so find that an adequate solution is obtained by taking

$$\Phi = \psi_1 = \frac{2hmXu}{\nu_2 (m_2 K)^{\frac{2}{s-1}} c^{\frac{s-5}{s-1}} I_1(s)} = \frac{2he\Xi c^{-\frac{s-5}{s-1}} u}{\Delta_1} \qquad \text{.........(607)},$$

where Δ_1 is given by equation (598).

304. The current i is equal to the flow of electricity per unit area of cross-section perpendicular to the axis of x, and so is given by

$$i = \iiint veuf(u, v, w, x, y, z)\, du dv dw$$

$$= \iiint veu\, Ae^{-hmc^2} \frac{2he\Xi c^{-\frac{s-5}{s-1}} u}{\Delta_1} du dv dw$$

$$= \frac{4\pi}{3} hve^2 \frac{(hm)^{\frac{3}{2}}}{\pi^{\frac{3}{2}}} \frac{\Gamma\left(\frac{2}{s-1} + 2\right)}{(hm)^{\frac{2}{s-1}+2}} \frac{\Xi}{\Delta_1}$$

$$= \frac{2}{3\sqrt{\pi}} \frac{e^2 \nu}{RT} \frac{\Gamma\left(\frac{2}{s-1} + 2\right)}{(hm)^{\frac{2}{s-1}+\frac{1}{2}}} \frac{\Xi}{\Delta_1} \qquad \text{.........................(608)}.$$

305. The current i is also equal to $\sigma\Xi$, where σ is the specific conductivity of the solid, so that

$$\sigma = \frac{2}{3\sqrt{\pi}} \frac{e^2 \nu}{RT\Delta_1} \frac{\Gamma\left(\frac{2}{s-1} + 2\right)}{(hm)^{\frac{2}{s-1}+\frac{1}{2}}} \qquad \text{.................(609)}.$$

The value of the thermal conductivity ϑ is obtained on putting $\beta = 0$ in equation (604). On comparison of the two coefficients, we obtain*

$$\frac{\vartheta}{\sigma} = \frac{2s}{s-1} \frac{R^2 T}{e^2 J} \qquad \text{.........................(610)}.$$

Diffusion of a light gas into a heavy one.

306. The analysis of this chapter will give an exact solution of the problem of the diffusion of a light gas into a heavy one, when the ratio m_1/m_2 is so small that it may be neglected, and may be expected to give a good

* For elastic spheres ($s = \infty$) this reduces to $\frac{\vartheta}{\sigma} = \frac{2R^2T}{e^2J}$, an equation given by Lorentz (*The Theory of Electrons*, p. 67 and note 29). Richardson (*The Electron Theory of Matter*, p. 421), by a method similar to that of Lorentz, obtains the more general equation $\frac{\vartheta}{\sigma} = \frac{2s}{s-1} \frac{R^2T}{e^2J}$. A formula similar to that of Richardson is also given by Bohr (*Studier over Metallernes Elektrontheori; afhandling for den filosofiske Doctorgrad*, Copenhagen, 1911, p. 53).

approximation when the ratio m_1/m_2 is quite small, as it is in many physical instances, such as the diffusion of hydrogen into air, oxygen or carbon-dioxide.

For a diffusion problem (cf. equation (585)) we take u_0, v_0, w_0 all constant, and $X = Y = Z = 0$. A sufficient solution of equation (585) is accordingly

$$\Phi = \psi_1 = -\frac{\frac{u}{\nu}\frac{\partial \nu}{\partial x}}{\Delta_1 c^{\frac{s-5}{s-1}}} \quad \text{.......................(611)},$$

where Δ_1 is given by equation (598), and the diffusion is assumed to take place in directions parallel to the axis of x.

The flow of molecules parallel to the axis of x, measured per unit area per unit time, is

$$\iiint \nu \left(\frac{hm}{\pi}\right)^{\frac{3}{2}} e^{-hmc^2}\left(u - \frac{u^2}{\nu\Delta_1}\frac{\partial \nu}{\partial x} c^{-\frac{s-5}{s-1}}\right) du\, dv\, dw$$

$$= -\frac{4\pi}{3\Delta_1}\left(\frac{hm}{\pi}\right)^{\frac{3}{2}} \frac{\partial \nu}{\partial x} \frac{\frac{1}{2}\Gamma\left(\frac{2}{s-1}+2\right)}{(hm)^{\frac{2}{s-1}+2}} \quad \text{......(612)}.$$

The coefficient of diffusion, say $\mathfrak{D}$, is the coefficient of $-\frac{\partial \nu}{\partial x}$ in this expression (cf. § 412), and so is given by

$$\mathfrak{D} = \frac{2}{3\sqrt{\pi}\Delta_1} \frac{\Gamma\left(\frac{2}{s-1}+2\right)}{(hm)^{\frac{2}{s-1}+\frac{1}{2}}} \quad \text{....................(613)}.$$

Comparing this with the coefficient of viscosity κ given by equation (595) we find

$$\mathfrak{D} = \frac{\zeta\kappa}{\rho},$$

where ζ is a pure number given by

$$\zeta = \frac{15(s-1)}{4s}\frac{I_2}{I_1} \quad \text{..........................(614)}.$$

Using the values of I_2/I_1 previously given (§ 301), we find that

when $s = 5$, $$\zeta = \frac{3I_2}{I_1} = 1{\cdot}5043 \quad \text{..........................(615)},$$

when $s = \infty$, $$\zeta = \frac{15}{4}\frac{I_2}{I_1} = 1{\cdot}25 \quad \text{..........................(616)}.$$

We shall return to a discussion of the problem of diffusion in a later chapter.

MIXTURE OF MOLECULES HAVING ANY MASS-RATIO.

307. The general solution, when the mass-ratio of the two kinds of molecules involved may have any mass-ratio whatever, has been worked out by Enskog*, following a method originally suggested by Lorentz†. The equation to be solved is now the general equation (562).

Consider, as in § 290, the tentative solution

$$\Phi = u\phi_1(c),$$

and let the value of I given by equation (563) be denoted by

$$I = I(u),$$

where $I(u)$, in addition to depending on u, depends on the various constants of the gas, m_1, m_2, h, etc. Similarly, let $\Phi = v\phi_1(c)$ lead to a value $I = I(v)$ and let $\Phi = w\phi_1(c)$ lead to a value $I = I(w)$.

Then, since I is linear in Φ, it is clear that a solution

$$\Phi = (lu + mv + nw)\,\phi_1(c) \quad \text{.......................(617)}$$

will lead to

$$I = lI(u) + mI(v) + nI(w) \quad \text{....................(618)}.$$

Now let l, m, n be regarded as direction-cosines, so that $lu + mv + nw$ will be the component velocity along a certain direction (l, m, n); it follows that the solution (617) will lead to a value of I given by

$$I = I(lu + mv + nw) \quad \text{.......................(619)},$$

where $I(lu + mv + nw)$ is the same function of $lu + mv + nw$ as $I(u)$ is of u. Comparing (618) and (619), we have

$$I(lu + mv + nw) - lI(u) - mI(v) - nI(w) = 0 \quad \text{........(620)}.$$

This equation must be true for all values of l, m, n and for all values of u, v, w for which c retains the same value. The condition that c shall retain the same value may be expressed in the differential form

$$udu + vdv + wdw = 0 \quad \text{.......................(621)}.$$

Differentiating (620) with respect to u, v, w, while keeping l, m, n and c constant, we obtain

$$(ldu + mdv + ndw)\,I'(lu + mv + nw) - lduI'(u) - mdvI'(v) - ndwI'(w) = 0,$$

where accents denote differential coefficients.

This equation must be true for all values of du, dv, dw which satisfy equation (621), so that we must have

$$\frac{l}{u}[I'(lu + mv + nw) - I'(u)] = \frac{m}{v}[I'(lu + mv + nw) - I'(v)]$$

$$= \frac{n}{w}[I'(lu + mv + nw) - I'(w)].$$

* "Kinetische Theorie der Vorgänge in mässig verdünnten Gäsen" (*Inaug. Dissertation*, Upsala, 1917).

† *Vorträge über die Kinetische Theorie der Materie und der Elektrizität* (Leipzig, 1914), p. 185.

Since this is true for all values of l, m, n we must have

$$I'(lu + mv + nw) = I'(u) = I'(v) = I'(w).$$

Each function must accordingly be a function of c and the constants of the gas only, so that we must have

$$I(u) = u\chi_1(c), \text{ etc.}$$

Thus we have shewn that a solution

$$\Phi = u\phi_1(c)$$

leads to a value of I_P of the form

$$I_P = u\chi_1(c).$$

308. By almost precisely similar reasoning*, we can extend the results obtained in §§ 291, 292, and shew that a solution

$$\Phi = (u^2 - \tfrac{1}{3}c^2)\,\phi_2(c)$$

leads to a value of I_P of the form

$$I_P = (u^2 - \tfrac{1}{3}c^2)\,\chi_2(c),$$

while a solution $$\Phi = uv\phi_2(c)$$

leads to $$I_P = uv\chi_2(c).$$

309. Thus, precisely as in § 294, the solution of the general steady-state equation will be

$$\Phi = \psi_1 + \psi_2 + B + Cmu + Dmv + Emw + Fmc^2 \;\ldots\ldots\ldots(622),$$

where ψ_1, ψ_2 have the forms

$$\psi_1 = \left\{u\left[2hmX - \frac{1}{\nu}\frac{\partial \nu}{\partial x} + \left(mc^2 - \frac{3}{2h}\right)\frac{\partial h}{\partial x}\right] + \ldots\right\}\phi_1(c) \;\ldots\ldots(623),$$

$$\psi_2 = \left\{(u^2 - \tfrac{1}{3}c^2)\frac{\partial u_0}{\partial x} + \ldots + uv\left(\frac{\partial v_0}{\partial x} + \frac{\partial u_0}{\partial y}\right) + \ldots\right\}\phi_2(c) \;\ldots\ldots(624).$$

Here $\phi_1(c)$, $\phi_2(c)$ are functions of c and the constants of the gas only. As we have already seen, they cannot be evaluated in finite terms. Their expansion for a series of powers of c^2 has been considered by Enskog†, who has also calculated numerical results‡. These are given later in the present book.

* Enskog, *l.c.* pp. 39, 40. † *l.c.* Chapters II and III. ‡ *l.c.* Chapter IV.

CHAPTER IX

PHENOMENA OF A GAS NOT IN A STEADY STATE (*continued*)

MAXWELL'S THEORY.

310. THE subject of the present chapter is primarily Maxwell's theory of the behaviour of a gas in which the molecules are supposed to be point centres of force, repelling according to the inverse fifth power of the distance*. Maxwell's original theory has been greatly improved and elaborated by Kirchhoff and Boltzmann, and an important generalisation has been made by Chapman†, who has extended the theory to molecules repelling according to any inverse power of the distance.

GENERAL EQUATIONS OF TRANSFER.

311. Let Q be any function of the velocity components of a single molecule, *e.g.* the x-component of its momentum, or its energy. We proceed to form general equations expressing the transfer of Q.

At any point x, y, z let $\bar{Q}$ be the mean value of Q, so that

$$\bar{Q} = \iiint f(u, v, w)\, Q\, du\, dv\, dw \quad \text{.....................(625).}$$

The number of molecules inside a fixed rectangular parallelepiped $dx\,dy\,dz$ at this point is $\nu\,dx\,dy\,dz$, and hence ΣQ, the aggregate amount of Q inside it, is given by

$$\Sigma Q = \nu \bar{Q}\, dx\,dy\,dz \quad \text{..........................(626).}$$

We now examine the various causes of change in ΣQ. In the first place some molecules will leave the element $dx\,dy\,dz$, taking a certain amount of Q with them. It has been already found in expression (406), that the total number of molecules of class A lost to the element $dx\,dy\,dz$ in time dt is

$$dx\,dy\,dz\,dt \left(u \frac{\partial}{\partial x} + v \frac{\partial}{\partial y} + w \frac{\partial}{\partial z}\right) (\nu f)\, du\,dv\,dw,$$

and hence the total amount of Q lost by motion into and out of the element is

$$dx\,dy\,dz\,dt \iiint \left[\left(u \frac{\partial}{\partial x} + v \frac{\partial}{\partial y} + w \frac{\partial}{\partial z}\right) (\nu f)\right] Q\, du\,dv\,dw \quad \text{......(627).}$$

* "On the Viscosity of Internal Friction of Air and other Gases," *Collected Works*, II. p. 1.

† *Phil. Trans.* 216 A. (1915), p. 279 and 217 A. (1916), p. 115.

If we write

$$\iiint uQf(u, v, w)\, du\, dv\, dw = \overline{uQ}, \text{ etc.},$$

so that $\overline{uQ}$ is the mean value of uQ averaged over all the molecules in the neighbourhood of the point x, y, z, expression (627) can be put in the form

$$dx\,dy\,dz\,dt\left[\frac{\partial}{\partial x}(\nu\overline{uQ}) + \frac{\partial}{\partial y}(\nu\overline{vQ}) + \frac{\partial}{\partial z}(\nu\overline{wQ})\right] \quad\text{.........(628).}$$

A second cause of change in ΣQ is the action of external forces on the molecules. For any single molecule, we have

$$\frac{\partial Q}{\partial t} = \frac{\partial Q}{\partial u}\frac{du}{dt} + \frac{\partial Q}{\partial v}\frac{dv}{dt} + \frac{\partial Q}{\partial w}\frac{dw}{dt} = \frac{1}{m}\left(X\frac{\partial Q}{\partial u} + Y\frac{\partial Q}{\partial v} + Z\frac{\partial Q}{\partial w}\right),$$

where X, Y, Z are the components of the external force acting on the molecule, so that in time dt the total value of ΣQ experiences an increase

$$dx\,dy\,dz\,dt\,\frac{\nu}{m}\left[X\overline{\left(\frac{\partial Q}{\partial u}\right)} + Y\overline{\left(\frac{\partial Q}{\partial v}\right)} + Z\overline{\left(\frac{\partial Q}{\partial w}\right)}\right] \quad\text{............(629),}$$

where again the bar over a quantity indicates an average taken over all molecules.

Lastly, ΣQ may be changed by collisions between molecules. If Q is any one of the quantities which have previously been denoted by $\chi_1, \chi_2, \dots \chi_5$, namely the mass, energy, and the three components of momentum of a molecule, there is no such change, but if Q is any other function of the velocities such changes will occur. In general let us denote the increase in ΣQ which is caused in the element $dx\,dy\,dz$ by collisions in time dt by

$$dx\,dy\,dz\,dt\,\Delta Q \quad\text{..............................(630).}$$

Expressions (628), (629) and (630) now contain between them the effect of all possible changes in ΣQ. The value of ΣQ is however given by expression (626), so that the change in ΣQ in time dt will be

$$\frac{d}{dt}(\nu\overline{Q})\, dx\,dy\,dz\,dt.$$

Comparing the two different values which have been obtained for this change, we have

$$\begin{aligned}\frac{d}{dt}(\nu\overline{Q}) = &-\left[\frac{\partial}{\partial x}(\nu\overline{uQ}) + \frac{\partial}{\partial y}(\nu\overline{vQ}) + \frac{\partial}{\partial z}(\nu\overline{wQ})\right]\\ &+\frac{\nu}{m}\left[X\overline{\left(\frac{\partial Q}{\partial u}\right)} + Y\overline{\left(\frac{\partial Q}{\partial v}\right)} + Z\overline{\left(\frac{\partial Q}{\partial w}\right)}\right] + \Delta Q \quad\text{.........(631).}\end{aligned}$$

312. If we put $Q = 1$ in this equation, so that $\overline{Q} = 1$, the equation becomes

$$\begin{aligned}\frac{d\nu}{dt} &= -\left[\frac{\partial}{\partial x}(\nu\overline{u}) + \frac{\partial}{\partial y}(\nu\overline{v}) + \frac{\partial}{\partial z}(\nu\overline{w})\right]\\ &= -\left[\frac{\partial}{\partial x}(\nu u_0) + \frac{\partial}{\partial y}(\nu v_0) + \frac{\partial}{\partial z}(\nu w_0)\right] \quad\text{...............(632),}\end{aligned}$$

the equation of continuity already obtained in equation (409).

If we multiply both sides of this equation by $\bar{Q}$, and subtract from corresponding sides of equation (631), we obtain as a new form for the general equation,

$$\nu \frac{d\bar{Q}}{dt} = \Sigma \left[\bar{Q} \frac{\partial}{\partial x}(\nu u_0) - \frac{\partial}{\partial x}(\nu \overline{uQ}) + \frac{\nu}{m} X \left(\overline{\frac{\partial Q}{\partial u}} \right) \right] + \Delta Q \quad \text{......(633)},$$

where Σ denotes summation with respect to x, y and z.

313. Let us now write $\quad u = u_0 + \mathsf{u}$, etc.

so that u, v, w are components of molecular velocity. Then

$$\bar{u} = u_0, \quad \bar{\mathsf{u}} = 0, \quad \overline{uQ} = \overline{u_0 Q + \mathsf{u}Q} = u_0 \bar{Q} + \overline{\mathsf{u}Q}.$$

Hence

$$\bar{Q} \frac{\partial}{\partial x}(\nu u_0) - \frac{\partial}{\partial x}(\nu \overline{uQ}) = \bar{Q} \frac{\partial}{\partial x}(\nu u_0) - \frac{\partial}{\partial x}(\nu u_0 \bar{Q}) - \frac{\partial}{\partial x}(\nu \overline{\mathsf{u}Q})$$

$$= -\nu u_0 \frac{\partial \bar{Q}}{\partial x} - \frac{\partial}{\partial x}(\nu \overline{\mathsf{u}Q}).$$

If Q is now expressed as a function of u_0, u, etc.

$$\frac{\partial Q}{\partial u} = \frac{\partial Q}{\partial (u_0 + \mathsf{u})} = \frac{\partial Q}{\partial u_0},$$

so that

$$\left(\overline{\frac{\partial Q}{\partial u}} \right) = \left(\overline{\frac{\partial Q}{\partial u_0}} \right) = \frac{\partial \bar{Q}}{\partial u_0}.$$

Making these substitutions, equation (633) becomes

$$\nu \frac{d\bar{Q}}{dt} = \Sigma \left[-\nu u_0 \frac{\partial \bar{Q}}{\partial x} - \frac{\partial}{\partial x}(\nu \overline{\mathsf{u}Q}) + \frac{\nu}{m} X \frac{\partial \bar{Q}}{\partial u_0} \right] + \Delta Q \quad \text{.........(634)},$$

or, again, if we write

$$\frac{D}{Dt} = \frac{d}{dt} + u_0 \frac{\partial}{\partial x} + v_0 \frac{\partial}{\partial y} + w_0 \frac{\partial}{\partial z} \quad \text{..................(635)},$$

so that $\frac{D}{Dt}$ denotes differentiation following an element of gas in its motion, the equation becomes

$$\nu \frac{D\bar{Q}}{Dt} = \Sigma \left[-\frac{\partial}{\partial x}(\nu \overline{\mathsf{u}Q}) + \frac{\nu}{m} X \frac{\partial \bar{Q}}{\partial u_0} \right] + \Delta Q \quad \text{...........(636)}.$$

Equations of transfer for a single gas.

314. When there is only one kind of gas present, a transformation of this equation can be effected.

Put $Q = u$ in equation (636), then since there is only one type of molecule $\Delta Q = 0$. Again we have $\bar{Q} = u_0$, so that

$$\frac{\partial \bar{Q}}{\partial u_0} = 1, \quad \frac{\partial \bar{Q}}{\partial v_0} = \frac{\partial \bar{Q}}{\partial w_0} = 0.$$

Also
$$\overline{\mathsf{U}Q} = \overline{\mathsf{U}\,(\mathsf{U} + u_0)} = \overline{\mathsf{U}^2},$$
$$\overline{\mathsf{V}Q} = \overline{\mathsf{V}\,(\mathsf{U} + u_0)} = \overline{\mathsf{UV}}, \text{ etc.}$$

The equation now becomes

$$\nu \frac{Du_0}{Dt} = -\left[\frac{\partial}{\partial x}(\nu\overline{\mathsf{U}^2}) + \frac{\partial}{\partial y}(\nu\overline{\mathsf{UV}}) + \frac{\partial}{\partial z}(\nu\overline{\mathsf{UW}})\right] + \frac{\nu}{m}X \quad\text{.........(637)},$$

which is identical with our previous equation of motion (417), and there are of course two similar equations obtained from equation (636) by putting $Q = v$ and $Q = w$ respectively.

From these three equations and the general equation (636) we can obtain an equation which does not contain X, Y or Z, and which is therefore true for a gas independently of the action of external forces. Effecting the elimination of X, Y and Z, this equation is found to be

$$\nu\left[\frac{D\bar{Q}}{Dt} - \frac{\partial\bar{Q}}{\partial u_0}\frac{Du_0}{Dt} - \frac{\partial\bar{Q}}{\partial v_0}\frac{Dv_0}{Dt} - \frac{\partial\bar{Q}}{\partial w_0}\frac{Dw_0}{Dt}\right]$$
$$= \Sigma\left[-\frac{\partial}{\partial x}(\nu\overline{\mathsf{U}Q}) + \frac{\partial\bar{Q}}{\partial u_0}\left\{\frac{\partial}{\partial x}(\nu\overline{\mathsf{U}^2}) + \frac{\partial}{\partial y}(\nu\overline{\mathsf{UV}}) + \frac{\partial}{\partial z}(\nu\overline{\mathsf{UW}})\right\}\right] + \Delta Q \quad\text{...(638)}.$$

In general $\bar{Q}$ will be a function of u_0, v_0, w_0 and of the mean values $\overline{\mathsf{U}^2}$, $\overline{\mathsf{UV}}$, etc. Clearly the bracket on the left-hand side will be the value of $D\bar{Q}/Dt$ calculated upon the assumption that $\overline{\mathsf{U}^2}$, $\overline{\mathsf{UV}}$, etc. are functions of the time, but that u_0, v_0, w_0 are independent of the time.

Special values for Q.

315. The equation just obtained is the general equation expressing the transfer of Q when there is only one kind of gas. Maxwell uses this same equation (638) for the investigation of the phenomena of viscosity and conduction of heat, regarding viscosity as a transfer of momentum and conduction of heat as a transfer of energy.

The values of Q must, however, be different for the two phenomena, and we shall proceed by finding the special forms assumed by equation (638) when Q has the requisite special values.

I. $Q = u^2$.

316. Let us first put

$$Q = u^2 = u_0^2 + 2u_0\mathsf{U} + \mathsf{U}^2.$$

Then
$$\bar{Q} = u_0^2 + \overline{\mathsf{U}^2},$$

and
$$\overline{\mathsf{U}Q} = \overline{\mathsf{U}^3} + 2u_0\overline{\mathsf{U}^2};\quad \overline{\mathsf{V}Q} = \overline{\mathsf{U}^2\mathsf{V}} + 2u_0\overline{\mathsf{UV}}, \text{ etc.}$$

Equation (638) now becomes

$$\nu\frac{D}{Dt}(\overline{\mathsf{U}^2}) = -\frac{\partial}{\partial x}\{\nu\overline{\mathsf{U}^3} + 2u_0\nu\overline{\mathsf{U}^2}\} - \frac{\partial}{\partial y}\{\nu\overline{\mathsf{U}^2\mathsf{V}} + 2u_0\nu\overline{\mathsf{UV}}\} - \frac{\partial}{\partial z}\{\ldots\}$$

$$+ 2u_0\left\{\frac{\partial}{\partial x}(\nu\overline{\mathsf{U}^2}) + \frac{\partial}{\partial y}(\nu\overline{\mathsf{UV}}) + \frac{\partial}{\partial z}(\nu\overline{\mathsf{UW}})\right\} + \Delta u^2$$

$$= -\frac{\partial}{\partial x}(\nu\overline{\mathsf{U}^3}) - \frac{\partial}{\partial y}(\nu\overline{\mathsf{U}^2\mathsf{V}}) - \frac{\partial}{\partial z}(\nu\overline{\mathsf{U}^2\mathsf{W}})$$

$$- 2\nu\left(\overline{\mathsf{U}^2}\frac{\partial u_0}{\partial x} + \overline{\mathsf{UV}}\frac{\partial u_0}{\partial y} + \overline{\mathsf{UW}}\frac{\partial u_0}{\partial z}\right) + \Delta u^2 \quad \ldots\ldots\ldots\ldots\ldots\ldots(639).$$

317. We shall require this equation in its present exact form when we come to discuss the conduction of heat. For other problems it is adequate to obtain a first approximation by neglecting deviations from Maxwell's Law of distribution of velocities. We accordingly take

$$\overline{\mathsf{U}^2} = \overline{\mathsf{V}^2} = \overline{\mathsf{W}^2} = q,$$

$$\overline{\mathsf{UV}} = \overline{\mathsf{VW}} = \overline{\mathsf{WU}} = 0.$$

The general equation (638) now reduces to the simpler form

$$\nu\left[\frac{D\overline{Q}}{Dt} - \frac{\partial\overline{Q}}{\partial u_0}\frac{Du_0}{Dt} - \frac{\partial\overline{Q}}{\partial v_0}\frac{Dv_0}{Dt} - \frac{\partial\overline{Q}}{\partial w_0}\frac{Dw_0}{Dt}\right]$$

$$= \Sigma\left[-\frac{\partial}{\partial x}(\nu\overline{\mathsf{U}Q}) + \frac{\partial\overline{Q}}{\partial u_0}\frac{\partial}{\partial x}(\nu q)\right] + \Delta Q \ldots\ldots\ldots(640),$$

while the special equation (639) becomes

$$\nu\frac{Dq}{Dt} = -2\nu q\frac{\partial u_0}{\partial x} + \Delta u^2 \quad \ldots\ldots\ldots\ldots\ldots\ldots\ldots(641),$$

and there are of course two similar equations for u^2 and v^2.

Let us now assume the molecules to be point-centres of force, so that their only kinetic energy is energy of translation; then from the conservation of energy, $\Delta(u^2 + v^2 + w^2)$ must vanish, so that on adding together the three equations of the type (641) we obtain

$$3\frac{Dq}{Dt} = -2q\left(\frac{\partial u_0}{\partial x} + \frac{\partial v_0}{\partial y} + \frac{\partial w_0}{\partial z}\right) \quad \ldots\ldots\ldots\ldots\ldots\ldots(642).$$

On elimination of $\frac{Dq}{Dt}$ between this and equation (641) we obtain

$$\nu q\left[2\frac{\partial u_0}{\partial x} - \frac{2}{3}\left(\frac{\partial u_0}{\partial x} + \frac{\partial v_0}{\partial y} + \frac{\partial w_0}{\partial z}\right)\right] = \Delta u^2 \ldots\ldots\ldots\ldots\ldots(643).$$

318. *Adiabatic Motion.* Incidentally we may notice that the equation of continuity (632) may be expressed in the form (cf. equation (635))

$$\frac{D\nu}{Dt} + \nu\left(\frac{\partial u_0}{\partial x} + \frac{\partial v_0}{\partial y} + \frac{\partial w_0}{\partial z}\right),$$

and this, in combination with (642), gives

$$\frac{1}{q}\frac{Dq}{Dt} - \frac{2}{3}\frac{1}{\nu}\frac{D\nu}{Dt} = 0.$$

On integration, it appears that $\frac{D}{Dt}(q\nu^{-\frac{2}{3}})$ is zero, so that as we follow an element of gas in its motion, $q\nu^{-\frac{2}{3}}$ remains constant. The value of q is however $\frac{1}{3}C^2$ or p/ρ, so that this result simply expresses that, following the motion of an element of gas, we must have

$$p\rho^{-\frac{5}{3}} = \text{constant} \quad \text{..........................(644).}$$

This is the particular case of the general adiabatic law of § 242, obtained by putting $\gamma = \frac{5}{3}$, this being the value appropriate to point-centres of force for which all the energy is translational.

II. $Q = uv$.

319. We next put $Q = uv$ in equation (640). We have

$$\overline{Q} = \overline{(u_0 + \mathsf{u})(v_0 + \mathsf{v})} = u_0 v_0, \quad \overline{\mathsf{u}Q} = v_0 q, \quad \overline{\mathsf{v}Q} = u_0 q, \quad \overline{\mathsf{w}Q} = 0,$$

$$\frac{d\overline{Q}}{du_0} = v_0, \quad \frac{d\overline{Q}}{dv_0} = u_0, \quad \frac{d\overline{Q}}{dw_0} = 0,$$

so that equation (640) becomes

$$0 = -\left[\frac{\partial}{\partial x}(\nu v_0 q) + \frac{\partial}{\partial y}(\nu u_0 q)\right] + v_0 \frac{\partial}{\partial x}(\nu q) + u_0 \frac{\partial}{\partial y}(\nu q) + \Delta(uv),$$

giving, upon simplification,

$$\nu q \left(\frac{\partial v_0}{\partial x} + \frac{\partial u_0}{\partial y}\right) = \Delta(uv) \quad \text{......................(645).}$$

III. $Q = u(u^2 + v^2 + w^2)$.

320. Lastly, in equation (640) we put

$$Q = u(u^2 + v^2 + w^2)$$
$$= (u_0 + \mathsf{u})(u_0^2 + v_0^2 + w_0^2 + 2u_0\mathsf{u} + 2v_0\mathsf{v} + 2w_0\mathsf{w} + \mathsf{u}^2 + \mathsf{v}^2 + \mathsf{w}^2),$$

so that
$$\overline{Q} = u_0(u_0^2 + v_0^2 + w_0^2) + 5u_0 q,$$

$$\overline{\mathsf{u}Q} = (3u_0^2 + v_0^2 + w_0^2)q + \overline{\mathsf{u}^2(\mathsf{u}^2 + \mathsf{v}^2 + \mathsf{w}^2)}, \quad \overline{\mathsf{v}Q} = 2u_0 v_0 q, \quad \overline{\mathsf{w}Q} = 2u_0 w_0 q,$$

$$\frac{\partial \overline{Q}}{\partial u_0} = 3u_0^2 + v_0^2 + w_0^2 + 5q, \quad \frac{\partial \overline{Q}}{\partial v_0} = 2u_0 v_0, \quad \frac{\partial \overline{Q}}{\partial w_0} = 2u_0 w_0.$$

From Maxwell's Law it is easily found that

$$\overline{\mathsf{u}^4} = \frac{3}{4h^2m^2} = 3q^2, \quad \overline{\mathsf{u}^2\mathsf{v}^2} = \overline{\mathsf{u}^2\mathsf{w}^2} = q^2, \quad \overline{\mathsf{u}^2(\mathsf{u}^2 + \mathsf{v}^2 + \mathsf{w}^2)} = 5q^2.$$

Hence on putting $Q = u(u^2 + v^2 + w^2)$ in equation (640), we obtain

$$5u_0\nu\frac{Dq}{Dt} = -\frac{\partial}{\partial x}\{\nu(3u_0^2 + v_0^2 + w_0^2 + 5q)q\} - \frac{\partial}{\partial y}(2u_0v_0\nu q) - \frac{\partial}{\partial z}(2u_0w_0\nu q)$$
$$+ (3u_0^2 + v_0^2 + w_0^2 + 5q)\frac{\partial}{\partial x}(\nu q) + 2u_0v_0\frac{\partial}{\partial y}(\nu q) + 2u_0w_0\frac{\partial}{\partial z}(\nu q)$$
$$+ \Delta u(u^2 + v^2 + w^2)$$
$$= -q\left[\frac{\partial}{\partial x}(3u_0^2 + v_0^2 + w_0^2 + 5q) + \frac{\partial}{\partial y}(2u_0v_0) + \frac{\partial}{\partial z}(2u_0w_0)\right]$$
$$+ \Delta u(u^2 + v^2 + w^2).$$

If we substitute for Dq/Dt from equation (642) and further simplify, this reduces to

$$2qu_0\left[2\frac{\partial u_0}{\partial x} - \frac{2}{3}\left(\frac{\partial u_0}{\partial x} + \frac{\partial v_0}{\partial y} + \frac{\partial w_0}{\partial z}\right)\right]$$
$$+ 2qv_0\left(\frac{\partial v_0}{\partial x} + \frac{\partial u_0}{\partial y}\right) + 2qw_0\left(\frac{\partial w_0}{\partial x} + \frac{\partial u_0}{\partial z}\right) + 5q\frac{\partial q}{\partial x} = \Delta u(u^2 + v^2 + w^2) \quad \ldots(646).$$

Calculations of ΔQ.

321. The various equations we have obtained depend on the values of ΔQ, where by definition ΔQ is such that the increase in ΣQ caused by collisions of all kinds is ΔQ per unit volume per unit time. To evaluate ΔQ we must return to the dynamics of collisions, which were worked out in detail in §§ 279—284 of the last chapter for the general law of force $m_1m_2Kr^{-s}$.

As before, let there be two kinds of molecules, of masses m_1 and m_2 respectively, and let ΔQ for the gas of the first kind be divided into two parts. Let

$$\Delta Q = \Delta_{11}Q + \Delta_{12}Q,$$

where $\Delta_{11}Q$, $\Delta_{12}Q$ denote the changes in ΣQ caused by collisions with molecules of the first and second kinds respectively.

Let $[Q]$ denote the change in the value of Q for a molecule of mass m_1 and velocity-components u, v, w produced by collision with a molecule of mass m_2 and velocity-components u', v', w'.

The number of collisions of this type is given by formula (516) on p. 210. Multiplying by $[Q]$, and integrating over all collisions which occur per unit time per unit volume, we find

$$\Delta_{12}Q = \iiiiiiii [Q]\,\nu_1\nu_2 f_1(u, v, w) f_2(u', v', w')\,du\,dv\,dw\,du'\,dv'\,dw'\,V p\,dp\,d\epsilon,$$

where all the symbols have the same meaning as before. Using the value for $p\,dp\,d\epsilon$ given in § 311, this becomes

$$\Delta_{12}Q = \nu_1\nu_2\{(m_1 + m_2)K\}^{\frac{2}{s-1}} \iiiiii f_1(u, v, w) f_2(u', v', w') V^{1-\frac{4}{s-1}} J_P$$
$$\times\, du\,dv\,dw\,du'\,dv'\,dw' \quad \ldots(647),$$

where $$J_P = \int_{a=0}^{a=\infty} \int_{\epsilon=0}^{\epsilon=2\pi} [Q]\, a\, da\, d\epsilon \quad \text{.....................(648).}$$

The value of $[Q]$ for any single collision is given at once by the equations of § 284, which can be written in the form

$$\bar{u} = u + \frac{m_2}{m_1 + m_2} [2(u' - u)\cos^2 \tfrac{1}{2}\theta' + \sqrt{V^2 - (u' - u)^2} \sin\theta' \cos(\epsilon - \omega_1)] \quad \text{......(649),}$$

$$\bar{v} = v + \frac{m_2}{m_1 + m_2} [2(v' - v)\cos^2 \tfrac{1}{2}\theta' + \sqrt{V^2 - (v' - v)^2} \sin\theta' \cos(\epsilon - \omega_2)] \quad \text{......(650),}$$

and a similar equation for $\bar{w}$. Here ω_1 is equal to zero, being introduced merely to maintain symmetry, while from equation (558),

$$\cos\omega_2 = -\frac{(u' - u)(v' - v)}{\sqrt{(V^2 - (u' - u)^2)(V^2 - (v' - v)^2)}} \quad \text{...........(651),}$$

and there is a similar equation for $\cos\omega_3$.

The value of $[Q]$ being obtained from these equations, we can without trouble evaluate the integral J_P given by equation (648).

The further integral (647) involves the laws of distribution f_1, f_2 and so cannot in general be evaluated unless these laws of distribution are known. In the last chapter it was seen that the laws of distribution assumed a specially simple form in the special case of $s = 5$.

LAW OF THE INVERSE FIFTH POWER.

322. Maxwell's original theory was confined entirely to the particular law of force $s = 5$. For this law of force the factor $V^{1-\frac{4}{s-1}}$ disappears entirely from the equation (647) for $\Delta_{12}Q$, and the equation becomes

$$\Delta_{12}Q = \nu_1\nu_2\sqrt{(m_1 + m_2)K} \iiiint\!\!\iint J_P (f_1\, du\, dv\, dw)(f_2\, du'\, dv'\, dw') \quad \text{...(652).}$$

We notice that the integral on the right is simply the value of J_P averaged over all molecules of the first kind and also over all molecules of the second kind. Owing to this simplification it will never be necessary to introduce actual expressions for the law of distribution.

We now proceed to calculate the values of ΔQ for certain values of Q when the law of force is the special law of the inverse fifth power.

Calculation of Δu.

323. Since Σu is unchanged by collisions between molecules of the same kind, it is clear that $\Delta_{11}u$ will vanish, and we shall have $\Delta u = \Delta_{12}u$.

For the value $Q=u$, $[Q]$ or $\bar{u}-u$ is given directly by equation (649). On calculating J_P by equation (648), the term in $\cos(\epsilon-\omega_1)$ disappears on integration with respect to ϵ, and we are left with

$$J_P = 2\pi\int_{\alpha=0}^{\alpha=\infty} \frac{2m_2}{m_1+m_2}(u'-u)\cos^2\tfrac{1}{2}\theta'\,\alpha d\alpha.$$

Following Maxwell, we write

$$4\pi\int_{\alpha=0}^{\alpha=\infty}\cos^2\tfrac{1}{2}\theta'\,\alpha d\alpha = A_1 \quad\text{.....................(653)},$$

the quantity A_1 being a pure number, and identical with the quantity denoted by $I_1(5)$ in the last chapter. Maxwell gives tables for the evaluation of A_1 by quadrature in his original paper, and finds

$$A_1 = 2{\cdot}6595 \quad\text{.............................(654)}.$$

We now have $$J_P = \frac{m_2}{m_1+m_2}A_1(u'-u),$$

and equation (652) gives immediately

$$\Delta u = \Delta_{12}u = \nu_1\nu_2 m_2\sqrt{\frac{K}{m_1+m_2}}\,A_1(u_{02}-u_{01}) \quad\text{........(655)},$$

where u_{02} is the average of u' for all the molecules of the second gas, and so is the mass-velocity of the second gas, and similarly for u_{01}.

The value of Δu is required for the problem of diffusion in which two gases are necessarily present. In the remaining calculations for ΔQ, which are needed for problems of viscosity and conduction of heat, we shall suppose that only one gas is present, so that we take $m_1=m_2=m$.

Calculation of Δu^2 and Δuv.

324. Putting $m_1=m_2$, the value of $[u^2]$ or $\bar{u}^2-u^2$ is found from equation (649) to be

$$[u^2] = \{u+(u'-u)\cos^2\tfrac{1}{2}\theta' + \tfrac{1}{2}\sqrt{V^2-(u'-u)^2}\sin\theta'\cos\epsilon\}^2 - u^2,$$

whence we obtain (equation (648))

$$J_P = 2\pi\int_0^\infty \{2u(u'-u)\cos^2\tfrac{1}{2}\theta' + (u'-u)^2\cos^4\tfrac{1}{2}\theta' + \tfrac{1}{8}(V^2-(u'-u)^2)\sin^2\theta'\}\,\alpha d\alpha$$

$$= 2\pi\int_0^\infty \{(u'^2-u^2)\cos^2\tfrac{1}{2}\theta' + \tfrac{1}{8}\{-2(u'-u)^2+(v'-v)^2+(w'-w)^2\}\sin^2\theta'\}\,\alpha d\alpha$$

$$\text{......(656)}.$$

Equation (647) now gives

$$\Delta u^2 = 2\pi\nu^2\sqrt{2mK}\,.\,\tfrac{1}{4}(-2\overline{U^2}+\overline{V^2}+\overline{W^2})\int_0^\infty \sin^2\theta'\,\alpha d\alpha \quad\text{......(657)}.$$

Maxwell writes $$\pi \int_0^\infty \sin^2 \theta' \alpha d\alpha = A_2 \quad \text{.......................(658)},$$

this being identical with the I_2 (5) of § 291, and finds

$$A_2 = 1{\cdot}3682 \quad \text{..............................(659)}.$$

The value of Δu^2 is now

$$\Delta u^2 = \tfrac{1}{2} \nu^2 \sqrt{2mK} A_2 (-2\overline{\mathrm{U}^2} + \overline{\mathrm{V}^2} + \overline{\mathrm{W}^2}) \quad \text{..............(660)}.$$

325. We also require Δuv, but this is more easily found by transformation of axes than by direct calculation.

Let us write $lx + my + nz$ instead of x, so that we write $lu + mv + nw$ instead of u. The left-hand member of equation (660) becomes

$$l^2 \Delta u^2 + 2lm \Delta uv + \ldots.$$

The bracket on the right-hand may be written

$$\overline{\mathrm{U}^2 + \mathrm{V}^2 + \mathrm{W}^2} - 3\overline{\mathrm{U}^2},$$

and therefore transforms into

$$(l^2 + m^2 + n^2) \{\overline{\mathrm{U}^2 + \mathrm{V}^2 + \mathrm{W}^2} - 3\,\overline{(l\mathrm{U} + m\mathrm{V} + n\mathrm{W})^2}\}.$$

As in § 318, we may equate coefficients of $2lm$, and obtain at once

$$\Delta uv = -\tfrac{3}{2} \nu^2 \sqrt{2mK}\, A_2 \overline{\mathrm{UV}} \quad \text{.....................(661)}.$$

Calculation of $\Delta \{u (u^2 + v^2 + w^2)\}$.

326. To evaluate $\Delta u (u^2 + v^2 + w^2)$, we need the complete system of three equations of the type of (649).

On putting $m_1 = m_2$, equation (649) becomes

$$\bar{u} = u + a + a' \cos (\epsilon - \omega_1) \quad \text{.....................(662)},$$

in which

$$a = (u' - u) \cos^2 \frac{\theta'}{2}, \quad a' = \tfrac{1}{2} \sqrt{V^2 - (u' - u)^2} \sin \theta',$$

and $\omega_1 = 0$. Similarly, from the two remaining equations,

$$\bar{v} = v + b + b' \cos (\epsilon - \omega_2) \quad \text{.....................(663)},$$

$$\bar{w} = w + c + c' \cos (\epsilon - \omega_3) \quad \text{.....................(664)},$$

where b, b' and c, c' are obtained from a, a' by replacing u by v and w respectively, and, by equation (651) and its companion,

$$\cos \omega_2 = \frac{(u' - u)(v' - v)}{4a'b'} \sin^2 \theta',$$

$$\cos \omega_3 = \frac{(u' - u)(w' - w)}{4a'c'} \sin^2 \theta'.$$

Squaring the system of three equations (662), (663) and (664), and adding corresponding sides,

$$\bar{u}^2 + \bar{v}^2 + \bar{w}^2 = \Sigma (u+a)^2 + 2\Sigma (u+a) a' \cos(\epsilon - \omega_1) + \Sigma a'^2 \cos^2(\epsilon - \omega_1),$$

so that

$$\begin{aligned} \bar{u}(\bar{u}^2 + \bar{v}^2 + \bar{w}^2) = (u+a)\Sigma(u+a)^2 + 2(u+a)\Sigma(u+a)a'\cos(\epsilon-\omega_1) \\ + (u+a)\Sigma a'^2\cos^2(\epsilon-\omega_1) \\ + a'\cos\epsilon\Sigma(u+a)^2 + 2a'\cos\epsilon\Sigma(u+a)a'\cos(\epsilon-\omega_1) \\ + a'\cos\epsilon\Sigma a'^2\cos^2(\epsilon-\omega_1). \end{aligned}$$

Hence, on integration with respect to ϵ,

$$\frac{1}{2\pi}\int_0^\epsilon [u(u^2+v^2+w^2)]\, d\epsilon$$

$$= -u\Sigma u^2 + (u+a)(\Sigma(u+a)^2 + \tfrac{1}{2}\Sigma a'^2) + 2a'\Sigma(u+a)a'\cos\omega_1 \quad \text{...(665)}.$$

To simplify this, we notice that

$$\Sigma a a' \cos\omega_1 = 0,$$

$$\Sigma(u+a)^2 = \Sigma u^2 + 2\Sigma u(u'-u)\cos^2\tfrac{1}{2}\theta' + V^2\cos^4\tfrac{1}{2}\theta',$$

$$\tfrac{1}{2}\Sigma a'^2 = \tfrac{1}{8}\sin^2\theta'\Sigma(V^2 - (u'-u)^2) = V^2\sin^2\tfrac{1}{2}\theta'\cos^2\tfrac{1}{2}\theta';$$

so that

$$\Sigma(u+a)^2 + \tfrac{1}{2}\Sigma a'^2 = \Sigma u^2 + \Sigma(u'^2 - u^2)\cos^2\tfrac{1}{2}\theta'.$$

It is now clear that the right-hand side of equation (665) can be expressed as the sum of two terms multiplied by $\cos^2\frac{1}{2}\theta'$ and $\sin^2\theta'$ respectively. Simplified as far as possible, we find for this expression the value

$$\begin{aligned} &(u'\Sigma u'^2 - u\Sigma u^2)\cos^2\tfrac{1}{2}\theta' \\ &\quad + \tfrac{1}{4}\sin^2\theta'\,[u(2\Sigma u'^2 - \Sigma u^2 - \Sigma uu') + u'(2\Sigma u^2 - \Sigma u'^2 - \Sigma uu')]. \end{aligned}$$

After integrating with respect to α, and averaging over all values of the velocities, the first line vanishes, while from the second we obtain, by the use of equation (647),

$$\begin{aligned} \Delta Q = \tfrac{1}{2} A_2 \nu^2 \sqrt{2mK} \\ \{4u_0(\overline{u^2+v^2+w^2}) - 2\overline{u(u^2+v^2+w^2)} - 2(u_0\overline{u^2} + v_0\overline{uv} + w_0\overline{uw})\}. \end{aligned}$$

Replacing u by $u_0 + \mathsf{U}$, etc., and also writing

$$\eta = \tfrac{3}{2}\sqrt{2mK}\,A_2 \quad \text{..........................(666)},$$

this equation becomes

$$\Delta u(u^2+v^2+w^2)$$

$$= \tfrac{4}{3}\eta\nu^2\{u_0(\overline{\mathsf{U}^2} + \overline{\mathsf{V}^2} + \overline{\mathsf{W}^2}) - 3(u_0\overline{\mathsf{U}^2} + v_0\overline{\mathsf{UV}} + w_0\overline{\mathsf{UW}}) + 3\overline{\mathsf{U}(\mathsf{U}^2+\mathsf{V}^2+\mathsf{W}^2)}\}\text{...(667)}.$$

Final Equations.

327. Substituting into this equation the value which was obtained for Δu^2 in equation (660), we have

$$\tfrac{1}{3}\eta\nu(\overline{\mathrm{V}^2} + \overline{\mathrm{W}^2} - 2\overline{\mathrm{U}^2}) = q\left\{2\frac{\partial u_0}{\partial x} - \frac{2}{3}\left(\frac{\partial u_0}{\partial x} + \frac{\partial v_0}{\partial y} + \frac{\partial w_0}{\partial z}\right)\right\} \quad \text{......(668)}.$$

Similarly, substituting the value obtained for Δuv from equation (661) into equation (645), we have

$$\eta\nu\overline{\mathrm{UV}} = -q\left(\frac{\partial v_0}{\partial x} + \frac{\partial u_0}{\partial y}\right) \quad \text{........................(669)},$$

and lastly, substituting the value of $\Delta u(u^2 + v^2 + w^2)$ just obtained into equation (646), we have, after simplification from equations (668) and (669),

$$\tfrac{2}{3}\eta\nu\overline{\mathrm{U}(\mathrm{U}^2 + \mathrm{V}^2 + \mathrm{W}^2)} = -5q\frac{\partial q}{\partial x} \quad \text{..................(670)}.$$

We have now obtained a sufficient amount of mathematical working material, and proceed to the discussion of physical phenomena.

Time of relaxation.

328. Let us in the first place consider a gas in which the law of distribution is initially some law other than that of Maxwell. Our equations enable us to determine the rate at which the gas approaches the steady state. We take the simplest case, and suppose that there is no mass-motion, so that $u_0 = v_0 = w_0 = 0$; we also suppose that the law of distribution is the same throughout the gas, so that $\overline{\mathrm{U}^2}$, $\overline{\mathrm{UV}}$, etc. are constants in space. With these suppositions equation (638) becomes

$$\nu\frac{\partial\overline{Q}}{\partial t} = \Delta Q \quad \text{.............................(671)},$$

expressing that the whole change in $\overline{Q}$ is caused by collisions. If we put

$$Q = u^2 - v^2 = \mathrm{U}^2 - \mathrm{V}^2,$$

we have, from equation (660),

$$\begin{aligned}\Delta Q &= -\tfrac{3}{2}\nu^2\sqrt{2mK}A_2(\overline{\mathrm{U}^2} - \overline{\mathrm{V}^2})\\ &= -\eta\nu^2(\overline{\mathrm{U}^2} - \overline{\mathrm{V}^2}),\end{aligned}$$

giving upon substitution in equation (671),

$$\frac{\partial}{\partial t}(\overline{\mathrm{U}^2} - \overline{\mathrm{V}^2}) = -\eta\nu(\overline{\mathrm{U}^2} - \overline{\mathrm{V}^2}) \quad \text{.....................(672)}.$$

Similarly, taking $Q = uv$, and inserting into equation (671) the value of Δuv given by equation (661), we obtain

$$\frac{\partial}{\partial t}(\overline{\mathrm{UV}}) = -\eta\nu(\overline{\mathrm{UV}}) \quad \text{..........................(673)}.$$

Thus $\overline{U^2} - \overline{V^2}$, $\overline{UV}$, etc. satisfy an equation of the form

$$\frac{\partial \phi}{\partial t} = -\eta \nu \phi,$$

of which the solution is

$$\phi = \phi_0 e^{-\eta \nu t},$$

shewing that ϕ decreases exponentially with the time, at such a rate that it is reduced to $1/e$ times its original value in a time $1/\eta\nu$. This time is called by Maxwell the "time of relaxation."

This time of relaxation measures the rate at which deviations from Maxwell's law of distribution will subside. A glance at equations (674) below will shew that it must also measure the rate at which inequalities of pressure must subside.

Numerical estimates for the value of η, as given by equation (666), are not available, so that it is not possible to find the absolute value of this "time of relaxation." We shall, however, be able to compare it with the known values of coefficients of viscosity, and shall find that it is extremely small (cf. § 330, below).

Viscosity.

329. We have already seen (§ 214) that the system of pressures at any point in the gas is given by the equations

$$\left. \begin{aligned} P_{xx} &= \rho \overline{U^2} \\ P_{yx} &= \rho \overline{UV}, \text{ etc.} \end{aligned} \right\} \quad \text{..........................(674).}$$

To arrive at these formulae we have taken $\varpi_{xx} = \varpi_{xy} = \ldots = 0$ in equations (420). This does not mean that we neglect the intermolecular forces which vary inversely as the fifth power of the distance, for we have already taken full account of these forces in supposing that two molecules are in collision as soon as these forces become appreciable; in neglecting the system of pressures ϖ_{xx}, ϖ_{xy}, etc. we are merely assuming that no forces exist other than those which vary inversely as the fifth power of the distance.

Let us write

$$p = \tfrac{1}{3}(P_{xx} + P_{yy} + P_{zz}) = \tfrac{1}{3}\rho\,(\overline{U^2 + V^2 + W^2}) \quad \text{...........(675),}$$

without at present attaching any physical interpretation to p. We have already supposed as a first approximation that $\overline{U^2} = \overline{V^2} = \overline{W^2} = q$, so that $p = \rho q$, and we may now replace q by p/ρ.

From equation (668) we have

$$\begin{aligned} P_{xx} = \rho \overline{U^2} &= \tfrac{1}{3}\rho\,(\overline{U^2 + V^2 + W^2}) - \tfrac{1}{3}\rho\,(\overline{V^2} + \overline{W^2} - 2\overline{U^2}) \\ &= p - \frac{p}{\eta\nu}\left\{2\frac{\partial u_0}{\partial x} - \frac{2}{3}\left(\frac{\partial u_0}{\partial x} + \frac{\partial v_0}{\partial y} + \frac{\partial w_0}{\partial z}\right)\right\} \quad \text{........(676),} \end{aligned}$$

and similarly, from equation (669),

$$P_{yx} = \rho \overline{\mathsf{UV}} = -\frac{p}{\eta\nu}\left(\frac{\partial v_0}{\partial x} + \frac{\partial u_0}{\partial y}\right) \quad \text{.................(677).}$$

These pressures agree exactly with those giving the components of pressure in a viscous gas (cf. equations (593) and (594) of § 298) if we take p to be the hydrostatic pressure, and suppose κ, the coefficient of viscosity, to be given by

$$\kappa = \frac{p}{\eta\nu} \quad \text{..............................(678).}$$

If we give to p its usual value νRT, this becomes

$$\kappa = \frac{RT}{\eta} \quad \text{..............................(679),}$$

so that κ is found to be independent of the density, and directly proportional to the temperature.

330. The physical discussion of this and other equations obtained from Maxwell's theory is reserved for later chapters. We may notice, however, in passing that equation (679) enables the value of η, and hence of the time of relaxation, to be deduced when κ is known.

For instance, the value of κ for air at 0° C. is ·000172, and $RT = 3{\cdot}69 \times 10^{-14}$, so that $\eta = 2{\cdot}15 \times 10^{-10}$, and the time of relaxation is

$$\frac{1}{\eta\nu} = \frac{1}{6 \times 10^9} \text{ seconds.}$$

It is, as we should expect, comparable with the time of describing a free path.

Conduction of Heat.

331. On summing three equations of the type of (639), each multiplied by m, we obtain

$$\rho\frac{D\overline{Q}}{Dt} = \Sigma\left[-\frac{\partial}{\partial x}\{\rho\overline{\mathsf{U}(\mathsf{U}^2+\mathsf{V}^2+\mathsf{W}^2)}\} - 2\rho\left(\overline{\mathsf{U}^2}\frac{\partial u_0}{\partial x} + \overline{\mathsf{UV}}\frac{\partial u_0}{\partial y} + \overline{\mathsf{UW}}\frac{\partial u_0}{\partial z}\right)\right] \text{...(680),}$$

where $Q = \mathsf{U}^2 + \mathsf{V}^2 + \mathsf{W}^2$.

This is the equation of transfer of Q; it is therefore the equation of transfer of energy, and this, in the Kinetic Theory, is the transfer of heat.

Writing $\overline{\mathsf{U}^2+\mathsf{V}^2+\mathsf{W}^2} = 3q$, the left-hand member becomes $3\rho\frac{Dq}{Dt}$. As regards the first term on the right-hand side, we have from equation (670)

$$\rho\overline{\mathsf{U}(\mathsf{U}^2+\mathsf{V}^2+\mathsf{W}^2)} = -\frac{15mq}{2\eta}\frac{\partial q}{\partial x}.$$

The remaining terms on the right-hand side, containing $\overline{U^2}$, $\overline{UV}$, etc., are given by equations (676) and (677). On substituting these values in equation (680), we obtain

$$3\rho \frac{Dq}{Dt} = \Sigma \frac{\partial}{\partial x}\left(\frac{15mq}{2\eta}\frac{\partial q}{\partial x}\right) - 2q\rho\left(\frac{\partial u_0}{\partial x} + \frac{\partial v_0}{\partial y} + \frac{\partial w_0}{\partial z}\right)$$
$$+ \frac{4mq}{\eta}\left\{\left(\frac{\partial u_0}{\partial x}\right)^2 + \left(\frac{\partial v_0}{\partial y}\right)^2 + \left(\frac{\partial w_0}{\partial z}\right)^2\right\}$$
$$- \frac{4mq}{3\eta}\left(\frac{\partial u_0}{\partial x} + \frac{\partial v_0}{\partial y} + \frac{\partial w_0}{\partial z}\right)^2 + \frac{2mq}{\eta}\Sigma\left(\frac{\partial w_0}{\partial y} + \frac{\partial v_0}{\partial z}\right)^2 \quad \text{...(681).}$$

In equation (642) we obtained a value for $\frac{Dq}{Dt}$ on the supposition that Maxwell's law was true at every point. The present equation is the generalisation of equation (642) for the more general case in which Maxwell's law is not assumed to hold.

If there is no mass-motion ($u_0 = v_0 = w_0 = 0$), the equation reduces to

$$3\rho \frac{dq}{dt} = \Sigma \frac{\partial}{\partial x}\left(\frac{15mq}{2\eta}\frac{\partial q}{\partial x}\right),$$

and since $mq = m\overline{U^2} = RT$, this is the equation of conduction of heat in a gas at rest. By comparison with Fourier's equation of conduction of heat,

$$C_v \rho \frac{dT}{dt} = \Sigma \frac{\partial}{\partial x}\left(\vartheta \frac{\partial T}{\partial x}\right),$$

we obtain for the coefficient of conduction of heat the value

$$\vartheta = \frac{5mq}{2\eta} C_v = \frac{5RT}{2\eta} C_v \quad \text{......................(682),}$$

or, introducing the coefficient of viscosity κ from equation (679),

$$\vartheta = \tfrac{5}{2}\kappa C_v{}^* \quad \text{............................(683).}$$

The physical significance of this equation will be discussed later.

Energy.

332. On substituting the values which have been found for ϑ, κ, etc., we find that equation (681) assumes the form

$$\rho C_v \frac{DT}{Dt} = \Sigma \frac{\partial}{\partial x}\left\{\vartheta \frac{\partial T}{\partial x}\right\} - \tfrac{2}{3}\rho C_v\left(\frac{\partial u_0}{\partial x} + \frac{\partial v_0}{\partial y} + \frac{\partial w_0}{\partial z}\right)$$
$$+ \frac{\kappa}{J}\left[2\Sigma\left(\frac{\partial u_0}{\partial x}\right)^2 + \Sigma\left(\frac{\partial w_0}{\partial y} + \frac{\partial v_0}{\partial z}\right)^2 - \tfrac{2}{3}\left(\frac{\partial u_0}{\partial x} + \frac{\partial v_0}{\partial y} + \frac{\partial w_0}{\partial z}\right)^2\right].$$

Obviously the term on the left-hand side is the increase of heat-energy of an element of the gas. On the right-hand side, the first term is the increase

* Maxwell, as the result of an arithmetical mistake, gave the numerical factor as $\frac{5}{3}$ in place of $\frac{5}{2}$. The error was pointed out by Boltzmann and Poincaré.

of heat which ordinary physics regards as due to conduction, the second term is that due to adiabatic expansion or compression, and the third term is that which ordinary physics attributes to the action of viscosity, being in fact twice the "dissipation function" of the viscous motion*.

To the Kinetic Theory, however, conduction of heat, change of temperature resulting from adiabatic motion and "heat generated by viscosity" are all equally resolved into the transfer of energy by molecules, so that to the Kinetic Theory the equation just obtained expresses nothing more than the conservation of this energy.

Diffusion.

333. In § 312, we obtained the general equation (633)

$$\nu \frac{d\bar{Q}}{dt} = \Sigma \left[\bar{Q} \frac{\partial}{\partial x}(\nu \bar{u}) - \frac{\partial}{\partial x}(\nu \overline{uQ}) + \frac{\nu}{m} X \left(\overline{\frac{\partial Q}{\partial u}} \right) \right] + \Delta Q \quad \text{......(684)},$$

where Σ denotes summation with respect to the three coordinate axes x, y and z.

This equation may be used to determine the value of the coefficient of diffusion. We shall suppose there are two kinds of gas, distinguished by the suffixes 1, 2, and these will be supposed to be diffusing into one another in a direction parallel to the axis of x. There will be an equation of the form of (684) for each kind of gas.

We may suppose that there are no externally applied forces, so that $X = 0$ in equation (684). The motion of diffusion may be supposed so slow that squares of the mass-velocity may be neglected, and since the motion is entirely parallel to the axis of x, we may put $\frac{\partial}{\partial y} = \frac{\partial}{\partial z} = 0$.

With these simplifications equation (684) reduces to

$$\nu \frac{d\bar{Q}}{dt} = \bar{Q} \frac{\partial}{\partial x}(\nu \bar{u}) - \frac{\partial}{\partial x}(\nu \overline{uQ}) + \Delta Q.$$

In this equation, we put $Q = u$, so that $\bar{Q} = u_0$, the mass-velocity parallel to Ox. Neglecting u_0^2, we may also put

$$\overline{uQ} = \overline{u^2} = \overline{U^2} = \frac{1}{2hm}.$$

Since h, which measures the temperature, is not supposed to vary with x, $\frac{\partial}{\partial x}(\nu \overline{uQ})$ reduces to $\frac{1}{2hm} \frac{\partial \nu}{\partial x}$. The term $\bar{Q} \frac{\partial}{\partial x}(\nu \bar{u})$ becomes $u_0 \frac{\partial}{\partial x}(\nu u_0)$, and this, being of the second order of small quantities, may be neglected. Further, in steady motion, the time-differential will vanish, and the equation reduces to

$$\frac{1}{2hm} \frac{\partial \nu}{\partial x} = \Delta u \quad \text{..............................(685)}.$$

* Cf. Lamb, *Hydrodynamics*, p. 518.

The value of Δu is not zero, because collisions with molecules of the second kind change the total momentum of the first gas. In equation (654) we obtained the value of Δu for the first gas in the form

$$\Delta u = \nu_1 \nu_2 m_2 \sqrt{\frac{K}{m_1 + m_2}} A_1 (u_{02} - u_{01}),$$

where u_{01}, u_{02} are the values of u_0 for the first and second gas respectively.

On substituting this value for Δu, equation (685), for the first and second gases, yields equations which can be put in the form

$$\frac{\partial \nu_1}{\partial x} = - \frac{\partial \nu_2}{\partial x} = 2h \nu_1 \nu_2 m_1 m_2 \sqrt{\frac{K}{m_1 + m_2}} A_1 (u_{02} - u_{01}) \quad \text{......(686)}.$$

334. The equation of equilibrium in the gas is $\frac{\partial p}{\partial x} = 0$, or $\frac{\partial}{\partial x} \left(\frac{\nu_1 + \nu_2}{2h} \right) = 0$ from which we deduce

$$\frac{\partial \nu_1}{\partial x} = - \frac{\partial \nu_2}{\partial x} \quad \text{.............................(687)}.$$

In order that the pressure may continue to maintain equilibrium, $\nu_1 + \nu_2$ must remain the same at all points of the gas throughout the whole time. Thus the total flow of molecules across any plane must be zero, and this requires

$$\nu_1 u_{01} + \nu_2 u_{02} = 0 \quad \text{..........................(688)}.$$

The flow of molecules of the first kind per unit area per unit time, namely $\nu_1 u_{01}$, is however equal to $- \mathfrak{D}_{12} \frac{\partial \nu_1}{\partial x}$, where $\mathfrak{D}_{12}$ is the coefficient of diffusion from the first gas into the second (cf. below § 412). Similarly of course

$$\nu_2 u_{02} = - \mathfrak{D}_{21} \frac{\partial \nu_2}{\partial x} \quad \text{..........................(689)},$$

whence equations (687) and (688) shew that $\mathfrak{D}_{12}$ and $\mathfrak{D}_{21}$ must be the same: there is only one mutual coefficient of diffusion between the two gases.

Instead of expressing $\mathfrak{D}_{21}$ in the unsymmetrical form of equation (689), we may, with the help of equations (686) and (688), express it in the form

$$\mathfrak{D}_{12} \frac{\partial \nu_1}{\partial x} (\nu_1 + \nu_2) = - \nu_1 u_{01} (\nu_1 + \nu_2) = \nu_1 \nu_2 (u_{02} - u_{01}) \quad \text{......(690)},$$

which is symmetrical.

Comparing the equation just found with equation (686) obtained from the dynamics of collision, we find at once

$$\mathfrak{D}_{12} = \frac{1}{2h m_1 m_2 A_1 (\nu_1 + \nu_2)} \sqrt{\frac{m_1 + m_2}{K}} \quad \text{..............(691)}.$$

For the diffusion of a single gas into itself, this becomes

$$\mathfrak{D} = \frac{1}{h\rho A_1 \sqrt{2mK}} = \frac{3A_2}{2h\rho A_1 \eta} \quad \text{....................(692)},$$

where η is introduced from equation (666). If we further introduce κ, the coefficient of viscosity, given by $\kappa = \frac{1}{2h\eta}$, the coefficient of diffusion is given by

$$\mathfrak{D} = \frac{3A_2}{A_1}\frac{\kappa}{\rho} \quad \text{.............................(693)},$$

or, introducing Maxwell's numerical values already given for A_1 and A_2,

$$\mathfrak{D} = 1\cdot5043\frac{\kappa}{\rho} \quad \text{...........................(694)}.$$

The physical discussion of these equations is reserved for Chapter XIII.

MORE GENERAL LAW OF FORCE.

335. As already mentioned, the theory of Maxwell just given, which applies only to imaginary molecules repelling according to the law r^{-5}, has been extended by Chapman* so as to apply to the general case of molecules repelling according to the law r^{-s}. Chapman's analysis is unfortunately too long to be given here even in outline, but some of its more interesting points may be noticed.

The main difficulty, as will be clear from what has already been said, centres round the circumstance that the factor $V^{1-\frac{4}{s-1}}$ remains in equation (647) for all values of s other than $s = 5$. It is consequently necessary to assume a definite law of distribution $f(u, v, w)$ before the quantity $\Delta_{12}Q$ given by equation (647) can be evaluated. The value of $f(u, v, w)$ cannot depend upon the special choice of axes made, so that f is necessarily invariant as regards orthogonal transformations of the axes of x, y, z. Now the only invariants possible are

$$C^2 \equiv U^2 + V^2 + W^2,$$

$$\epsilon = \frac{\partial u_0}{\partial x} + \frac{\partial v_0}{\partial y} + \frac{\partial w_0}{\partial z},$$

$$\zeta = \left(U\frac{\partial}{\partial x} + V\frac{\partial}{\partial y} + W\frac{\partial}{\partial z}\right)h,$$

$$\eta = U^2\frac{\partial u_0}{\partial x} + V^2\frac{\partial v_0}{\partial y} + W^2\frac{\partial w_0}{\partial z} + VW\left(\frac{\partial w_0}{\partial y} + \frac{\partial v_0}{\partial z}\right) + \ldots$$

and others involving derivatives of the second order with respect to x, y, z. For gases in which temperature and density gradients are small, these second order derivatives may be neglected; indeed the ordinary assumption that definite coefficients of viscosity, conduction and diffusion exist, will be found to involve the assumption that all derivatives of second and higher orders are

* *Phil. Trans.* 216 A (1915), p. 279. See also footnote to p. 219.

negligible. Thus it is permissible at the outset to assume a distribution of velocities of the form

$$f(u, v, w) = F_1(c^2) + \epsilon F_2(c^2) + \zeta F_3(c^2) + \eta F_4(c^2).$$

This is obviously of the general form already obtained in § 309. Chapman assumes expansions of F_1, F_2, F_3, F_4 in powers of $\mathbf{c}^2$, and calculates the values of $\Delta_{12} Q$ for various values of Q from equation (647). The substitution of these values into equation (631) leads to a system of equations which determine the coefficients in the series F_1, F_2, F_3, F_4, and the subsequent analysis is similar in all essentials to that of Maxwell already given.

The numerical results obtained by Chapman are stated and discussed in Chapters XI—XIII below.

CHAPTER X

THE FREE PATH

336. In the two preceding chapters we have seen how problems of viscosity, conduction of heat, diffusion, etc. can be reduced to a problem of the dynamics of collisions. We shall in subsequent chapters see how these same problems can be treated by a study of the problems associated with the free path. For this and other reasons, the present chapter is devoted to problems connected with the free path in a gas, the molecules being assumed for this purpose to be elastic spheres.

Length of Mean Free Path.

337. In § 33 we gave a calculation of the mean free path in a gas. We shall now give a more detailed investigation applicable to the free paths of molecules in a mixture of gases, the molecules of the different gases being of different sizes, and shall at the same time examine the correlation between the velocity of a molecule and its probable free path.

We shall suppose the constants of the molecules of different types to be distinguished by suffixes, those of the first type having a suffix unity ($\nu_1, m_1, \sigma_1, \ldots$), and so on. We shall require a system of symbols to denote the distances apart at collision of the centres of two molecules of different kinds. Let these be S_{11}, S_{12}, S_{23}, etc., S_{pq} being the distance of the centres of two molecules of types p, q when in collision. Obviously

$$S_{12} = \tfrac{1}{2}(\sigma_1 + \sigma_2), \quad S_{11} = \sigma_1, \text{ etc. } \quad \ldots\ldots\ldots\ldots\ldots\ldots(695).$$

Maxwell's Mean Free Path.

338. As in expression (48) the number of collisions per unit time between molecules of types 1 and 2 and of classes A and B respectively (pp. 18, 19) is

$$\nu_1\nu_2 f_1(u, v, w) f_2(u', v', w')\, VS_{12}^2 \cos\theta\, du\,dv\,dw\,du'\,dv'\,dw'\,d\omega \ldots(696),$$

where V is the relative velocity, and $d\omega$ the element of solid angle to within which the line of centres is limited.

Replacing f_1, f_2 by the values appropriate to the steady state, and carrying out the integration with respect to $d\omega$, the number of collisions per unit time is found to be

$$\pi\nu_1\nu_2 \left(\frac{h^3 m_1^{\frac{3}{2}} m_2^{\frac{3}{2}}}{\pi^3} e^{-h(m_1c^2 + m_2c'^2)}\right) VS_{12}^2\, du\,dv\,dw\,du'\,dv'\,dw' \ldots\ldots(697),$$

this expression being exactly analogous to expression (50) previously obtained.

Let the velocity-components u, v, w, u', v', w' now be replaced by new variables, $\mathbf{u}, \mathbf{v}, \mathbf{w}, \alpha, \beta, \gamma$ given by

$$\mathbf{u} = \frac{m_1 u + m_2 u'}{m_1 + m_2}, \text{ etc.}; \ \alpha = u' - u, \text{ etc.} \quad \text{(698)},$$

so that $\mathbf{u}, \mathbf{v}, \mathbf{w}$ are the components of velocity of the centre of gravity of the two molecules, and α, β, γ are as usual the components of the relative velocity V. We may put

$$\mathbf{u}^2 + \mathbf{v}^2 + \mathbf{w}^2 = \mathbf{c}^2, \quad \alpha^2 + \beta^2 + \gamma^2 = V^2,$$

so that
$$m_1 c^2 + m_2 c'^2 = (m_1 + m_2)\, \mathbf{c}^2 + \frac{m_1 m_2}{m_1 + m_2} V^2.$$

We readily find that

$$\frac{\partial(\mathbf{u}, \alpha)}{\partial(u, u')} = 1,$$

so that $du\,du' = d\mathbf{u}\,d\alpha$, and expression (697) may be replaced by

$$\pi \nu_1 \nu_2 \left(\frac{h^3 m_1^{\frac{3}{2}} m_2^{\frac{3}{2}}}{\pi^3} e^{-h\left[(m_1+m_2)\mathbf{c}^2 + \frac{m_1 m_2}{m_1+m_2} V^2\right]} \right) V S_{12}^2 \, d\mathbf{u}\, d\mathbf{v}\, d\mathbf{w}\, d\alpha\, d\beta\, d\gamma \quad \text{(699)}.$$

339. On integrating with respect to all possible directions in space for the velocity $\mathbf{c}$ of the centre of gravity, we may replace $d\mathbf{u}\,d\mathbf{v}\,d\mathbf{w}$ by $4\pi \mathbf{c}^2 d\mathbf{c}$ while similarly, integrating with respect to all possible directions for V, we may replace $d\alpha\,d\beta\,d\gamma$ by $4\pi V^2 dV$. We accordingly obtain for the number of collisions per unit volume per unit time for which $\mathbf{c}$, V lie within specified small ranges $d\mathbf{c}\,dV$,

$$16 \nu_1 \nu_2 h^3 m_1^{\frac{3}{2}} m_2^{\frac{3}{2}} S_{12}^2 e^{-h\left[(m_1+m_2)\mathbf{c}^2 + \frac{m_1 m_2}{m_1+m_2} V^2\right]} \mathbf{c}^2 V^3 d\mathbf{c}\, dV \quad \text{(700)}.$$

Integrating from $\mathbf{c} = 0$ to $\mathbf{c} = \infty$, the number of collisions for which V lies between V and $V + dV$ is found to be

$$4 \nu_1 \nu_2 \sqrt{\frac{\pi h^3 m_1^3 m_2^3}{(m_1 + m_2)^3}} S_{12}^2 e^{-\frac{h m_1 m_2}{m_1 + m_2} V^2} V^3 dV \quad \text{(701)},$$

and again integrating this expression from $V = 0$ to $V = \infty$, the total number of collisions per unit volume per unit time between molecules of types 1 and 2 is found to be

$$2 \nu_1 \nu_2 S_{12}^2 \sqrt{\frac{\pi}{h}\left(\frac{1}{m_1} + \frac{1}{m_2}\right)} \quad \text{(702)}.$$

340. This formula gives the number of free paths of the ν_1 molecules of the first type in unit volume, which are terminated per unit time by molecules of the second type. When the difference between the two types of molecules is ignored, it reduces to twice expression (53) already found, the reason for the multiplying factor 2 being that already explained on p. 36.

Expression (702) divided by ν_1 will represent the mean chance of collision per unit time for a molecule of type 1 with a molecule of type 2. Hence the total mean chance of collision per unit time for a molecule of type 1 is

$$2\Sigma \nu_s S_{1s}^2 \sqrt{\frac{\pi}{h}\left(\frac{1}{m_1}+\frac{1}{m_s}\right)} \quad \text{.....................(703).}$$

The mean time interval between collisions is of course the reciprocal of this.

The total distance described by ν_1 molecules of the first kind per unit time is

$$\nu_1 \bar{c}_1 = \frac{2\nu_1}{\sqrt{\pi h m_1}} \text{..............................(704),}$$

while the total number of free paths described by these ν_1 molecules is equal to ν_1 times expression (703). By division, the mean free path for molecules of the first type is found to be

$$\lambda_1 = \frac{1}{\pi \Sigma \nu_s S_{1s}^2 \sqrt{1+\frac{m_1}{m_s}}} \quad \text{.......................(705).}$$

When there is only one kind of gas present, this reduces to formula (56) already found.

An Alternative Calculation of the Mean Free Path.

341. There is another way of estimating the length of the mean free path, as has already been indicated in § 33. According to this method of calculation, the mean free path is not taken to be the mean of all the paths described in unit time, but the mean of all the paths described from a given instant to the next collision.

Let us fix our attention on a molecule of the first type, moving with velocity c. The chance of collision per unit time with a molecule of the second type having a specified velocity c' is equal to the probable number of molecules of this second kind in a cylinder of base πS_{12}^2, and of height V, where V is the relative velocity.

The second molecule is supposed to have a velocity c'. Let θ, ϕ be angles determining the direction of this velocity, θ measuring the angle between its direction and that of c, and ϕ being an azimuth. The number of molecules of the second kind per unit volume for which c', θ, ϕ lie within small specified ranges $dc', d\theta', d\phi'$ is

$$\nu_2 \left(\frac{hm_2}{\pi}\right)^{\frac{3}{2}} e^{-hm_2c'^2} c'^2 \sin\theta \, d\theta \, d\phi \, dc'.$$

The result of integrating with respect to ϕ is obtained by replacing $d\phi$ by 2π, and on multiplying this by $\pi S_{12}^2 V$, we obtain for the number of molecules of the second kind which lie within the cylinder of volume $\pi S_{12}^2 V$ and which are such that c', θ lie within a range dc', $d\theta$,

$$2\nu_2 S_{12}^2 \sqrt{\pi h^3 m_2^3}\, V e^{-hm_2c'^2} c'^2 \sin\theta\, d\theta\, dc \quad \text{..............(706)}.$$

When c, c' are given, the value of V depends on θ, being given by

$$V^2 = c^2 + c'^2 - 2cc' \cos\theta,$$

whence we obtain by differentiation, keeping c, c' constant,

$$V dV = cc' \sin\theta\, d\theta.$$

Thus expression (706) can be replaced by

$$2\nu_2 S_{12}^2 \sqrt{\pi h^3 m_2^3}\, e^{-hm_2c'^2} \frac{c'}{c}\, dc' V^2 dV \quad \text{..................(707)}.$$

We first integrate with respect to V, keeping c and c' constant. We have

$$\int V^2 dV = \tfrac{1}{3}[V^3],$$

in which the limits for V are $c + c'$ and $c \sim c'$, so that

$$\int V^2 dV = \tfrac{2}{3}c(c^2 + 3c'^2) \text{ when } c' > c,$$
$$= \tfrac{2}{3}c'(c'^2 + 3c^2) \text{ when } c' < c.$$

Thus the result of integrating expression (707) with respect to V is

$$\text{when } c' > c,\ \tfrac{4}{3}\nu_2 S_{12}^2 \sqrt{\pi h^3 m_2^3}\, e^{-hm_2c'^2} c'(c^2 + 3c'^2)\, dc' \quad \text{......(708)},$$

$$\text{when } c' < c,\ \tfrac{4}{3}\nu_2 S_{12}^2 \sqrt{\pi h^3 m_2^3}\, e^{-hm_2c'^2} \frac{c'^2}{c}(c'^2 + 3c^2)\, dc' \quad \text{......(709)}.$$

If we now integrate this quantity with respect to c' from $c' = 0$ to $c' = \infty$ (using the appropriate form according as c' is greater or less than c), we obtain for the aggregate chance per unit time of a collision between a given molecule of the first type moving with velocity c, and a molecule of the second type,

$$\tfrac{4}{3}\nu_2 S_{12}^2 \sqrt{\pi h^3 m_2^3} \left[\int_c^\infty c'(c^2 + 3c'^2)\, e^{-hm_2c'^2} dc' + \int_0^c \frac{c'^2(c'^2 + 3c^2)}{c} e^{-hm_2c'^2} dc'\right] \quad \text{.........(710)}.$$

342. The former of the two integrals inside the square bracket can be evaluated directly, and is found to be equal to

$$(2hm_2c^2 + \tfrac{2}{3}) \frac{e^{-hm_2c^2}}{h^2 m_2^2}.$$

The second integral cannot be evaluated in finite terms. If, however, we replace $hm_2c'^2$ by y^2, the integral becomes

$$\frac{1}{c\sqrt{h^5 m_2^5}} \int_0^{c\sqrt{hm_2}} y^2(y^2 + 3hm_2c^2)\, e^{-y^2} dy,$$

which, after continued integration by parts with regard to y^2, reduces to

$$\frac{1}{c\sqrt{h^5 m_2^5}}\left\{-e^{-hm_2c^2}c\sqrt{hm_2}\,(2hm_2c^2+\tfrac{3}{4})+\tfrac{3}{4}(2hm_2c^2+1)\int_0^{c\sqrt{hm_2}}e^{-y^2}dy\right\}.$$

The sum of the two integrals in expression (710) is accordingly

$$\frac{3}{4c\sqrt{h^5 m_2^5}}\left[c\sqrt{hm_2}\,e^{-hm_2c^2}+(2hm_2c^2+1)\int_0^{c\sqrt{hm_2}}e^{-y^2}dy\right]\ \ldots(711).$$

If we introduce a function* $\psi(x)$ defined by

$$\psi(x)=xe^{-x^2}+(2x^2+1)\int_0^x e^{-y^2}dy\ \ldots\ldots\ldots\ldots\ldots(712),$$

expression (711) may be expressed in the form

$$\frac{3}{4c\sqrt{h^5 m_2^5}}\psi(c\sqrt{hm_2}),$$

and hence if we denote expression (710) by Θ_{12}, its value is found to be

$$\Theta_{12}=\frac{\sqrt{\pi}\nu_2 S_{12}^2}{hm_2c}\psi(c\sqrt{hm_2})\ \ldots\ldots\ldots\ldots\ldots\ldots(713).$$

With this definition of Θ_{12} we see that when a molecule of the first kind is moving with a velocity c, the chance that it collides with a molecule of the second kind in time dt is $\Theta_{12}dt$.

343. If we change the suffix 2 into 1 wherever it occurs, we obtain an expression Θ_{11} for the chance per unit time that a molecule of the first kind moving with velocity c shall collide with another molecule of the same kind.

By addition, the total chance per unit time that a molecule of the first kind moving with velocity c shall collide with a molecule of any kind is

$$\Sigma\Theta_{1s}=\Theta_{11}+\Theta_{12}+\Theta_{13}+\ldots\ \ldots\ldots\ldots\ldots\ldots\ldots(714).$$

In unit time the molecule we are considering describes a distance c, hence the chance of collision per unit length of path is

$$\frac{1}{c}\Sigma\Theta_{1s}\ \ldots\ldots\ldots\ldots\ldots\ldots\ldots\ldots\ldots(715).$$

The mean free path λ_c, for molecules of the first kind moving with velocity c, is accordingly

$$\lambda_c=\frac{c}{\Sigma\Theta_{1s}}\ \ldots\ldots\ldots\ldots\ldots\ldots\ldots\ldots(716).$$

Tait's Free Path.

344. When there is only one kind of gas, equation (716) assumes the form

$$\lambda_c=\frac{c}{\Theta}=\frac{hmc^2}{\sqrt{\pi}\nu\sigma^2\psi(c\sqrt{hm})}\ \ldots\ldots\ldots\ldots\ldots\ldots(717),$$

* The value of $\int_0^x e^{-y^2}dy$ cannot be expressed in simpler terms, so that $\psi(x)$ as defined by equation (712) is already in its simplest form. Tables for the evaluation of $\psi(x)$ are given in appendix B.

and from this formula we can without difficulty calculate Tait's expression for the mean free path, defined as explained in § 33. For, in a single gas, there is at any instant a fraction

$$\sqrt{\frac{h^3m^3}{\pi^3}}\,4\pi e^{-hmc^2}c^2dc$$

of the whole number of molecules moving with velocity c, and therefore, on the average, starting to describe distances c/Θ each before collision. Hence Tait's mean free path (λ_T) is given by

$$\lambda_T=\int_0^\infty\frac{c}{\Theta}\sqrt{\frac{h^3m^3}{\pi^3}}\,4\pi e^{-hmc^2}c^2dc=\frac{1}{\pi\nu\sigma^2}\int_0^\infty\frac{4x^4e^{-x^2}dx}{\psi(x)}\quad\ldots\ldots(718).$$

This integral can only be evaluated by quadrature. The evaluation has been performed by Tait* and Boltzmann†, who agree in assigning to it the value 0·677, leading to the value for λ_T which has already been given in § 33.

Maxwell's Free Path.

345. We can also deduce Maxwell's formula for the free path from the results obtained in § 343.

Out of all the molecules of the first kind, a fraction

$$4\pi\left(\frac{hm_1}{\pi}\right)^{\frac{3}{2}}e^{-hm_1c^2}c^2dc$$

will be moving with a velocity between c and $c+dc$, and the chance of collision per unit time for each of these molecules is $\Sigma\Theta_{1s}$. Hence the average chance of collision per unit time, for all molecules of the first kind, is

$$4\pi\left(\frac{hm_1}{\pi}\right)^{\frac{3}{2}}\int_0^\infty e^{-hm_1c^2}c^2(\Sigma\Theta_{1s})\,dc,$$

or, from equation (713),

$$4\sum_s\frac{\nu_sS_{1s}^2(hm_1)^{\frac{3}{2}}}{hm_s}\int_0^\infty e^{-hm_1c^2}c\psi(c\sqrt{hm_s})\,dc\quad\ldots\ldots\ldots\ldots(719).$$

Putting $c\sqrt{hm_s}=x$, this becomes

$$4\sum_s\frac{\nu_sS_{1s}^2(hm_1)^{\frac{3}{2}}}{(hm_s)^2}\int_0^\infty e^{-\frac{m_1}{m_s}x^2}x\psi(x)\,dx\quad\ldots\ldots\ldots\ldots\ldots(720).$$

Upon substitution for $\psi(x)$ from equation (712), the integral in this formula becomes the sum of two integrals

$$\int_0^\infty e^{-x^2\left(1+\frac{m_1}{m_s}\right)}x^2dx+\int_0^\infty\int_0^x x(2x^2+1)\,e^{-y^2-\frac{m_1}{m_s}x^2}dx\,dy\quad\ldots(721).$$

* *Edin. Trans.* XXXIII. p. 74 (1886).

† *Wiener Sitzungsberichte*, XCVI. p. 905 (1887); *Gastheorie*, I. p. 73.

The first of these integrals has for its value

$$\frac{\sqrt{\pi}}{4}\left(1+\frac{m_1}{m_s}\right)^{-\frac{3}{2}} \quad \text{.........................(722)},$$

while on writing $y = Kx$, the second integral

$$= \int_0^\infty \int_0^1 x^2(2x^2+1)\, e^{-x^2\left(K^2+\frac{m_1}{m_s}\right)} dx\, dK$$

$$= \int_0^1 \left[\frac{3\sqrt{\pi}}{4}\left(K^2+\frac{m_1}{m_s}\right)^{-\frac{5}{2}} + \frac{\sqrt{\pi}}{4}\left(K^2+\frac{m_1}{m_s}\right)^{-\frac{3}{2}}\right] dK$$

$$= \frac{\sqrt{\pi}}{4}\left\{\frac{3m_s^2}{m_1^2}\left(1+\frac{m_1}{m_s}\right)^{-\frac{1}{2}} - \frac{m_s^2}{m_1^2}\left(1+\frac{m_1}{m_s}\right)^{-\frac{3}{2}} + \frac{m_s}{m_1}\left(1+\frac{m_1}{m_s}\right)^{-\frac{1}{2}}\right\}.$$

Adding this to expression (722) and simplifying, we obtain as the value of expression (721)

$$\frac{\sqrt{\pi}}{2}\frac{m_s^2}{m_1^2}\sqrt{1+\frac{m_1}{m_s}}.$$

This is the value of the integral which occurs in expression (720). The whole expression is therefore equal to

$$2\sum_s \nu_s S_{1s}^2 \sqrt{\frac{\pi}{h}\left(\frac{1}{m_1}+\frac{1}{m_s}\right)} \quad \text{....................(723)},$$

which is identical with expression (703), as of course it ought to be.

Formula (723) gives the mean chance of collision per unit time for a single molecule of the first kind, and every collision terminates a free path of this molecule. The total number of free paths described by all molecules of the first kind per unit time per unit volume is therefore

$$\mathfrak{n}_1 = 2\Sigma \nu_1 \nu_s S_{1s}^2 \sqrt{\frac{\pi}{h}\left(\frac{1}{m_1}+\frac{1}{m_s}\right)} \quad \text{..................(724)}.$$

The distance described per unit time by the ν_1 molecules of the first kind in a unit volume is, as before, given by expression (701), so that the mean free path, λ_1, of all molecules of the first kind is

$$\lambda_1 = \frac{2\nu_1}{\mathfrak{n}_1\sqrt{\pi h m_1}} = \frac{1}{\pi\Sigma \nu_s S_{1s}^2 \sqrt{1+\frac{m_1}{m_s}}} \quad \text{..............(725)},$$

agreeing with formula (705).

Problems connected with the Free Path.

Dependence of Free Path on Velocity.

346. The way in which λ_c depends on the value of c is of some interest. The formula expressing λ_c as a function of c is, however, too complex to convey much definite meaning to the mind, and we are therefore compelled to fall back on numerical values. The following table, which is taken from Meyer's *Kinetic Theory of Gases* (p. 429), gives the ratio of λ_c (equation (717)) to Maxwell's mean free path λ (equation (56)) for different values of c, from $c=0$ to $c=\infty$.

$c/\bar{c}$	hmc^2	λ_c/λ	λ/λ_c
0		0	∞
0·25		0·3445	2·9112
0·5		0·6411	1·5604
0·627	$\frac{1}{2}$	0·7647	1·3111
0·886	1	0·9611	1·0407
1·0		1·0257	·9749
1·253	2	1·1340	·8819
1·535	3	1·2127	·8247
1·772	4	1·2572	·7954
2		1·2878	·7765
3		1·3551	·7380
4		1·3803	·7244
5		1·3923	·7182
6		1·3989	·7149
∞		1·4142	·7071

Probability of a Free Path of given length.

347. It is of interest to find the probability that a molecule shall describe a free path of given length.

Let $f(l)$ denote the probability that a molecule moving with a velocity c shall describe a free path at least equal to l. After the molecule has described a distance l, the chance of collision within a further distance dl is, by formula (715), equal to dl/λ_c. Hence the chance that a molecule shall describe a distance l, and then a further distance dl, without collision is

$$f(l)(1-dl/\lambda_c).$$

This must however be the same thing as $f(l+dl)$ or

$$f(l)+\frac{\partial f(l)}{\partial l}dl.$$

Equating these expressions we have

$$\frac{\partial f(l)}{\partial l}=-\frac{f(l)}{\lambda_c},$$

of which the solution is

$$f(l)=e^{-l/\lambda_c} \quad\text{.............................(726)},$$

the arbitrary constant of the integration being determined by the condition that $f(0)=1$.

By differentiation, the probability that a molecule moving with a velocity c shall describe a free path of length between l and $l+dl$ is

$$e^{-l/\lambda_c}\frac{dl}{\lambda_c} \quad\text{.................................(727).}$$

348. It is clear from the form of these expressions that free paths which are many times greater than the mean free path will be extremely rare. For instance, the probability that a molecule moving with velocity c shall describe a path greater than n times λ_c is $f(n\lambda_c)$ or, by equation (726), e^{-n}. Thus only one molecule in 148 describes a path as great as $5\lambda_c$, only one in 22,027 a path as great as $10\lambda_c$, only one in $2{\cdot}7\times10^{43}$ a path as great as $100\lambda_c$, and so on.

The foregoing results apply only to molecules moving with a given velocity c. At any given instant the fraction of the whole number of molecules which have described a distance greater than l since their last collision will be

$$\sqrt{\frac{\pi^3}{h^3m^3}}\int_0^\infty 4\pi c^2 e^{-hmc^2-l/\lambda_c}\,dc \quad\text{..................(728).}$$

This function is not easy to calculate in any way. As the result of a rough calculation by quadrature, I have found that through the range of values for l in which its value is appreciable, it does not ever differ by more than about 1 per cent. from $e^{-1{\cdot}04l/\lambda}$, which is the value for molecules moving with velocity $1/\sqrt{hm}$.

Law of Distribution of Velocities in Collision.

349. In many physical problems, it is important to consider the distribution of relative velocities, ratios of velocities, etc. in the different collisions which occur. We attempt to obtain expressions for various laws of distribution of this type.

In formulae (708) and (709) we obtained expressions for the chance per unit time that a molecule of type 1 moving with velocity c should collide with a molecule of type 2 moving with a velocity between c' and $c'+dc'$.

The number of molecules of type 1 per unit volume moving with a velocity between c and $c+d$ is

$$4\pi\nu_1\left(\frac{hm}{\pi}\right)^{\frac{3}{2}}e^{-hm_1c^2}\,c^2dc.$$

Multiplying expressions (708) and (709) by this number we obtain as the total number of collisions per unit time per unit volume between molecules having specified velocities within ranges dc, dc',

when $c'>c$,

$$\tfrac{16}{3}\nu_1\nu_2S_{12}{}^2h^3m_1{}^{\frac{3}{2}}m_2{}^{\frac{3}{2}}e^{-h(m_1c^2+m_2c'^2)}\,c^2c'\,(c^2+3c'^2)\,dc\,dc' \quad\text{......(729),}$$

when $c'<c$,

$$\tfrac{16}{3}\nu_1\nu_2S_{12}{}^2h^3m_1{}^{\frac{3}{2}}m_2{}^{\frac{3}{2}}e^{-h(m_1c^2+m_2c'^2)}\,cc'^2\,(3c^2+c'^2)\,dc\,dc' \quad\text{......(730).}$$

350. We proceed next to find the number of collisions in which the velocities c, c' stand in a given ratio to one another. Let $c = \kappa c'$, and let the variables in expressions (729) and (730) be changed from c, c' to κ, c'. Clearly the differential $dcdc'$ becomes $c'd\kappa dc'$, and the two expressions become

when $\kappa > 1$,

$$\tfrac{16}{3}\nu_1\nu_2 S_{12}^2 h^3 m_1^{\frac{3}{2}} m_2^{\frac{3}{2}} e^{-hc'^2(m_1\kappa^2+m_2)}\,\kappa(3\kappa^2+1)\,d\kappa c'^6 dc' \quad \text{......(731)},$$

when $\kappa < 1$,

$$\tfrac{16}{3}\nu_1\nu_2 S_{12}^2 h^3 m_1^{\frac{3}{2}} m_2^{\frac{3}{2}} e^{-hc'^2(m_1\kappa^2+m_2)}\,\kappa^2(\kappa^2+3)\,d\kappa c'^6 dc' \quad \text{......(732)}.$$

On integrating these expressions with respect to c' from $c' = 0$ to $c' = \infty$, we obtain the number of collisions for which κ, the ratio of the velocities, lies within a given range $d\kappa$. The numbers are readily found to be

$$\text{when } \kappa > 1, \qquad 5\nu_1\nu_2 S_{12}^2\left(\frac{\pi m_1^3 m_2^3}{h}\right)^{\frac{1}{2}}\frac{\kappa(3\kappa^2+1)}{(m_1\kappa^2+m_2)^{\frac{7}{2}}}\,d\kappa \quad \text{..............(733)},$$

$$\text{when } \kappa < 1, \qquad 5\nu_1\nu_2 S_{12}^2\left(\frac{\pi m_1^3 m_2^3}{h}\right)^{\frac{1}{2}}\frac{\kappa^2(\kappa^2+3)}{(m_1\kappa^2+m_2)^{\frac{7}{2}}}\,d\kappa \quad \text{..............(734)}.$$

The total number of collisions per unit time per unit volume may of course be derived by integrating this quantity from $\kappa = 0$ to $\kappa = \infty$. It is found to be

$$5\nu_1\nu_2 S_{12}^2\left(\frac{\pi m_1^3 m_2^3}{h}\right)^{\frac{1}{2}}\left\{\int_0^1 \frac{\kappa^2(\kappa^2+3)}{(m_1\kappa^2+m_2)^{\frac{7}{2}}}\,d\kappa + \int_1^\infty \frac{\kappa(3\kappa^2+1)}{(m_1\kappa^2+m_2)^{\frac{7}{2}}}\,d\kappa\right\}$$

$$= 2\nu_1\nu_2 S_{12}^2\sqrt{\frac{\pi}{h}\left(\frac{1}{m_1}+\frac{1}{m_2}\right)} \quad \text{......(735)},$$

which agrees, as it ought, with formula (702).

351. The law of distribution of κ in different collisions can be obtained by dividing expressions (733) and (734) by the total number of collisions (735). This law of distribution is found to be

$$\text{when } \kappa > 1, \qquad \frac{5}{2}\frac{m_1^2 m_2^2}{(m_1+m_2)^{\frac{1}{2}}}\frac{\kappa(3\kappa^2+1)}{(m_1\kappa^2+m_2)^{\frac{7}{2}}} \quad \text{.....................(736)},$$

$$\text{when } \kappa < 1, \qquad \frac{5}{2}\frac{m_1^2 m_2^2}{(m_1+m_2)^{\frac{1}{2}}}\frac{\kappa^2(\kappa^2+3)}{(m_1\kappa^2+m_2)^{\frac{7}{2}}} \quad \text{.....................(737)}.$$

The law of distribution of values of κ when the molecules are similar is obtained on taking $m_1 = m_2$. We must notice however that if we simply put $m_1 = m_2$ in expressions (736) and (737) each collision is counted twice, once as having a ratio of velocities κ and once as having a ratio of velocities $1/\kappa$. It seems simplest to define the value of κ for a collision in this case as the ratio of the greater to the smaller velocity, so that κ is always greater than unity, and we then obtain the law of distribution by putting $m_1 = m_2$ in expression (736) and multiplying by two so that each collision shall only count once. The law of distribution is found to be

$$\frac{5\kappa(3\kappa^2+1)}{\sqrt{2}(1+\kappa^2)^{\frac{7}{2}}}d\kappa \quad \text{..........................(738)},$$

of which the value when integrated from $\kappa = 1$ to $\kappa = \infty$ is unity, as it ought to be.

Persistence of Velocity after Collision.

352. The next problem will be to examine the average effect of a collision as regards reversal or deflection of path. We shall find that in general a collision does not necessarily reverse the velocity in the original direction of motion, or even reduce it to rest: there is a marked tendency for the original velocity to persist to some extent after collision. It is obviously of the utmost importance to form an estimate of the extent to which this persistence of velocity occurs.

Persistence of Velocity when the molecules are similar elastic spheres.

353. Let us begin by considering two molecules of equal mass colliding with velocities c, c'. In fig. 18 let OP and OQ represent these velocities, and let R be the middle point of PQ. Then we can resolve the motion of the two molecules into

(i) a motion of the centre of mass of the two, the velocity of this motion being represented by OR, and

(ii) two equal and opposite velocities relative to the centre of mass, these being represented by RP and RQ.

Imagine a plane RTS drawn through R parallel to the common tangent to the spheres at the moment of impact, and let P', Q' be the images of P, Q

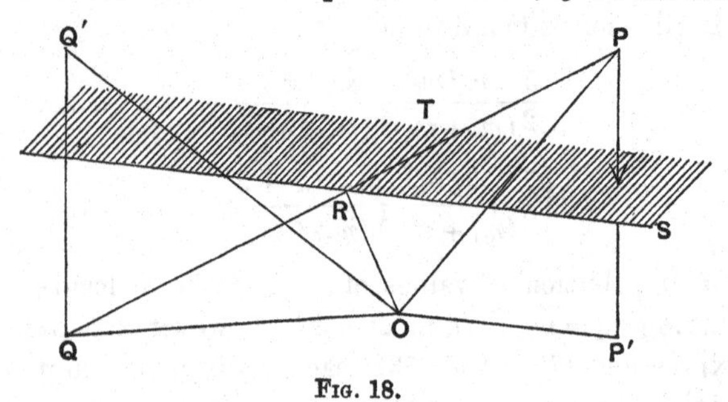

FIG. 18.

in this plane. Then clearly RP' and RQ' represent the velocities relatively to the centre of gravity after impact, so that OP' and OQ' represent the actual velocities in space.

354. Let us now suppose the molecules to be elastic spheres of uniform diameter σ. In fig. 19 let the directions of motion relative to the centre of gravity before impact be AB, DE, and let those after impact be BC, EF. Then the line of centres bisects each of the angles ABC, DEF. Let us call each of these angles ϕ, measured so as to be acute in the figure. Imagine the point E surrounded by a circle of radius σ of which the plane is perpendicular to the direction AB. Then in order that a collision may take place,

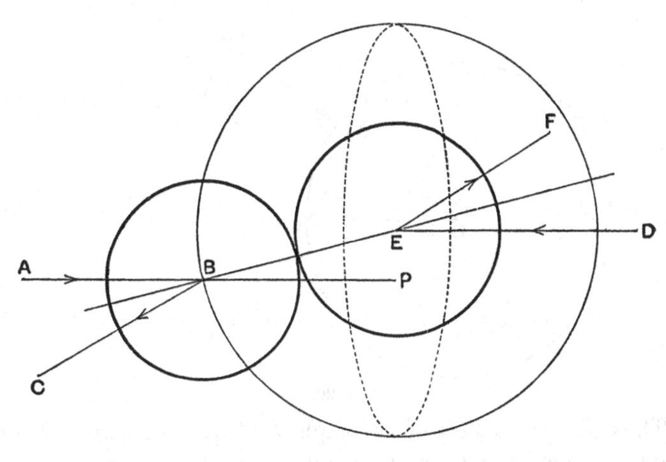

Fig. 19.

the line AB produced must cut the plane of this circle at some point P inside the circle. Also, all positions of P inside this circle are equally probable, so that the probability that the distance EP shall lie between r and $r + dr$ is $2r\,dr/\sigma^2$. Since $r = \sigma \sin \frac{1}{2}\phi$, this may be written $\sin \frac{1}{2}\phi \cos \frac{1}{2}\phi\, d\phi$. This, then, is the probability that ϕ shall lie between ϕ and $\phi + d\phi$, and therefore that the angle which EF makes with DE shall be between ϕ and $\phi + d\phi$. The expression found is, however, equal to $\frac{1}{2} \sin \phi\, d\phi$ and therefore to that part of the area of a unit sphere for which the radius makes an angle between ϕ and $\phi + d\phi$ with a given line. It follows that all directions for EF are equally probable, a result which was first given by Maxwell in 1859*. Hence in fig. 18 all directions of RQ', RP' are equally probable, so that the "expectation" of the component of velocity of either molecule after impact in any direction is equal to the component of OR in that direction.

355. Let us now average over all possible directions for the velocity of the second molecule, keeping the magnitude of this velocity constant. In fig. 20 let OP, OQ as before represent the velocities of the two colliding molecules, and let R be the middle point of PQ, so that OR represents the velocity of the centre of gravity of the two molecules. We have to average the components of the velocity OR over all positions of Q which lie on a

* *Collected Works*, I. p. 378.

sphere having O for centre. It is at once obvious that the average component of OR in any direction perpendicular to OP is zero. We have, therefore, only to find the component in the direction OP, say ON. We must not suppose all directions for OQ to be equally likely, for (cf. Chapter II) the probability of collision with any two velocities is proportional to the relative

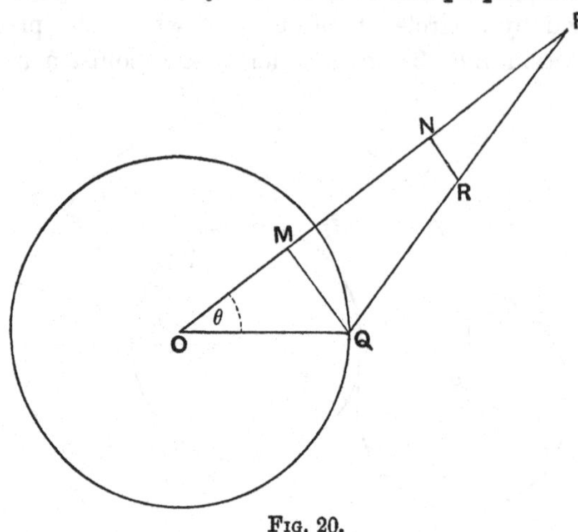

FIG. 20.

velocity. Thus the probability of the angle POQ lying between θ and $\theta + d\theta$ is not simply proportional to $\sin\theta d\theta$, but is proportional to $PQ \sin\theta d\theta$, for PQ represents the relative velocity. The average value of the component ON is therefore

$$\overline{ON} = \frac{\int_0^\pi ON \,.\, PQ \sin\theta d\theta}{\int_0^\pi PQ \sin\theta d\theta} \quad \text{.....................(739).}$$

Let us now write $\quad OP = c, \quad OQ = c', \quad PQ = V,$

so that
$$V^2 = c^2 + c'^2 - 2cc' \cos\theta \quad \text{.......................(740).}$$

Then $\quad ON = \frac{1}{2}(OP + OM) = \frac{1}{2}(c + c' \cos\theta) = \frac{1}{4c}(3c^2 + c'^2 - V^2)$ ...(741).

By differentiation of relation (740), we have $VdV = cc' \sin\theta d\theta$, so that equation (739) becomes

$$\overline{ON} = \frac{\int (3c^2 + c'^2 - V^2) V^2 dV}{4c \int V^2 dV} = \frac{3c^2 + c'^2}{4c} - \frac{\int V^4 dV}{4c \int V^2 dV} \quad \text{......(742),}$$

the limits of integration being from $V = c' \sim c$ to $V = c' + c$.

Performing the integration we find that

when $c > c'$,
$$\overline{ON} = \frac{15c^4 + c'^4}{10c(3c^2 + c'^2)} \quad \text{.......................(743),}$$

when $c < c'$,
$$\overline{ON} = \frac{c(5c'^2 + 3c^2)}{5(3c'^2 + c^2)} \quad \text{..........................(744).}$$

We notice that these expressions are necessarily positive for all values of c and c', so that whatever the velocities of the two colliding molecules may be, the "expectation" of the velocity of the first molecule after collision is definitely in the same direction as the velocity before collision. Naturally the same also is true of the second molecule.

If we denote $\overline{ON}$, the "expectation" of velocity after collision of the first molecule in the direction of OP, by α, then the ratio α/c may be regarded as a measure of the persistence of the velocity of the first molecule.

Formulae (743) and (744) give the values of α, and hence of the persistence α/c. It is at once seen that the values of α/c depend only on the ratio c/c', and not on the values of c and c' separately. If, as before (§ 350), we denote c/c' by κ, the values of the persistence are

when $\kappa > 1$, $$\frac{\alpha}{c} = \frac{15\kappa^4 + 1}{10\kappa^2 (3\kappa^2 + 1)} \quad \text{.........................(745)},$$

when $\kappa < 1$, $$\frac{\alpha}{c} = \frac{3\kappa^2 + 5}{5(\kappa^2 + 3)} \quad \text{.............................(746)}.$$

356. These expressions are too intricate to convey much meaning as they stand. The following table gives numerical values of the persistence α/c corresponding to different values of κ, the ratio of velocities:

$\frac{c}{c'} =$	∞	4	2	$1\frac{1}{2}$	1	$\frac{2}{3}$	$\frac{1}{2}$	$\frac{1}{4}$	0
$\frac{\alpha}{c} =$	·500	·492	·473	·441	·400	·368	·354	·339	·333
$\frac{c'}{c} =$	0	$\frac{1}{4}$	$\frac{1}{2}$	$\frac{2}{3}$	1	$1\frac{1}{2}$	2	4	∞

It now appears that the persistence is a fraction which varies from $33\frac{1}{3}$ to 50 per cent., according to the ratio of the original velocities. From the values given, it is clear that we are likely to obtain fairly accurate results if we assume, for purposes of rough approximation, that the persistence is always equal to 40 per cent. of the original velocity.

357. By averaging over all possible values of the ratio κ we can obtain an exact value for the mean persistence averaged over all collisions.

Each collision involves two molecules of which the rôles are entirely interchangeable. Let us agree to speak of the molecule of which the initial velocity is the greater as the first molecule, so that c/c' or κ is always greater than unity.

The persistences of the velocities of the two molecules involved in any one collision are respectively

$$\frac{15\kappa^4 + 1}{10\kappa^2 (3\kappa^2 + 1)} \text{ and } \frac{3 + 5\kappa^2}{5(3\kappa^2 + 1)},$$

the first of these expressions being given directly by formula (745), while the second is immediately obtained by writing $1/\kappa$ for κ in expression (746).

The mean persistence of the two molecules concerned in this collision, being the mean of the two expressions just found, is

$$\frac{25\kappa^4 + 6\kappa^2 + 1}{20\kappa^2(3\kappa^2 + 1)} \quad \text{..........................(747).}$$

A few numerical values of this quantity are found to be:

$\kappa =$	1	$1\frac{1}{4}$	$1\frac{1}{2}$	2	3	4	∞
mean persistence =	·400	·401	·404	·413	·415	·416	·417

The law of distribution of values of κ in the different collisions which occur has been found in formula (738) to be

$$\frac{5\kappa(3\kappa^2 + 1)}{\sqrt{2}(1 + \kappa^2)^{\frac{7}{2}}} d\kappa \quad \text{..........................(748).}$$

Multiplying together expressions (747) and (748) and integrating from $\kappa = 1$ to $\kappa = \infty$, we obtain for the mean persistence of all velocities after collision

$$\int_1^\infty \frac{25\kappa^4 + 6\kappa^2 + 1}{4\sqrt{2}\kappa(1 + \kappa^2)^{\frac{7}{2}}} d\kappa = \frac{1}{4} + \frac{1}{4\sqrt{2}} \log_e(1 + \sqrt{2}) = \cdot 406.$$

Thus the average value of the persistence is very nearly equal to $\frac{2}{5}$, the value when the molecules collide with exactly equal velocities.

Persistence when molecules have different masses.

358. The calculations just given apply only when the molecules are all similar. Let us now examine what value is to be expected for the persistence when the molecules have different masses and sizes, but are still supposed to be elastic spheres.

Consider a collision between two molecules of masses m_1, m_2; let their velocities be a, b respectively as before, and let their relative velocity be V.

It is clear upon examination of §§ 353, 354, that all the analysis of these sections will be directly applicable to the present case, the only alteration being that S_{12} (cf. § 337) must replace σ as the distance BE between the centres. Thus Maxwell's result is still true; all directions are equally likely for the velocities after impact relative to the centre of gravity, and the expectation of any component of velocity after collision is exactly that of the common centre of gravity.

We may accordingly proceed to average exactly as in § 355. But if OR in fig. 20 represents the velocity of the centre of gravity, R will no longer be the middle point of PQ; it will divide PQ in such a way that $m_1RP = m_2RQ$. Thus in place of relations (741) we have

$$ON = \frac{m_1OP + m_2OM}{m_1 + m_2} = \frac{m_1c + m_2c' \cos\theta}{m_1 + m_2}$$

$$= \frac{m_1 - m_2}{m_1 + m_2}c + \frac{m_2}{m_1 + m_2}(c + c'\cos\theta) \quad \text{.........(749).}$$

So long as c and c' are kept constant we can average this exactly as before. The first term $\frac{m_1 - m_2}{m_1 + m_2}c$, being constant, is not affected by averaging, while the average value of the second term $\frac{m_2}{m_1 + m_2}(c + c'\cos\theta)$ is equal to $\frac{2m_2}{m_1 + m_2}$ times the average of the term $\frac{1}{2}(c + c'\cos\theta)$ already found in § 355.

Hence, if $\left(\frac{\alpha}{c}\right)_e$ denotes the value of the persistence $\frac{\alpha}{c}$ when the two masses are equal, we have, in the general case in which the masses are unequal,

$$\frac{\alpha}{c} = \frac{m_1 - m_2}{m_1 + m_2} + \frac{2m_2}{m_1 + m_2}\left(\frac{\alpha}{c}\right)_e \quad \text{....................(750).}$$

This gives the persistence of the velocity c of the molecule of mass m_1, the values of $\left(\frac{\alpha}{c}\right)_e$ being given by the table on p. 263. The persistence is of course a function of the two quantities m_1/m_2 and c'/c.

359. If we assume as a rough approximation that the value of $\left(\frac{\alpha}{c}\right)_e$ is equal to ·400 regardless of the ratio of velocities κ, then equation (750) reduces to the approximate formula

$$\frac{\alpha}{c} = \frac{m_1 - \frac{1}{5}m_2}{m_1 + m_2} \quad \text{..............................(751),}$$

which of course depends only on m_1/m_2. This formula, however, must not be applied when the ratio m_1/m_2 is either very large or very small.

When m_1/m_2 is very small, c will be large compared with c' in practically all collisions, so that κ is very great and the appropriate value to assume for $\left(\frac{\alpha}{c}\right)_e$ is $\frac{1}{2}$, this corresponding to $\kappa = \infty$. From equation (750) we now obtain the approximate formula

$$\frac{\alpha}{c} = \frac{m_1}{m_1 + m_2} \cdots \left(\frac{m_1}{m_2} \text{ small}\right) \quad \text{....................(752).}$$

We notice that in the limit when m_1 vanishes in comparison with m_2, the persistence vanishes. Indeed this can easily be seen directly: at a collision the light molecule simply bounces off the heavy molecule; all directions can be seen to be equally likely by the method of § 354, and therefore the persistence is *nil*.

At the opposite extreme, when m_1/m_2 is very large, the appropriate values to assume are $\kappa = 0$ and $\left(\frac{\alpha}{c}\right)_e = \frac{1}{3}$. The approximate formula derived from equation (750) is now

$$\frac{\alpha}{c} = \frac{m_1 - \frac{1}{3}m_2}{m_1 + m_2} \ldots \left(\frac{m_1}{m_2} \text{ large}\right) \quad \ldots\ldots\ldots\ldots\ldots\ldots (753).$$

In the limit when m_2 vanishes, the persistence becomes equal to unity. This also can be seen directly: the heavy molecule merely knocks the light molecule out of its way, and passes on with its velocity unaltered.

360. In place of these approximate formulae, we can obtain a formula accurate for all values of m_1/m_2 by averaging the exact equation (750).

When $\kappa > 1$, we have (by formula (745))

$$\left(\frac{\alpha}{c}\right)_e = \frac{15\kappa^4 + 1}{10\kappa^2(3\kappa^2 + 1)},$$

while the law of distribution of the κ's is (by formula (736))

$$\frac{5}{2} \frac{m_1^2 m_2^2}{(m_1 + m_2)^{\frac{1}{2}}} \frac{\kappa(3\kappa^2 + 1)}{(m_1\kappa^2 + m_2)^{\frac{7}{2}}} d\kappa.$$

When $\kappa < 1$, the corresponding quantities are (formulae (746) and (737))

$$\frac{3\kappa^2 + 5}{5(\kappa^2 + 3)} \text{ and } \frac{5}{2} \frac{m_1^2 m_2^2}{(m_1 + m_2)^{\frac{1}{2}}} \frac{\kappa^2(\kappa^2 + 3)}{(m_1\kappa^2 + m_2)^{\frac{7}{2}}} d\kappa.$$

Hence the value of $\left(\frac{\alpha}{c}\right)_e$ averaged over all collisions is given by

$$\overline{\left(\frac{\alpha}{c}\right)_e} = \frac{5m_1^2 m_2^2}{2(m_1 + m_2)^{\frac{1}{2}}} \left[\int_0^1 \frac{(3\kappa^2 + 5)\kappa^2 d\kappa}{5(m_1\kappa^2 + m_2)^{\frac{7}{2}}} + \int_1^\infty \frac{(15\kappa^4 + 1) d\kappa}{10\kappa(m_1\kappa^2 + m_2)^{\frac{7}{2}}}\right]$$

$$= \frac{1}{4\mu^3\sqrt{(1 + \mu^2)}} \log(\sqrt{1 + \mu^2} + \mu) + \frac{2\mu^8 + 5\mu^6 + 3\mu^4 - \mu^2 - 1}{4\mu^2(1 + \mu^2)^3} \quad \ldots (754),$$

where $\mu^2 = m_2/m_1$. On substituting this into equation (750) we obtain a formula giving the average persistence of velocity for any ratio of masses.

361. It is readily verified that when $\mu = 1$ the value of this average persistence is ·406, as already found in § 357.

When μ is very small, formula (754) reduces to

$$\left(\frac{\alpha}{c}\right)_e = \frac{1}{3} + \frac{2}{15}\mu^2 - \frac{4}{35}\mu^4 + \text{terms in } \mu^6, \text{ etc.},$$

and similarly when μ is large, the expansion is

$$\left(\frac{\alpha}{c}\right)_e = \frac{1}{2} - \frac{1}{4\mu^2} + \frac{\log_e \mu}{4\mu^4} + \text{terms in } \frac{1}{\mu^6}, \text{ etc.}$$

From formulae (754) and (750), the following values can be calculated:

$\frac{m_2}{m_1}=$	0	$\frac{1}{10}$	$\frac{1}{5}$	$\frac{1}{2}$	1	2	5	10	∞
$\overline{\left(\frac{\alpha}{c}\right)_e}=$	·333	·335	·339	·360	·406	·432	·491	·498	·500
$\overline{\left(\frac{\alpha}{c}\right)}=$	1·000	·879	·779	·573	·406	·243	·152	·086	·000

These figures shew that the persistence is always positive but may have any value whatever, according to the ratio of the masses of the molecules.

362. For laws of force between molecules different from that between elastic spheres, the persistence of velocity will obviously be different from what it is for elastic spheres. Clearly, however, everything will depend on our definition of a collision. If we suppose that a very slight interaction is sufficient to constitute a collision, then the mean free path will be very short, while the persistence will be nearly equal to unity. If, on the other hand, we require large forces to come into play before calling a meeting of two molecules a collision, then the free path will be long, but the persistence will be small, or possibly even negative. In fact, the variations in the persistence of velocities just balance the arbitrariness of the standard we set up in defining a collision. This being so, it will be understood that the conception of persistence of velocities is hardly suited for use in cases where a collision is not a clearly defined event.

CHAPTER XI

VISCOSITY

363. In Chapters VIII and IX we developed a purely mathematical theory, which was found to lead to an explanation of the phenomena of viscosity, conduction of heat and diffusion of gases. This theory, although mathematically perfect, did not go far towards revealing the physical mechanism underlying the phenomena.

There is another method of treating these problems, in which we follow as closely as possible the physical processes which result in the phenomena. This method we now proceed to examine. It does not lead to results possessing the same mathematical exactness as the former method: its importance lies rather in its disclosure of the physical mechanism at work. Briefly, the three phenomena under consideration are regarded as *transport phenomena*—viscosity is a transport of momentum, conduction of heat is a transport of energy, and diffusion is a transport of mass. The mechanism of transport is provided by the free path; a molecule describing a free path of length λ is in effect transporting certain amounts of momentum, energy and mass through a distance λ. If the gas were in a steady state each such transport would be exactly balanced by an equal and opposite transport in the reverse direction, and the net transport would always be *nil*. But if the gas is not in a steady state there will always be an unbalanced residue, and this want of balance results in the phenomena we wish to study.

General Equations of Viscosity.

364. We begin by discussing the motion in a gas in which the mass-velocity varies from point to point. At the particular point considered, let us choose our axes so that the mass-velocity is parallel to the axis of x, while the surfaces of equal velocity are parallel to the plane of xy; thus in the neighbourhood of the point the mass-velocity is a function of z only.

Let us write μ for mu, the momentum of any single molecule in the direction of the x-axis. The mean value of μ at any point will be denoted by $\bar{\mu}$, where of course $\bar{\mu}$ varies from point to point in the gas. At the particular point considered we have chosen the direction of our axes so that the gas is arranged, as regards the distribution of $\bar{\mu}$, in a series of layers parallel to the plane of xy, so that $\bar{\mu}$ is a function of z only. We proceed to attempt to calculate the amount of μ which is transported by the molecular motion across any one of the planes $z =$ constant.

The physical principle underlying the calculation can easily be explained. To fix our ideas, let us suppose that the average value of μ increases as z increases, that the planes $z =$ constant are horizontal and that z increases as we move upwards as in fig. 21. The molecules will cross the planes $z =$ constant in both directions. Those which cross any plane, say $z = z_0$, in the downward direction will, however, be coming from regions in which the average value of μ per molecule is greater than it is over the plane $z = z_0$, and will therefore, on the average, possess a value of μ in excess of that appropriate to the plane $z = z_0$. In the same way, those molecules which cross this plane in the upward direction will, on the average, possess a value of μ smaller than that appropriate to the plane $z = z_0$. Since, however, there is no mass motion parallel to the axis of z, the number of molecules which cross the plane $z = z_0$ in one direction is exactly equal to the number which cross it in the opposite direction. There is, therefore, more momentum carried through the plane $z = z_0$ in the downward direction than in the upward direction. In other words, there is a downward transport of momentum.

365. As regards any single molecule which meets the plane $z = z_0$ at any point P, the amount of μ carried across the plane $z = z_0$ will of course depend, in actual fact, upon the whole past history of the molecule before reaching the point P. We are going to conduct our preliminary calculations upon the supposition that the history of the molecule previous to the last collision before meeting P, say at Q, is immaterial. This would be justifiable if, on the average, all past histories previous to the point Q were equally probable. This, unfortunately, is not so when the molecules are elastic spheres. From the persistence of velocities investigated in the last chapter, it follows that a molecule which is known to have arrived at P from Q has probably started originally from some point further away from P than the point Q. Since, however, the amount of the persistence depends in general on the particular molecular structure assumed, it will be simplest to neglect it altogether at first, and subsequently correct our results for it as far as possible.

366. Consider a molecule meeting the plane $z = z_0$ in P, having previously come from a collision at Q. Let the velocity components of the molecule be u, v, w, and let the velocity be regarded as consisting of two parts:

(i) a velocity u_0, of components $u_0, 0, 0$, equal to the mass-velocity of the gas at P;

(ii) a velocity c, of components $u - u_0, v, w$, the molecular-velocity of the molecule relatively to the gas at P.

In fig. 21, let QP be the actual path described in the gas before the molecule arrives at P. Let RP represent the distance travelled by the gas

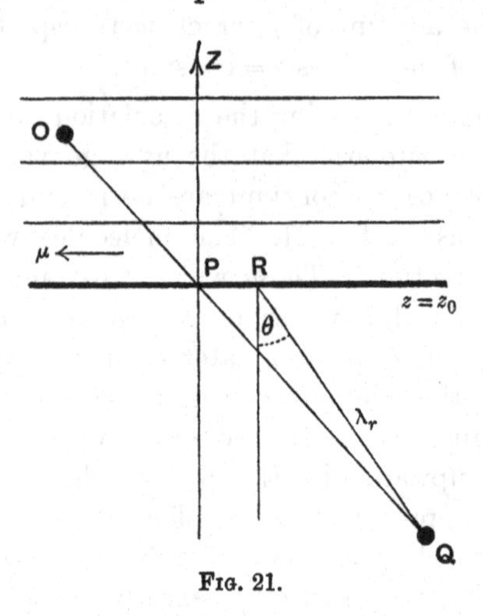

FIG. 21.

in the same interval of time, owing to its mass-velocity u_0, 0, 0. Then QR will represent the path described by the molecule relative to the mass-motion of the surrounding gas. Let the length QR be denoted by λ_r, and let this make an angle θ with the axis of z.

We are working upon the hypothesis that the expectation of μ for the molecule in question is that appropriate to the point Q. We shall therefore take it to be the mean value of μ at the point Q, of which the z coordinate is

$$z_0 - \lambda_r \cos\theta \qquad (755).$$

Since $\bar{\mu}$, the mean value of μ, is a function of z only, we can denote the value of $\bar{\mu}$ over the plane $z = \zeta$ by $\bar{\mu}(\zeta)$, and the value of $\bar{\mu}$ at Q will be

$$\bar{\mu}(z_0 - \lambda_r \cos\theta).$$

Since λ_r is small compared with the scale of variation in $\bar{\mu}$, this expression may be written as

$$\bar{\mu}(z_0) - \lambda_r \cos\theta \left(\frac{\partial\bar{\mu}}{\partial z}\right) \qquad (756).$$

This is the expectation of μ for any molecule which crosses the plane $z = z_0$, having a relative molecular-velocity c inclined at an angle θ to the axis of z.

Since all directions of this molecular-velocity may be regarded as equally probable, the probability of θ lying between θ and $\theta + d\theta$ is proportional to $\sin\theta\, d\theta$.

The number of molecules per unit volume which have relative molecular-velocities for which c and θ lie within specified small ranges dc, $d\theta$ may therefore be taken to be

$$\tfrac{1}{2}\nu f(c) \sin\theta \, d\theta \, dc,$$

where of course $\int_0^\infty f(c)\,dc = 1$.

The number of molecules having a velocity satisfying these conditions, which cross a unit area of the plane $z = z_0$ in time dt, is equal to the number which at any instant occupy a cylinder of base unity in the plane $z = z_0$ and of height $c\cos\theta\,dt$, and is therefore

$$\tfrac{1}{2}\nu c f(c) \cos\theta \sin\theta \, d\theta \, dc \, dt \quad \text{.....................(757)}.$$

Each molecule, on the average, carries with it the amount of momentum given by expression (756). The total momentum transferred across unit area of the plane by the molecules now under discussion is therefore

$$\tfrac{1}{2}\nu c f(c) \left\{ \bar{\mu}(z_0) - \overline{\lambda_r} \cos\theta \left(\frac{\partial \bar{\mu}}{\partial z}\right) \right\} \cos\theta \sin\theta \, d\theta \, dc \, dt,$$

where $\overline{\lambda_r}$ denotes the mean value of λ_r, and is therefore the mean value of the path QR relative to the moving gas of all molecules which move with a velocity c relative to the gas. Clearly $\overline{\lambda_r}$ is the same as the λ_c of § 343, and will therefore be replaced by λ_c.

On integrating the expression just found with respect to θ, we shall obtain the total transfer of momentum by all molecules with velocities between c and $c + dc$, whatever their direction. The limits for θ are 0 to π, values of θ from 0 to $\frac{1}{2}\pi$ covering molecules which cross the plane from below, and values of θ from $\frac{1}{2}\pi$ to π those which cross the plane from above. The result of this integration is

$$\tfrac{1}{3}\nu c f(c) \lambda_c \left(\frac{\partial \mu}{\partial z}\right) dc \, dt,$$

a negative sign indicating that the transfer is from above to below. On further integrating from $c = 0$ to $c = \infty$, we obtain for the total transport of momentum across unit area of the plane in time dt,

$$\tfrac{1}{3}\nu \left(\frac{\partial \bar{\mu}}{\partial z}\right) \int_0^\infty c\lambda_c f(c) \, dc \, dt = \tfrac{1}{3}\nu \left(\frac{\partial \bar{\mu}}{\partial z}\right) \overline{c\lambda_c} \quad \text{...........(758)},$$

where $\overline{c\lambda_c}$ denotes the mean value of $c\lambda_c$ averaged over all the molecules of the gas.

In the foregoing argument it might perhaps be thought that $\overline{\lambda_r}$ ought to be replaced by $\frac{1}{2}\lambda_c$ instead of by λ_c. For if QO (fig. 21) is the whole free path described before collision occurs, there is no reason why PO should be less than PQ, so that the probable value of PQ might be thought to be $\frac{1}{2}\lambda$.

The fallacy in this reasoning becomes obvious on considering that in selecting free paths at random by choosing points on these free paths, the longer free paths have a greater chance of being chosen than the shorter ones, the chance of any path being chosen being in fact exactly proportional to the length of the path. The average path chosen in this way, accordingly, will be much longer than that calculated in § 33. To see that λ_c is the right value to assign to $\overline{\lambda_r}$, we notice that after a molecule has left P, its chances of collision are exactly the same whether it has just undergone collision at P or has come undisturbed from Q. Hence $\overline{PO} = \lambda_c$, and therefore, by a similar argument, $\overline{PQ} = \lambda_c$.

A simple example taken from Boltzmann's *Vorlesungen** will perhaps elucidate the point further. In a series of throws with a six-faced die the average interval between two throws of unity is of course five throws. But starting from any instant the average number of throws since a unit throw last occurred will be five, and similarly, working back from any instant, the average number of throws since a unit throw occurred is also five.

367. We can conveniently suppose that

$$\overline{c\lambda_c} = \bar{c}l \quad\text{.................................(759),}$$

where l is a new quantity, which is of course the mean free path of a molecule, this mean being taken in a certain way. The way in which the mean has to be taken is not the same as any of the ways in which it was taken in the last chapter, so that we do not obtain an accurate result, in the case of elastic spheres, by replacing l by any of the known values of the mean free path. At the same time the mean values calculated in different ways will not greatly differ from one another, and as our present calculation is at best one of approximation, we shall be content for the moment to suppose l to be identical with the mean free path, however calculated. The extent of the error involved in this procedure will be examined later.

368. We have shewn that the aggregate transfer of momentum per unit of time across a unit area of a plane parallel to the plane of xy is

$$\tfrac{1}{3}\nu\bar{c}l\frac{\partial\bar{\mu}}{\partial z} \quad\text{.................................(760).}$$

Across the plane $z + dz$ the similar transfer is

$$\tfrac{1}{3}\nu\bar{c}l\left(\frac{\partial\bar{\mu}}{\partial z} + dz\frac{\partial^2\bar{\mu}}{\partial z^2}\right),$$

so that the gain of momentum to the layer between the planes z and $z + dz$ is

$$\tfrac{1}{3}\nu\bar{c}l\frac{\partial^2\bar{\mu}}{\partial z^2}dz.$$

Since μ is equal to mu, we may replace $\bar{\mu}$ by mu_0 and this expression assumes the form

$$\tfrac{1}{3}\nu\bar{c}lm\frac{\partial^2 u_0}{\partial z^2}dz \quad\text{.............................(761).}$$

* I. p. 72.

Also if we have a viscous fluid of coefficient of viscosity κ moving with the mass-velocity of the gas, of which the components are u_0, 0, 0, the force per unit area of the z plane in the direction of the axis of x, acting upon the layer of fluid enclosed by the planes z and $z + dz$, is

$$\kappa \frac{\partial u_0}{\partial z} \quad \text{.................................(762).}$$

Similarly that on the plane $z + dz$, acting in the other direction, is

$$\kappa \left(\frac{\partial u_0}{\partial z} + dz \frac{\partial^2 u_0}{\partial z^2} \right) \quad \text{..........................(763).}$$

The rate of increase of momentum per unit area of the layer between these two planes must be the resultant of the two forces (762) and (763), namely

$$\kappa \frac{\partial^2 u_0}{\partial z^2} dz.$$

This will be identical with expression (761) if

$$\kappa = \tfrac{1}{3} \nu \bar{c} l m \quad \text{.............................(764).}$$

We therefore see that our gas will behave exactly like a viscous fluid, of which the coefficient of viscosity is given by equation (764). If we replace $m\nu$ by ρ, this takes the simple form

$$\kappa = \tfrac{1}{3} \rho \bar{c} l \quad \text{............................(765).}$$

369. From the results of our analysis we can now obtain an insight into the molecular mechanics of viscosity in the case of a gas. Let us imagine two molecules, with velocities u, v, w and $-u, v, w$, penetrating from a layer at which the mass-velocity is 0, 0, 0 to one at which it is u_0, 0, 0. By the time the molecules have reached this second layer we must suppose that their velocities are divided into two parts, namely,

$$u - u_0,\ v,\ w \ \text{ and } \ u_0,\ 0,\ 0$$

for the first, and

$$-u - u_0,\ v,\ w \ \text{ and } \ u_0,\ 0,\ 0$$

for the second. The first part in each case will represent molecular-motion, and the second part will represent mass-motion. Now in equation (32), we saw that the total energy of the gas could be regarded as the sum of the energies of the molecular and mass-motions. The sum of the energies of the molecular-motions of the two molecules now under discussion is, however,

$$\tfrac{1}{2} m \left[(u - u_0)^2 + v^2 + w^2 \right] + \tfrac{1}{2} m \left[(-u - u_0)^2 + v^2 + w^2 \right],$$

which can be written

$$m (u^2 + v^2 + w^2) + m u_0^2.$$

The first term is equal to the energy of the molecular-motion of the two molecules at the start; the second term represents an increase which must be regarded as gained at the expense of the mass-motion of the gas. Thus the phenomenon of viscosity in gases consists essentially in the degradation of the energy of mass-motion into energy of molecular-motion; it is therefore accompanied by a rise of temperature in the gas.

Corrections when Molecules are assumed to be Elastic Spheres.

370. From want of definite knowledge of the molecular structure two errors have been introduced into our calculations. In the first place we have neglected the persistence of velocities after collision, and in the second place we have ignored the difference between two different ways of estimating the mean free path. If the molecules are assumed to be elastic spheres, it is possible to estimate the amount of error introduced by both these simplifications.

We may begin by an exact calculation of $\overline{c\lambda_c}$, to replace the assumption of equation (759). The quantity required is

$$\overline{c\lambda_c} = \int_0^\infty f(c)\,\lambda_c c\,dc \qquad \text{.......................(766)},$$

where λ_c is the same as the λ_c of § 343. Substituting the value given for λ_c by equation (717), and putting

$$f(c) = \sqrt{\frac{h^3 m^3}{\pi^3}}\, 4\pi c^2 e^{-hmc^2},$$

we find $$\overline{c\lambda_c} = \int_0^\infty \frac{4(hm)^{\frac{3}{2}} c^5 e^{-hmc^2}}{\pi\nu\sigma^2 \psi(c\sqrt{hm})}\,dc = \frac{4}{\pi\sqrt{hm}\,\nu\sigma^2}\int_0^\infty \frac{x^5 e^{-x^2}\,dx}{\psi(x)}.$$

Thus if l is defined by equation (759), we must take

$$l = \frac{\overline{c\lambda_c}}{\bar{c}} = \overline{c\lambda_c}\sqrt{2hm} = \frac{8}{\sqrt{2}\,\pi\nu\sigma^2}\int_0^\infty \frac{x^5 e^{-x^2}\,dx}{\psi(x)}.$$

The integral can only be evaluated by quadrature. Tables for its evaluation are given by Tait*. The integral has also been evaluated by Boltzmann†, whose result agrees to three significant figures with that obtained by Tait.

Using this value for the integral, it is found that

$$l = 1{\cdot}051\,\frac{1}{\sqrt{2}\,\pi\nu\sigma^2} \qquad \text{.......................(767)}.$$

The value of l, calculated accurately for our present purpose, accordingly differs only by about 5 per cent. from Maxwell's mean free path calculated in § 33.

* *Collected Works*, II. pp. 152 and 178.

† *Wiener Sitzungsber.* LXXXIV. p. 45 (1881).

371. We turn now to the more serious error which has been introduced by ignoring the persistence of velocities. In a gas of which the molecules are elastic spheres, we have found that this persistence is measured by a numerical factor which is always intermediate between $\frac{1}{2}$ and $\frac{1}{3}$, and of which the mean value, averaged over all collisions, is ·406.

If, on the average, each particle has described a path of which the projection on the axis of z is ζ, with a velocity of which the component parallel to the direction of the axis of z is w, then, on tracing back the motion, we know that as regards the previous path of each molecule the expectation of average velocity parallel to the axis of z is θw, where θ measures the persistence. The expectation of the projection of this path on the axis of z may therefore be taken to be $\theta\zeta$. Similarly, the expectation of the projection of each of the paths previous to these may be taken to be $\theta^2\zeta$, and so on.

It follows that if we trace the motion a sufficient distance back, each molecule must be supposed to have come, not from a distance ζ measured along the axis of z, but from a distance

$$\zeta + \theta\zeta + \theta^2\zeta + \ldots = \frac{\zeta}{1-\theta} \quad \text{.....................(768).}$$

We must not, however, assume that each molecule on arriving at the plane $z = z_0$ has, on the average, a value of μ appropriate to the plane $z = z_0 + \frac{\zeta}{1-\theta}$. For the molecule has not travelled a distance $\frac{\zeta}{1-\theta}$ undisturbed, and at each collision a certain amount of its excess of momentum will have been shared with the colliding molecule. Of the various simple assumptions possible, the most obvious one to make is that at each collision, the excess of momentum above that appropriate to the point at which the collision takes place is halved, half going to the colliding molecule and half remaining with the original molecule. Making this assumption, it is clear that the excess of momentum to be expected is not that due to having travelled undisturbed a distance equal to that given by expression (768), but a distance

$$\zeta + \tfrac{1}{2}\{\theta\zeta + \tfrac{1}{2}[\theta^2\zeta + \tfrac{1}{2}(\theta^3\zeta + \ldots)]\} = \frac{\zeta}{1-\frac{1}{2}\theta} \quad \text{...........(769).}$$

Taking $\theta = $ ·406, this becomes $1·255\zeta$.

372. It follows that the persistence of velocities, when the molecules are elastic spheres, can roughly be allowed for by supposing the free path in the viscosity formula to be the mean free path multiplied by a factor 1·255. Combining this with expression (767), we find that the free path in the viscosity formula must be taken to be

$$l = \frac{1·317}{\sqrt{2}\pi\nu\sigma^2}.$$

A better approximation can be obtained by inserting the factor $1/(1-\frac{1}{2}\theta)$ in the integrand of § 370 before integration, the value of θ being obtained from the table on p. 263. As the result of a rough integration by quadratures, I find

$$l = \frac{1 \cdot 382}{\sqrt{2}\pi\nu\sigma^2} \quad \text{..........................(770).}$$

Hence for a gas, in which the molecules are elastic spheres, the viscosity coefficient (equation (765)) is given approximately by

$$\kappa = \tfrac{1}{3}\rho\bar{c}l = \cdot 461 \frac{m\bar{c}}{\sqrt{2}\pi\sigma^2} \quad \text{.....................(771).}$$

373. This formula, although undoubtedly better than formula (765), is still only an approximation. It would doubtless be possible to improve on the rough assumptions just made, and so obtain results still closer to the truth.

This, however, seems unnecessary, since exact numerical results are obtainable by the mathematical methods of Chapters VIII and IX. Chapman, following Maxwell's method, arrived in his first paper* at the formula

$$\kappa = \frac{5\pi}{32} \frac{m\bar{c}}{\sqrt{2}\pi\sigma^2} = \cdot 491 \frac{m\bar{c}}{\sqrt{2}\pi\sigma^2} \quad \text{...................(772).}$$

This, as already explained†, could only be regarded as an approximation, since the function Φ was assumed to be limited to terms of the third degree. In a later paper‡, Chapman examined the error of this approximation by taking successive further approximations, and found that the successive further approximations in turn require that the numerical factor ·491 should be multiplied by 1·01485, 1·01588, 1·01606. These numbers clearly converge rapidly to a number which, to three places of decimals, may be taken to be 1·016, and we may therefore suppose that the true value of κ is 1·016 times that given by equation (772), or

$$\kappa = \cdot 499 \frac{m\bar{c}}{\sqrt{2}\pi\sigma^2} \quad \text{..........................(773).}$$

Enskog, in the paper already referred to§, also takes formula (772) as a first approximation, and calculates two further approximations for all inverse powers of the distance. For elastic spheres, his last approximation agrees with the second approximation of Chapman, the correcting factor being 1·0159. Thus formula (773) may be regarded as fully established.

* *Phil. Trans.* 211 A (1911), p. 433.

† See footnote, p. 219.

‡ *Phil. Trans.* 216 A (1915), p. 279.

§ See footnote to p. 229.

Variation of κ with Density.

374. Equation (773) shews that theoretically κ is independent of the density of the gas, when the molecules are assumed to be elastic spheres.

Indeed, whatever structure we assume for the molecules of the gas, it is clear that l will, to a first approximation, vary inversely as the number of molecules per unit volume of the gas. Hence equation (765) gives a value of κ which is independent of ν, and we obtain Maxwell's law:

The coefficient of viscosity of a gas is independent of its density.

In spite of its apparent improbability, this law was predicted by Maxwell on purely theoretical grounds, and its subsequent experimental confirmation has constituted one of the most striking triumphs of the Kinetic Theory.

Some of the physical consequences of this law are interesting, and occasionally surprising. For instance, according to the well-known law of Stokes, the final steady velocity of a sphere falling through a viscous fluid is given by

$$v = \frac{g(M - M_0)}{6\pi a\kappa},$$

where a, M are the radius and mass of the sphere, and M_0 the mass of fluid displaced. Since κ is, by Maxwell's law, independent of the density, it follows that, within the limits within which Stokes's law is true, the final velocity of a sphere falling through air or any other gas, will be independent of the density of the gas, or more strictly will depend on the density of the gas only through the term $M - M_0$, which will differ only inappreciably from M. Thus, a small sphere will fall as rapidly through a dense gas as through a rare gas. Again the air-resistance experienced by a pendulum ought to be independent of the density of the air, so that the oscillations of a pendulum ought to die away as rapidly in a rare gas as in a dense gas, as was found to be the case by Boyle in 1660*.

At the same time it ought to be mentioned that Maxwell's law is by no means completely confirmed by experiment. Meyer† gives a detailed and full account of a variety of experiments which have been designed and performed in order to test this law, and concludes that the divergences from the law found by experiment are not sufficiently great to invalidate the law within the limits of pressure from 1 to $\frac{1}{60}$ atmosphere‡.

The most interesting example of such experiments is perhaps found in a set by Maxwell himself§. He suspended three parallel and coaxal circular

* Thomson and Poynting, *Properties of Matter*, p. 218.

† *Kinetic Theory of Gases*, p. 181.

‡ See also Winkelmann's *Handbuch der Physik* (IIte aufl.) I. pp. 1399, 1406.

§ *Phil. Trans.* 156 (1866), p. 249.

discs horizontally on a common axis by a torsion thread in such a way that they could oscillate between four parallel fixed discs. If the law in question were true, the oscillations ought, as in Boyle's pendulum experiment, to die away at the same rate whether the air were dense or rare, or at least ought only to vary by a small difference of the nature of that found above in discussing Stokes's law. The following table shews the values of the logarithmic decrement found experimentally by Maxwell, and also the values calculated by him on the assumption of constancy of the coefficient of viscosity:

Pressure (inches of mercury)	Logarithmic decrement	
	Observed	Calculated
0·54	0·157	0·156
5·68	0·156	0·156
20·09	0·152	0·153
29·29	0·153	0·154

At considerably higher pressures, Maxwell's law fails altogether for certain gases. For instance, the following values give the viscosity of carbon-dioxide at high pressures* and at a temperature of 32·6° C.

Pressure (atmospheres)	Density	κ
60·3	0·170	0·000189
69·9	0·240	0·000214
74·6	0·310	0·000241
76·6	0·380	0·000273
77·2	0·450	0·000315
77·6	0·520	0·000367
78·2	0·590	0·000426
80·7	0·660	0·000496
88·5	0·730	0·000575
107·3	0·800	0·000678

A similar dependence of viscosity on density has also been observed in hydrogen at moderate pressures by Kamerlingh Onnes, Dorsman and Weber†.

* Experiments by Warburg and v. Babo (*Wied. Ann.* XVII. (1882), p. 390, and *Berlin. Sitzungsber.* (1882), p. 509). The numbers here given are those of Warburg after correction by Brillouin (*Leçons sur la viscosité des fluides*, 1907).

† *Communications from the Leiden Phys. Laboratory*, 134 *a* (1913).

At the other limit of excessively small pressure, a remarkable departure from Maxwell's law may also occur, owing to the free path becoming comparable with, or even greater than, the dimensions of the vessel in which the experiment is conducted. If the molecule has not room to describe a free path equal to the theoretical free path assumed in § 366, the resulting formula obtained for the viscosity must obviously fail. If l cannot, from the arrangement of the apparatus, be greater than some value l_0, then κ (cf. equation (765)) cannot be greater than $\frac{1}{3}\rho\bar{c}l_0$, and so ought to vanish with ρ. This is found to be the case. Crookes* measured the viscosities of gases at pressures of only a few thousandths of a millimetre of mercury, and obtained values much smaller than those at higher pressures, which tended to vanish altogether as the density of the gas vanished.

Variation of κ with Temperature.

375. Since $\bar{c}$ is proportional to the square root of the absolute temperature, it appears from formula (773) that if the molecules were true elastic spheres, the value of κ would be proportional to the square root of the temperature.

As a matter of fact, it is found that κ varies a good deal more rapidly than this as the temperature increases. The divergence between experiment and the theoretical value obtained on the assumption that the molecules are elastic spheres is, however, one that could have been predicted. This assumption is, at best, only an approximation, and we must continually examine what deviations are to be expected from the results to which it leads.

The peculiarity of a system of elastic spheres is that the motion remains geometrically the same if the velocity of each sphere is increased in the same ratio. Thus in order to determine the motion of two spheres after collision it is only necessary to know the directions of motion before collision, and the *ratio* of their velocities; we are not concerned with the actual values of these velocities. When, on the other hand, we suppose the molecules to be surrounded by fields of force, this ceases to be true: the paths after collision do not depend solely on the *ratio* of the velocities, but on the absolute magnitudes of these velocities. For instance, in fig. 22 let $OPQR$, $O'P'Q'R'$ be the paths described when two molecules surrounded by fields of force meet one another, the figure being drawn, for the sake of simplicity, for the case in which the two velocities are equal and opposite. If now we suppose the molecules moving along the same lines before encounter, but each with double its former velocity, the paths will be different. For obviously the higher velocity will carry each molecule further into the other's field of force before the centres of the two molecules reach their shortest distance apart, so that the path described, instead of being $OPQR$, will be, let us say, $OPST$.

* *Phil. Trans.* 172, p. 387.

We can, however, by an obvious geometrical construction find the size of two spheres which would describe paths having the same deflections as *OPQR* or *OPST*. If we perform the construction for the two paths of fig. 22 we find that the size of the sphere for the former path described with small velocity is greater than for the latter path described with large velocity. Thus, if we attempt to represent our molecules by spheres, the size of these spheres must be supposed to decrease as the mean molecular velocity increases, and therefore as the temperature rises.

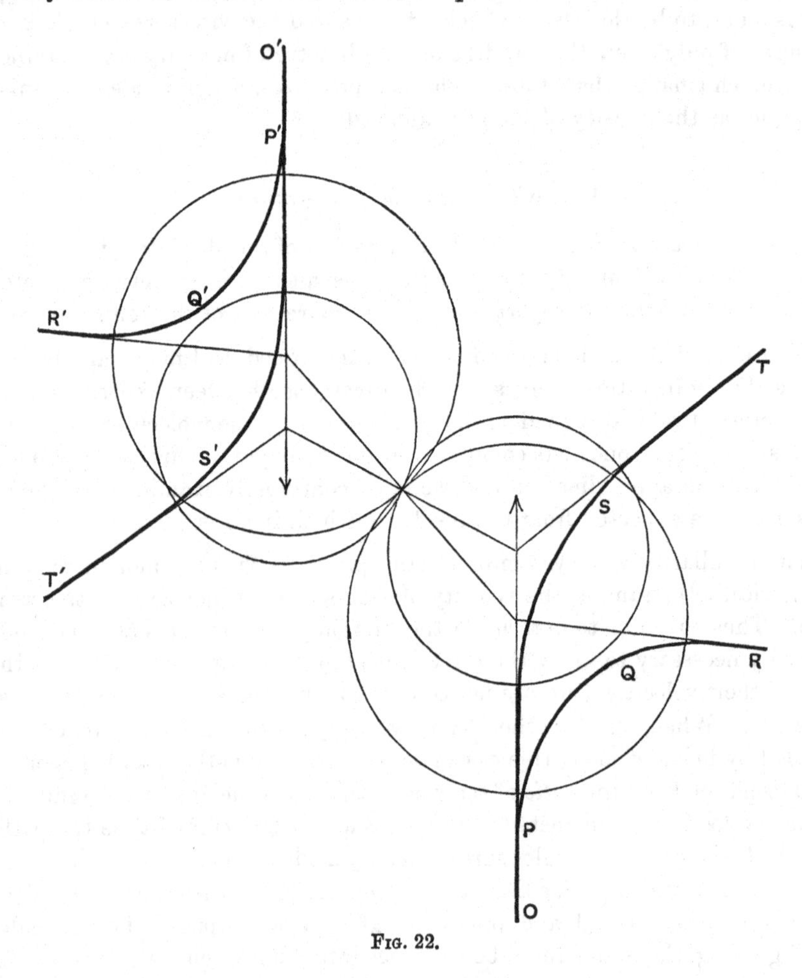

FIG. 22.

376. Thus in formula (773), κ must be supposed to depend on the temperature both through the factor $\bar{c}$ in the numerator, and also through the factor σ^2 in the denominator. The value of κ will accordingly not vary as the square root of the temperature, but will vary with the temperature more rapidly than this.

In Meyer's *Kinetic Theory of Gases**, there will be found a full account of experiments to test the variation of κ with temperature. The following table gives the variation of κ with temperature found by Schultze† for the gases helium and argon, together with the corresponding values of $\frac{1}{2}\sigma$ calculated from them by use of formula (773).

Gas	Temperature	κ (observed)	$\frac{1}{2}\sigma$ (calculated)
Helium......	15·3° C.	·0001969	$1{\cdot}08 \times 10^{-8}$
"	99·6° C.	·0002348	$1{\cdot}04 \times 10^{-8}$
"	184·6° C.	·0002699	$1{\cdot}02 \times 10^{-8}$
Argon	14·7° C.	·0002208	$1{\cdot}81 \times 10^{-8}$
"	99·7° C.	·0002733	$1{\cdot}74 \times 10^{-8}$
"	183·7° C.	·0003224	$1{\cdot}68 \times 10^{-8}$

377. A table expressing the variation of σ with the temperature will give some information, although slight, as to the field of force surrounding the molecules. For the calculated value of σ is, roughly speaking, the average distance of closest approach of the centres of two molecules in collision, so that the mutual potential energy of two molecules at a distance σ is, on the average, equal to the kinetic energy of the velocities along the line of centres before collision.

From formula (52), the average value of V^2, the square of the relative velocity before collision, is $\frac{8}{3}C^2$. Thus the square of the velocity of each molecule relatively to the centre of gravity of the two colliding molecules will be, on the average, $\frac{2}{3}C^2$. The probability that the direction of this velocity makes an angle with the line of centres, which lies between θ and $\theta + d\theta$, is $2 \sin\theta \cos\theta \, d\theta$, so that the average square of the relative velocity along the line of centres $V \cos\theta$ is

$$\frac{2}{3}C^2 \int_0^{\pi/2} 2 \sin\theta \cos^3\theta \, d\theta = \frac{1}{3}C^2.$$

The kinetic energy which has been destroyed by the intermolecular field of force when the molecules are, on the average, at their point of closest approach at distance σ apart is therefore $\frac{1}{3}mC^2$ or RT.

Thus we may say that the mutual potential energy of two molecules at a distance σ apart will be RT, where T is the temperature corresponding to the value of σ in question. The force of repulsion between two molecules at a distance σ is accordingly $-R\dfrac{dT}{d\sigma}$.

* § 85, English translation, p. 215 *et seq.*

† *Ann. d. Phys.* VI. (1901), p. 302 for helium, and V. (1901), p. 140 for argon.

For instance if the law of force is μr^{-s}, we must have

$$\frac{\mu}{\sigma^s} = -R\frac{dT}{d\sigma},$$

giving on integration

$$\sigma = \left[\frac{\mu}{RT(s-1)}\right]^{\frac{1}{s-1}} \quad\text{.......................(774).}$$

In this argument σ has been taken to be the distance of closest approach of two molecules at an encounter, and when the orbits are at all curved, this is not quite the same thing as the diameter of the equivalent sphere obtained by a construction such as that of fig. 22. Thus equation (774) will give a value of σ which will be in error by a numerical multiplier. This multiplier will of course vary for different values of s. It will reduce to unity for elastic spheres, and will differ most from this for the smallest values of s.

378. It will be remembered that in § 172 we found that molecules with a law of force μr^{-s} could be regarded as elastic spheres for the purpose of calculating the pressure, if σ were supposed given by

$$\sigma = \left[\frac{\mu}{RT(s-1)}\right]^{\frac{1}{s-1}} \sqrt[3]{\Gamma\left(1-\frac{3}{s-1}\right)} \quad\text{...........(775),}$$

which agrees with (774) except for the numerical factor. We see that molecules which are really point centres of force may be treated as elastic spheres, both as regards pressure and viscosity, but the spheres must be of different sizes in the two cases.

When s is nearly infinite—*i.e.* when the molecules are very hard—formulae (774) and (775) become identical, but for smaller values of s, the divergence between them becomes very considerable. The lowest value for s which can be supposed to occur for any gas is probably about $s = 5$ (cf. § 380, below), and when $s = 5$,

$$\sqrt[3]{\Gamma\left(1-\frac{3}{s-1}\right)} = \sqrt[3]{\Gamma\left(\frac{1}{4}\right)} = 1{\cdot}5363 \quad\text{...............(776).}$$

Thus for such a gas as carbon-dioxide, for which $s = 5{\cdot}2$, we may expect a difference of as much as 50 per cent. between the values of σ calculated from viscosity and Boyle's law.

In such a case as this, however, the calculation from Boyle's law fails because b, which from equation (341) ought to vary as $T^{-\frac{3}{s}}$, is supposed, in evaluating b experimentally, to remain independent of the temperature.

379. Whatever the value of the numerical multiplier may be, it appears that $\frac{1}{\sigma^2}$ will vary as $T^{\frac{2}{s-1}}$, so that κ will vary as T^n, where

$$n = \tfrac{1}{2} + \frac{2}{s-1} \quad\text{..............................(777).}$$

It is of interest to notice, as was first pointed out by Lord Rayleigh*, that this result could have been obtained purely from a consideration of physical dimensions, without any exact analysis or detailed study of the mechanism of viscosity.

For κ, the coefficient of viscosity, can only depend on the following quantities: m and C which measure the mass and mean velocity of the molecules, μ which measures the distance at which their action on one another reaches a certain intensity (replacing the "size" of the molecules, which has now become meaningless), and ν the number of molecules per cubic centimetre. It is clear, as in § 374, that κ must be independent of ν, so that κ must be expressible as a function of m, C and μ.

The physical dimensions of κ, m, C and μ are as follows:

$$\begin{array}{llll} \kappa & \text{is of} & \text{dimensions} & ML^{-1}T^{-1}, \\ m & \text{,,} & \text{,,} & M, \\ C & \text{,,} & \text{,,} & LT^{-1}, \\ \mu & \text{,,} & \text{,,} & ML^{s+1}T^{-2}. \end{array}$$

Hence κ must be proportional to

$$(m^{s+1} C^{s+3} \mu^{-2})^{\frac{1}{s-1}},$$

this being the only way of combining m, C and μ, so as to get a quantity of the same physical dimensions as κ. If κ is observed to vary as the nth power of the absolute temperature, and therefore as the $2n$th power of C, we have the relation

$$2n = \frac{s+3}{s-1} \qquad \text{.............................(778),}$$

which is the same relation as is given by equation (777).

380. For a great number of substances, it is found as a matter of experiment that κ varies approximately as a power of T, being represented with very tolerable accuracy by the formula

$$\kappa = \kappa_0 \left(\frac{T}{273 \cdot 1}\right)^n \qquad \text{..........................(779),}$$

where κ_0 is of course the coefficient of viscosity at 0° C.

The molecules of such substances may be regarded as point centres of force, repelling according to the law μ/r^s, where s is given by equation (778).

The following table gives the values of n observed for various substances, together with the values of s calculated from relation (778). An instance of the closeness of agreement between formula (779) and observation will be found below (§ 383).

* *Proc. Roy. Soc.* LXVI. p. 68.

VALUES OF n AND s.

Gas	Authority*	Value of n (observed)	Value of s (calculated)
Hydrogen	1	·681	12·05
	4	·70	11·0
Helium	1	·681	12·05
	2	·6852	11·80
	3	·647	14·6
Nitrogen..............	4	·74	9·3
Carbon-monoxide ...	4	·74	9·3
Air	1	·754	8·87
Oxygen	1	·782	8·09
	4	·80	7·7
Argon	1	·815	7·36
	2	·8227	7·19
Nitrous oxide.........	4	·93	5·6
Carbon-dioxide	4	·98	5·2

381. *Sutherland's Formula.* The supposition that the molecular force falls off as an inverse power of the distance leads to formula (774) which requires σ to vanish absolutely at high temperatures. It seems more probable that a molecule possesses a hard kernel which is not penetrated by other molecules no matter how violent the collision between them may be.

Sutherland's formula† is based upon certain physical assumptions which amount to assuming that the effective value of σ at temperature T is

$$\sigma^2 = \sigma_\infty^2\left(1+\frac{C}{T}\right) \qquad (780),$$

where C, σ_∞ are constants, σ_∞ being the value of σ when $T=\infty$, and therefore being the diameter of the hard kernel of the molecule, while C is the temperature at which $\sigma^2 = 2\sigma_0^2$.

If l is the free path at temperature T and l_0 its value at 0° C., we must have

$$\frac{l}{l_0} = \frac{\sigma^2{}_{\theta=0}}{\sigma^2} = \frac{1+\dfrac{C}{273\cdot 1}}{1+\dfrac{C}{T}},$$

* Authorities:

1. Lord Rayleigh, *Proc. Roy. Soc.* LXVI. p. 68, and *Collected Scientific Papers*, IV. pp. 452 and 481.
2. Schultze, *Ann. d. Phys.* V. p. 163, and VI. p. 310.
3. Kamerlingh Onnes and Sophus Weber, *Communications from the Leiden Phys. Laboratory*, 134 *b*, p. 18.
4. von Obermayer, *Wiener Sitzungsber.* LXXIII. (2), p. 433.

† *Phil. Mag.* [5], XXXVI. (1893), p. 507.

which is Sutherland's formula for the free path. From this it at once follows that

$$\frac{\kappa}{\kappa_0} = \frac{\bar{c}l}{(\bar{c}l)_{\theta=0}} = \left(\frac{T}{273\cdot1}\right)^{\frac{1}{2}} \frac{1 + \dfrac{C}{273\cdot1}}{1 + \dfrac{C}{T}},$$

or

$$\kappa = \kappa_0 \left(\frac{T}{273\cdot1}\right)^{\frac{3}{2}} \frac{C + 273\cdot1}{C + T},$$

and this is Sutherland's formula for the viscosity at temperature T.

382. For many gases this formula meets with very considerable success in measuring the variation of viscosity with temperature. As an illustration may be given the following tables, taken from a paper by Breitenbach*, in which the observed and calculated values of the viscosity are compared.

Ethylene
($\kappa_0 = \cdot00009613$, $C = 225\cdot9$)

Temperature	κ (observed)	κ (calculated)
$-21\cdot2°$ C.	·0000891	·0000890
15·0	1006	1012
99·3	1278	1278
182·4	1530	1519
302·0	1826	1833

Carbon-dioxide
($\kappa_0 = \cdot00013879$, $C = 239\cdot7$)

Temperature	κ (observed)	κ (calculated)
$-20\cdot7°$ C.	·0001294	0001284
15·0	1457	1462
99·1	1861	1857
182·4	2221	2216
302·0	2682	2686

The following are the values for C found by different observers:

Helium	$C = 80\cdot3$ (Schultze), 78·2 (Schmitt).
Argon	$C = 169\cdot9$ (Schultze), 174·6 (Schmitt).
Krypton	$C = 142$ (Rankine).
Xenon	$C = 252$ (Rankine).
Hydrogen	$C = 72\cdot2$ (Rayleigh), 71·7 (Breitenbach), 79 (Sutherland), 83 (Schmitt).
Nitrogen	$C = 110\cdot6$ (Bestelmeyer), 113 (Schmitt).
Carbon-monoxide	$C = 100$ (Sutherland).
Air	$C = 111\cdot3$ (Rayleigh), 119·4 (Breitenbach), 113 (Sutherland).
Nitric oxide	$C = 195$ (Sutherland).
Oxygen	$C = 127$ (Sutherland), 138 (Schmitt).
Chlorine	$C = 199$ (Sutherland).
Nitrous oxide	$C = 260$ (Sutherland).
Carbon-dioxide	$C = 239\cdot7$ (Breitenbach), 277 (Sutherland).
Ethylene	$C = 225\cdot9$ (Breitenbach), 272 (Sutherland).
Methyl chloride	$C = 454$ (Breitenbach).

* *Ann. d. Phys.* VI. p. 168.

383 On the other hand Kamerlingh Onnes* finds very definitely that the viscosity of helium at low temperatures cannot be represented by Sutherland's formula with anything like the accuracy given by the simpler formula (779). This is shewn in the following table: the first column gives the values of κ observed for helium, the second column gives the values calculated from formula (779) on taking $\kappa_0 = \cdot0001887$, $n = \cdot647$, while the third column gives values of κ calculated by Sutherland's formula, taking $C = 78{\cdot}2$.

VISCOSITY OF HELIUM.

Temperature	κ (observed)	$\kappa_0\left(\frac{T}{273{\cdot}1}\right)^{\cdot647}$	κ (calculated, Sutherland)
183·7° C.	·0002681	·0002632	·0002682
99·8	2337	2309	2345
18·7	1980	1970	1979
17·6	1967	1965	1974
−22·8	1788	1783	1771
−60·9	1587	1603	1563
−70·0	1564	1558	1513
−78·5	1506	1515†	1460
−102·6	1392	1389	1317
−183·3	09186	09185	0745
−197·6	08176	08213	0628
−198·4	08132	08155	0621
−253·0	03498	03489	0135
−258·1	02946	02887	0092

Very similar results have also been obtained for hydrogen by Kamerlingh Onnes, Dorsman and Weber‡.

The general failure of Sutherland's formula to represent viscosity at low temperatures has been noticed and discussed by Schmitt, Bestelmeyer, Vogel and others§.

GENERAL FORMULA FOR THE COEFFICIENT OF VISCOSITY.

Law of Force μr^{-s}.

384. We have seen that molecules attracting according to the law μr^{-s} may be treated as elastic spheres having an effective diameter given, except

* Kamerlingh Onnes and Sophus Weber, *Communications from the Leiden Phys. Laboratory*, 134 *b*, p. 18.

† This entry, which was obviously wrong in the original table, has been recalculated.

‡ *Communications from the Leiden Phys. Laboratory*, 134 *a* (1913).

§ For references see Chapman, *Phil. Trans.* 216 A (1915), p. 342.

for a multiplying constant, by equation (774), and on substituting this value for σ into equation (765) or (773) it appears that the coefficient of viscosity must be given by an equation of the form

$$\kappa = A\sqrt{mRT}\left[\frac{RT(s-1)}{\mu}\right]^{\frac{2}{s-1}} \quad \text{.................(781)},$$

where A is a numerical constant.

Chapman*, using the method already explained, has determined the value of this constant. To a first approximation he found its value to be

$$A = \frac{5\sqrt{\pi}}{8I_2(s)\,\Gamma\left(4-\frac{2}{s-1}\right)} \quad \text{...................(782)},$$

where $I_2(s)$ is the number defined by equation (575). With this value for A, equation (781) reduces to Maxwell's exact formula (679) when $s = 5$, and to Chapman's approximate formula (772) when $s = \infty$.

In his second paper, Chapman carries the calculations to a second approximation, and finds that the value of A given by equation (782) must be multiplied by a factor which increases continuously from unity when $s = 5$ to 1·01485 when $s = \infty$. This last number of course agrees with Chapman's second approximation for elastic spheres already given in § 373. This indicates that the error in using approximation (782) for A is never more than about $1\frac{1}{2}$ per cent., and as this is smaller than experimental errors of observation, it is hardly worth carrying the approximation further. Before leaving this question, it may be remarked that Enskog† has given the factor by which A must be multiplied when the molecules repel as the inverse sth power of the distance in the form

$$1 + \frac{3(s-5)^2}{2(s-1)(101s-113)} + \dots.$$

This of course reduces to unity when $s = 5$ and has the value $1\frac{3}{202}$ or 1·01485, when $s = \infty$, thus agreeing with Chapman's value already quoted.

Viscosity in a Mixture of Gases.

385. As the proportions of two kinds of gas in a mixture change from 1 : 0 to 0 : 1, the coefficient of viscosity of the mixture will of course also change, starting from the coefficient of viscosity κ_1 of the first gas, and ending at the coefficient of viscosity of the second gas κ_2. But for certain pairs of gases, it is found that the change is not a continuous one, and the coefficient of viscosity of the mixture κ_{12} may for certain proportions of the mixture have a value greater than either of the coefficients of viscosity κ_1, κ_2 of the pure gases.

* See footnote on p. 219.

† *l.c. ante* (see footnote to p. 229).

The theoretical investigation of viscosity in a mixture of gases is very complicated. Formulae for the coefficients of viscosity of mixture have been given by Maxwell, Kuenen, Chapman and Enskog, and in every case the theory predicts a maximum value for a certain ratio of the gases in accordance with observation. Maxwell's investigation* deals only with molecules repelling as the inverse fifth power of the distance; Kuenen† deals with elastic spheres, the formulae being corrected for the phenomenon of "per-

Gas	Molec. Weight	κ (observed)	Authority‡	Assumed κ at 0° C.	$\frac{1}{2}\sigma$ (calculated) (cms.)
Hydrogen............	H_2 : 2	·00008822 (23° C.)	1	·0000857	$1{\cdot}36 \times 10^{-8}$
		·00008574 (0° C.)	2		
Helium..............	He : 4	·0001969 (15° C.)	3	·000189	$1{\cdot}09 \times 10^{-8}$
		·0001887 (0° C.)	4		
Water-vapour......	H_2O : 18	·0000904 (0° C.)	5	·0000904	$2{\cdot}29 \times 10^{-8}$
Carbon-monoxide..	CO : 28	·000163 (0° C.)	6	·000163	$1{\cdot}90 \times 10^{-8}$
Ethylene	C_2H_4 : 28	·0000961 (0° C.)	2	·0000961	$2{\cdot}78 \times 10^{-8}$
Nitrogen	N_2 : 28	·00017648 (23° C.)	1	·000167	$1{\cdot}89 \times 10^{-8}$
		·0001674 (0° C.)	4		
Air	—	·00018227 (23° C.)	7	·000172	$1{\cdot}87 \times 10^{-8}$
		·0001725 (0° C.)	8		
Nitric oxide.........	NO : 30	·0001794	9	·000179	$1{\cdot}88 \times 10^{-8}$
Oxygen..............	O_2 : 32	·00020423 (23° C.)	1	·000192	$1{\cdot}81 \times 10^{-8}$
		·0001926 (0° C.)	4		
Argon	Ar : 40	·0002208 (15° C.)	10	·000211	$1{\cdot}83 \times 10^{-8}$
		·000211 (0° C.)	4		
Carbon-dioxide ...	CO_2 : 44	·0001388 (0° C.)	2	·000139	$2{\cdot}31 \times 10^{-8}$
Nitrous oxide	N_2O : 44	·0001353 (0° C.)	8	·000135	$2{\cdot}35 \times 10^{-8}$
Methyl chloride ...	CH_3Cl : 50	·0000988 (0° C.)	2	·0000988	$2{\cdot}83 \times 10^{-8}$
Ethyl chloride......	C_2H_5Cl : 64	·0000935	6	·0000935	$3{\cdot}09 \times 10^{-8}$
Chlorine	Cl_2 : 71	·0001287	6	·000129	$2{\cdot}70 \times 10^{-8}$
Benzene	C_6H_6 : 78	·0000700	8	·0000700	$3{\cdot}75 \times 10^{-8}$
Krypton	Kr : 83	·000246 (16° C.)	11	·000238	$2{\cdot}07 \times 10^{-8}$
Xenon	Xe : 130	·000222 (15° C.)	11	·000214	$2{\cdot}44 \times 10^{-8}$

* *Collected Works*, II. p. 1.

† *Konink. Akad. Wetenschappen, Amsterdam, Proc.* 16 (1914), p. 1162, and 17 (1915), p. 1068. *Communications from the Leiden Phys. Laboratory*, Supplements 36 *a* and 38.

‡ Authorities:

1. Kia-Lok Yen, *Phil. Mag.* XXXVIII. (1919), p. 582.
2. Breitenbach, *Ann. d. Phys.* V. (1901), p. 166.
3. Schultze, *Ann. d. Phys.* VI. (1901), p. 310.
4. K. Schmitt, *Ann. d. Phys.* XXX. (1909), p. 398. This paper contains a summary of results obtained in recent years at Halle, by the pupils of Prof. Dorn.
5. Puluj, *Wiener Sitzungsber.* 1878.
6. *Recueil de Constantes physiques.* The values are those deduced from the transpiration experiments of Graham (*Phil. Trans.* 1846, 8).
7. Millikan, *Proc. Nat. Acad. Sci.* III. (1917), p. 233.
8. *Recueil de Constantes physiques* (mean value).
9. Vogel (Halle), quoted by Eucken, *Phys. Zeits.* XIV. (1913), p. 324.
10. Schultze, *Ann. d. Phys.* V. (1901), p. 140.
11. Rankine, *Proc. Roy. Soc.* 83 A (1910), p. 516, and 84 A (1910), p. 181.

sistence of velocity" explained in Chapter X; while Chapman* and Enskog† discuss the viscosity of gas-mixture by following the general methods which have already been explained in Chapters IX and VIII respectively. Unfortunately the requisite analysis is too lengthy to be reproduced here; reference must be made to the original papers.

DETERMINATION OF SIZE OF MOLECULES.

386. We have already, in § 376, had an instance of the calculation of molecular radii from the coefficients of viscosity. When the coefficient of viscosity of any gas has been determined by experiment, it is possible to regard equation (773) as an equation for σ, and so obtain the molecular radius on the supposition that the molecules may be regarded as elastic spheres.

In the table on p. 288 are given the coefficients of viscosity of various gases, and the values of $\frac{1}{2}\sigma$, calculated from equation (773).

It is at once seen that the values of $\frac{1}{2}\sigma$ in the last column are at least of the same order of magnitude as the values calculated from the deviations from Boyle's Law in § 179. The reason why still better agreement cannot be expected will be clear from what has already been said in § 378.

* *Phil. Trans.* 216 A (1915), p. 279, and 217 A (1916), p. 115; *Proc. Royal Soc.* 93 A (1916), p. 1.

† *Kinetische Theorie der Vorgänge in mässig verdünnten Gases*, Inaug. Dissertation, Upsala, 1917.

CHAPTER XII

CONDUCTION OF HEAT

Elementary Theory.

387. An elementary theory of conduction can be developed in the same way as the elementary theory of viscosity, given at the beginning of the last chapter.

Let a gas be supposed arranged in layers of equal temperature parallel to the plane of xy. Let $\bar{E}$ denote the mean energy of a molecule at any point in the gas, so that $\bar{E}$ will be a function of z.

Let us fix our attention on the molecules which cross a unit area of the plane $z = z_0$. Some molecules will cross this unit area after having come a distance l from their last collision in a direction making an angle θ with the axis of z. The last collision of these molecules must accordingly have taken place in the plane

$$z = z_0 - l \cos \theta.$$

We may suppose that the mean energy of these molecules is that appropriate to this plane, and this may be taken to be (cf. formula (756))

$$\bar{E} - l \cos \theta \frac{\partial \bar{E}}{\partial z} \quad \text{.........................(783)},$$

where $\bar{E}$ is evaluated at $z = z_0$.

The number of molecules which cross the unit area in question in a direction making an angle between θ and $\theta + d\theta$ with the axis of z per unit time is (cf. formula (757))

$$\tfrac{1}{2} \nu \bar{c} \cos \theta \sin \theta d\theta,$$

and if we assume that each of these has an average amount of energy given by formula (783), the total flow of energy across the unit area of the plane will be

$$\int_{\theta=0}^{\theta=\pi} \left(\bar{E} - l \cos \theta \frac{\partial \bar{E}}{\partial z} \right) \tfrac{1}{2} \nu \bar{c} \cos \theta \sin \theta d\theta = -\tfrac{1}{3} \nu \bar{c} l \frac{\partial \bar{E}}{\partial z} \quad \text{......(784)}.$$

If $\bar{E}$ had been independent of z, this flow of energy would of course have been *nil*, for as much would have crossed the plane in one direction as in the other. But if $\bar{E}$ increases with z, the molecules which cross the plane in the direction of z decreasing, since they come from regions in which z is greater, carry more energy than those crossing the plane in the reverse direction, and so there is a resulting flow of energy in the direction of z decreasing.

If ϑ is the coefficient of conduction of heat, the flow of heat across unit area of the plane $z = z_0$ in the direction of z increasing is $-\vartheta \frac{\partial T}{\partial z}$, so that the flow of energy is $-J\vartheta \frac{\partial T}{\partial z}$, where J is the mechanical equivalent of heat.

Equating this to expression (784),

$$J\vartheta \frac{\partial T}{\partial z} = \tfrac{1}{3}\nu\bar{c}l \frac{\partial \bar{E}}{\partial z} = \tfrac{1}{3}\nu\bar{c}l \frac{d\bar{E}}{dT}\frac{\partial T}{\partial z},$$

from which it follows that the value of ϑ is

$$\vartheta = \frac{1}{3}\frac{\nu\bar{c}l}{J}\frac{d\bar{E}}{dT} \qquad \text{.........................(785).}$$

From equation (465) we have the relation

$$C_v = \frac{1}{Jm}\frac{d\bar{E}}{dT} \qquad \text{.........................(786),}$$

where C_v is the specific heat at constant volume, and again, from equation (764), if κ is the coefficient of viscosity,

$$\kappa = \tfrac{1}{3}\nu\bar{c}lm \qquad \text{.........................(787).}$$

Using these relations, equation (785) becomes

$$\vartheta = \kappa C_v \qquad \text{.........................(788).}$$

MEYER'S THEORY.

Correction when Molecules are Elastic Spheres.

388. We found that the first formula obtained for the coefficient of viscosity was true as regards order of magnitude, but required correction by multiplication by a numerical factor substantially different from unity. So also here, we shall find that strict analysis leads to a value of ϑ which differs very appreciably from that given by equation (788), although again the only difference will lie in the occurrence of a numerical multiplier. We proceed to apply analysis, as rigorously as possible, to the case of conduction of heat in a gas of which the molecules are elastic spheres. The solution which follows is substantially that given in Meyer's *Kinetic Theory of Gases*. The main difference arises out of the fact that Meyer neglects certain terms expressing the variation of collision-frequency, although these terms are of the same order of magnitude as terms retained, while I have found it possible to give the more complete investigation in which these terms are taken into account.

We consider any element $dxdy$ of the plane $z=z_0$, and with the centre of this element as origin, we take spherical polar coordinates r, θ, ϕ, the line $\theta=0$ being parallel to the axis of z.

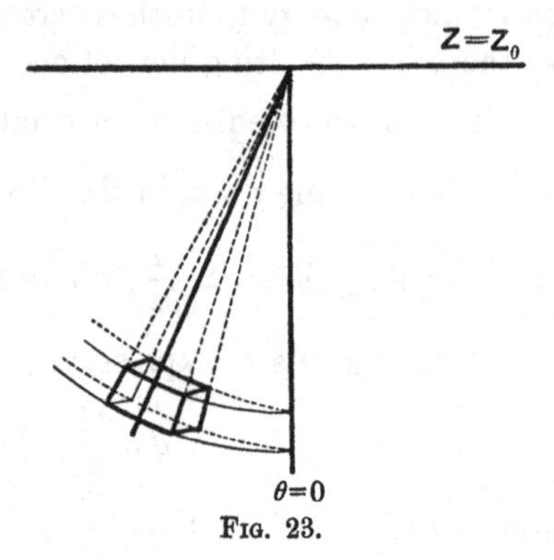

FIG. 23.

The curvilinear element of volume for which r, θ, ϕ lie between r and $r+dr$, θ and $\theta+d\theta$, ϕ and $\phi+d\phi$ is the volume

$$dv = r^2 \sin\theta\, d\theta\, d\phi\, dr.$$

We begin by considering the possibility of a molecule undergoing collision in the element dv, leaving it with a velocity c in such a direction as to pass through the small area $dxdy$, and describing a free path which reaches at least as far as the element $dxdy$ without collision.

Since the whole motion is reversible, the number of collisions in which one of the molecules has a velocity between c and $c+dc$ after collision in the element dv is exactly equal to that of the collisions in which one of the molecules has a velocity within these limits before collision.

The number of molecules which at any instant are moving with a velocity between c and $c+dc$ in the element dv is

$$\nu r^2 \sin\theta\, d\theta\, d\phi\, dr \left(\frac{hm}{\pi}\right)^{\frac{3}{2}} 4\pi c^2 e^{-hmc^2}\, dc \quad \text{..............(789)},$$

so that the number of collisions experienced by these molecules in time dt is, by § 342, equal to

$$\Theta\, dt \times (\text{expression (789)}) \quad \text{......................(790)},$$

where

$$\Theta = \frac{\sqrt{\pi}\,\nu\sigma^2}{hmc}\, \psi(c\sqrt{hm}) \quad \text{........................(791)}.$$

This, then, is the number of molecules which, in time dt, experience a collision in the element dv, and leave it with a velocity between c and $c+dc$. All directions are equally likely. The element $dxdy$ subtends a solid angle

$$dx\,dy\,\frac{\cos\theta}{r^2}$$

at the element dv, so that the chance that, if a molecule escapes collision, it will pass through the element $dxdy$, is

$$dx\,dy\,\frac{\cos\theta}{4\pi r^2} \quad \text{..............................(792)}.$$

Multiplying expressions (790) and (792) together, we obtain for the number of molecules which leave the element dv in time dt with a velocity between limits c and $c+dc$, in a direction suitable for passing through the element of area $dxdy$,

$$\nu\, dx\, dy\, dt \sin\theta \cos\theta\, d\theta\, d\phi\, dr \left(\frac{hm}{\pi}\right)^{\frac{3}{2}} \Theta c^2 e^{-hmc^2}\, dc \quad \text{......(793).}$$

In this expression the quantities Θ, h, r are to be evaluated in the element dv, and therefore for the value $z = z_0 - r\cos\theta$. We may accordingly put

$$\Theta = \Theta_{z=z_0} - r\cos\theta \left(\frac{\partial\Theta}{\partial z}\right)_{z=z_0}$$

and give similar values to h, r. The expression then becomes

$$\nu\, dx\, dy\, dt \sin\theta \cos\theta\, d\theta\, d\phi\, dr \left(\frac{hm}{\pi}\right)^{\frac{3}{2}} \Theta c^2 e^{-hmc^2}\, dc$$

$$\times \left[1 - \frac{r\cos\theta}{\nu}\frac{d\nu}{dz} - \frac{3}{2}\frac{r\cos\theta}{h}\frac{dh}{dz} - \frac{r\cos\theta}{\Theta}\frac{d\Theta}{dz} + mc^2 r\cos\theta \frac{dh}{dz}\right] \quad \text{...(794),}$$

in which all quantities are evaluated for the plane $z = z_0$.

389. We next calculate the probability of a molecule describing the free path r from dv to $dxdy$ without collision.

We cannot use the analysis of § 347 to determine this probability, for in the present problem the state of the gas varies from point to point as we proceed along the path of the molecule. If, however, we denote by $f(l)$ the fraction of the whole which travel a distance at least equal to l without collision, we obtain, just as in § 347, the differential equation

$$\frac{\partial f(l)}{\partial l} = -\frac{f(l)}{\lambda_c} = -\frac{f(l)\,\Theta}{c} \quad \text{..................(795),}$$

in which Θ is now evaluated at a point at distance l along the path. The solution of this equation is

$$f(l) = e^{-\frac{1}{c}\int_0^l \Theta\, dl} \quad \text{........................(796).}$$

If the path is small compared with the scale of variation of conditions in the gas, we may to a sufficient approximation write the value of Θ at a distance l along the path in the form

$$\Theta = \Theta_{l=0} + l\frac{\partial\Theta}{\partial l},$$

so that

$$\int_0^l \Theta\, dl = l\Theta_{l=0} + \tfrac{1}{2} l^2 \frac{\partial\Theta}{\partial l},$$

and the right-hand side is equal to l times the value of Θ at a distance $\frac{1}{2}l$ along the path, at which the value of z is $z - \frac{1}{2}r\cos\theta$. Thus

$$\int_0^l \Theta\, dl = l\left(\Theta_{z=z_0} - \tfrac{1}{2} r\cos\theta \frac{d\Theta}{dz}\right),$$

giving on substitution into equation (796),

$$f(r) = e^{-r\Theta/c}\left(1 + \frac{1}{2c} r^2 \cos\theta \frac{d\Theta}{dz}\right).$$

390. By multiplication of this expression and (794) we find, as the number of particles which cross the plane $z = z_0$ per unit area per unit time with velocities between c and $c + dc$, having started from the element dv,

$$d\mathfrak{n} = \nu \sin\theta \cos\theta \, d\theta \, d\phi \left(\frac{hm}{\pi}\right)^{\frac{3}{2}} \Theta c^2 e^{-hmc^2} e^{-r\Theta/c} (1 - Fr\cos\theta)\, dc\, dr,$$

where $$F = \frac{1}{\nu}\frac{d\nu}{dz} + \frac{3}{2}\frac{1}{h}\frac{dh}{dz} + \frac{1}{\Theta}\frac{d\Theta}{dz} - mc^2\frac{dh}{dz} - \frac{r}{2c}\frac{d\Theta}{dz},$$

and all quantities are evaluated in the plane $z = z_0$.

We first integrate this expression with respect to θ and ϕ so as to obtain the total number of molecules which cross the plane $z = z_0$ from the side $z < z_0$ with velocities between c and $c + dc$, coming from a distance intermediate between r and $r + dr$. In this integration, the limits for ϕ are from 0 to 2π, those for θ are from 0 to $\frac{1}{2}\pi$ only. With these limits of integration,

$$\iint \sin\theta \cos\theta \, d\theta \, d\phi = \pi, \quad \iint \sin\theta \cos^2\theta \, d\theta \, d\phi = \tfrac{2}{3}\pi.$$

Thus if we denote the number in question by $dn_{r,c}$, we have

$$dn_{r,c} = \int_{\theta=0}^{\theta=\pi/2}\int_{\phi=0}^{\phi=2\pi} d\mathfrak{n} = 2\pi\nu\left(\frac{hm}{\pi}\right)^{\frac{3}{2}} \Theta c^2 e^{-hmc^2} e^{-r\Theta/c}\left(\tfrac{1}{2} - \tfrac{1}{3}Fr\right) dc\, dr.$$

We can now integrate this over all values of r, the limits being $r = 0$ to $r = \infty$. The number so obtained, say dn_c, will be the total number of molecules crossing unit area of the plane $z = z_0$ in the direction of z increasing, having velocities lying between c and $c + dc$.

The quantity Fr depends on $\frac{d\Theta}{dz}$ through the terms $\left(\frac{r}{\Theta} - \frac{r^2}{2c}\right)\frac{d\Theta}{dz}$, and on multiplying by $e^{-r\Theta/c}$ and integrating, these terms destroy one another. Omitting these terms, F becomes independent of r, so that

$$\int_0^\infty e^{-r\Theta/c}\,\Theta\, dr = c, \quad \int_0^\infty Fre^{-r\Theta/c}\,\Theta\, dr = \frac{Fc}{\Theta},$$

and we finally obtain

$$dn_c = 2\pi\nu\left(\frac{hm}{\pi}\right)^{\frac{3}{2}} e^{-hmc^2} c^2 dc\left[\frac{c}{2} - \frac{c^2}{3\nu\Theta}\frac{d\nu}{dz} - \frac{c^2}{2h\Theta}\frac{dh}{dz} + \frac{mc^4}{3\Theta}\frac{dh}{dz}\right].$$

This expression gives the flow, in the direction of z increasing, of molecules having velocities intermediate between c and $c + dc$. The corresponding flow in the opposite direction, say dn_c', will be given by an expression which will be exactly similar except that the signs of all the differential coefficients with respect to z must be changed. Thus δn_c, the excess flow in the direction of z increasing, of molecules with velocities intermediate between c and $c + dc$, will be given by

$$\delta n_c = dn_c - dn_c'$$
$$= -\frac{2\pi\nu}{\Theta}\left(\frac{hm}{\pi}\right)^{\frac{3}{2}}\left[\frac{2}{3\nu}\frac{d\nu}{dz} + \frac{1}{h}(1 - \tfrac{2}{3}hmc^2)\frac{dh}{dz}\right]e^{-hmc^2}c^4 dc.$$

On substituting the value of Θ from equation (791), and writing x for $c\sqrt{hm}$, this becomes

$$\delta n_c = \frac{4}{3\pi\sigma^2\sqrt{hm}\,\psi(x)}\left[(x^2 - \tfrac{3}{2})\frac{1}{h}\frac{dh}{dz} - \frac{1}{\nu}\frac{d\nu}{dz}\right] \quad\text{.........(797)},$$

and the total flow of molecules of all velocities is found by integrating this expression from $c = 0$ to $c = \infty$.

391. Heat can be transferred either by conduction or by convection. If we wish to deal, as in the present case, with conduction only, we must introduce the condition that there is to be no convection. This simply requires that there shall be no transfer of mass, and therefore that expression (797), integrated over all values of c, shall vanish. This condition becomes

$$\frac{1}{\nu}\frac{d\nu}{dz}\int_0^\infty \frac{x^5 e^{-x^2}}{\psi(x)}dx = \frac{1}{h}\frac{dh}{dz}\int_0^\infty \frac{(x^2 - \frac{3}{2})\,x^5 e^{-x^2}}{\psi(x)}dx \quad\text{.........(798)}.$$

The integrals have been calculated by W. Conrau, and the result is given in Meyer's *Kinetic Theory of Gases* (p. 464)*. It appears that equation (798) can be written in the form

$$\frac{1}{\nu}\frac{d\nu}{dz} = 0{\cdot}71066\,\frac{1}{h}\frac{dh}{dz} \quad\text{.......................(799)}.$$

392. A condition which ought also to be satisfied† is that the pressure shall be the same throughout the gas in order that equilibrium may be maintained. The condition for this is $\frac{d}{dz}\left(\frac{\nu}{2h}\right) = 0$, or

$$\frac{1}{\nu}\frac{d\nu}{dz} = \frac{1}{h}\frac{dh}{dz} \quad\text{.............................(800)},$$

and this is at once seen to be inconsistent with equation (799).

The origin of this inconsistency is easily found if we remember that the analysis of this chapter has assumed Maxwell's Law to be true throughout the gas, while in Chapter VIII it was conclusively shewn that Maxwell's Law could not hold throughout the gas. Our analysis is in point of fact defective because we have based it all on the assumption of an erroneous law of

* See also footnote to p. 299 below.

† See Meyer's *Kinetic Theory of Gases*, or Sommerfeld's paper in *Vorträge über die Kinetische Theorie der Materie und der Elektrizität*, § 6, p. 158.

distribution. The amount of divergence between equations (799) and (800) gives a rough indication of the amount of error introduced by the erroneous law of distribution.

393. As the divergence revealed in this way is not excessive, and moreover as there is no better way available, we may proceed by assuming the relation expressed by equation (799) to be true.

Assuming this relation, equation (797) becomes

$$\delta n_c = \frac{4}{3\pi\sigma^2\sqrt{hm}\,\psi(x)}[x^2 - 2{\cdot}21066]\frac{1}{h}\frac{dh}{dz}x^5 e^{-x^2}dx \quad\text{......(801).}$$

Each of the molecules counted in δn_c carries kinetic energy of translation $\frac{1}{2}mc^2$ or $x^2/2h$ across the plane $z = z_0$. The amount of heat energy transferred by this kinetic energy of translation is accordingly $x^2/2hJ$ per molecule. If the total transfer of heat energy arising in this way is denoted by Γ_t we obtain, on integration,

$$\Gamma_t = \sum_{x=0}^{x=\infty}\frac{x^2}{2hJ}\delta n_c \quad\text{..............................(802)}$$

$$= \frac{2}{3h^2J\pi\sigma^2\sqrt{hm}}I\frac{dh}{dz} \quad\text{......................(803),}$$

where
$$I = \int_0^\infty \frac{[x^2 - 2{\cdot}21066]\,x^7}{\psi(x)}e^{-x^2}dx \quad\text{.................(804).}$$

394. We may suppose, as in § 241, that the average, for all molecules, of the ratio of internal to translational energy is β, but we are not entitled to assume that the transfer of internal energy is equal to β times that of translational energy.

For the internal energy to be expected in a molecule coming from collision in the element dv is not β times $\frac{1}{2}mc^2$, where c is the velocity of the molecule, but is β times $\frac{1}{2}\overline{mc^2}$, where $\overline{mc^2}$ is the average value of mc^2 in the element dv. The internal energy to be expected is therefore $\frac{3}{4}\beta/h$, where h is evaluated in the element dv, or

$$\frac{3}{4}\frac{\beta}{h} + \frac{3}{4}r\cos\theta\frac{\beta}{h^2}\frac{dh}{dz},$$

in which all quantities are evaluated in the plane $z = z_0$.

If Γ_i is the transfer of internal energy, we shall have

$$\Gamma_i = \Sigma\left(\frac{3}{4}\frac{\beta}{h} + \frac{3}{4}r\cos\theta\frac{\beta}{h^2}\frac{dh}{dz}\right)dn \quad\text{.............(805),}$$

where dn is the number evaluated in § 390, and the summation is equivalent to integration with respect to θ, ϕ, r and c. Unfortunately these integrations cannot be effected in finite terms.

395. As a very rough approximation, we may assume

$$\Gamma_i = \beta\Gamma_t \quad \text{.............................(806)},$$

and more generally we may assume

$$\Gamma_i = \theta\beta\Gamma_t \quad \text{.............................(807)},$$

where θ is a pure number, of which the value cannot be calculated.

We can see that θ must be less than unity, in the following way. The process of conduction consists essentially in molecules coming from elements such as dv and bringing to the plane $z = z_0$ amounts of energy appropriate, not to the plane $z = z_0$, but to the element dv. If the free path were vanishingly small, this process could not occur; indeed, we may notice that ϑ, as given by equation (785), vanishes when $l = 0$. The molecules which describe very short free paths are therefore useless as carriers of energy, and the molecules which are efficient carriers are those which describe long free paths. But it has been seen (§ 346) that on the whole the molecules which describe long free paths are the fast moving molecules. These molecules have an amount of translational energy which is above the average, but only have their average share of internal energy.

Thus the mechanism of conduction is specially favourable to the transfer of translational energy, but is not specially favourable to the transfer of internal energy. Molecules which have an especially large amount of translational energy to carry are *ipso facto* particularly efficient as carriers, but the same is not true of molecules which possess a specially large amount of internal energy. The former kind of energy is accordingly the more readily transferred, whence it results that θ is less than unity.

396. It has for a long time been recognised that the problem of the conduction of heat must be complicated by the difference in the conditions of transfer of translational and internal energy; indeed, if this had not been expected on theoretical grounds, a consideration of the values obtained experimentally for the coefficient of conduction would soon shew that we had to deal with something of this kind (cf. § 403, below).

Stefan* and Boltzmann† made the assumption that the translational energy of the molecules was passed on from place to place with greater speed than the remaining energy. Since the internal energy travelled the less rapidly, the conductivity would be less for a gas with much internal energy, than for one in which the energy was mainly translational. Meyer seems to have accepted these ideas in the first edition of his *Kinetic Theory of Gases* (1877), but repudiated them in the second edition‡ (1899). It would

* *Wien. Sitzungsber.* LXXII. [2] (1875), p. 74.

† *Ibid.* p. 458, *Pogg. Ann.* CLVII. (1876), p. 457.

‡ English translation, p. 285.

seem, however, to be fairly certain that the two kinds of energy must travel at precisely the same rate—otherwise there could be no regular propagation of sound in a gas. The considerations brought forward in § 395 merely shew that the internal energy plays a smaller part in the process of conduction than the translational energy. The facts appear to be as assumed by Stefan and Boltzmann, although the explanation of them that we have arrived at is different from theirs.

397. As in § 301, the total transfer of heat energy of all kinds must be

$$-\vartheta \frac{dT}{dz} \text{ or } \frac{\vartheta}{2h^2 R} \frac{dh}{dz}.$$

Hence, using equations (806) and (807), we must have

$$\frac{\vartheta}{2h^2 R} \frac{dh}{dz} = \Gamma_t + \Gamma_i = (1 + \theta\beta)\, \Gamma_t \quad \text{..............(808)},$$

giving, on comparison with equation (803),

$$\vartheta = \frac{4(1 + \theta\beta) R}{3J\pi\sigma^2 \sqrt{hm}} I \quad \text{..........................(809)}.$$

398. If we agree for the present to neglect the difference between θ and unity, then, by use of the formulae

$$\bar{c} = \frac{2}{\sqrt{\pi h m}}, \quad C_v = \tfrac{3}{2}(1 + \beta) \frac{R}{Jm},$$

equation (809) may be replaced by

$$\vartheta = \frac{4Im}{9\sqrt{\pi}\sigma^2} \bar{c} C_v \quad \text{..........................(810)}.$$

The value of κ obtained before correction for persistence of velocities (equations (765) and (767)) was

$$\kappa = \tfrac{1}{3}\rho\bar{c}l = \tfrac{1}{3}\rho\bar{c}\left[\frac{1{\cdot}051}{\sqrt{2}\pi\nu\sigma^2}\right] \quad \text{......................(811)}.$$

By comparison with equation (810),

$$\vartheta = \frac{4}{3} \frac{I\sqrt{2\pi}}{1{\cdot}051} \kappa C_v \quad \text{..........................(812)},$$

giving the multiplying factor by which equation (788) must be corrected when the molecules are treated as elastic spheres, but persistence of velocities is neglected.

399. We proceed to the evaluation of the integral I given by equation (804). Let us write

$$I_n = \int_0^\infty \frac{x^n}{\psi(x)} e^{-x^2} dx.$$

The value of I_n can be obtained by quadrature from values for $\psi(x)$, and tables of I_n to the upper limit 4 are given by Tait*. The values of the complete integrals have been sent me by Prof. L V. King†, who finds that $I_9 = 1\cdot4625$ and $I_7 = \cdot4631$, so that $I = I_9 - 2\cdot21066\, I_7 = \cdot4387$.

Substituting this value, equation (812) becomes

$$\vartheta = \epsilon\kappa C_v \quad\text{.............................(813)},$$

where $\epsilon = 1\cdot395$.

400. So far no correction has been introduced for the persistence of velocities. We are, however, trying to correct equation (788), namely,

$$\vartheta = \kappa C_v,$$

by determining the numerical multiplier. In other words, we are trying to evaluate the ratio ϑ/κ. Now ϑ and κ are each proportional to the mean free path, and are therefore both affected in the same way by the persistence of velocities. It is therefore clear that the fraction ϑ/κ will be approximately unaffected by this persistence.

Hence, since we have used the value of κ which was obtained before correction, we can use the value of ϑ given by equation (813) without further correction, and shall obtain approximately accurate results.

As in § 372, a better result can be obtained by putting the correcting factor $1/(1-\frac{1}{2}\theta)$ inside the integral (804), using the values for θ given on p. 263. I have evaluated the integral obtained in this way by quadrature, but find that the only effect of this extra refinement is to increase the value of ϵ to $1\cdot497$.

401. Although Meyer's method is of value in throwing light on the physical processes at work in conduction of heat, yet it cannot be said to be very successful in approximating to the value of the numerical multiplier ϵ.

The value of ϵ has been calculated for monatomic gases by Chapman, using the method of successive approximations explained in the last chapter (cf. § 373). In his first paper he obtained the first approximation $\epsilon = 2\cdot500$ for all laws of force of the form μr^{-s}, this including of course the special case $s = 5$ studied by Maxwell, for which the value $\epsilon = 2\cdot500$ is known to be exact (cf. § 331). In his second paper he finds that further approximations alter this value by less than one per cent. of its value. The greatest error in the

* *Trans. Roy. Soc. Edinburgh*, XXXIII. (1886), p. 74.

† In a letter of date Dec. 17, 1908. Prof. King has checked his values by an independent computation, which also confirms the calculation of Conrau used in equation (799). I am indebted also to Prof. King for drawing my attention to an inaccurate equation which I had unfortunately copied from Meyer's book into the first edition of the present book, namely the equation $\vartheta = 1\cdot6027\,\kappa C_v$, which appears to be the result of a faulty calculation by Meyer. I have myself evaluated the integral I independently, and obtained a value agreeing very closely with that sent me by Prof. King.

first approximation is found to occur in the case of elastic spheres, for which the value of ϵ is 2·522.

Enskog* has obtained results identical with those of Chapman, and has further given the general formula for monatomic molecules repelling as the inverse sth power of the distance

$$\epsilon = \frac{5}{2} \frac{1 + \dfrac{(s-5)^2}{4(s-1)(11s-13)} + \dots}{1 + \dfrac{3(s-5)^2}{2(s-1)(101s-113)} + \dots},$$

which reduces to Maxwell's exact value $\epsilon = \frac{5}{2}$ when $s = 5$.

Experimental Values.

402. We proceed to examine the relation between ϑ and κ which is found experimentally.

Monatomic Gases. The following table of recent determinations of $\vartheta/\kappa C_v$, the quantity we have denoted by ϵ, is given by Enskog†:

Helium‡	at	0° C.,	$\epsilon = 2·40$,
	„	− 191·6° C.,	$\epsilon = 2·23$,
	„	− 252·1° C.,	$\epsilon = 2·02$.
Argon‡	at	0° C.,	$\epsilon = 2·49$,
	„	182·5° C.,	$\epsilon = 2·57$.
Neon§	at	10° C.,	$\epsilon = 2·501$.

Other investigators have found similar, although not identical values. Thus Schwarze‖ in 1903 found $\epsilon = 2·507$¶ for helium and $\epsilon = 2·501$ for argon, while Hercus and Laby** give $\epsilon = 2·31$ for helium and $\epsilon = 2·47$ for argon. Thus the theoretical law seems to be confirmed to within the limits of experimental error except at low temperatures.

Polyatomic Gases. In the table on p. 301 the first column gives the observed values of ϑ for a number of polyatomic gases. The values of κ in the third column are observed values taken direct from the table on p. 288. It has been more difficult to assign values to C_v. When no direct experimental determination is available, it is possible to use either of the formulae of Chapter VII,

$$C_v = C_p - \frac{R}{Jm}, \quad C_v = \frac{R}{Jm(\gamma - 1)} \quad \dots\dots\dots\dots\dots\dots (814).$$

* See footnote to p. 229. † *l.c.* p. 104.

‡ Eucken, *Phys. Zeit.* XII. (1911), p. 1101, XIV. (1913), p. 324.

§ Bannawitz, *Ann. d. Phys.* XLVIII. (1915), p. 577.

‖ *Ann. d. Phys.* XI. (1903), p. 303.

¶ Eucken (*Phys. Zeit.* XIV. p. 328) states that Schwarze gives too high a value for helium owing to miscalculation of the value of C_v.

** *Proc. Roy. Soc.* XCV. A (1918), p. 190.

The values of C_v for H_2, N_2, O_2 and CO_2 have been calculated from the first formula, using the values given in the table on p. 190; for the remaining cases, the second formula has been used, except where otherwise stated.

403. It appears that there is no uniformity in the values of $\vartheta/\kappa C_v$. An inspection of the values obtained shews, however, that $\vartheta/\kappa C_v$ is greatest for monatomic gases, and least for gases in which the molecules are of most complex structure (ethylene, carbon-dioxide, etc.). In other words $\vartheta/\kappa C_v$ is largest when $\beta = 0$, and smallest when β is large. This suggests that the want of uniformity may come largely from the ignoring of the factor θ, by which β ought to have been multiplied.

VALUES OF ϑ AND OF $\vartheta/\kappa C_v$.

Gas	ϑ (obs.)	Authority*	κ (p. 288)	C_v	$\vartheta/\kappa C_v$ (obs.)	$\frac{1}{4}(9\gamma - 5)$
Hydrogen	·0003970	1	·0000857	2·42	1·91	1·90
Helium	·0003360	1	·000189	·746†	2·38	2·44
Carbon-monoxide..	·00005425	1	·000163	·177	1·88	1·91
Nitrogen	·0000566	1	·000167	·178‡	1·91	1·91
Ethylene	·0000407	1	·0000961	·274§	1·55	1·55
Air	·0000566	1	·000172	·172‖	1·91	1·91
Nitric oxide.........	·0000555	1	·000179	·167	1·86	1·88
Oxygen..............	·0000570	1	·000192	·156	1·90	1·90
Argon	·00003894	2	·000211	·0745¶	2·49	2·44
Carbon-dioxide ...	·0000337	1	·000139	·156	1·55	1·72
Nitrous oxide	·0000351	3	·000135	·148	1·76	1·73

It will be remembered that the factor θ was required by the circumstance that the moving molecules formed carriers which were more efficient for the transport of translational than of internal energy.

* Authorities:

1. Eucken, *Phys. Zeitschrift*, XIV. (1913), p. 324.
2. Schwarze, *Ann. d. Phys.* XI. (1903), p. 303.
3. Value assumed by Eucken (*l.c.*). This is the mean of determinations by Winkelmann and Wüllner.

† Determined by Vogel, and quoted by Eucken. The value of C_v for helium given by the first of formulae (814) is, however, ·767.

‡ Calculated from the first of formulae (814). Eucken takes C_v= ·177, Pier gives C_v= ·175.

§ The mean of values given by Winkelmann (*Pogg. Ann.* CLIX. (1876), p. 177) and Wüllner (*Wied. Ann.* IV. (1878), p. 321).

‖ Direct experimental value.

¶ The value assumed by Eucken. Schwarze uses the value C_v= ·0740, based upon an experimental determination of C_p by Dittenberger (*Halle Diss.* 1897). Pier gives C_v= ·0746. The theoretical value given by formula (814) is ·0767.

Neglecting this circumstance, we found, in § 387, the simple formula

$$\vartheta = \kappa C_v = \tfrac{3}{2}(1+\beta)\frac{R}{Jm}\kappa \quad \text{.....................(815)},$$

this last form being obtained by substituting the value of C_v from equation (472). We have seen that when the energy is wholly translational ($\beta = 0$), the value of ϑ given by this formula must be multiplied by (approximately) $\frac{5}{2}$. Eucken* has, however, suggested that the simpler formula (815) may be accurate for the transport of internal energy, for which (cf. § 395) there is no correlation between the velocity of the molecule and the amount of internal energy carried.

Combining these two contributions, we arrive at the formula

$$\begin{aligned}\vartheta &= \tfrac{3}{2}\left(\tfrac{5}{2}+\beta\right)\frac{R}{Jm}\kappa \\ &= \frac{\frac{5}{2}+\beta}{1+\beta}\kappa C_v \\ &= \tfrac{1}{4}(9\gamma - 5)\kappa C_v \quad \text{..........................(816)},\end{aligned}$$

the last two forms being obtained on substituting the values of C_v and γ from equations (472) and (474). According to this equation the ratio $\vartheta/\kappa C_v$ ought to have a value $\frac{1}{4}(9\gamma - 5)$ which depends on the ratio of the specific heats. In the last column of the table on p. 301 the values of $\frac{1}{4}(9\gamma - 5)$ are given, and are seen to agree very well with the observed values of $\vartheta/\kappa C_v$.

It is worthy of remark that Boltzmann† proposed a theory according to which the value of $\vartheta/\kappa C_v$ was to be $\frac{15}{4}(\gamma - 1)$, but this is obviously not in accordance with observation‡.

CONDUCTION OF HEAT AND ELECTRICITY IN SOLIDS.

Conduction of Heat.

404. In 1900 Drude§ propounded a theory of conduction of heat in solids, according to which the process is exactly similar to that in gases which we have just been considering, except that the carriers of the heat-energy are the free electrons in the metals.

According to the simplest form of this theory, the coefficient of conduction of heat in a solid will be given by equation (785), namely

$$\vartheta = \frac{1}{3}\frac{v\bar{c}l}{J}\frac{d\bar{E}}{dT} \quad \text{............................(817)},$$

* *Phys. Zeits.* XIV. (1913), p. 324.
† *Pogg. Ann.* CLVII. (1876), p. 457.
‡ See Chapman, *Phil. Trans.* 211 A, p. 465.
§ *Ann. d. Phys.* I. (1900), p. 566.

in which all the quantities refer to the free electrons in the solid, so that ν is the number of free electrons per unit volume, l is their average free path as they thread their way through the solid, and so on. Since an electron is believed to have no energy of rotation or internal motion, we may take $\bar{E} = \frac{3}{2}RT$, and the formula for ϑ becomes

$$\vartheta = \frac{1}{2J}\nu\bar{c}lR \quad\text{.............................(818).}$$

This formula is only of the roughest nature; when the complicated physical conditions are taken fully into account, it must be replaced by the more complex formula (604) obtained by exact analysis in Chapter VIII.

Conduction of Electricity.

405. Drude's theory supposes that the free electrons also act as carriers in the conduction of electricity. If there is an electric force Ξ in the direction of the axis of x, each electron will be acted on by a force Ξe, and so will gain momentum in the direction Ox at a rate Ξe per unit time. The time required to describe an average free path l, with average velocity $\bar{c}$, will be $l/\bar{c}$, so that in describing such a free path, the electron will acquire an additional momentum in the direction of the axis of x equal to $\Xi el/\bar{c}$.

Since the mass of the electron is very small compared with that of the atom or molecule with which it collides, we may suppose (cf. § 359) that there is no persistence of velocities after collisions, so that an electron starts out from collision with a velocity for which all directions are equally likely, and, in describing its free path, superposes on to this a velocity

$$\frac{\Xi el}{m\bar{c}}$$

parallel to the axis of x. It follows that at any instant the free electrons have an average velocity u_0, parallel to the axis of x, given by

$$u_0 = \frac{1}{2}\frac{\Xi el}{m\bar{c}}.$$

Across unit area perpendicular to the axis of x, there will be a flow of electrons at the rate νu_0 per unit time, and these will carry a current i given by

$$i = \nu e u_0 = \frac{1}{2}\frac{\Xi \nu e^2 l}{m\bar{c}}.$$

The coefficient of electric conductivity σ is defined by the relation $i = \sigma\Xi$, and is therefore equal to the coefficient of Ξ in the above equation. To the order of accuracy to which we are now aspiring, $\bar{c}$ may be supposed to be the velocity of each electron, so that we may put $\frac{1}{2}m(\bar{c})^2 = \frac{3}{2}RT$, and the conductivity is given by

$$\sigma = \frac{\nu e^2 l\bar{c}}{6RT} \quad\text{.............................(819).}$$

This is Drude's formula for electric conductivity. Numerically it can obviously only give an approximation of the roughest kind*, and must be replaced by the exact formula (609) when exact numerical values are required.

Ratio of the two Conductivities.

406. *The Wiedemann-Franz law.* By comparison of equations (818) and (819), we obtain

$$\frac{\vartheta}{\sigma} = 3\left(\frac{R}{e}\right)^2 \frac{T}{J}.$$

This is Drude's approximate formula for ϑ/σ. The exact formula of Richardson and Bohr, obtained in § 305, was

$$\frac{\vartheta}{\sigma} = \frac{2s}{s-1}\left(\frac{R}{e}\right)^2 \frac{T}{J} \quad \text{.........................(820).}$$

From these equations it appears that:

at a given temperature, the ratio of the electric and thermal conductivities must be the same for all substances.

This is the law of Wiedemann and Franz, announced by them as an empirical discovery in 1853†.

The Law of Lorenz. From equation (820) it also follows that:

the ratio of the thermal and electric conductivities must be proportional to the absolute temperature,

a law put forward on theoretical grounds by Lorenz in 1872‡.

Comparison with Experiment.

407. For elastic spheres ($s = \infty$) equation (820) reduces to

$$\frac{\vartheta}{\sigma} = 2\left(\frac{R}{e}\right)^2 \frac{T}{J} \quad \text{.........................(821),}$$

the formula originally given by Lorentz§. Using the numerical values given in §§ 8 and 151, we find that this equation becomes

$$\frac{\vartheta}{\sigma T} = 1{\cdot}777 \times 10^8 \text{ in electromagnetic units.}$$

For instance, at 18° C. ($T = 291{\cdot}1$), the value of ϑ/σ ought to be $5{\cdot}17 \times 10^{10}$. The values of ϑ/σ have been determined experimentally by

* Various attempts have been made to obtain more accurate values for the numerical multiplier in Drude's theory; see in particular N. Bohr, *Studier over Metallernes Elektrontheorie*, p. 54, and W. F. G. Swann, *Phil. Mag.* XXVII. (1914), p. 441. But nothing short of the full analysis of Chap. VIII is likely to lead to an accurate result.

† *Pogg. Ann.* LXXXIX. (1853), p. 497.

‡ *Pogg. Ann.* CXLVII. (1872), p. 429 and *Wied. Ann.* XIII. (1882), p. 422.

§ See footnote to p. 227.

Jäger and Diesselhorst* for a large number of substances at 18° C. and at 100° C. They find for example,

for three samples of copper, $\vartheta/\sigma = 6{\cdot}76,\ 6{\cdot}65,\ 6{\cdot}71 \times 10^{10}$,
„ silver, $\vartheta/\sigma = 6{\cdot}86 \times 10^{10}$,
„ two samples of gold, $\vartheta/\sigma = 7{\cdot}27,\ 7{\cdot}09 \times 10^{10}$.

More elaborate experiments, covering a wide range of temperature, have been conducted by Lees†. A sample of his results is given in the following table, which gives values of $\frac{\vartheta}{\sigma T}$ at different temperatures.

VALUES OF $\frac{\vartheta}{\sigma T} \times 10^{-8}$.

	From the experiments of Lees					From experiments of Jäger and Diesselhorst	
Temperature ...	− 170° C.	− 60° C.	− 50° C.	0° C.	18° C.	18° C.	100° C.
Copper	1·85	2·17	2·26	2·30	2·32	2·29	2·32
Silver	2·04	2·29	2·36	2·33	2·33	2·36	2·37
Zinc	2·20	2·39	2·40	2·45	2·43	2·31	2·33
Lead	2·55	2·54	2·52	2·53	2·51	2·46	2·51
Steel	3·34	3·09	3·10	3·06	3·05	3·10	3·09
Manganine	5·94	4·16	3·58	3·41	3·34	3·14	2·97

408. These numbers shew that the observed values of ϑ/σ, although of the same order of magnitude as those predicted by theory, are in every case somewhat too large.

It is more difficult to test the values for ϑ and σ separately which are predicted by theory, since the theoretical formulae for these coefficients separately contain the quantities ν and l, for which it is difficult to form a reliable numerical estimate. But such evidence as is available shews quite definitely that the formulae for ϑ and σ separately do not shew anything like so good an agreement with observation as that shewn by the formula for their ratio ϑ/σ.

* *Berlin. Sitzungsber.* XXXVIII. (1899), p. 719, and *Abhand. d. Phys.-Tech. Reichsanstalt*, III. (1900), p. 369.

† C. H. Lees, "The effects of low temperatures on the Thermal and Electrical conductivities of certain approximately pure metals and alloys" (Bakerian Lecture, 1908), *Phil. Trans.* 208 A, p. 381.

Even at ordinary temperatures the value predicted for σ by equation (819) can only be reconciled with the observed values of σ by assigning to the product νl a value considerably greater than is consistent with other available evidence as to the values of ν and l. At low temperatures this difficulty becomes much greater. For instance Kamerlingh Onnes* has found that at very low temperatures the specific resistance of many metals is reduced to only an infinitesimal fraction of the resistance at 0° C.; at helium temperatures (*i.e.* below about 4° absolute) the resistance may be only of the order of 10^{-11} times that at ordinary temperatures. At these very low temperatures the value of $\bar{c}$ in formula (819) will be decreased to about one-tenth of its value at 0° C., while we cannot suppose that any very great change occurs in the value of ν. It accordingly appears that the 10^{11}-fold increase in the conductivity would require a 10^{10}-fold increase in l, if the theory on which formula (819) is based were true. This requires that the electrons shall describe free paths measured in metres or even in kilometres, a requirement which it is quite impossible to reconcile with the known facts of the structure of matter.

The need for these long free paths can be seen in a very direct manner. In one experiment described by Kamerlingh Onnes, a coil of lead wire was placed in liquid helium, and was found to have a resistance equal to 5×10^{-11} times its resistance at 0° C. The ends of the coil were then fused together and a current started in the coil by magnetic induction. It was found that the "time of relaxation" of this current—*i.e.* the time required to fall to $1/e$ times its initial strength—was of the order of a day, whereas under ordinary conditions it would have been about $\frac{1}{70,000}$ second. In an observation lasting one hour no perceptible decrease of the current could be noticed. If the current consisted of free electrons in motion, each free electron in this time would have had to describe a path of about 30,000 kilometres without its motion being seriously checked by collisions.

Wien† has suggested a modification of Drude's theory, based on Planck's quantum-theory, which attempts to remove this difficulty. An alternative suggestion, also based on the conceptions of the quantum-theory, will be referred to in Chapter XVII below. But it would be beyond the scope of this book to examine into these theories in detail: their relation to the conceptions of the dynamical theory of gases is only slight.

* Experiments with liquid helium, *Konink. Akad. Wetenschappen, Amsterdam, Proc.* XXIII. (1914), p. 12.

† *Berliner Sitzungsber.* VII. (1913), p. 184.

CHAPTER XIII

DIFFUSION

ELEMENTARY THEORIES.

Meyer's Theory.

409. THE difficulties in the way of an exact mathematical treatment of diffusion are similar to those which occurred in the problems of viscosity and heat conduction. Following the method adopted in discussing these earlier problems, we shall begin by giving a simple, but mathematically inexact, treatment of the question.

We imagine two gases diffusing through one another in a direction parallel to the axis of z, the motion being the same at all points in a plane perpendicular to the axis of z. The arrangement of the gases is accordingly in layers perpendicular to this axis. Let the mass-velocity of the whole gas in the direction of z increasing be w_0, and let the molecular densities of the two gases be ν_1, ν_2. Then ν_1, ν_2 and w_0 are functions of z only.

We assume that, as far as the order of approximation required in the problem, the mass-velocity of the gas is small compared with its molecular-velocity, and we also assume that the linear scale of variation of either gas is great compared with the average mean free path of a molecule. We shall also, to obtain a rough first approximation, assume that Maxwell's law of distribution of velocities obtains at every point, and that h is the same for the two gases.

410. The number of molecules of the first kind, which cross the plane $z = z_0$ per unit area per unit time in the direction of z increasing, is

$$\left(\frac{hm_1}{\pi}\right)^{\frac{3}{2}} \iiint \nu_1 e^{-hm_1[u^2+v^2+(w-w_0)^2]}\, w\,du\,dv\,dw \quad \ldots\ldots\ldots(822),$$

in which the limits are from $-\infty$ to $+\infty$ as regards u and v, and from 0 to ∞ as regards w.

Since these molecules do not all come from the same point, ν_1 must, in accordance with the principles already explained, be evaluated at the point from which they started after their last collision. Those which move so as to make an angle θ with the axis of z may be supposed, on the average, to come from a point of which the z coordinate is $z_0 - \lambda \cos\theta$, and at this point the value of ν_1 may be taken to be

$$\nu_1 = \nu_1(z_0) - \lambda \cos\theta \left(\frac{\partial \nu_1}{\partial z}\right)_{z_0} \quad \text{......................(823).}$$

We may now divide the integral (822) into two integrals corresponding to the two terms of the right-hand side of equation (823).

The value of the first is

$$\left(\frac{hm_1}{\pi}\right)^{\frac{3}{2}} \iiint \nu_1(z_0)\, e^{-hm_1[u^2+v^2+(w-w_0)^2]}\, w\, du\, dv\, dw$$

$$= \nu_1(z_0)\left(\frac{hm_1}{\pi}\right)^{\frac{3}{2}} \iiint e^{-hm_1(u^2+v^2+\mathsf{W}^2)} (\mathsf{W}+w_0)\, du\, dv\, d\mathsf{W} \quad \text{......(824),}$$

in which W, as usual, stands for $w - w_0$. The limits of integration are from $-\infty$ to $+\infty$ for u and v, and from $\mathsf{W} = -w_0$ to ∞. We have

$$\int_{-\infty}^{+\infty} e^{-hm_1 u^2} du = \int_{-\infty}^{+\infty} e^{-hm_1 v^2} dv = \left(\frac{\pi}{hm}\right)^{\frac{1}{2}},$$

and, as far as the first power of w_0,

$$\int_{-w}^{+\infty} e^{-hm_1 \mathsf{W}^2} \mathsf{W}\, d\mathsf{W} = \frac{1}{2hm_1}, \quad \int_{-w_0}^{+\infty} e^{-hm_1 \mathsf{W}^2} w_0\, d\mathsf{W} = \tfrac{1}{2} w_0 \left(\frac{\pi}{hm_1}\right)^{\frac{1}{2}}.$$

Hence we obtain, as the value of expression (824),

$$\tfrac{1}{2}\nu_1(z_0)\left(\frac{1}{\sqrt{\pi hm_1}} + w_0\right) = \tfrac{1}{2}\nu_1(z_0)\left(\tfrac{1}{2}\bar{c}_1 + w_0\right) \quad \text{...........(825),}$$

where $\bar{c}_1$, as in § 30, denotes the mean molecular-velocity of all the molecules of the first kind, and is given by equation (44).

The second integral required for the evaluation of expression (822) is

$$\lambda\left(\frac{\partial \nu_1}{\partial z}\right)_{z_0}\left(\frac{hm_1}{\pi}\right)^{\frac{3}{2}} \iiint e^{-hm_1[u^2+v^2+(w-w_0)^2]}\, w \cos\theta\, du\, dv\, dw \quad \text{...(826).}$$

Owing to the presence of the multiplier $\lambda\left(\frac{\partial \nu_1}{\partial z}\right)_{z_0}$, this expression is already a small quantity of the first order, so that in evaluating it we may put $w_0 = 0$. Replacing $\cos\theta$ by w/c, it becomes

$$\lambda\left(\frac{\partial \nu_1}{\partial z}\right)_{z_0}\left(\frac{hm_1}{\pi}\right)^{\frac{3}{2}} \iiint e^{-hm_1 c^2} \frac{w^2}{c}\, du\, dv\, dw,$$

in which the integral is taken over all values of u and v, and over all positive values of w.

This expression is easily evaluated by noticing that it has just half the value it would have if taken over all values of u, v, w, and is therefore equal to $\tfrac{1}{2}\lambda\left(\frac{\partial \nu_1}{\partial z}\right)_{z_0}$ times the average value of $\frac{w^2}{c}$ taken over all molecules, in a gas having no mass-motion. This average value is equal to one-third of the average of $\frac{u^2+v^2+w^2}{c}$ or c, and is therefore $\frac{1}{3}\bar{c}_1$. Hence the value of expression (826) is

$$\tfrac{1}{6}\lambda\left(\frac{\partial \nu_1}{\partial z}\right)_{z_0}\bar{c}_1.$$

Combining this with expression (825), we find as the total value of expression (822),

$$\tfrac{1}{2}\nu_1(\tfrac{1}{2}\bar{c}_1+w_0)-\tfrac{1}{6}\lambda\frac{\partial \nu_1}{\partial z}\bar{c}_1,$$

in which all quantities are to be evaluated in the plane $z=z_0$.

This is the total flow of molecules across unit area of the plane $z=z_0$, in the direction of z increasing. The corresponding flow in the opposite direction is

$$\tfrac{1}{2}\nu_1(\tfrac{1}{2}\bar{c}_1-w_0)+\tfrac{1}{6}\lambda\frac{\partial \nu_1}{\partial z}\bar{c}_1.$$

411. The rate of increase of the number of molecules of the first kind on the positive side of the plane $z=z_0$, measured per unit time per unit area, is the difference of these two expressions. Denoting this quantity by Γ_1, we have

$$\Gamma_1=\nu_1 w_0-\tfrac{1}{3}\lambda_1\frac{\partial \nu_1}{\partial z}\bar{c}_1 \quad\text{..........................(827).}$$

Similarly, for the rate of increase of molecules of the second kind,

$$\Gamma_2=\nu_2 w_0-\tfrac{1}{3}\lambda_2\frac{\partial \nu_2}{\partial z}\bar{c}_2 \quad\text{..........................(828).}$$

Eliminating w_0 from these equations, we obtain

$$\Gamma_1\nu_2-\Gamma_2\nu_1=\tfrac{1}{3}\nu_1\lambda_2\bar{c}_2\frac{\partial \nu_2}{\partial z}-\tfrac{1}{3}\nu_2\lambda_1\bar{c}_1\frac{\partial \nu_1}{\partial z} \quad\text{..............(829).}$$

412. The pressure must be constant throughout the gas, so that we must have

$$\nu_1+\nu_2=\text{cons.},$$

whence, by differentiation with respect to z,

$$\frac{\partial \nu_1}{\partial z}+\frac{\partial \nu_2}{\partial z}=0.$$

Moreover if the flow is steady, $\nu_1+\nu_2$ must not vary with the time, so that the total flow of molecules over every plane must be zero, and this requires that

$$\Gamma_1+\Gamma_2=0.$$

Equation (829) now becomes

$$-\Gamma_1=\Gamma_2=\frac{1}{3}\frac{\nu_1\lambda_2\bar{c}_2+\nu_2\lambda_1\bar{c}_1}{\nu_1+\nu_2}\frac{\partial \nu_1}{\partial z} \quad\text{..................(830).}$$

The number of molecules of the first kind in a layer of unit cross-section between the planes $z = z_0$ and $z = z_0 + dz$, is $\nu_1 dz$; the rate at which this quantity increases is $\frac{d\nu_1}{dt} dz$, but is also found to be $-\frac{\partial \Gamma_1}{\partial z} dz$, by calculating the flow across the two boundary planes. Hence we have

$$\frac{d\nu_1}{dt} = -\frac{\partial \Gamma_1}{\partial z},$$

and on using the value of Γ_1 provided by equation (830), and neglecting small quantities of the second order, this becomes

$$\frac{d\nu_1}{dt} = \mathfrak{D}_{12} \frac{\partial \nu_1{}^2}{\partial z^2} \quad \text{.............................(831)},$$

where

$$\mathfrak{D}_{12} = \frac{1}{3} \frac{\nu_1 \lambda_2 \bar{c}_2 + \nu_2 \lambda_1 \bar{c}_1}{\nu_1 + \nu_2} \quad \text{.......... ....(832)}.$$

Equation (831) is the well-known equation of diffusion, $\mathfrak{D}_{12}$ being the coefficient of diffusion of the two gases. Hence the coefficient of diffusion is given by formula (832). Clearly it is symmetrical as regards the physical properties of the two gases, but depends on the ratio ν_1/ν_2 in which they are mixed.

The foregoing analysis is essentially the same as that given by Meyer in his *Kinetic Theory of Gases*, and formula (832) is generally known as Meyer's formula* for the coefficient of diffusion.

Some special cases of this formula may be noticed.

Coefficient of Self-diffusion.

413. If we consider diffusion between two gases in which the molecules are approximately of equal size and weight, and agree to neglect the differences in size and weight, we may take λ and c to be the same for each gas, and so obtain

$$\mathfrak{D} = \tfrac{1}{3} \lambda \bar{c} \quad \text{.................................(833)}.$$

Comparing this with the value of the coefficient of viscosity (equation (765))

$$\kappa = \tfrac{1}{3} \lambda \bar{c} \rho,$$

we obtain the relation

$$\mathfrak{D} = \frac{\kappa}{\rho} \quad \text{.....................................(834)}.$$

The quantity $\mathfrak{D}$ obtained in this way may also be regarded as the coefficient of self-diffusion or interdiffusivity of a single gas. It measures the rate at which selected molecules of a homogeneous gas diffuse into the remainder.

* The actual value of $\mathfrak{D}_{12}$ given by Meyer (*Kinetic Theory of Gases*, p. 255, English trans.) is $\frac{3}{8}\pi$ times that given by formula (832). Meyer's formula has, however, attempted to take into account a correction which is here reserved for later discussion (§ 415). Meyer does not claim that his correction is exact.

Dependence on Proportions of Mixture.

414. In the special case to which formula (833) applies, the value of $\mathfrak{D}$ is independent of ν_1 and ν_2, but formula (832) shews that in general $\mathfrak{D}_{12}$ ought to vary with the proportions of the mixture. In the limiting case in which $\nu_1/\nu_2 = 0$, we have

$$\mathfrak{D}_{12} = \tfrac{1}{3}\lambda_1 \bar{c}_1 = \frac{2}{3\sqrt{\pi h}\,(m_1 + m_2)\,\pi \nu S_{12}^2}\left(\frac{m_2}{m_1}\right)^{\frac{1}{2}},$$

and there is a similar formula for the case of $\nu_2/\nu_1 = 0$, in which m_1 and m_2 are interchanged.

Thus the coefficients of diffusion in these two cases stand in the ratio

$$\frac{\mathfrak{D}_{\nu_1=0}}{\mathfrak{D}_{\nu_2=0}} = \frac{m_2}{m_1} \quad \text{.........................(835)},$$

shewing that the value of $\mathfrak{D}$ ought according to Meyer's formula to vary greatly with the proportions of the mixture. The predicted variation will be greatest for molecules of very uneven mass. For example for the diffusion of H_2—CO_2, the extreme variation would be 22 to 1, for A—He it would be 10 to 1, and so on.

As we shall see later, the observed variation of $\mathfrak{D}_{12}$ with ν_1/ν_2 is nothing like as great as is predicted by this formulae, but we shall now proceed to correct the formulae for persistence of velocities, and shall find that the corrected equations predict a much smaller dependence of $\mathfrak{D}_{12}$ on ν_1/ν_2.

Correction to Meyer's Theory when the Molecules are Elastic Spheres.

415. As was the case with the corresponding formulae for viscosity and conduction of heat, the approximate formulae which have been obtained can be improved by a correction of the numerical multiplier.

We shall consider first the correction to be applied to the simple formula for self-diffusion, namely

$$\mathfrak{D} = \tfrac{1}{3}\lambda\bar{c} \quad \text{..............................(836)}$$

$$= \frac{\kappa}{\rho} \quad \text{..................................(837)}.$$

As before, there have been two sources of error introduced into these approximate equations, the first arising from the assumption that λ is the same for all velocities, and the second from neglect of the persistence of velocities.

As regards the first, it is clear that in expression (826), λ must be replaced by λ_c and taken under the sign of integration. Hence instead of λ in the final result, we must have l, where

$$l = \frac{\int_0^\infty \lambda_c e^{-hmc^2} c^3 dc}{\int_0^\infty e^{-hmc^2} c^2 dc} = \frac{\overline{\lambda_c c}}{\bar{c}} \quad \text{.....................(838).}$$

This however is exactly the same as the l of the viscosity formula, of which the value was found in § 370 to be

$$\frac{1{\cdot}051}{\sqrt{2}\pi\nu\sigma^2}.$$

Hence this correction affects $\mathfrak{D}$ and κ exactly similarly, multiplying each by 1·051, but does not affect equation (837).

416. We now examine the effect of the persistence of velocities. We found in § 371, that when a molecule arrives at the plane $z = z_0$ in a given direction, the expectation of the distance it has travelled in that direction is not λ, but $k\lambda$, where

$$k = \frac{1}{1-\theta}.$$

Here θ is the persistence of velocities at a collision between two molecules of equal mass, of which the value was found in § 357 to be ·406. Thus the expectation of the molecule belonging to the one gas or the other is not that appropriate to a distance λ back, but to a distance $k\lambda$, and the effect of "persistence" is therefore to multiply the value of $\mathfrak{D}$ given in equation (836) by a factor k. Also, as we saw in § 371, the effect of persistence on the coefficient of viscosity is to multiply the simple expression $\frac{1}{3}\lambda\bar{c}\rho$ by a factor $1/(1-\frac{1}{2}\theta)$.

The values of $\mathfrak{D}$ and κ, both corrected for persistence, accordingly become

$$\mathfrak{D} = \frac{1}{3(1-\theta)}\lambda\bar{c},$$

$$\kappa = \frac{1}{3(1-\frac{1}{2}\theta)}\lambda\bar{c}\rho,$$

so that the corrected form of equation (837) must be

$$\mathfrak{D} = \frac{1-\frac{1}{2}\theta}{1-\theta}\frac{\kappa}{\rho}.$$

Putting $\theta = {\cdot}406$, the value found in § 357, this becomes

$$\mathfrak{D} = 1{\cdot}34\frac{\kappa}{\rho} \quad \text{..............................(839).}$$

It is of interest to examine into the origin of the difference between the effect of persistence of velocities on diffusion on the one hand, and on viscosity and conduction of heat on the other. Diffusion, it will be seen, is a transport of a *quality*, while viscosity and heat-conduction are transports of *quantities*. The difference rests ultimately upon the circumstance that qualities remain unaltered by collisions, whereas quantities do not.

417. The effect of persistence when the molecules are not of equal mass is more difficult to estimate.

When the molecules were equal, the expectation of the distance a molecule had come was increased by persistence from λ to

$$\lambda + \theta\lambda + \theta^2\lambda + \theta^3\lambda + \ldots = \frac{\lambda}{1-\theta} \qquad (840).$$

When the molecules are of unequal masses, the persistence will be different at different collisions, and instead of expression (840), we shall have one of the form

$$\lambda + p\lambda + pq\lambda^2 + pqr\lambda^3 + \ldots \qquad (841),$$

where $p, q, r, \ldots$ are the different persistences at the various collisions. Suppose we are considering the motion of a molecule of mass m_1 in a mixture of molecules of masses m_1, m_2, mixed in the proportion ν_1/ν_2. Then of the quantities $p, q, r, \ldots$ a certain proportion, say β, of the whole will have an average value $\theta = \cdot 406$, these representing collisions with other molecules of the first kind, while the remainder, a proportion $1-\beta$ of the whole, will have an average value which we shall denote by θ_{12}, this being the persistence for a molecule of the first kind colliding with one of the second kind.

Let P denote expression (841), and let s denote $\beta\theta + (1-\beta)\theta_{12}$, this being the expectation of each of the quantities $p, q, r, \ldots$. We have

$$P = \lambda + p\lambda + pq\lambda^2 + pqr\lambda^3 + \ldots,$$

$$Ps = \quad s\lambda + ps\lambda^2 + pqs\lambda^3 + \ldots,$$

and hence, by subtraction,

$$P(1-s) = \lambda + (p-s)\lambda + p(q-s)\lambda^2 + pq(r-s)\lambda^3 + \ldots.$$

Clearly the expectation of the right-hand side is λ, for the expectations of $p-s, q-s, r-s, \ldots$ are all zero. Hence the expectation of P is

$$P = \frac{\lambda}{1-s} = \frac{\lambda}{1-\{\beta\theta + (1-\beta)\theta_{12}\}} \qquad (842).$$

Accordingly the effect of persistence in this mixture of gases is to increase λ to a value of which the expectation is that on the right-hand side of equation (842).

In § 345 we found for the mean chance of collision per unit time for a molecule of the first kind, moving in a mixture of two kinds of gas,

$$2\nu_1\sigma_1^2\sqrt{\frac{2\pi}{hm_1}} + 2\nu_2 S_{12}^2\sqrt{\frac{\pi}{h}\left(\frac{1}{m_1}+\frac{1}{m_2}\right)}.$$

In this the first term represents collisions with molecules of the first kind, and the second represents collisions with molecules of the second kind. The ratio of these two terms is therefore exactly the ratio $\beta : 1-\beta$, and we have

$$\frac{\beta}{\sqrt{2}\,\pi\nu_1\sigma_1^2} = \frac{1-\beta}{\pi\nu_2 S_{12}^2\sqrt{\left(1+\frac{m_1}{m_2}\right)}}.$$

Each fraction is equal to

$$\frac{1}{\sqrt{2}\pi\nu_1\sigma_1^2 + \pi\nu_2 S_{12}^2\sqrt{\left(1+\frac{m_1}{m_2}\right)}},$$

and this again is equal to λ_1 by equation (840).

Using this value for β, we find for the value of expression (842), the free path of a molecule of the first kind increased by persistence,

$$P_1 = \frac{1}{(1-\theta)\sqrt{2}\pi\nu_1\sigma_1^2 + (1-\theta_{12})\,\pi\nu_2 S_{12}^2\sqrt{\left(1+\frac{m_1}{m_2}\right)}} \quad \text{......(843)},$$

and for the corresponding quantity for the second molecule,

$$P_2 = \frac{1}{(1-\theta)\sqrt{2}\pi\nu_2\sigma_2^2 + (1-\theta_{21})\,\pi\nu_1 S_{12}^2\sqrt{\left(1+\frac{m_2}{m_1}\right)}} \quad \text{......(844)}.$$

On replacing λ_1, λ_2 in equation (832) by their enhanced values, as given above, we find as the form of Meyer's equation, after correction for persistence of velocities,

$$\mathfrak{D}_{12} = \frac{1}{3}\frac{\nu_1 P_2\bar{c}_2 + \nu_2 P_1\bar{c}_1}{\nu_1+\nu_2} \quad \text{......................(845)}.$$

418. In this formula the value of θ is always ·406; the value of θ_{12} depends, as was seen in § 359, on the ratio of the two masses. It was found that θ_{12} was of the form

$$\theta_{12} = \frac{m_1 - \alpha_{12}m_2}{m_1+m_2},$$

where α was a small positive number, depending on the ratio of the masses, but lying always between 0 and $\frac{1}{4}$, and equal to ·188 for equal masses.

When ν_1 is small, we have, instead of the limiting form given in § 414,

$$\mathfrak{D}_{12} = \frac{1}{3}\frac{\nu_2 P_1\bar{c}_1}{\nu_1+\nu_2} = \frac{2}{3(1-\theta_{12})\,\pi(\nu_1+\nu_2)S_{12}^2}\sqrt{\frac{m_2}{\pi h m_1(m_1+m_2)}}$$

$$= \frac{2}{3(1+\alpha_{12})\,\pi(\nu_1+\nu_2)S_{12}^2}\sqrt{\frac{1}{\pi h}\left(\frac{1}{m_1}+\frac{1}{m_2}\right)} \quad \text{......(846)}.$$

The limiting form, when ν_2 is small, is the same except that α_{12} is replaced by α_{21}. Thus the ratio of the extreme values of $\mathfrak{D}$ as ν_1/ν_2 varies is

$$\frac{\mathfrak{D}_{\nu_1=0}}{\mathfrak{D}_{\nu_2=0}}=\frac{1+\alpha_{21}}{1+\alpha_{12}},$$

instead of the ratio $m_2:m_1$ found from Meyer's formula (835). Since the extreme values possible for α are 0 and $\frac{1}{3}$, it appears that the greatest range possible for $\mathfrak{D}$ is at most one of 4 : 3.

Thus, when persistence of velocities is taken into account, Meyer's formula yields values which do not vary greatly with the proportion $\nu_1:\nu_2$ of the mixture*.

The Stefan-Maxwell Theory.

419. Another theory of diffusion was put forward by Stefan† and Maxwell‡, based upon physical principles which will now be explained.

It will be noticed that equation (831) is of the same form as the well-known equation of conduction of heat: it indicates a progress or spreading out of the gas of the first kind, similar to the progress and spreading out of heat in a problem of conduction. The larger $\mathfrak{D}$ is, the more rapidly this progress takes place; $\mathfrak{D}$ is largest when the free paths are longest, and vice versa. Long free paths mean rapid diffusion, as we should expect.

Now the formula for the mean path λ_1 in a mixture of two gases was found in § 340 to be

$$\lambda_1=\frac{1}{\sqrt{2}\pi\nu_1\sigma_1^2+\sqrt{\left(1+\frac{m_1}{m_2}\right)}\,\pi\nu_2 S_{12}^2}\quad\text{..............(847)},$$

where S_{12} is the arithmetic mean of the diameters of the two kinds of molecules. The larger the denominator in this expression, the smaller λ_1 will be, and so the slower the process of diffusion. Both terms in the denominator of expression (847) accordingly contribute something towards hindering the process of diffusion.

The second of these terms arises from collisions of the molecules of the first kind with molecules of the second kind, and that these collisions should

* This was pointed out in a valuable paper by Kuenen ("The diffusion of Gases according to O.,E. Meyer," Supp. no. 28 to the *Communications from the Phys. Lab. of Leiden*, Jan. 1913). Kuenen took α uniformly equal to ·188, its value when $m_1=m_2$, and assumed the number of collisions to be in the ratio $\nu_1^2:\nu_1\nu_2$, so that his result is different from mine, but the principle was essentially the same. In a later paper by the same author (*Communications from the Phys. Lab. of Leiden*, Supp. 38) the mass difference was taken into account.

† *Wiener Sitzungsberichte*, LXIII: [2] (1871), p. 63, and LXV. (1872), p. 323.

‡ *Coll. Scientific Papers*, I. p. 392, and II. p. 57 and p. 345. See also Boltzmann, *Wiener Sitzungsberichte*, LXVI. [2] (1872), p. 324, LXXVIII. (1878), p. 733, LXXXVI. (1882), p. 63, and LXXXVIII. 1883), p. 835. Also *Vorlesungen über Gastheorie*, I. p. 96.

hinder diffusion is intelligible enough. But it is not so clear how collisions of the molecules of the first kind with one another, represented by the first term in the denominator of expression (847), can hinder the process of diffusion. When molecules of the same kind collide, their average forward motion will, from the conservation of momentum, remain unaffected by the collision, and it is not easy to see how the process of diffusion has been hindered by the collision.

420. If we entirely neglected collisions between molecules of the same kind, we should have free paths given by the equations

$$\lambda_1 = \frac{1}{\sqrt{\left(1+\frac{m_1}{m_2}\right)}\pi\nu_2 S_{12}^2}; \quad \lambda_2 = \frac{1}{\sqrt{\left(1+\frac{m_2}{m_1}\right)}\pi\nu_1 S_{12}^2} \quad \text{......(848)}$$

in place of equation (847). Using these values for the free paths, equation (832) becomes

$$\mathfrak{D}_{12} = \frac{m_1^{\frac{1}{2}}\bar{c}_2 + m_2^{\frac{1}{2}}\bar{c}_1}{3\pi(\nu_1+\nu_2)S_{12}^2(m_1+m_2)^{\frac{1}{2}}}$$

$$= \frac{2}{3\pi(\nu_1+\nu_2)S_{12}^2}\sqrt{\frac{1}{\pi h}\left(\frac{1}{m_1}+\frac{1}{m_2}\right)} \quad \text{............(849),}$$

or, in terms of the molecular-velocities*,

$$\mathfrak{D}_{12} = \frac{1}{3\pi\nu S_{12}^2}\sqrt{\bar{c}_1^2+\bar{c}_2^2} \quad \text{.........................(850).}$$

421. If the two kinds of molecules are of equal mass and size, formula (850) becomes

$$\mathfrak{D} = \frac{2}{3}\frac{\bar{c}}{\sqrt{2}\pi\nu\sigma^2} \quad \text{..........................(852),}$$

which may be contrasted with Meyer's uncorrected formula (833).

* Meyer, using the value of $\mathfrak{D}_{12}$ already explained (see footnote to p. 310), obtains a value for $\mathfrak{D}_{12}$ on Maxwell's theory equal to $\frac{3}{8}\pi$ times this, namely

$$\mathfrak{D}_{12} = \frac{1}{8\nu S_{12}^2}\sqrt{\bar{c}_1^2+\bar{c}_2^2} \quad \text{....................................(851),}$$

and this same value is given by Maxwell (*l.c. ante* and *Nature*, VIII. (1873), p. 298). On the other hand, Stefan (*Wiener Sitzungsber.* LXVIII. (1872), p. 323), Langevin (*Ann. de Chimie et de Physique*, [8], v. (1905), p. 245), and Chapman in his first paper (*Phil. Trans.* 211 A, p. 449) all arrived at the formula

$$\mathfrak{D}_{12} = \frac{3}{32\nu S_{12}^2}\sqrt{\bar{c}_1^2+\bar{c}_2^2},$$

which differs from (851) by a factor $\frac{3}{4}$. Chapman and Langevin both extended their method to the general law of force μr^{-s}; their method was somewhat similar to that of Maxwell as given in Chap. IX (§ 358), but they assumed Maxwell's law of distribution to hold, so that their results were only exact for the case of $s=5$, for which their result agrees with Maxwell's formula (694).

Using Chapman's corrected formula (773) for the coefficient of viscosity,

$$\kappa = \cdot 499 \frac{\rho \bar{c}}{\sqrt{2}\pi \nu \sigma^2},$$

equation (852) may be put in the form

$$\mathfrak{D} = 1\cdot 336 \frac{\kappa}{\rho} \quad \text{.........................(853)},$$

agreeing almost exactly with equation (839) which was obtained on correcting Meyer's formula for persistence of velocities.

It is more difficult to compare equation (849) with equation (845) which was obtained by correcting Meyer's formula for persistence of velocities. We at once notice the outstanding difference between the two, namely that Meyer's formula depends on the ratio ν_1/ν_2, whereas Maxwell's formula (849) does not. In the limiting case of $\nu_1 = 0$, Meyer's formula reduces to formula (846), which is identical with Maxwell's formula (849) divided by the factor $(1 + \alpha_{12})$, where α_{12} is the small number defined in § 418. Thus it appears that the formulae approximate closely, although they naturally cannot agree exactly, as one predicts slight variation with ν_1/ν_2, while the other predicts none at all.

Exact General Formulae.

422. The formulae we have so far obtained are only approximate formulae for the very special case of molecules which may be treated as elastic spheres.

In two special cases, we have obtained perfectly rigorous values for $\mathfrak{D}_{12}$. In § 306 (equation (613)) we found the solution

$$\mathfrak{D}_{12} = \frac{\frac{2}{3}\Gamma\left(\frac{2}{s-1}+2\right)}{\nu_2 (h m_1 m_2 K)^{\frac{2}{s-1}} I_1(s)} \sqrt{\frac{1}{\pi h m_1}} \quad \text{..............(854)}$$

for the case in which $\nu_1/\nu_2 = 0$ and $m_1/m_2 = 0$.

Also in § 334 (equation (691)) we obtained Maxwell's solution

$$\mathfrak{D}_{12} = \frac{1}{2 h m_1 m_2 (\nu_1 + \nu_2) A_1} \sqrt{\frac{m_1 + m_2}{K}} \quad \text{..............(855)}$$

for the case in which $s = 5$, independently of the values of ratios ν_1/ν_2 and m_1/m_2.

The two solutions (854) and (855) are both perfectly rigorous and exact, and so must be expected to agree in the case which they cover in common, namely the case of $s = 5$, $\nu_1/\nu_2 = 0$, $m_1/m_2 = 0$, as it is easily verified that they do. They are two special cases of a general solution, which as we have seen, cannot be expressed in finite terms.

423. When the molecules are treated as point centres of force repelling as the sth power of the distance, the methods of Chapman and Enskog, already explained, allow the value of $\mathfrak{D}_{12}$ to be calculated, by successive approximations, to any desired degree of accuracy.

A first approximation, arrived at by the assumption that the components of velocity u, v, w relative to the velocity of mass-motion were distributed according to Maxwell's law, had been given by Langevin* in 1905; the same formula was given independently by Chapman† in 1917. To a first approximation the value of $\mathfrak{D}_{12}$ is found to be given by

$$\mathfrak{D}_{12} = \frac{3\left[\pi (m_1 + m_2)\right]^{\frac{1}{2}}}{8 (\nu_1 + \nu_2) \left[h m_1 m_2 K\right]^{\frac{2}{s-1} + \frac{1}{2}} I_1(s) \Gamma\left(3 - \frac{1}{s-1}\right)} \quad \text{......(856).}$$

When $s = 5$, this is exact, and reduces to Maxwell's formula (855). In other cases, the exact value of $\mathfrak{D}_{12}$ will be obtained on multiplying the above approximate value of $\mathfrak{D}_{12}$ by a numerical multiplier.

Self-Diffusion.

424. Chapman gives the following table of the values of this multiplier obtained from calculations carried as far as a second approximation for special values of s in the case in which the two kinds of molecules are exactly similar (self-diffusion).

	$s = 5$	$s = 9$	$s = 17$	$s = \infty$
Multiplier	$= 1{\cdot}000$	$1{\cdot}004$	$1{\cdot}008$	$1{\cdot}015$.

It appears that the multiplier increases steadily as s increases from 5 upwards, and reaches its maximum of 1·015 when $s = \infty$ (elastic spheres). In the special case of elastic spheres, Chapman supposes that the multiplying factor, to a third approximation, would have the value 1·017, and that this is the value to which successive approximations are converging, to an accuracy of one part in a thousand. Assuming this, Chapman finds that the accurate value of the coefficient of self-diffusion for elastic spheres‡ is

$$\mathfrak{D} = \frac{0{\cdot}1520}{4 \nu \sigma^2 (hm)^{\frac{1}{2}}} \quad \text{.........................(857)}$$

$$= 1{\cdot}200 \frac{\kappa}{\rho} \quad \text{.........................(858).}$$

Pidduck§, following a method originated by Hilbert‖, based on the transformation of Boltzmann's characteristic equation (518), had previously arrived

* *Annales de Chimie et de Physique*, [8], v. (1905), p. 245.

† *Phil. Trans.* 217 A (1917), p. 166.

‡ *l.c.* p. 172.

§ *Proc. Lond. Math. Soc.* xv. (1915), p. 89.

‖ *Math. Annalen*, LXXII. (1912), p. 562.

at the formula (857) for $\mathfrak{D}$, the only difference being that the number in the numerator, calculated to three places of decimals only, is given as 0·151.

Diffusion in a Mixture.

425. Both Chapman* and Enskog† have shewn how to obtain, by successive approximations, formulae for the general value of $\mathfrak{D}_{12}$ in a mixture of gases. Chapman starts by taking the value given by equation (856) as a first approximation. Denoting this value by $(\mathfrak{D}_{12})_0$, the general value of $\mathfrak{D}_{12}$ is put in the form

$$\mathfrak{D}_{12} = \frac{(\mathfrak{D}_{12})_0}{1 - \epsilon_0} \qquad \text{..............................(859)},$$

where ϵ_0 is a small quantity, to be evaluated by successive approximations. As we have already seen, the value of ϵ_0 is zero for Maxwellian molecules repelling as the inverse fifth-power of the distance. In other cases the formulae, even when carried only to a second approximation, are extremely complicated, and the reader who wishes to study them in detail is referred to the original memoirs‡.

The principal interest of these general formulae lies in the amount of dependence of $\mathfrak{D}_{12}$ on the proportion of the mixture which is predicted by them. We have already seen that the approximate Maxwell-Stefan theory (§ 420) predicted that $\mathfrak{D}_{12}$ would be independent of this proportion, whereas the theory of Meyer predicted very great dependence when the molecules of the two kinds of gas were of very unequal mass. This latter dependence, it is true, was greatly reduced when "persistence of velocities" was taken into account, but still remained quite appreciable.

For the ratio of $\mathfrak{D}_{12}$ in the two extreme cases of $\nu_1/\nu_2 = 0$ and $\nu_1/\nu_2 = \infty$, Enskog§ gives the formula (accurate to a second approximation) for the case in which the molecules may be treated as elastic spheres,

$$\frac{\mathfrak{D}_{\nu_1=0}}{\mathfrak{D}_{\nu_2=0}} = \frac{1 + \dfrac{m_2^2}{12m_2^2 + 16m_1m_2 + 30m_1^2}}{1 + \dfrac{m_1^2}{12m_1^2 + 16m_1m_2 + 30m_2^2}}.$$

This may be compared with the value m_2/m_1 predicted by Meyer's uncorrected formula (835), and with our formula (§ 418) obtained by correcting Meyer's formula for persistence of velocities. To take a definite instance, it will be found that when $m_1/m_2 = 10$, the predicted values are as follows:

Meyer	10·000
„ (corrected for persistence)	1·324
Chapman-Enskog	1·072

* *Phil. Trans.* 217 A (1917), p. 166.

† *l.c. ante* (see footnote to p. 229).

‡ Chapman, *l.c.* equations (13·07), (13·28), *et ff.*; Enskog, *l.c.* equation (168), *et ff.*

§ *l.c.* p. 103.

Comparison with Experiment.

426. As a matter of convenience, it will be well to consider first the experimental evidence bearing on the question as to how far the coefficient of diffusion depends on the proportion in which the gases are mixed. A series of experiments* have been made at Halle to test this question, a summary of which will be found in a paper by Lonius†.

Experiments were made on the pairs of gases H_2—O_2, H_2—N_2 and N_2—O_2 by Jackmann, on H_2—O_2 and H_2—CO_2 by Deutsch, and on He—A by Schmidt and Lonius. On every theory which has been considered, we should expect the greatest variation of $\mathfrak{D}_{12}$ with ν_1/ν_2 to occur when the ratio of the masses of the molecules differs most from unity. The following table‡ gives the values obtained for $\mathfrak{D}_{12}$ with different values of ν_1/ν_2 for the two pairs of gases for which this inequality of masses is greatest.

Pair of Gases (1, 2 respectively)	$\frac{\nu_1}{\nu_2}$	$\mathfrak{D}_{12}$ (observed)	Observer	$\mathfrak{D}_{12}$ (calculated) (Chapman)
H_2—CO_2	3	0·21351	Deutsch	0·212
	1	0·21774	”	0·222
	$\frac{1}{3}$	0·22772	”	0·226
He—A	2·65	0·24418	Lonius	0·248
	2·26	0·24965	”	0·250
	1·66	0·25040	Schmidt	0·251
	1	0·25405	”	0·254
	·477	0·25626	Lonius	0·257
	·311	0·26312	”	0·259

In the last column are given the values calculated by Chapman from his theoretical formulae. In these calculations an absolute value of $\mathfrak{D}_{12}$ is assumed such as to make the mean of the calculated values of $\mathfrak{D}_{12}$ for each pair of gases equal to the mean of the observed values.

It will be seen that the degree of variation in $\mathfrak{D}_{12}$ predicted by the formula of Chapman is at least of the same order of magnitude as that actually observed. The superiority of Chapman's formulae over the two others already discussed is shewn in the following table, which gives a comparison between the extreme values of $\mathfrak{D}_{12}$ for He—A observed, and those predicted by these various formulae. (All values of $\mathfrak{D}_{12}$ are multiplied by a factor chosen so as to make $\mathfrak{D}_{12} = 1$ when $\nu_1 = \nu_2$.)

* R. Schmidt, *Ann. d. Physik*, XIV. (1904), p. 801, and the following Inaug.-Dissertations: R. Schmidt (1904), O. Jackmann (1906), R. Deutsch (1907), and Lonius (1909).

† *Ann. d. Phys.* XXIX. (1909), p. 664. See also Chapman, *Phil. Trans.* 211 A, p. 478.

‡ Lonius, *l.c.* p. 676.

ν_1/ν_2	$\mathfrak{D}_{12}$ (obs.)	$\mathfrak{D}_{12}$ (calculated)		
		(Chapman)	(Meyer; corrected)	(Meyer)
2·65	·961	·976	·910	·548
1·00	1·000	1·000	1·000	1·000
0·311	1·036	1·021	1·110	1·526

From the foregoing discussion, it will have been noticed that the actual variation of $\mathfrak{D}_{12}$ with the proportion of the mixture is, in any case, very slight. Consequently, throughout the remainder of the chapter we shall be content to disregard the dependence of $\mathfrak{D}_{12}$ on the ratio ν_1/ν_2.

Coefficient of Self-Diffusion.

427. The simplest formula to test numerically is that for self-diffusion, but the coefficient of self-diffusion of a gas into itself is not a quantity which admits of direct experimental determination.

A convenient plan, adopted by Lord Kelvin, is to take a set of three gases for which the coefficients $\mathfrak{D}_{12}$, $\mathfrak{D}_{23}$, $\mathfrak{D}_{31}$ are known. All the quantities in formula (849) are then known with great accuracy except only S_{12}. Hence from the three values of $\mathfrak{D}_{12}$, $\mathfrak{D}_{23}$, $\mathfrak{D}_{31}$ we can calculate S_{12}, S_{23}, S_{31} and so deduce values of σ_1, σ_2, σ_3. Instead of comparing these values with other determinations of σ_1, σ_2 and σ_3, Lord Kelvin inserted them into formula (852) and so obtained the coefficients of self-diffusion of the three gases in question.

Lord Kelvin* gives the following values of coefficients of interdiffusivity of four gases, calculated from the experimental determinations of Loschmidt.

Gases
H_2 —(1)
O_2 —(2)
CO —(3)
CO_2—(4)

Pairs of Gases	$\mathfrak{D}_{11}$	Pairs of Gases	$\mathfrak{D}_{22}$
(12, 13, 23)	1·32	(12, 13, 23)	·193
(12, 14, 24)	1·35	(12, 14, 24)	·190
(13, 14, 34)	1·26	(23, 24, 34)	·183
Mean	1·31	Mean	·189

Pairs of Gases	$\mathfrak{D}_{33}$	Pairs of Gases	$\mathfrak{D}_{44}$
(12, 13, 23)	·169	(12, 14, 24)	·106
(13, 14, 34)	·175	(13, 14, 34)	·111
(23, 24, 34)	·178	(23, 24, 34)	·109
Mean	·174	Mean	·109

* *Baltimore Lectures*, p. 295.

The agreement *inter se* of the values obtained by different sets of three gases gives a striking confirmation of the theory, except of course as regards the numerical multiplier which does not affect the values obtained for $\mathfrak{D}_{11}$, $\mathfrak{D}_{22}$, etc.

428. It remains to test the numerical multiplier. The calculations of Chapman, Enskog and Pidduck combine in predicting the relation (approximately)

$$\mathfrak{D} = 1{\cdot}200 \frac{\kappa}{\rho}$$

for elastic spheres, while Maxwell's theory given in Chapter VIII predicted the relation (exactly)

$$\mathfrak{D} = 1{\cdot}504 \frac{\kappa}{\rho}$$

for molecules repelling according to the inverse fifth-power of the distance.

In the following table, the first column gives the value assumed for κ in the table of p. 288, the second column gives ρ, the third gives the value of $\mathfrak{D}$ calculated from Loschmidt's experiments, and the fourth gives the value of $\mathfrak{D}\rho/\kappa$.

Gas	κ (p. 288)	ρ	$\mathfrak{D}$ (p. 321)	$\frac{\mathfrak{D}\rho}{\kappa}$
Hydrogen	·0000857	·0000899	1·31	1·37
Oxygen	·000192	·001429	·189	1·40
Carbon-monoxide ...	·000163	·001250	·174	1·34
Carbon-dioxide	·000139	·001977	·109	1·50

It at once appears that $\mathfrak{D}\rho/\kappa$ has in each case a value intermediate between the two values 1·200 and 1·504 predicted by theory for elastic spheres and inverse fifth-power molecules. Not only is this so, but the values of $\mathfrak{D}\rho/\kappa$ vary between these limits in a manner which accords well with the knowledge we already have as to the laws of force (μr^{-s}) in the different gases concerned, as the following figures shew:

	Value of s (p. 284)	$\frac{\mathfrak{D}\rho}{\kappa}$
Theory	∞	1·200
Hydrogen	12	1·37
Carbon-monoxide ...	9·3	1·34
Oxygen	7·9	1·40
Carbon-dioxide	5·2	1·50
Theory	5·0	1·504

It is somewhat remarkable that the values of $\mathfrak{D}\rho/\kappa$ for hydrogen and carbon-monoxide, in which the molecules are comparatively "hard" (in the sense of § 375), approximate closely to the value 1·34 predicted by Meyer's corrected theory (§ 416) and also by the Stefan-Maxwell theory (§ 421). This suggests that for ordinary natural molecules the two simpler theories may represent the processes at work with as great accuracy as the more elaborate theories of Chapman and Enskog so long as we remain in ignorance of the exact molecular structure which ought to be assumed in the latter.

Coefficient of Diffusion for Elastic Spheres.

429. In the following table are given the observed values* of $\mathfrak{D}_{12}$ for a number of pairs of gases in which the molecules are comparatively hard, having values of s greater than 8 in the table of p. 284. The table gives also the values of S_{12} calculated from them by formula (850) (using values of $\bar{c}$ given on p. 119), and, in the last column, the values of S_{12} calculated from the coefficient of viscosity as on p. 288.

The agreement between the two sets of values of S_{12} is as good as could reasonably be expected, providing a corresponding confirmation of

Gases	$\mathfrak{D}_{12}$ (observed)	S_{12} (calc. from $\mathfrak{D}_{12}$)	S_{12} (calc. from viscosity)
Hydrogen—Air	0·661	$3{\cdot}23\times10^{-8}$	$3{\cdot}23\times10^{-8}$
,, —Oxygen	0·679	3·18	3·17
Oxygen—Air	0·1775	3·69	3·68
,, —Nitrogen	0·174	3·74	3·70
Carbon-monoxide—Hydrogen...	0·642	3·28	3·26
,, —Oxygen......	0·183	3·65	3·71

formula (850). When one or both of the two kinds of molecules involved is softer than those in the foregoing table the agreement is still good although less striking than that found above, as is shewn in the table on page 324.

Thermal Diffusion.

430. Maxwell's treatment of diffusion for molecules repelling as the inverse fifth-power of the distance was based upon the general equation of transfer (§ 333)

$$\frac{\partial}{\partial x}(\nu\overline{uQ}) - \overline{Q}\frac{\partial}{\partial x}(\nu\bar{u}) = \Delta Q \text{ ....(860)},$$

* The values used are taken from the Smithsonian tables (1910 edition).

Gases	$\mathfrak{D}_{12}$ (observed)	S_{12} (calc. from $\mathfrak{D}_{12}$)	S_{12} (calc. from viscosity)
Carbon-dioxide—Hydrogen	0·538	$3·56 \times 10^{-8}$	$3·67 \times 10^{-8}$
„ —Air.....................	0·138	4·03	4·18
„ —Carbon-monoxide...	0·136	4·09	4·21
Nitrous oxide—Hydrogen	0·535	3·57	3·71
„ —Carbon-dioxide	0·0983	4·53	4·66
Ethylene—Hydrogen	0·486	3·75	4·14
„ —Carbon-monoxide	0·101	4·99	4·68

which is applicable to any steady motion parallel to the axis of x. On putting $Q = u$ and assuming the inverse fifth-power law, this equation becomes

$$\frac{1}{2hm_1}\frac{\partial \nu_1}{\partial x} = \nu_1 \nu_2 m_2 \sqrt{\frac{K}{m_1 + m_2}} A_1 (u_{02} - u_{01}) \text{ ....(861)},$$

and a comparison with the general equation of diffusion enables us at once to determine the coefficient of diffusion.

The success of the method depended on the assumption of the law of the inverse fifth-power. Under this law Δu was found to be proportional to $u_{02} - u_{01}$ and the equation of diffusion followed at once. Under any other law Δu is not proportional to $u_{02} - u_{01}$; the value of ΔQ depends on the law of distribution of velocities whatever value is given to Q, and the equation of diffusion no longer follows on putting $Q = u$.

To obtain the equation of diffusion from equation (860) in the general case, it is necessary to assume Q to be equal to u multiplied by a series of powers of c. The resulting equation* is found to be of the same general type as (861), except for the important difference that the left-hand member includes terms in $\partial T/\partial x$ and $\partial p/\partial x$ in addition to the term in $\partial \nu/\partial x$.

If T and p do not vary with x, these additional terms disappear. But their presence in the general equation indicates that a process of diffusion is necessarily going on in any gas in which T and p vary from point to point, exception being made of the special case in which the molecules repel according to the exact inverse fifth-power of the distance. These phenomena were first predicted on purely theoretical grounds by Chapman (1917) in his paper already quoted; shortly afterwards they were predicted independently by Enskog†.

* For details see Chapman, *Phil. Trans.* 217 A (1917), pp. 124, 181. The analysis given by Chapman in this paper is invalidated by an error of algebra, as was pointed out by Enskog (*Arkiv f. Mat. Astron. und Fysik* XVI. (1921)), but the numerical consequences are not serious. For a corrected discussion see Chapman and Hainsworth, *Phil. Mag.* 48 (1924), p. 593.

† "Kinetische Theorie der Vorgänge in mässig verdünnten Gäsen" (Inaug. Dissertation, Upsala, 1917).

The phenomenon of "pressure-diffusion" does not appear to possess any great importance physically. That of "thermal-diffusion" is of importance because numerically its magnitude is comparable with that of ordinary diffusion. Let us imagine that we have a tube or cylinder, originally filled with a uniform mixture of two gases, and let the two ends be kept permanently at different temperatures. As the result of thermal diffusion, currents will be set up in the tube, the molecules of the heavier gas tending to diffuse in the direction of decreasing temperature and vice versa. There is a limit to the inequality of composition of the mixture which can be established by this means, for the inequality brings into play ordinary diffusion which acts in the opposite direction and tends to restore uniformity of composition. Thus a steady state will ultimately be reached in which the proportion of the mixture will vary gradually as we pass along the tube.

431. This predicted variation in the proportion of the mixture was first discovered experimentally by F. W. Dootson*. In a typical experiment a tube with a bulb at each end was filled with a mixture of hydrogen and carbon-dioxide in approximately equal proportions. One bulb was then kept for four hours at a steady temperature of 230° C., the other being kept water-cooled at 10° C. At the end of the four hours samples were drawn off from the two bulbs and analysed, with the following results:

Hot bulb (230° C.): 44·9 per cent. H_2; 55·1 per cent. CO_2.
Cold bulb (10° C.): 41·3 per cent. H_2; 58·7 per cent. CO_2.

The effect detected here is of the sign and order of magnitude predicted by theory. In actual amount it is rather less than half the amount predicted on the supposition that the molecules behave like elastic spheres. The theoretical effect for elastic spheres is, however, greater than that for any other type of molecule, and it will be remembered that it vanishes altogether for molecules repelling according to the Maxwellian law μr^{-5}. Thus the effect detected by experiment was about what was to be expected, and as later and more accurate experiments† have given very similar results, the theory may be regarded as being verified.

The steady state phenomenon we have been considering is of interest in that it depends greatly upon the law of force between molecules. It seems possible that it may in time lead to powerful methods for the investigation of molecular fields of force. Chapman has suggested also that it may prove to be of value for the separation of gases of equal molecular weight‡ (e.g. C_2H_4 and N_2), and also possibly for the separation of isotopes §.

* *Phil. Mag.* XXXIII. (1917), p. 248.
† Ibbs, *Proc. Roy. Soc.* 99 A (1921), p. 385, and 107 A (1925), p. 470; Elliott and Masson, *Proc. Roy. Soc.* 108 A (1925), p. 378.
‡ *Phil. Mag.* XXXIV. (1917), p. 146.
§ *Phil. Mag.* XXXVIII. (1919), p. 182.

CHAPTER XIV

THE EVIDENCE OF THE KINETIC THEORY AS TO THE SIZE OF MOLECULES

Evaluation of Molecular radius from Free Path Phenomena.

432. In the last three chapters we have considered the free path phenomena of viscosity, conduction of heat and diffusion, and have found for the three corresponding coefficients formulae involving in every case the quantity σ, the diameter of the molecule of the gas in question. Thus we have three phenomena from which the molecular diameter may be calculated.

The values of σ which can be deduced from the phenomenon of viscosity have already been calculated and exhibited in the table on p. 288. A similar set of values can be deduced from the observed values of ϑ, the coefficient of conduction of heat given on p. 301. To do this, it is necessary to make some definite assumption as to the transfer of internal molecular energy, and the assumption which has been made is that already explained in § 403. Finally, it is possible to obtain a third set from the coefficients of diffusion given in the tables on pp. 323, 324, although the procedure here is rather more complicated than in the two former cases. In these tables we have thirteen observations from which to determine the eight molecular diameters involved. A least-square solution would be laborious, and of little real value since obviously some of the values given for $\mathfrak{D}_{12}$ have a much greater observational value than others. The following simple plan has therefore been followed.

The values of the molecular diameters of the three gases hydrogen, oxygen and air have been determined solely from the first three entries in the first table. The value of σ for nitrogen can then be obtained from the fourth entry. The two remaining entries then give the two somewhat discordant values 1·92 and 1·83 for the radius of the molecule of carbon-monoxide, and I have assumed the true value to be the mean of these, namely 1·87.

The three first entries in the second table then give for the radius of the CO_2 molecule the values 2·20, 2·16, 2·22, and I have assumed the true value to be the mean of these, namely 2·19. In a similar way I have taken mean values for the radii of the molecules of nitrous oxide and ethylene.

The three sets of values obtained in this way are exhibited in the following table. The agreement of the different entries *inter se* is surprisingly good considering the assumptions which have been introduced. Rather naturally, it is less good for gases with soft molecules than for those with hard. The entries in the fourth column give the mean of the entries in the preceding columns, and the entries in the last column give the mean free path calculated by Maxwell's formula (56).

VALUES OF MOLECULAR RADIUS AND MEAN FREE PATH.

Gas	Value of $\frac{1}{2}\sigma \times 10^8$ calculated from			Mean value of $\frac{1}{2}\sigma \times 10^8$	Mean free path (cms.)
	Viscosity	Conduction of heat	Diffusion		
Hydrogen	1·36	1·36	1·36	1·36	$11{\cdot}25 \times 10^{-6}$
Helium	1·09	1·10	—	1·10	17·2
Water-vapour	2·29	—	—	2·29	4·0
Carbon-monoxide	1·90	1·91	1·87	1·89	5·8
Ethylene	2·78	2·78	2·75	2·77	2·7
Nitrogen	1·89	1·89	1·92	1·90	5·8
Air	1·87	1·87	1·87	1·87	5·9
Nitric oxide	1·88	1·89	—	1·88	5·9
Oxygen	1·81	1·81	1·82	1·81	6·4
Argon	1·83	1·82	—	1·82	6·3
Carbon-dioxide	2·31	2·42	2·19	2·31	3·9
Nitrous oxide	2·35	2·33	2·27	2·32	3·9
Methyl chloride	2·83	—	—	2·83	2·6
Ethyl chloride	3·09	—	—	3·09	2·2
Chlorine	2·70	—	—	2·70	2·9
Benzene	3·75	—	—	3·75	1·5
Krypton	2·07	—	—	2·07	4·9
Xenon	2·44	—	—	2·44	3·5

Evaluation of $\frac{1}{2}\sigma$ from Deviations from Boyle's Law.

433. In each of these phenomena the molecular diameter has entered through the free path. In Chapter VI the molecular diameter was estimated from the observed deviations from Boyle's Law, and the diameter then entered through the total volume occupied by all the molecules in a given space. The two sets of values obtained for $\frac{1}{2}\sigma$ are as follows:

Gas	$\frac{1}{2}\sigma$ from Boyle's Law	$\frac{1}{2}\sigma$ from free path
Hydrogen	$1{\cdot}27 \times 10^{-8}$	$1{\cdot}36 \times 10^{-8}$
Helium	0·99	1·10
Nitrogen	1·78	1·90
Air	1·66	1·87
Carbon-dioxide	1·71	2·31

Although these numbers agree tolerably well, the agreement must to some extent be regarded as accidental. For, as we saw in § 378, the two sets of values of σ do not really measure the same quantity. Of the five gases in the table, the two which have the hardest molecules are hydrogen and helium, and for each of these we may suppose that $s=12$ approximately (cf. p. 284). Thus the correcting factor which was found to be necessary in § 378 has for these gases the value $\sqrt[3]{\Gamma(\frac{8}{11})}$ or 1·079. We must divide the values for $\frac{1}{2}\sigma$ obtained from Boyle's Law by this number and find for the quantity

$$\frac{1}{2}\left[\frac{\mu}{RT(s-1)}\right]^{\frac{1}{s-1}}$$

the corrected values 1·18 for hydrogen and 0·91 for helium.

The values of $\frac{1}{2}\sigma$ obtained from free path phenomena attempt to measure this quantity directly, but the values obtained are now seen to differ from those just found by about 13 per cent. for hydrogen and about 22 per cent. for helium. For the softer molecules, such as that of carbon-dioxide, the agreement is naturally very much worse. We have found for the CO_2 molecule a correcting factor 1·5363 (cf. equation (776)), so that the entry 1·71 for $\frac{1}{2}\sigma$ ought to be reduced to 1·10. But as mentioned in § 377 there are other corrections to be made before we can start comparing the two values of $\frac{1}{2}\sigma$.

The general result, however, emerges quite clearly that the values for $\frac{1}{2}\sigma$ obtained from Boyle's Law are uniformly smaller than those obtained from free path phenomena. A simple geometrical interpretation of this can be given. In measuring the deviations from Boyle's Law we are virtually measuring the volume of a molecule, while in measuring the free path in a gas we are measuring the cross-section of the same molecule. The figures obtained suggest that the mean radius of the molecule, regarded as a solid, is less than the mean average radius of the cylinders circumscribing its various cross-sections. But this, simply as a matter of geometry, must necessarily be the case if the molecule has any shape except that of a sphere.

To illustrate this suppose that the molecules of a gas were constituted of coin-shaped discs of radius a and small thickness h, and therefore of total volume $\pi a^2 h$. Except for a correction necessitated by the fact that these molecules would not behave like elastic spheres at collision, the deviations from Boyle's Law would lead to a value of σ such that $\frac{1}{6}\pi\sigma^3 = \pi a^2 h$ or $\sigma = (6a^2h)^{\frac{1}{3}}$, while measurements on the free path would lead to the value $\sigma = a$. If h is small, the value of $(6a^2h)^{\frac{1}{3}}$ will of course be very much smaller than a.

The difference between the two sets of values found for $\frac{1}{2}\sigma$ may accordingly be interpreted as indicating that the molecules are not really spherical. The comparative closeness of the two sets of values for hydrogen and helium suggests, however, that for these gases the assumption of spherical molecules will give a tolerably good approximation to the truth. For the softer molecules such as carbon-dioxide it is perhaps safest not to attempt to draw any conclusions, in view of the difficulties already explained in § 378. It need hardly be said that what we are concerned with in these evaluations of $\frac{1}{2}\sigma$ is the extension of the field of force surrounding the molecule, and not the size of the material structure out of which this field of force originates. We shall see how this latter structure can be measured in § 437.

434. Some investigators avoid the difficulties arising out of the "softness" of the molecules, by assuming Sutherland's formula (§ 381) to hold, and comparing the values of σ deduced from Boyle's Law with the values of the diameter σ_∞ of the hard kernel. For instance Eucken*, in a very interesting paper, has given the following comparison, amongst others, of diameters of molecules calculated from deviations from Boyle's Law and from free paths:

Gas	$\frac{1}{2}\sigma$ from Boyle's Law	$\frac{1}{2}\sigma_\infty$ from free path
Nitrogen............	$1{\cdot}540 \times 10^{-8}$	$1{\cdot}527 \times 10^{-8}$
Oxygen	1·447	1·467
Argon	1·427	1·417
Carbon-dioxide ...	1·602	1·601

It must be remarked that Eucken uses values of Van der Waals' quantity b which are deduced from the critical data, and so may differ very widely from true observational values (cf. § 190). For this reason Eucken's values of $\frac{1}{2}\sigma$ deduced from Boyle's Law differ considerably from those given on p. 327, and little seems to be gained by comparing $\frac{1}{2}\sigma$ evaluated from the critical temperature with $\frac{1}{2}\sigma_\infty$ which refers to temperature $T = \infty$.

Evaluation of $\frac{1}{2}\sigma$ from Densities in the Solid and Liquid States.

435. A further estimate of the molecular radius can be formed by a consideration of the maximum density of the substance when in the solid or liquid state. This method, however, only enables us to calculate an upper limit to the molecular radius.

For instance, Dewar† found the density of solid hydrogen at $13{\cdot}2°$ absolute to be ·0763. The mass of a cubic centimetre of solid hydrogen is

* *Phys. Zeitsch.* XIV. (1913), p. 331. Eucken also calculates values for three other gases—Helium, Hydrogen and Benzene, but the values agree nothing like so well as those given here. The author states that the material for the calculation of $\frac{1}{2}\sigma$ from Boyle's Law is uncertain in the case of the gases helium and hydrogen. See also Chapman, *Phil. Trans.* 211 A, p. 481.

† *Proc. Roy. Soc.* LXXIII. (1904), p. 251.

accordingly ·0763 grammes, while the mass of each molecule (cf. § 8) is known to be $3{\cdot}32 \times 10^{-24}$ grammes. The number of molecules in a cubic centimetre is accordingly $2{\cdot}30 \times 10^{22}$. If the molecules of hydrogen are regarded as hard spheres of diameter σ, these molecules, if packed as closely as possible, would occupy a volume

$$2{\cdot}30 \times 10^{22} \times \frac{\sigma^3}{\sqrt{2}}.$$

This volume, then, is certainly less than a cubic centimetre. Or, what comes to the same thing, the value of the molecular radius of hydrogen is certainly less than the value of $\frac{1}{2}\sigma$ which makes the above expression equal to a cubic centimetre. This value is

$$\tfrac{1}{2}\sigma = 1{\cdot}97 \times 10^{-8}.$$

In this way we obtain a superior limit to the value of $\frac{1}{2}\sigma$ for hydrogen. The similar limits for other gases can be obtained in the same way, and we arrive at the following table:

Substance	Temp. (C.)	Density	Upper limit of $\frac{1}{2}\sigma$	Value of $\frac{1}{2}\sigma$ (p. 327)
Hydrogen (solid)*	−259·9°	0·0763	$1{\cdot}97 \times 10^{-8}$	$1{\cdot}36 \times 10^{-8}$
Helium (liquid)†	−271·6	0·1456	2·00	1·10
Water	4·0	1·0000	1·73	2·29
Carbon-monoxide‡	−205·0	·08558	2·12	1·89
Ethylene§	−21	0·414	2·68	2·77
Nitrogen (solid)*	−252·5	1·0265	1·99	1·90
Oxygen (solid)*	−252·5	1·4256	1·86	1·81
Argon (liquid)‡	−189	1·423	2·02	1·82
Carbon-dioxide (solid)‖	−79	1·53	2·03	2·31
Nitrous oxide§	−20·6	1·003	2·32	2·32
Methyl chloride¶	−20	0·983	2·46	2·83
Ethyl chloride**	0	0·925	2·65	3·09
Chlorine††	−80	1·6602	2·31	2·70
Benzene‡‡	0	0·899	2·94	3·75
Krypton§§	−169	2·15	2·23	2·07
Xenon§§	−140	3·52	2·25	2·44

* Dewar, *Proc. Roy. Soc.* LXXIII. (1904), p. 251.
† Kamerlingh Onnes (1911). ‡ Baly and Donnan (1902).
§ Cailletet and Mathias, *Journ. de Phys.* [2], v. (1886), p. 555.
‖ Behn (1900). ¶ Vincent and Delachanal, *Comptes Rendus*, LXXXVII. (1878), p. 987.
** Darling, *Ann. Chem.* 160 (1871), p. 214, and *Recueil de Constantes Physiques.*
†† Knietsch, *Ann. Chem.* 259 (1890), p. 100. ‡‡ *Recueil de Constantes Physiques*, p. 146.
§§ Ramsey and Travers (1900).

Considering the general crudeness of the supposition upon which we have been working, the comparative agreement of these two sets of figures cannot be regarded as otherwise than satisfactory. In regarding all molecules as spherical, regardless of their true shape, we run the risk of an error in $\frac{1}{2}\sigma$ comparable with the value of the quantity itself. That discrepancies of this order of magnitude occur in the two sets of figures cannot, therefore, be regarded as a matter for surprise. At the same time the circumstance that some of the values for $\frac{1}{2}\sigma$ given in the last column are greater than their upper limits is one which demands explanation.

436. If the molecules were known to be hard spheres, the circumstance just mentioned would be perfectly incomprehensible. But when we allow for deviations from perfect hardness the difficulty is at once removed. For the upper limit is obviously an upper limit to the radius of the hard kernel of the molecule, and so must not be compared with the value of $\frac{1}{2}\sigma$ at 0° C., which is considerably larger than that of the hard kernel.

Assuming the relation between the radius of the hard kernel ($\frac{1}{2}\sigma_\infty$) and the effective radius of the molecule to be that given by Sutherland's formula (§ 381), it is easy to deduce values of $\frac{1}{2}\sigma_\infty$ from the observed values of $\frac{1}{2}\sigma$ at 0° C. A set of such values is given in the following table, together with the values of the upper limit, now regarded as an upper limit for $\frac{1}{2}\sigma_\infty$, for comparison.

Gas	Assumed value for C	Value of $\frac{1}{2}\sigma_\infty$ (calc.)	Upper limit to $\frac{1}{2}\sigma_\infty$
Hydrogen	76	$1{\cdot}21 \times 10^{-8}$	$1{\cdot}97 \times 10^{-8}$
Helium..................	79	0·97	2·00
Carbon-monoxide ...	100	1·63	2·12
Ethylene	249	2·01	2·68
Nitrogen	112	1·60	1·99
Oxygen..................	132	1·48	1·86
Argon	169	1·43	2·02
Carbon-dioxide	240	1·68	2·03
Nitrous oxide	260	1·66	2·32
Methyl chloride	454	1·74	2·46
Chlorine	199	2·06	2·31
Krypton	142	1·68	2·23
Xenon	252	1·78	2·25

It will be noticed that the upper limit is now in every case above the value for $\frac{1}{2}\sigma_\infty$.

Evaluations of $\frac{1}{2}\sigma$ from Dielectric Constant.

437. Mention may also be made here of an interesting determination of the size of molecules which does not depend upon the kinetic theory at all. Regarding molecules as spheres which are perfect conductors of electricity*, the dielectric capacity K of a gas which contains N molecules per cubic centimetre is found to be given by

$$K - 1 = \tfrac{1}{2}\pi N\sigma^3.$$

On assigning to N the value $2{\cdot}705 \times 10^{19}$ it is possible to calculate σ directly K is known. When K is not known by direct experiment, we may assume Maxwell's relation $K = \mu^2$, where μ is the refractive index for light as compared with a vacuum. For a gas, K is very nearly equal to unity, so that the relation becomes $K = 2\mu - 1$. In the following table many of the values of K are calculated in this way.

Gas	$\frac{K}{K_0}$ (observed)	Method†	Lower limit to $\frac{1}{2}\sigma_\infty$	$\frac{1}{2}\sigma_\infty$ (p. 331)
Hydrogen	1·000264	E	$0{\cdot}92 \times 10^{-8}$	$1{\cdot}21 \times 10^{-8}$
Helium	1·0000724	O	0·60	0·97
Carbon-monoxide ...	1·000692	E	1·27	1·63
Ethylene	1·001385	E	1·61	2·01
Nitrogen	1·000594	O	1·21	1·60
Air	1·000588	E	1·20	1·57
Oxygen	1·000543	O	1·18	1·48
Argon	1·000568	O	1·19	1·43
Carbon-dioxide	1·000965	E	1·41	1·68
Nitrous oxide	1·001082	E	1·47	1·66
Methyl chloride	1·001732	O	1·72	1·74
Ethyl chloride	1·002346	O	1·90	—
Chlorine	1·001536	O	1·65	2·06
Benzene	1·003382	O	2·15	—
Krypton	1·000850	O	1·36	1·68
Xenon	1·001378	O	1·60	1·78

* Mossotti's *Hypothesis*. See Maxwell, *Elect. and Mag.* (3rd edition), p. 70, or Jeans, *Elect. and Mag.* (5th edition), p. 130.

† METHODS: E—Electric method by direct measurement of K. The values given are in each case the mean of two determinations: Boltzmann, *Wiener Sitzungsber.* LXIX. p. 795, and Klemenčič, *Wiener Sitzungsber.* XCI. p. 712.

O—Optical method, the value of K being calculated from the observed refractive index. The values of μ are taken from Cuthbertson and Metcalfe, *Phil. Trans.* 207 A (1907), p. 135, Travers, *Study of Gases*, p. 296, and the *Recueil de Constantes Physiques*.

It will be noticed that the value of $\frac{1}{2}\sigma$, obtained in this manner, is the radius of the electrical structure of the atom, this being supposed for simplicity to be spherical. The values obtained for $\frac{1}{2}\sigma$ may accordingly be regarded as lower limits for the quantity $\frac{1}{2}\sigma_\infty$ which we have already had under consideration. The values obtained for this lower limit are given in the above table, together with the values of $\frac{1}{2}\sigma_\infty$ obtained on p. 331, for comparison.

It is satisfactory to find that these values of $\frac{1}{2}\sigma_\infty$ lie in every case between the lower limit just obtained, and the upper limit obtained on p. 330. The limits are in some cases quite wide, as for instance with helium, but it must be remembered that we have arrived at the limits by assuming the helium atom to be spherical, whereas it probably consists of three coplanar electric charges. The widest limits occur for the gases of lowest molecular weight, except for a tendency for monatomic gases to have wider limits than others. This is seen from the following table, giving the ratio of the limits for the various gases:

Gas:	He	Ar	Kr	Xe
Ratio:	3·3	1·7	1·6	1·5

Gas:	H_2	CO, C_2H_4, N_2	O_2 N_2O	CO_2, CH_3Cl, C_2H_5Cl, Cl_2, C_6H_6
Ratio:	2·1	1·7	1·6	1·4

Evaluation of $\frac{1}{2}\sigma$ from Crystalline Structure.

438. Finally, it is possible to measure the distances apart of the centres of the atoms in crystal structures with great accuracy. From measures of this kind, W. L. Bragg* has determined the following radii for the atoms of the commoner elements which occur in gases:

Carbon $\frac{1}{2}\sigma = 0·77 \times 10^{-8}$ cms. Nitrogen $\frac{1}{2}\sigma = 0·65 \times 10^{-8}$ cms.
Oxygen $\frac{1}{2}\sigma = 0·65 \times 10^{-8}$ cms. Chlorine $\frac{1}{2}\sigma = 1·05 \times 10^{-8}$ cms.

The radii of the atoms of the inert gases cannot be determined in this way, but from a consideration of their positions in the periodic table of elements, Bragg deduces the following radii:

Neon $\frac{1}{2}\sigma = 0·65 \times 10^{-8}$ cms. Krypton $\frac{1}{2}\sigma = 1·17 \times 10^{-8}$ cms.
Argon $\frac{1}{2}\sigma = 1·02 \times 10^{-8}$ cms. Xenon $\frac{1}{2}\sigma = 1·35 \times 10^{-8}$ cms.

These radii are substantially lower than those calculated from free path phenomena, shewing that the atoms are packed more closely in crystalline solids than in gaseous collisions.

* *Phil. Mag.* XL. (1920), p. 169.

CHAPTER XV

AEROSTATICS AND PLANETARY ATMOSPHERES

439. IN the present chapter we shall apply the principles and results of the kinetic theory to a discussion of problems connected with the atmosphere, both of this and other planets. The problems dealt with consist of various problems of aerostatics, and an investigation into the question of dissipation of planetary atmospheres.

AEROSTATICS.

Atmosphere in Isothermal (Conductive) Equilibrium.

440. An atmosphere is essentially a mixture of gases of different kinds under the influence of a permanent field of force, namely that of gravitation. The potential of this field of force at a height z above the surface of the planet may be taken to be gz.

For the densities of the different kinds of gas at a height z, we now have, from § 113, the equations

$$\rho = \rho_0 e^{-2hmgz} \quad \text{.........................(862)},$$

$$\rho' = \rho_0' e^{-2hm'gz} \quad \text{.........................(863)}.$$

These equations give the densities of the constituents of the atmosphere at different heights, $\rho_0, \rho_0', \ldots$ clearly being the densities at the planet's surface. We neglect variations in the value of g, and also the rotation of the planet.

These equations are independent in the sense that each is concerned with one and only one of the different constituents of the atmosphere. The equations therefore contain the mathematical expression of the law formulated by Dalton for an atmosphere in isothermal equilibrium:

An atmosphere in isothermal equilibrium may be regarded as the aggregate of a number of atmospheres, one for each constituent gas, the law of density in each atmosphere being the same as if it alone was present.

Considering, for simplicity, two constituents only, the ratio in which they are mixed at any height z is seen to be

$$\frac{\rho}{\rho'} = \frac{\rho_0}{\rho_0'} e^{-2hgz(m-m')} = \frac{\rho_0}{\rho_0'} e^{-\frac{gz}{T}\frac{m-m'}{R}} \quad \text{...............(864)},$$

and numerical values can be obtained on inserting the values of R/m, R/m' from the table on p. 119.

For instance, if two kinds of gas are the oxygen and nitrogen in the earth's atmosphere, we find that the index of the exponential becomes equal to about ·01 at a height of five kilometres. Thus the proportions of oxygen and nitrogen would, on the suppositions we have made, change at the slow rate of about one per cent. in every five kilometres.

Atmosphere in Adiabatic (Convective) Equilibrium.

441. The atmospheric distribution which has just been investigated, a distribution in which the temperature is constant throughout, while the density of each component gas falls off exponentially with the height above the earth's surface, is the law which would undoubtedly become established if the earth's atmosphere were left at rest for a sufficient time.

Under actual conditions, however, the earth's atmosphere is incessantly being agitated by currents and storms, so that there is a continual mechanical transference of air from one part of the atmosphere to another. From this circumstance, coupled with the fact that the conduction of heat in gases is very slow, it follows that the atmosphere is never permitted to assume the equilibrium distribution which has just been discussed. As the density of an element is changed by its enforced motion from one layer of the atmosphere to another, the temperature also tends to change, but before the temperature has adjusted itself by conduction to the temperature of its new surroundings the element finds itself again moved away. Hence it happens that the factor which determines the distribution of the atmosphere is not the equalisation of temperature necessary to a permanent state, but is the condition that an element of gas, on being moved from one place to another, shall take up the requisite pressure and volume in its new position without any loss or gain of heat by conduction taking place. The law connecting the pressure and volume in the atmosphere must accordingly be the adiabatic law found in § 242.

442. The general equation of equilibrium of the atmosphere is

$$\frac{\partial p}{\partial z} = -g\rho \quad \text{.................................(865)},$$

and if, in accordance with the adiabatic law, we write

$$p = k\rho^\gamma \quad \text{................................(866)},$$

we obtain

$$k\gamma\rho^{\gamma-1}\frac{\partial \rho}{\partial z} = -g\rho \quad \text{..........................(867)}.$$

The integral of this equation is

$$\frac{k\gamma}{\gamma-1}(\rho_0^{\gamma-1} - \rho^{\gamma-1}) = gz \quad \text{..........................(868)},$$

where ρ_0 is the density at $z=0$. This is the law according to which the density falls off with the height. Since by equation (866), T is proportional to $\rho^{\gamma-1}$, it follows that equation (868) can be expressed in the form

$$\frac{T_0-T}{z}=\text{a constant} \qquad (869),$$

where T_0 is the temperature at $z=0$. Thus the temperature decreases as we move upwards in the atmosphere, the amount of decrease being proportional to the height.

The process of diffusion, as well as that of conduction, being very slow in gases, it follows that the constituent gases of an atmosphere in convective equilibrium ought to occur in approximately the same proportion at all heights. This is found by experiment to be true of the atmosphere of the earth. Frankland* has found that the proportions of nitrogen and oxygen in our atmosphere are the same for all altitudes up to 14,000 feet. As has already been remarked, there would be a variation of about one per cent. at this height in an atmosphere in conductive equilibrium.

On putting $\rho=0$ in equation (868) we obtain

$$z=\frac{k\gamma\rho_0^{\gamma-1}}{g(\gamma-1)} \qquad (870),$$

from which it appears that there is a superior limit to the height of an atmosphere in convective equilibrium. Since, by equation (866), $p=k\rho^\gamma$, this limiting height may be written in the form

$$\frac{p_0\gamma}{g\rho_0(\gamma-1)} \qquad (871),$$

where p_0, ρ_0 are the pressure and density at the earth's surface. Substituting numerical values, this height is found to be about 29 kilometres.

Our atmosphere, then, if supposed to be in convective equilibrium throughout, would have to be regarded as a layer of gas of uniform composition throughout, having a height of about 29 kilometres, the temperature decreasing uniformly as we ascend.

On substituting numerical values, we find for the constant of the right-hand side of equation (869) a value of approximately 10° C. per kilometre. In practice, however, the problem is one of extreme complexity, owing in part to the irregularities of the earth's surface which prevent the surfaces of equal temperature from being strictly spherical. It is found that this theoretical estimate of the temperature gradient is only approximately confirmed by observation, the observed value being more like 5° C. per kilometre†.

* *Journ. Chem. Soc.* XIII. p. 22.

† See Dines, *Phil. Trans.* 211 A (1912), p. 253, or Gold, *Proc. Roy. Soc.* LXXXII. A (1909), p. 43. The *Comité météorologique international* has adopted the rate of 1° per 200 metres for reductions of temperature observations to sea level.

THE OUTER ATMOSPHERE.

Outer Atmosphere in Conductive Equilibrium.

443. When we examine in detail the *molecular* mechanism by which the adiabatic law is maintained in an atmosphere we find that there must be a limit beyond which adiabatic equilibrium is impossible. For at the free surface which is predicted from the supposition that the adiabatic law obtains throughout, the density would be zero, and therefore the mean free path infinite. Hence there would be molecules arriving at this surface from layers of gas inside it, with finite velocities and infinitesimal probabilities of collision. The majority of these molecules would of course pass outside the free surface predicted by the simpler theory, in a manner somewhat similar to that in which molecules escape from the free surface of a liquid and form a vapour.

These molecules would form what may be described as an "outer" atmosphere. In this atmosphere the density is very small, so that collisions are rare, and the majority of molecules will simply describe orbits under the earth's gravitation, undisturbed by collisions, and will finally fall back again into the "inner" or adiabatic atmosphere. This at any rate is true of those molecules which start with velocities such that they describe elliptic orbits under the earth's attraction. Others, starting with greater velocities, will describe parabolic or hyperbolic orbits, and these may be regarded as lost altogether to the earth's atmosphere. We shall return to the consideration of these losses later.

444. A brief calculation will shew that the isothermal layer predicted by these considerations would be expected to start only a few metres from what would otherwise have been the outer limit of the atmosphere; and so long as we consider only the molecular mechanism of the atmosphere, there seems to be no reason why the isothermal layer should extend further than this. The question assumes a very different aspect when the radiation of the different layers of the atmosphere is taken into account. It is obvious that an atmosphere in which the outer layers were at, or near to, the absolute zero of temperature could not be in permanent equilibrium. For these outer layers, being at this low temperature, would lose no heat by radiation, and would be continually gaining heat by absorption of radiation from the warm inner layers of the atmosphere. For any atmospheric arrangement to be permanent, the radiation and absorption of each element must be equal.

The mathematical theory of an atmosphere in radiation equilibrium has been worked out by Gold*. Regarding the atmosphere as consisting of two shells, the inner in the adiabatic state, and the outer in the isothermal, it is

* "The Isothermal Layer of the Atmosphere, and Atmospheric Radiation," *Proc. Roy. Soc.* LXXXII. A (1909), p. 43.

shewn that for an atmosphere of uniform constitution, the adiabatic state could not extend to a height greater than that given by $p = \frac{1}{2}p_0$, where p_0 is the surface pressure. When the atmosphere is not supposed to be uniform, the height to which the adiabatic layer can extend is increased. Using an approximate formula to represent the varying amounts of water-vapour at different heights, it is shewn that the adiabatic layer must extend to a height greater than that given by $p = \frac{1}{2}p_0$ (namely $z = 5\frac{1}{2}$ kilometres), but cannot extend to a height greater than that given by $p = \frac{1}{4}p_0$ (namely $z = 10\frac{1}{2}$ kilometres).

445. The existence of an isothermal layer above the adiabatic region of the atmosphere has been amply confirmed by observation. The phenomenon was first observed by M. Teisserenc de Bort, and was communicated to the Société Française de Physique in 1899*; it has now become a matter of every day observation. The average height at which the isothermal layer is encountered is between 10 and 11 kilometres in European latitudes. The altitude is greater near the equator and less near the poles; it is generally less in cyclonic weather than in anticyclonic†. Occasionally the phenomenon is found to be one of temperature-inversion, the temperature reaching a minimum at the boundary of the two layers, and increasing as the isothermal layer is further penetrated. In an observation over Strassburg in 1905, the temperature recorded at 26 kilometres was 20° C. higher than that at 14 kilometres.

In general we shall obtain a fair approximation to average conditions by assuming that the temperature is adiabatic up to a height of $10\frac{1}{2}$ kilometres, at which the pressure is $\frac{1}{4}p_0$, that at this point the temperature is $-54°$ C. ($= 219°$ abs.), and that beyond this the atmosphere is in isothermal equilibrium.

446. According to the simple theory of which the result is expressed by equation (862), there can be no upper limit to the height of the outer atmosphere in isothermal equilibrium, for the value of ρ, as given by this equation, does not vanish at any finite height.

It must, however, be remembered that in arriving at equation (862), no account was taken of the rotation of the planet, or of the variation of the value of gravity at heights above the earth's surface. When there is found to be no limit to the height of the atmosphere, the neglect of these disturbing agencies becomes inadmissible.

* See also Teisserenc de Bort, *Comptes Rendus*, 134 (April 1902), 138 (Jan. 1904), and 145 (July 1907) and Dines, *Phil. Trans.* 211 A, pp. 263, 268.

† For fuller details see *The Computer's Handbook* (Government publication, issued by the Meteorological Office), especially diagram facing II. p. 55; W. H. Dines, "The characteristics of the Free Atmosphere" (*Geophysical Memoirs*, No. 13, London, 1919); McAdie, *Principles of Aerography*, chap. v. especially pp. 47, 49.

For a planet rotating about the axis of z with angular velocity ω, the law of distribution of density may be taken to be

$$\rho = Ae^{-2hmV} \quad\text{.............................(872)},$$

where

$$V = -\tfrac{1}{2}\omega^2(x^2+y^2) - \frac{ga^2}{r} \quad\text{......................(873)},$$

the term $\frac{ga^2}{r}$ representing the gravitational potential at a distance r from the earth's centre. Thus the equation of the surfaces of equal density may be taken to be

$$\omega^2(x^2+y^2) + \frac{2ga^2}{r} = \text{constant} \quad\text{....................(874)}.$$

The form of these surfaces was first studied by Edward Roche*; a discussion of his results is given in a paper by Prof. G. H. Bryan†. Sufficient information for our present purpose will be obtained by examining the distribution of density in the equatorial plane of the planet. Replacing x^2+y^2 by r^2, we have in the equatorial plane

$$\rho = Ae^{hm\left[\omega^2 r^2 + \frac{2ga^2}{r}\right]} \quad\text{......................(875)}.$$

Differentiating, we find that $d\rho/dr$ vanishes when

$$\omega^2 r = \frac{ga^2}{r^2} \quad\text{.............................(876)}.$$

There is therefore a single minimum of density, and the position of this minimum is the same for each constituent atmosphere, being in fact the series of points at which the apparent centrifugal force exactly balances the gravitational attraction of the planet. As we pass outwards the density decreases until this minimum is reached, and afterwards increases continually.

Helmert's value for $\omega^2 a/g$, the ratio of the apparent centrifugal force to gravity at the earth's equator, is $\frac{1}{288\cdot38}$, so that the minimum of density ought to occur at a distance from the earth's centre given by

$$r = \sqrt[3]{288\cdot38}\,a = 6\cdot607\,a.$$

At a point so far from the earth's surface as this, the density is so small that it may be treated as insignificant.

Constitution of the Outer Atmosphere.

447. Throughout the whole of the outer atmosphere the law of distribution must be that expressed by equation (872). We can now carry out calculations similar to those of § 442, but having reference to the outer

* *Mémoires Acad. Sci. Montpellier.*

† "The Kinetic Theory of Planetary Atmospheres," *Phil. Trans.* 196 A, p. 12.

atmosphere. It will not be found necessary to carry these calculations to a height above the earth's surface so great that the rotational term becomes of any importance.

At a height z above the surface of the planet, the value of V is

$$V = -\frac{ga^2}{r} = -\frac{ga^2}{a+z},$$

so that equation (872) becomes

$$\rho = Ae^{2hmga\left(\frac{a}{a+z}\right)}.$$

Also ρ_0, the value of ρ at the lower boundary of the outer atmosphere, is given by

$$\rho_0 = Ae^{2hmga},$$

since we may neglect the thickness of the inner (convective) atmosphere in comparison with the earth's radius.

The elimination of A leads to

$$\rho = \rho_0 e^{-2hmga\left(\frac{z}{a+z}\right)} \qquad \text{.......................(877).}$$

Also at the inner surface the proportion in which the different constituents occur is the same as throughout the inner atmosphere, and therefore the same as at the earth's surface.

448. The following table gives the constitution by volume of the atmosphere at the earth's surface:

Hydrogen*......	about 1	in 10,000
Helium†.........	1	„ 250,000
Neon†	1	„ 80,000
Nitrogen.........	78·03	„ 100
Oxygen	20·99	„ 100
Argon‡	0·94	„ 100
Krypton§	1	„ 2,000,000
Xenon§	1	„ 17,000,000

Traces of Carbon-dioxide and Ammonia are also present in the atmosphere in varying quantities.

* This is the proportion given by the *Recueil de Constantes Physiques*, as also by the *Computer's Handbook* of the British Meteorological Office. Claude, *Comptes Rendus*, 148 (1909), p. 1454, gives the proportion as less than 1 in 1,000,000 by weight; Lord Rayleigh (*Phil. Mag.* III. (1902), p. 416) gives less than 1 in 30,000 by volume. Some authorities believe that there is no appreciable amount of free hydrogen in the atmosphere. Important evidence bearing on this question will be found in a paper by Chapman and Milne, *Royal Meteorological Society*, June 16, 1920.

† W. Ramsay, *Proc. Roy. Soc.* LXXX. (1908), p. 599.

‡ Moissan, *Comptes Rendus*, 137, p. 600.

§ W. Ramsay, *Proc. Roy. Soc.* LXXI. (1903), p. 421, and LXXX. (1908), p. 599.

We can now study the way in which the composition of the upper atmosphere varies at different heights by a use of formula (877). The density at the bottom of the outer atmosphere must be a matter of uncertainty, but as a rough and convenient approximation, let us suppose that it corresponds to 10^{19} molecules per cubic centimetre. The number of molecules of different kinds at this level can be at once deduced from the table just given, and then. from formula (877), we can calculate the following table*, giving the number of molecules per cubic centimetre at different heights in the isothermal atmosphere. The table is calculated for the assumed temperature $T = -54°$ C. = 219° absolute, and z is measured from the *top* of the adiabatic atmosphere assumed to be at 10·5 kilometres above sea level.

The table shews at a glance how the heavier gases tend to sink to the bottom of the isothermal atmosphere, while the lighter ones rise to the top. As we ascend in the outer atmosphere, the proportion of any light to any heavier constituent gas must continually increase, so that the proportion of a light gas, however rare at the bottom, must necessarily exceed that of all heavier gases after a sufficient height.

NUMBER OF MOLECULES PER CUBIC CENTIMETRE IN THE OUTER ATMOSPHERE.

Gas	Molecular weight	Number of molecules per c.c. at height z (in kilometres)				
		$z=0$	$z=20$	$z=80$	$z=160$	$z=800$
Hydrogen...	2	100×10^{13}	80×10^{13}	4300×10^{11}	1820×10^{11}	3×10^{11}
Helium......	4	4×10^{13}	$2{\cdot}6\times10^{13}$	73×10^{11}	13×10^{11}	10^{6}
Neon.........	20	$12{\cdot}5\times10^{13}$	$1{\cdot}4\times10^{13}$	$0{\cdot}3\times10^{11}$	$0{\cdot}5\times10^{7}$	0
Nitrogen ...	28	$780{,}300\times10^{13}$	$42{,}900\times10^{13}$	520×10^{11}	35×10^{7}	0
Oxygen......	32	$209{,}900\times10^{13}$	$7{,}000\times10^{13}$	25×10^{11}	$0{\cdot}3\times10^{7}$	0
Argon	40	$9{,}400\times10^{13}$	139×10^{13}	$0{\cdot}04\times10^{11}$	10^{2}	0
Krypton ...	83	$0{\cdot}5\times10^{13}$	10^{9}	0	0	0
Xenon	130	$0{\cdot}06\times10^{13}$	10^{5}	0	0	0
Total......		10^{19}	5×10^{17}	5×10^{14}	2×10^{14}	3×10^{11}

For instance hydrogen passes nitrogen and all other gases at about 75 kilometres up, at 80 kilometres hydrogen forms 87 per cent. of the whole atmosphere, and at 800 kilometres practically the whole atmosphere is hydrogen.

* This table is from a paper by the present writer (*Bull. of the Mount Weather Observatory*, II. [6] (1910)). A similar table was first given by Hann in 1903 (*Meteorolog. Zeitschrift* (1903), p. 122), and will be found reproduced in the *Recueil de Constantes Physiques*, p. 688. See also Humphreys, *Bull. of the Mount Weather Observatory*, II. [2] (1910), and Wegener, *Phys. Zeitschrift*, XII. (1911), p. 170.

The proportion of helium nowhere attains to any great value. At about 95 kilometres, helium exceeds nitrogen in amount, but is itself already exceeded about eighty-fold by hydrogen*.

The three remaining monatomic gases are all heavier than air, and so nowhere exceed the small proportions which they contribute at the base.

Collisions in the outer atmosphere.

449. At a height of 800 kilometres, the atmosphere is practically all hydrogen. The value of ν here is about 3×10^{11}, and assuming a molecular diameter $\sigma = 2{\cdot}7 \times 10^{-8}$, we find a free path of about 10 metres. At this height, then, molecular collisions are still comparatively frequent.

At a height equal to four times this, the density is reduced by about 10^{-6}, and the free path is accordingly about 10,000 kilometres. Thus it appears that at this height the chance of collision for a molecule is practically negligible, and the atmosphere may be supposed to consist of molecules in free flight, undisturbed by collisions. It will be noticed that even at this height there are still about 300,000 molecules per cubic centimetre.

When the free path of a molecule is sufficiently short, it may be regarded as a straight line, but the molecules we are now considering are in flight for so long that gravity will produce a very appreciable curvature of their paths. The paths of some of these molecules will be approximately parabolic, the molecules behaving like projectiles discharged at their last collision. Others may rise to such heights that the variations in the values of gravity become perceptible, and their orbits must be treated as ellipses: these molecules, while in flight, form in effect a series of infinitesimal satellites to the earth. A small minority of the molecules, which happen to have acquired very high velocities by a series of unusually violent collisions, will describe hyperbolic orbits, and unless they meet with another collision, will be lost to the earth's atmosphere for ever.

In this way we see that there must be a continual loss to the atmospheres both of the earth and of other planets. The amount of this loss we may now try to estimate.

THE RATE OF LOSS OF PLANETARY ATMOSPHERES.

450. Imagine a sphere of radius R drawn in the atmosphere of a planet concentric with the planet's surface, this sphere being of such radius that collisions outside it are very infrequent, but the radius being left otherwise undetermined for the present.

The gravitational potential at the surface of this sphere will be ga^2/R, so that a molecule arriving at the sphere with a velocity c will describe an elliptic or a hyperbolic orbit according as $c^2 <$ or $> 2ga^2/R$.

* If, however, hydrogen is absent from the upper atmosphere as Chapman and Milne suggest, then the upper atmosphere is preponderatingly helium at all heights above about 95 kilometres (see footnote to p. 340).

The number of molecules which cross unit area of this sphere in an outward direction in unit time with a velocity greater than $\sqrt{2ga^2/R}$ will be

$$\nu\left(\frac{hm}{\pi}\right)^{\frac{3}{2}}\iiint e^{-hm(u^2+v^2+w^2)}\,w\,du\,dv\,dw \quad \text{............(878)},$$

where ν is the density of molecules at the sphere $r=R$, w is the component of velocity normal to the sphere, and the integration is taken for all values of u, v, w which are such that w is positive, and

$$u^2+v^2+w^2>\frac{2ga^2}{R}.$$

Each molecule counted in expression (878) is describing a hyperbolic orbit, and if the value of R is supposed so great that collisions outside the sphere of radius R are very infrequent, then each of these molecules may be supposed to be permanently lost to the planet's atmosphere.

To integrate expression (878) put

$$u=c\sin\theta\cos\phi,\quad v=c\sin\theta\sin\phi,\quad w=c\cos\theta,$$

then the limits of integration are from $\theta=0$ to $\frac{1}{2}\pi$, from $\phi=0$ to 2π, and from $c=\sqrt{2ga^2/R}$ to ∞. We find for the value of expression (878),

$$\nu\left(\frac{hm}{\pi}\right)^{\frac{3}{2}}\pi\int_{c=\sqrt{2ga^2/R}}^{c=\infty}e^{-hmc^2}c^3\,dc=\frac{\nu}{2\sqrt{\pi hm}}e^{-hm\frac{2ga^2}{R}}\left(1+hm\frac{2ga^2}{R}\right).$$

As in equation (877), we may take the value of ν to be

$$\nu=\nu_0 e^{-2hmga\frac{R-a}{R}} \quad \text{........................(879)},$$

where ν_0 is the molecular density at the base of the isothermal atmosphere. On substituting this value for ν, we obtain for the loss per unit area per unit time over the sphere of radius R,

$$\frac{\nu_0}{2\sqrt{\pi hm}}e^{-2hmga}\left(1+hm\frac{2ga^2}{R}\right) \quad \text{..................(880)}.$$

Comparing this with formula (318), we may notice that the loss is exactly what it would be if gas from the base of the isothermal layer were streaming freely into space, without any resistance, through a series of orifices of total area equal to

$$e^{-2hmga}\left(1+hm\frac{2ga^2}{R}\right)$$

times the surface of the sphere of radius R.

451. The expression obtained in this way is not independent of R, as we might at first have expected it to be. The reason for this is as follows. In the complete atmosphere, supposing it to be constituted according to the law expressing the steady state throughout, there will be a number of molecules describing orbits which never pass within a sufficiently small distance from the planet's centre for the chance of collision to be appreciable. Some of these describe hyperbolic or parabolic orbits, travelling from infinity past the planet to infinity again without collision. Now if p is the distance of the apse of any orbit from the planet's centre, it is clear that a molecule describing this orbit will be counted in expression (880) as escaping from the planet's atmosphere if $R > p$, but not if $R < p$. We should therefore expect expression (880) to increase with R, as is in fact seen to be the case.

It is questionable whether molecules of the kind just considered ought to be supposed to exist in an actual atmosphere. The analysis by which the specification of the steady state is arrived at takes no account of the length of time required for the establishment of this steady state. In the present instance the steady state implies the arrival of molecules which have described hyperbolic and parabolic orbits from infinity. It is therefore obvious that it will require infinite time to establish such a steady state.

On the other hand, molecules which are supposed to describe orbits in the regions in which no collisions occur have no influence on the rest of the atmosphere and may therefore be removed without disturbing the equilibrium of the remainder of the atmosphere. In nature these molecules cannot be supposed to exist. They would be counted in our estimate of the escape of molecules from the atmosphere by taking R great. We shall therefore obtain the most accurate results by taking R as small as possible, and the error would vanish altogether if we could reduce the sphere of radius R to such a size that collisions might be regarded as frequent everywhere inside it.

This we are prohibited from doing, because we have already supposed R to be so great that collisions outside the sphere of radius R are very infrequent. It must, nevertheless, be noticed that the *order of magnitude* of expression (880) is determined solely by the exponential e^{-2hmga}, so that the value of $2hmga$ determines whether the escape of molecules is appreciable or not. This criterion, as we should expect, is independent of R.

In the case of a rotating atmosphere, we found that there must be supposed to be a complete atmosphere extending to infinity, lying outside the region in which practically no collisions occur. This atmosphere can be treated in the same way in which individual molecules coming from infinity have been treated. It can be supposed to be removed bodily without disturbing the equilibrium of the remainder of the atmosphere.

452. Formula (880) gives the number of molecules which are lost per unit area per unit time from the planet's atmosphere, ν_0 being the number of molecules per unit volume at the base of the isothermal layer. Hence the time required for the planet to lose an amount equal to a layer one centimetre thick of the gas in question at the base of its isothermal atmosphere will be given by

$$t_0 = \frac{2\sqrt{\pi h m}}{1 + 2hmga^2/R} e^{2hmga} \text{ seconds} \quad \text{.................(881).}$$

For the earth, the actual mean radius is 6370 km., the radius of the base of the isothermal atmosphere is about 6380 km., while we have seen that R can certainly be taken to be less than (6380 + 3200) km., or about $\frac{3}{2}a$. We are only concerned with the evaluation of t_0 as regards order of magnitude, and to this degree of accuracy the distinction between R and a may be disregarded, both for the earth and the other planets. Putting $R/a = 1$, and replacing $2hm$ by its value $3/C^2$, formula (881) becomes approximately

$$t_0 = \frac{4{\cdot}34}{C\left(1+\frac{3ga}{C^2}\right)} e^{\frac{3ga}{C^2}} \text{ seconds} \quad \text{.................(882).}$$

Imagine that the total amount of gas of the kind under consideration is equal to that in a layer of thickness H and density equal to that of this gas at the base of the isothermal layer. At this base, the partial pressure of the gas in question must be approximately $\nu_0 mgH$, and is also equal to $\frac{1}{3}\nu_0 mC^2$. Thus the value of H must be approximately $\frac{1}{3}C^2/g$, and the time t_1 required for the whole outer atmosphere to stream away, if its present rate of loss were kept up, would be

$$t_1 = \frac{C^2}{3g} t_0 = \frac{1{\cdot}45C}{g\left(1+\frac{3ga}{C^2}\right)} e^{\frac{3ga}{C^2}} \text{ seconds} \quad \text{..............(883).}$$

For the earth, $a = 6{\cdot}37 \times 10^8$ cms., $g = 981$, so that $ga = 6{\cdot}36 \times 10^{11}$. The value of C for hydrogen at $-53°$ C., the temperature we have assumed for the isothermal atmosphere of the earth, is $1{\cdot}65 \times 10^5$, so that

$$\frac{3ga}{C^2} = 70{\cdot}0 \quad \text{..............................(884).}$$

Hence we find that for hydrogen in the earth's atmosphere, $t_0 = 9{\cdot}4 \times 10^{23}$ seconds or about 3×10^{17} years, while t_1 is about $2{\cdot}8 \times 10^{24}$ years. This represents a quite inappreciable rate of dissipation, even when measured by astronomical standards.

The question of the rate of loss of an atmosphere has recently been discussed by J. E. Jones* and E. A. Milne†. Although both writers treat the question with a higher degree of mathematical refinement than that of the above investigation, their numerical results do not differ to any important extent from those given by our simpler investigation.

453. It will have been noticed that formulae (882) and (883) are very sensitive to variations in temperature, owing to the presence of the exponential factor. If we had assumed a temperature of 550° abs. (277° C.) instead of one of $-54°$ C. for the earth's isothermal atmosphere, the value of C^2 would be $2\frac{1}{2}$ times its former value, and the index of the exponential (cf. equation

* *Trans. Cambridge Phil. Soc.* XXII. (1923), p. 534.

† *Trans. Cambridge Phil. Soc.* XXII. (1923), p. 483.

(884)) would be 28 in place of 70, giving values of t_0 and t_1 about e^{-42} or 10^{-18} times those found above. The actual value of t_1 is now about 10 million years, and this would represent appreciable dissipation on the astronomical scale of time.

It accordingly seems certain that at present our atmosphere is retaining its hydrogen, and *à fortiori* all the heavier gases, but the loss of a hydrogen atmosphere is readily understood if we are at liberty to contemplate an epoch of time in which the temperature of the outer atmosphere was greater than about 277° C.

454. Whenever a constituent of an atmosphere is still in existence, the values of t_0 and t_1 must be of astronomical orders of magnitude, so that the index $3ga/C^2$ must be a fairly large number. Thus we may, to within our present accuracy, neglect unity in comparison with this number in formulae (882) and (883), and replace them by the approximate formulae

$$t_0 = \frac{1{\cdot}45C}{ga} e^{\frac{3ga}{C^2}}, \quad t_1 = \frac{C^3}{2g^2a} e^{\frac{3ga}{C^2}} \quad \text{.................(885).}$$

The application of these formulae is naturally not limited to the earth. The masses and radii of the other planets and their satellites are known with fair accuracy. From these g can be estimated, and hence we obtain, for any planet, a relation between t_1, the time of dissipation, and C, the velocity of molecules of any kind in the atmosphere of the planet. The following table gives the values of M, a and g for various members of the solar system. In the last three columns are given the values of C which correspond to $t_1 = 1000$ years, $t_1 = 1{,}000{,}000$ years and $t_1 = 1000$ million years. These times may roughly be supposed to represent astronomically rapid, moderate and imperceptible dissipation respectively.

Planet	Mass (Earth = 1)	Radius (Earth = 1)	Gravity at surface (Earth = 1)	Value of molecular velocity C		
				$t_1 = 1000$ years	$t_1 = 10^6$ years	$t_1 = 10^9$ years
Sun	333,432	109·05	27·9	$1{\cdot}6 \times 10^7$	$1{\cdot}4 \times 10^7$	$1{\cdot}3 \times 10^7$
Mercury	0·056	0·37	0·41	$1{\cdot}1 \times 10^5$	$1{\cdot}1 \times 10^5$	$0{\cdot}9 \times 10^5$
Venus	0·817	0·966	0·88	$2{\cdot}7 \times 10^5$	$2{\cdot}4 \times 10^5$	$2{\cdot}1 \times 10^5$
Earth	1·000	1·000	1·000	$2{\cdot}9 \times 10^5$	$2{\cdot}6 \times 10^5$	$2{\cdot}3 \times 10^5$
Moon	0·012	0·273	0·165	$6{\cdot}1 \times 10^4$	$5{\cdot}4 \times 10^4$	$4{\cdot}8 \times 10^4$
Mars	0·108	0·54	0·37	$1{\cdot}3 \times 10^5$	$1{\cdot}2 \times 10^5$	$1{\cdot}0 \times 10^5$
Jupiter	318	11·14	2·53	$1{\cdot}6 \times 10^6$	$1{\cdot}4 \times 10^6$	$1{\cdot}2 \times 10^6$
Sat. I	0·005	0·31	0·05	4×10^4	3×10^4	3×10^4
Sat. II	0·007	0·28	0·09	5×10^4	4×10^4	4×10^4
Sat. III ...	0·028	0·47	0·13	7×10^4	6×10^4	6×10^4
Sat. IV......	0·013	0·40	0·08	5×10^4	5×10^4	4×10^4
Saturn	95·22	9·4	1·06	$9{\cdot}0 \times 10^5$	$8{\cdot}1 \times 10^5$	$7{\cdot}1 \times 10^5$
Titan	0·02	0·37	0·14	7×10^4	6×10^4	6×10^4
Uranus	14·58	4·0	0·92	$5{\cdot}6 \times 10^5$	$4{\cdot}9 \times 10^5$	$4{\cdot}4 \times 10^5$
Neptune	17·26	4·3	0·95	$5{\cdot}8 \times 10^5$	$5{\cdot}1 \times 10^5$	$4{\cdot}6 \times 10^5$

This table ought to be used in conjunction with one such as the following, which gives the values of C at various temperatures for different possible constituents of planetary atmospheres:

VALUES OF C AT DIFFERENT TEMPERATURES.

Gas	Temperature		
	$-100°$ C.	$0°$ C.	$300°$ C.
Hydrogen	$1{\cdot}47\times10^5$	$1{\cdot}84\times10^5$	$2{\cdot}66\times10^5$
Helium	$1{\cdot}04\times10^5$	$1{\cdot}31\times10^5$	$1{\cdot}90\times10^5$
Water-vapour.........	$4{\cdot}9\times10^4$	$6{\cdot}1\times10^4$	$8{\cdot}8\times10^4$
Nitrogen	$3{\cdot}9\times10^4$	$4{\cdot}9\times10^4$	$7{\cdot}1\times10^4$
Oxygen	$3{\cdot}7\times10^4$	$4{\cdot}6\times10^4$	$6{\cdot}7\times10^4$
Argon	$3{\cdot}3\times10^4$	$4{\cdot}1\times10^4$	$5{\cdot}9\times10^4$
Carbon-dioxide	$3{\cdot}1\times10^4$	$3{\cdot}9\times10^4$	$5{\cdot}7\times10^4$

These predictions of the Kinetic Theory appear to be in accordance with the facts in every instance. The tables explain at once the existence of atmospheres on Venus, the earth, and all the superior planets.

Theory would lead us to expect an atmosphere on Venus very similar in composition to that on our earth. There appears to be quite conclusive evidence that an atmosphere of some sort exists on Venus, although it has not yet been found possible to determine its constitution. What evidence is available is interpreted by H. N. Russell as shewing that the atmosphere of Venus is so permeated with particles of vapour as to be translucent rather than transparent, and this leaves the question of constitution unsolved.

Mars ought to retain water-vapour and all heavier gases with certainty, but the retention of helium must remain open to question in our present ignorance of the Martian temperature, while hydrogen could not possibly be retained. Lowell and Slipher claim to have found spectroscopic evidence of the existence of water-vapour on Mars, while Campbell argues from the smallness of atmospheric absorption that the atmosphere of Mars cannot at most have a density as great as a quarter of that of our own atmosphere.

Jupiter, Saturn, Uranus and Neptune ought clearly to retain all constituents of their atmospheres, including hydrogen. Not only this, but these planets ought to have retained hydrogen even if, in the past, their atmospheres had been at temperatures about ten times as great as those which we must now assign to them. Spectroscopic evidence indicates that all of these planets have very dense atmospheres, in which hydrogen is almost certainly a prominent constituent.

The critical molecular velocities for the moon are about a fifth of those for the earth, so that, if the temperature conditions had always been the same as for the earth, the moon ought to have retained gases having molecular weights equal to 25 times those retained by the earth. But any atmosphere of this kind on the moon must probably have been very thin, and the resulting high temperatures on the illuminated side of the moon would probably soon result in a loss of whatever atmosphere there was. An atmosphere has been observed on Titan, for which the critical velocities are about the same as for the moon, but this is explicable in view of its greater distance from the sun, and the same consideration is probably adequate to account for the suspected atmospheres on two of Jupiter's satellites. Mercury is believed to be devoid of atmosphere, although its critical velocities are higher than those for any of these satellites. The high temperature resulting from its proximity to the sun provides an adequate explanation.

Free electrons have such high velocities that they ought to escape freely from everything, including the sun. If this occurred the sun and all the planets would have become positively charged until a state of electrical equilibrium was obtained, in which planets and satellites with varying positive charges of electricity moved through a space which was negatively charged by the presence of free electrons. This possibility leads to interesting fields of speculation, but there is as yet no evidence that anything of the kind happens to any great extent. The question has been fully discussed by Milne in the paper already quoted*.

455. The escape of "atmospheres" from the surfaces of wholly gaseous stars presents a further problem, the atmospheres here being of course merely the outer layers of the stars themselves. Our formulae (885) shew that the question of escape or non-escape turns on the value of ga, the gravitational potential at the surface of the star. For all known stars ga has such a value that the present rate of escape is inappreciable†. The problem has a certain cosmogonic interest; the cosmogonist can exclude from his consideration all configurations which would be merely transitory on account of the escape of their outer layers. For instance the satellites of Mars and the smaller satellites of Saturn and Jupiter which are even now too small to retain an atmosphere, could not exist for long in the gaseous form, so that unless these bodies are mere relics of far larger masses, we must conclude that they were all born in either the solid or liquid state‡.

* *Trans. Cambridge Phil. Soc.* XXII. (1923), p. 483.

† *l.c.* § 13 (p. 507).

‡ J. H. Jeans, *Problems of Cosmogony and Stellar Dynamics*, p. 281.

CHAPTER XVI

STATISTICAL MECHANICS AND THE PARTITION OF ENERGY IN CONTINUOUS MEDIA

GENERAL THEORY.

456. IN Chapter V we considered the statistical mechanics of a general dynamical system. The theory there obtained was applied in particular to dynamical systems consisting of a great number of similar particles—molecules or atoms. In this application there was always some difficulty and uncertainty arising out of our ignorance of the exact structure of the molecules or atoms under consideration.

We proceed now to apply the same theory to the motion of continuous media. Three media will be of interest, namely a gas, the luminiferous ether (on the supposition that one exists) and an elastic solid. In these applications of the theory, our former uncertainty as to the mechanics of the system under discussion disappears, for, to a first approximation at least, the dynamics of each medium is known. The degrees of freedom of the medium represent a capacity for transmitting wave-motions, and we shall find that the number of these degrees of freedom can be easily determined.

Degrees of freedom in continuous media.

457. In each of the three media just mentioned, the possible motions are all determined by an equation of the form

$$\frac{d^2\phi}{dt^2} = a^2\nabla^2\phi \qquad (886),$$

and this is known to represent wave-motion propagated with a velocity a*. In this equation, ϕ is a scalar quantity, or a component of a vector, having different meanings according to the problem in hand. If the medium is gaseous, ϕ must be the velocity-potential†, while in the ether all the six components of electric and magnetic force satisfy equations of the form of (886)‡. In an elastic solid, the equation is satisfied by the dilatation Δ and by the three components of rotation $\varpi_1, \varpi_2, \varpi_3$, these quantities being defined by the equations

$$\Delta = \frac{\partial u}{\partial x} + \frac{\partial v}{\partial y} + \frac{\partial w}{\partial z}; \quad \varpi_1 = \frac{1}{2}\left(\frac{\partial w}{\partial y} - \frac{\partial z}{\partial v}\right), \text{ etc.},$$

where u, v, w are the components of the displacement at any point§.

* See for instance, Jeans, *Electricity and Magnetism* (5th edition), §§ 578–580.

† Lord Rayleigh, *Theory of Sound*, II. chap. XIII.

‡ Jeans, *Electricity and Magnetism* (5th edition), § 577.

§ Love, *Theory of Elasticity*, chap. V.

In the two latter cases, the different sets of solutions are not independent. If X, Y, Z the components of electric force in the ether satisfy equation (886), then the components of magnetic force must necessarily satisfy the same equation, and moreover of the three solutions X, Y, Z, only two are independent, since X, Y, Z are connected by the relation

$$\frac{\partial X}{\partial x}+\frac{\partial Y}{\partial y}+\frac{\partial Z}{\partial z}=0.$$

In the elastic solid solutions, $\varpi_1, \varpi_2, \varpi_3$ are connected by the relation

$$\frac{\partial \varpi_1}{\partial x}+\frac{\partial \varpi_2}{\partial y}+\frac{\partial \varpi_3}{\partial z}=0.$$

Thus only two of these three sets of solutions are independent, to which the Δ solution must be added, making in all three sets of independent solutions.

458. For simplicity, let the medium under consideration be supposed limited to a rectangular volume, extending from $x=0$ to $x=\alpha$, from $y=0$ to $y=\beta$, and from $z=0$ to $z=\gamma$.

Let the value of ϕ at time $t=0$ be denoted by ϕ_0, this being of course a function of x, y and z only. By Fourier's theorem, the value of ϕ_0 at every point inside the volume $\alpha\beta\gamma$ can be expressed in the form

$$\phi_0=\Sigma\Sigma\Sigma A_{lmn}\cos\frac{l\pi x}{\alpha}\cos\frac{m\pi y}{\beta}\cos\frac{n\pi z}{\gamma}$$

$$+\Sigma\Sigma\Sigma B_{lmn}\sin\frac{l\pi x}{\alpha}\cos\frac{m\pi y}{\beta}\cos\frac{n\pi z}{\gamma}+\dots \quad \dots\dots(887).$$

In this equation the triple summation is to be taken over all positive integral values of l, m and n from 0 to ∞. The coefficients A_{lmn}, B_{lmn}, ... are given by

$$A_{lmn}=\frac{8}{\alpha\beta\gamma}\iiint\phi_0\cos\frac{l\pi x}{\alpha}\cos\frac{m\pi y}{\beta}\cos\frac{n\pi z}{\gamma}\,dx\,dy\,dz,$$

$$B_{lmn}=\frac{8}{\alpha\beta\gamma}\iiint\phi_0\sin\frac{l\pi x}{\alpha}\cos\frac{m\pi y}{\beta}\cos\frac{n\pi z}{\gamma}\,dx\,dy\,dz,$$

etc. and there are six other sets of coefficients (say C, D, E, F, G, H), the total of eight corresponding to all possible arrangements of sines and cosines of

$$\frac{l\pi x}{\alpha},\ \frac{m\pi y}{\beta}\ \text{and}\ \frac{n\pi z}{\gamma}.$$

Let the rate of increase in ϕ at time $t=0$ be denoted by $\dot{\phi}_0$. The quantity $\dot{\phi}_0$ can also be expanded in series similar to the right-hand side of equation (887). Let the coefficients in this expansion of $\dot{\phi}_0$ be A'_{lmn}, B'_{lmn}, etc.

Knowing the initial values of ϕ and $\dot{\phi}$, the complete solution of equation (886) can be written down. It is readily seen to be

$$\phi = \Sigma\Sigma\Sigma \cos\frac{l\pi x}{\alpha}\cos\frac{m\pi y}{\beta}\cos\frac{n\pi z}{\gamma}\left(A_{lmn}\cos pt + A'_{lmn}\frac{\sin pt}{p}\right)$$

$$+\Sigma\Sigma\Sigma \sin\frac{l\pi x}{\alpha}\cos\frac{m\pi y}{\beta}\cos\frac{n\pi z}{\gamma}\left(B_{lmn}\cos pt + B'_{lmn}\frac{\sin pt}{p}\right) + \ldots \ldots (888),$$

where, in order that equation (886) may be satisfied, we must have

$$p^2 = a^2\pi^2\left(\frac{l^2}{\alpha^2} + \frac{m^2}{\beta^2} + \frac{n^2}{\gamma^2}\right) \ldots\ldots (889).$$

On combining the terms which have the same values of l, m, n in equation (888), it is found that ϕ can be expressed as a sum of terms of the form

$$\phi = \Sigma K \cos\left(pt \pm \frac{l\pi x}{\alpha} \pm \frac{m\pi y}{\beta} \pm \frac{n\pi z}{\gamma} - \epsilon\right) \ldots\ldots (890),$$

where the summation is over all values of $\pm l$, $\pm m$, $\pm n$, and the constants K and ϵ are of course different for each set of values. Put in this form, it is clear that the solution represents sets of plane waves, travelling in different directions. From equation (889) it follows that all waves are propagated with the same velocity a.

459. Questions of great importance arise out of the classification of these waves according to frequency and wave-length. It will be remembered that the values of l, m, n are necessarily integral. We may now imagine different values of l, m, n represented in a three-dimensional space having ξ, η, ζ for rectangular coordinates, and the different sets of integral values will occur at the rate of one per unit volume in this space. The number of sets for which p is less than some assigned value p_0 will by equation (889) be equal to the number of points having integral coordinates (l, m, n) in this space, such that

$$\frac{l^2}{\alpha^2} + \frac{m^2}{\beta^2} + \frac{n^2}{\gamma^2} < \frac{p_0^2}{a^2\pi^2},$$

and therefore to the number of such points which lie inside the ellipsoid whose equation is

$$\frac{\xi^2}{\alpha^2} + \frac{\eta^2}{\beta^2} + \frac{\zeta^2}{\gamma^2} = \frac{p_0^2}{a^2\pi^2}.$$

The volume of this ellipsoid is $\frac{4}{3}\frac{\alpha\beta\gamma}{\pi^2 a^3}p_0^3$, and this is accordingly the number of sets of values of l, m, n for which $p < p_0$. By differentiation, the number of sets for which p lies within a range from p to $p + dp$ is

$$\frac{4\alpha\beta\gamma}{\pi^2 a^3}p^2 dp \ldots\ldots (891).$$

460. We have so far been dealing only with abstract solutions of the equation (886). Before applying these results to a definite problem, we must eliminate such solutions as do not satisfy the boundary conditions of the particular problem in hand.

In a sound problem, we are dealing with the vibrations of a gaseous medium; ϕ is the velocity-potential, and the boundary condition to be satisfied is that $\partial\phi/\partial n$ shall vanish all over the boundary. Applied to the particular volume under consideration, this requires that $\partial\phi/\partial x$ shall vanish when $x=0$ and when $x=\alpha$, and similar conditions must be satisfied for y and z. The effect of this restriction is to limit the solution (887) for ϕ_0 to terms in A; all the solutions in $B, C, \ldots H$ disappear because they do not satisfy the boundary conditions. Thus the complete solution (888) reduces to

$$\phi = \Sigma\Sigma\Sigma \cos\frac{l\pi x}{\alpha}\cos\frac{m\pi y}{\beta}\cos\frac{n\pi z}{\gamma}\left(A_{lmn}\cos pt + A'_{lmn}\frac{\sin pt}{p}\right) \ldots(892).$$

The number of separate free vibrations is now only one-eighth of that previously estimated, and the law of distribution according to values of p will, from formula (891), be

$$\frac{\alpha\beta\gamma}{2\pi^2 a^3} p^2 dp \quad \ldots\ldots\ldots\ldots\ldots\ldots(893).$$

461. In a light or radiation problem, ϕ may be supposed to be any one of three components of electric force or of magnetic force. For definiteness, let us suppose ϕ to be identical with X, the x-component of electric force.

In order that the system may be a conservative one, there must be no possibility of energy passing through the walls of the containing vessel. These walls must accordingly be thought of as perfect reflectors, and therefore as perfect conductors. One boundary condition must clearly be that X shall vanish over the planes $y=0$, $y=\beta$ and $z=0$, $z=\gamma$, and this requires that all the terms in ϕ_0 (equation (887)) which contain cosines of $m\pi y/\beta$ or $n\pi z/\gamma$ must vanish. We are left with the solution

$$X = \Sigma\Sigma\Sigma \cos\frac{l\pi x}{\alpha}\sin\frac{m\pi y}{\beta}\sin\frac{n\pi z}{\gamma}\left(E_{lmn}\cos pt + E'_{lmn}\frac{\sin pt}{p}\right)$$
$$+ \Sigma\Sigma\Sigma \sin\frac{l\pi x}{\alpha}\sin\frac{m\pi y}{\beta}\sin\frac{n\pi z}{\gamma}\left(H_{lmn}\cos pt + H'_{lmn}\frac{\sin pt}{p}\right) \ldots(894).$$

From this value of X, the values of the remaining components of electric and magnetic force can be written down, and it is at once found that all the other boundary conditions are satisfied*. Thus formula (894) contains the solution of the problem; it contains only a quarter as many constants as the general solution (888) in which the boundary conditions were disregarded.

* Cf. Jeans, *Electricity and Magnetism* (5th edition), § 593.

Formula (891), which expressed the law of distribution before the boundary conditions were taken into account, must accordingly be divided by 4, and we obtain for the law of distribution of values of p in the light problem,

$$\frac{\alpha\beta\gamma}{\pi^2 a^3} p^2 dp \qquad\qquad (895).$$

462. Any one term in the solution for ϕ can exist by itself, and the number of separate terms will accordingly be the same as the number of separate free vibrations. It will be noticed that the number of free vibrations in the ether (cf. formula (893)) is double that in a gas. This could have been foreseen from the circumstance that a sound-wave is determined by one vector, namely the displacement in the direction of propagation, while a light-wave is determined by two vectors which determine the intensity and direction of propagation and of polarisation.

In an elastic solid waves of both kinds can coexist. We have a normal wave of compression analogous to a sound-wave, and a tangential wave of distortion analogous to a light-wave. Thus the number of separate free vibrations in an elastic solid medium is equal to the sum of the corresponding numbers in a gas and in the ether. It must however be remembered that in an elastic solid there are two different velocities of propagation, say a_1 for waves of compression and a_2 for waves of distortion.

463. In the formulae obtained, the factor $\alpha\beta\gamma$ the volume of the enclosure enters as a multiplier, so that the number of free vibrations per unit volume is the same whatever the volume of the enclosure. We have only proved this to be true for rectangular enclosures, but in physical applications the wave-length of the vibration will always be very small in comparison with the dimensions of the enclosure, so that we should expect that the shape of the enclosure would become unimportant just as its size has been seen to do. That this is actually the case has been formally proved by Weyl*.

Collecting our results, we find that the number of free vibrations per unit volume in different media is that given in the table below. The first column gives the number of vibrations classified according to frequency of vibration. If λ is the wave-length of any vibration of frequency $p/2\pi$, we have $\lambda = 2\pi a/p$, so that $d\lambda = 2\pi a p^{-2} dp$. On transforming the variable from p to λ, we obtain the corresponding numbers for vibrations classified according to wave-length, and these are given in the last column of the table.

* H. Weyl, *Math. Annalen*, LXXI. (1912), p. 441. See also papers by the same author: *Journal für die reine und angewandte Mathematik*, 141 (1912), pp. 1 and 163; 143 (1913), p. 177, and *Rendiconti del Circolo Mat. di Palermo*, XXXIX. (1915), p. 1.

Medium	Vibrations per unit volume of the medium classified according to	
	Frequency $\frac{p}{2\pi}$	Wave-length λ
Gas	$\frac{p^2dp}{2\pi^2a^3}$	$4\pi\lambda^{-4}d\lambda$
Ether	$\frac{p^2dp}{\pi^2a^3}$	$8\pi\lambda^{-4}d\lambda$
Elastic solid	$\frac{p^2dp}{\pi^2}\left(\frac{1}{2a_1^3}+\frac{1}{a_2^3}\right)$	$12\pi\lambda^{-4}d\lambda$

464. Before leaving this subject, it may be noticed that in any medium whatever, the number of free vibrations per unit volume, of wave-length between λ and $\lambda + d\lambda$ (where λ is supposed large compared with the scale of structure of the medium, if it is coarse-grained, and small compared with the size of the medium), must necessarily be of the form

$$C\lambda^{-4}d\lambda,$$

where C is a numerical constant. For no other type of formula would be possible, consistently with the physical dimensions of the quantities involved.

The whole problem accordingly reduces to the determination of the multiplier C, which of course can differ from one medium to another. And for the three media we have had under consideration, the three values of C must be in the ratio 1 : 2 : 3 for the reasons already stated—in a gas there is only the one set of normal vibrations, in the ether there are only transverse vibrations, but there are two independent transverse vibrations, in different planes of polarisation, for each normal vibration in the gas; and finally in the elastic solid there are both normal and transverse vibrations.

Statistical Mechanics.

465. When once the number of independent free vibrations has been counted up, the problem of determining the partition of energy in a continuous medium becomes an exceedingly simple one. For each vibration is known from the dynamics of the medium, as expressed in the general equation (886) and its solution (888), to consist of a simple harmonic motion, so that if ϕ_1 is the coordinate of any vibration, the corresponding energy will be of the form

$$E_1 = \tfrac{1}{2}a_1\dot{\phi}_1^2 + \tfrac{1}{2}b_1\phi_1^2 \quad \text{.........................(896)},$$

where a_1, b_1 are constants. Each vibration accordingly contributes two squared terms to the energy of the whole system. Let λ be supposed so small that the number of vibrations of wave-length between λ and $\lambda + d\lambda$, namely

$C\lambda^{-4}d\lambda$, is a large number. Then the total energy of the corresponding $2C\lambda^{-4}d\lambda$ squared terms must, by the general theorem of equipartition of energy (§ 100), be

$$CRT\lambda^{-4}d\lambda \quad \text{.............................(897)},$$

this energy being of course at the rate RT per vibration.

On inserting the appropriate value for C, this formula will give the partition of energy according to wave-length in the medium under consideration.

Analysis of Energy in a Gas.

466. We have so far thought of the heat-energy of a gas as residing in molecular motion, but formula (897) regards it as the energy of trains of waves of sound. Similarly the heat of a solid is usually thought of as a manifestation of random agitations of the molecules, but formula (897) shews that it also may be regarded as the energy of regular elastic solid vibrations (cf. § 3).

Before proceeding further, it will be useful to make a somewhat detailed study of the relation between these two widely different ways of regarding the energy of a medium. For simplicity we shall confine our attention to the case of a gas.

Imagine the whole volume $\Omega = \alpha\beta\gamma$ of the gas divided into n rectangular cells each of volume ω, and of edges $\delta\alpha$, $\delta\beta$, $\delta\gamma$ parallel to the edges of the large rectangular volume, δ here being regarded as a small fraction. The total number of cells n is also equal to $1/\delta^3$.

The molecular arrangement in the gas is sufficiently known from the investigation of Chap. III. The molecules are distributed absolutely at random between the n cells, and each velocity component is distributed according to Maxwell's Law.

Let us consider the arrangement of positions first. An arrangement in which the molecules are placed at exactly equal distances apart, in some regular geometrical order, is of course a possible arrangement, but is no more typical of the normal state than would be a motion in which each molecule had exactly the same velocity. So also an arrangement in which each of the n cells into which the volume is divided contains the same number N/n of molecules is possible, but is not typical of the normal state.

As in § 45, let us consider as typical an arrangement in which

$$a_1, a_2, a_3, \dots a_n,$$

the numbers of molecules in the respective cells, are given by

$$a_1 = \frac{N}{n} + \alpha_1, \quad a_2 = \frac{N}{n} + \alpha_2, \text{ etc.} \quad \text{..................(898)},$$

where the α's are small compared with N/n, and

$$\alpha_1 + \alpha_2 + \ldots + \alpha_n = 0 \quad \text{.......................(899)}.$$

Let the sth cell have as the coordinates of its point farthest from the origin $p\delta\alpha$, $q\delta\beta$, $r\delta\gamma$, and let a_s also be denoted by a_{pqr}.

Consider an arrangement in which the distribution of molecules is such that in every cell

$$a_{pqr} = \xi_{lmn} \cos lp\,\delta\pi \cos mq\,\delta\pi \cos nr\,\delta\pi \quad \text{...........(900)},$$

where ξ_{lmn} is a constant. This is not a typical arrangement, for there is regularity in it. As we pass along any line parallel to the axis of x, $p\delta$ varies from 0 to 1, so that sin $lp\,\delta\pi$ varies harmonically, with a wave-length such that there are l half-periods within the length α inside the rectangle. Similarly the arrangement in any chain of cells parallel to the axis of y is harmonic, there being m half-periods inside the rectangle, and similarly for the axis of z. When the cells become small compared with the wave-length, $\cos lp\,\delta\pi$ may be replaced by $\cos \frac{l\pi x}{\alpha}$, and so on, whence it appears that equation (900) expresses a distribution of density in a single free vibration such as is specified by equation (892).

In general the values of a_{pqr} for the different cells will not be of the form (900), but whatever the values of a_{pqr}, they may always be expressed by Fourier's theorem as the sum of a number of terms of this form. Thus we may suppose the most general typical distribution to be of the form

$$a_{pqr} = \sum_l \sum_m \sum_n \xi_{lmn} \cos lp\,\delta\pi \cos mq\,\delta\pi \cos nr\,\delta\pi \quad \text{...........(901)},$$

where the summation is from $l = 1$ to $l = 1/\delta$, and similarly for m and n.

If, as before, probability is measured as a fraction in the appropriate generalised space, then the probability that $a_1, a_2, \ldots$ shall have the values assigned to them by equations (898) is, by equation (73), equal to θ_a given by

$$\theta_a = n^{\frac{1}{2}n} (2\pi N)^{-\frac{1}{2}(n-1)} e^{-NK_a},$$

where (cf. equation (82))

$$NK_a = \frac{1}{2}\frac{n}{N}\Sigma a_s^2 \quad \text{..........................(902)}.$$

The probability that $\alpha_1, \alpha_2, \alpha_3, \ldots$ shall lie within a small range $d\alpha_1 d\alpha_2 d\alpha_3 \ldots$, $d\alpha_1$ being small compared with N/n, and so on, is therefore

$$\theta_a d\alpha_1 d\alpha_2 d\alpha_3 \ldots \quad \text{..........................(903)}.$$

Formula (903) gives the law of distribution of the α's, and therefore of the a's. Let us now change the variables from the α's to the ξ's, the relation between the two sets of variables being that given by equations (901). Since these equations are linear, we have at once

$$d\alpha_1 d\alpha_2 \ldots = \Delta\, d\xi_1 d\xi_2 \ldots,$$

where Δ the modulus of transformation is a constant.

Squaring equations (901) and adding corresponding sides,

$$\sum_1^n \alpha^2 = \sum_1^n \xi^2 \left(\sum_l \sum_m \sum_n \cos^2 lp\,\delta\pi \cos^2 mq\,\delta\pi \cos^2 nr\,\delta\pi\right)$$

$$= \sum_1^n \frac{1}{8\delta^3}\xi^2 = \tfrac{1}{8}n\sum_1^n \xi^2 \quad \text{.................................(904)},$$

so that equation (902) becomes

$$NK_a = \frac{1}{16}\frac{n^2}{N}(\xi_1^2 + \xi_2^2 + \ldots).$$

The law of distribution (903) now becomes

$$n^{\frac{1}{2}n}(2\pi N)^{-\frac{1}{2}(n-1)}\Delta e^{-\frac{n^2}{16N}(\xi_1^2+\xi_2^2+\ldots)}\,d\xi_1 d\xi_2 \ldots .$$

Thus the law of distribution of each ξ is the same, namely

$$Ce^{-\frac{n^2\xi^2}{16N}}\,d\xi,$$

where C is a constant, and we have

$$\overline{\xi_1^2} = \overline{\xi_2^2} = \ldots = \frac{8N}{n^2} \quad \text{.......................(905)}.$$

The next step is to find the potential energy of the trains of waves $\xi_1, \xi_2, \ldots$. If p_0 is the equilibrium pressure, and if s denotes the condensation ($\delta\rho/\rho$) at any point, the potential energy measured from the equilibrium configuration is

$$V = \tfrac{1}{2}p_0\iiint s^2\,dx\,dy\,dz = \tfrac{1}{2}p_0\sum_1^n\left(\frac{\alpha n}{N}\right)^2\omega.$$

Substituting for $\Sigma\alpha^2$ from equation (904), this becomes

$$V = \tfrac{1}{16}p_0\Omega\left(\frac{n}{N}\right)^2(\xi_1^2 + \xi_2^2 + \ldots)$$

$$= \tfrac{1}{16}RT\frac{n^2}{N}(\xi_1^2 + \xi_2^2 + \ldots),$$

and it now follows from equations (905) that the average value of each term in this expression for V is $\tfrac{1}{2}RT$.

Thus we have seen how, in the typical normal state, the randon arrangement of the molecules as regards position may be resolved into regular trains of waves. And, as we should expect from general statistical mechanics, the amplitudes of these waves are seen to be distributed according to Maxwell's Law, and the mean potential energy of each wave is $\tfrac{1}{2}RT$.

467. The random *motions* of the molecules can be resolved into trains of waves in a similar way*; the final result of this analysis can be seen from the general principles of wave-motion, which indicate that the mean kinetic energy of each wave must be equal to its mean potential energy and therefore to $\tfrac{1}{2}RT$.

* For details of the actual analysis, see *Phil. Mag.* XVII. (1909), p. 239.

468. To sum up, it appears that the potential energy in a gas, which originates from the molecules not being perfectly evenly spaced, can be expressed as the potential energy of trains of waves in the form

$$2V = \beta_1 \phi_1^2 + \beta_2 \phi_2^2 + \dots,$$

while the corresponding kinetic energy, which arises from the random agitation of the molecules, can be expressed as the kinetic energy of the same waves in the form

$$2T = \alpha_1 \dot{\phi}_1^2 + \alpha_2 \dot{\phi}_2^2 + \dots.$$

Further, the average energy of each term is the same, being given by

$$\overline{\alpha_1 \dot{\phi}_1^2} = \overline{\alpha_2 \dot{\phi}_2^2} = \dots = \overline{\beta_1 \phi_1^2} = \overline{\beta_2 \phi_2^2} = \dots = \frac{1}{2h} = RT,$$

and hence the average value of the energy of any vibration, say E_1 as given by equation (896), must also be RT.

Thus a detailed study of the mechanism of molecular energy leads to exactly the same result as was obtained in § 465 from the general principles of statistical mechanics. It has also given us a further insight into the physical meaning of this result.

469. We arrived at formula (897) by supposing λ to be small, but except in a perfectly structureless medium there is a limit set to the smallness of λ by the coarse-grainedness of the medium. For instance, in a gas at normal pressure, the distance apart of adjacent molecules is about 3×10^{-7} cms., so that it is meaningless to consider a wave of sound in this gas for which the wave-length is smaller than 3×10^{-7} cms. But there will always be some range, say from λ_0 (small) to λ_1 (large) over which formula (897) is applicable.

On integration the total energy of waves of wave-length between these limits is found to be

$$\int_{\lambda=\lambda_0}^{\lambda=\lambda_1} \mathrm{C}\, RT\lambda^{-4}\, d\lambda = \tfrac{1}{3}\mathrm{C}\, RT\left(\frac{1}{\lambda_0^3} - \frac{1}{\lambda_1^3}\right).$$

As a rough approximation, we may suppose that λ_1^{-3} is negligible in comparison with λ_0^{-3}, and also that all the energy of the medium can be expressed as the energy of waves of wave-length greater than λ_0, where λ_0 is a length comparable with the scale of coarse-grainedness of the medium. The total energy per unit volume of the medium is accordingly

$$\tfrac{1}{3}\mathrm{C}\, RT\lambda_0^{-3} \quad \dots\dots\dots\dots(906).$$

This energy represents the potential and kinetic energies of the wave-motion in the medium; it takes no account of the internal energy, if any, of the molecules. Let β denote, as usual, the ratio of the internal energy of the molecules to their energy of translation, then the sum of these two energies, say E, will be given by

$$E = \tfrac{1}{3}(1+\beta)\,\mathrm{C}\, RT\lambda_0^{-3} \dots\dots\dots\dots(907).$$

The specific heat at constant volume of unit volume of the medium is, however, given by

$$C_v \rho = \frac{1}{J}\frac{dE}{dT} = \frac{1}{3}\frac{R}{J}(1+\beta)\,\mathrm{C}\lambda_0^{-3} = 1\cdot 07 \times 10^{-24} \times (1+\beta)\,\mathrm{C}\lambda_0^{-3},$$

so that the value of λ_0 for any medium can be found from a knowledge of the specific heat.

For instance, for air at normal pressure, $C_v = \cdot 172$, $\rho = \cdot 00129$, $\beta = \frac{2}{3}$, $\mathrm{C} = 4\pi$, so that $\lambda_0 = 4\cdot 5 \times 10^{-7}$ cms. This is comparable with the average molecular distance $3\cdot 3 \times 10^{-7}$ cms., as of course it ought to be. If the scale of molecular structure in a gas had not been known, it could have been roughly determined in this way.

470. In a perfectly continuous or structureless medium we may take $\lambda_0 = 0$. The only structureless medium which we need consider is the luminiferous ether. For this $\mathrm{C} = 8\pi$, and the law of distribution of radiant energy (cf. formula (897)) will be

$$8\pi RT\lambda^{-4}\,d\lambda \quad \text{.............................(908).}$$

According to the classical mechanics this is the formula which ought to give the partition of energy in ether in temperature equilibrium at temperature T. It ought to be obeyed for all wave-lengths from a great wave-length λ_1 comparable with the dimensions of the medium up to the very shortest wave-lengths. It was first given by Lord Rayleigh* in 1900, being suggested only as a theoretical formula which might perhaps be found to agree with experiment for very long waves; there was never any question of its being expected to give the partition of energy for all waves.

On putting $\lambda_0 = 0$, formula (906) shews that the energy of a structureless medium would be infinite except when $T = 0$, the specific heat C_v of the medium becoming infinite. Thus, excluding the impossible case of a system having infinite energy, we see that the temperature of a perfectly structureless medium ought to be invariably zero. Whenever an exchange of energy takes place between the medium and a material system placed in it, the medium must always gain energy and the rest of the system must always lose energy. The final state can therefore only be one in which all of the energy of the material system has been transmitted to the medium, and both are at zero temperature, the only exception being when a state is previously reached such that further transfers of energy cannot take place.

We see that the presence of a medium of this kind leads to exactly the same result as was previously obtained in Chapter V (§ 84) by supposing the material system to lose energy in a way which could be represented by the existence of a dissipation-function. We now see that the presence of a continuous medium, with an insatiable capacity for energy arising from its

* See Lord Rayleigh, *Phil. Mag.* [5] XLIX. (1900), p. 539, and *Nature*, LXXII. (1905), pp. 54, 243. See also J. H. Jeans, *Phil. Mag.* [6] X. (1905), p. 91.

infinite degrees of freedom, can be exactly represented by the existence of a dissipation-function; a final state is only reached when the material system can no longer transmit energy to this medium.

471. Physical illustrations of the ideas which have just been developed will occur to everyone. A pendulum, suspended and set in motion inside a closed vessel containing air, will continue to lose its energy of vibration until it comes to rest. Its energy will have been absorbed, primarily, in setting up waves in the air, but ultimately these may be regarded as equivalent to a random molecular motion. Strictly speaking, the air is not an absolutely continuous medium: its capacity for energy is not infinite, whence it results that the pendulum is not absolutely reduced to rest. In its final state it has a certain very small motion, of the nature of a Brownian movement, and of amount appropriate to a "particle" of its size, namely a motion such that its kinetic energy is on the average equal to that of one molecule of the air surrounding it.

A further instance of these principles will be found in the state of things considered at the end of § 3 (p. 3).

472. If we suppose an ether to exist, precisely similar results must necessarily follow for this ether if we make the assumption that the classical laws of dynamics are obeyed. All available evidence goes to shew that, if an ether exists, there can be nothing of the nature of coarse-grainedness in the structure of the ether. Even if the ether has a structure, this structure must certainly be very much more fine-grained than that of a gas, being indeed so fine-grained as to have escaped observation altogether. Thus we should expect the loss of energy of a material system to continue until all its energy had been transmitted to the ether.

Consider for definiteness a mass of iron, say at 0° C., placed inside an enclosure whose walls are perfectly impervious to radiation, so that the iron and the ether in the enclosure form a self-contained system. According to our theoretical result, the iron ought to lose energy continually until its temperature becomes close to the absolute zero, and the whole energy of the system has passed into the ether. But so far from this actually happening experimental evidence indicates with considerable certainty that a final state is rapidly attained in which the iron remains approximately at 0° C., and the partition of energy in the ether is in thermodynamical equilibrium with the matter. In this state, the energy of the ether is of density $3{\cdot}93 \times 10^{-5}$ ergs per cubic centimetre, while that of the iron is of the order of 8×10^{9} ergs per cubic centimetre. Thus, almost all the energy resides in the comparatively few degrees of freedom of the iron, while only an infinitesimal amount passes into the enormously greater number of degrees of freedom in the ether. Nothing of the nature of equipartition of energy appears to hold.

Here we see, in its most vivid form, the complete contradiction between the result of observation and the theoretical results obtained from the assumptions that an ether exists and that the classical laws of dynamics are obeyed. Analysis which will now be given will shew that the contradiction arises out of the classical laws of dynamics and not out of the assumed existence of an ether. For, without assuming that an ether exists at all, we shall deduce the law of distribution of radiant energy (908) from the classical dynamics alone.

Radiation from Resonators.

473. Consider first the exchange of energy between a material resonator, as for instance a Hertzian vibrator, and a field of radiant energy surrounding it.

Let the resonator be regarded as a dynamical system, obeying the classical laws. Let its kinetic and potential energies be of the forms

$$2L = a\dot{\phi}^2,$$
$$2V = b\phi^2,$$

in which a and b are constants. When the resonator oscillates free from external disturbance, its equation of motion is

$$a\ddot{\phi} + b\phi = 0 \quad\text{.............................(909)},$$

and the solution of this equation is

$$\phi = A\cos pt + B\sin pt \quad\text{.......................(910)},$$

where A, B are constants, and p is such that $ap^2 = b$.

For a motion which is influenced by external agencies, such as the interaction with the assumed field of radiation, the equation of motion will be of the form

$$a\ddot{\phi} + b\phi = U \quad\text{.............................(911)}.$$

The value of U during any interval $t=0$ to $t=\tau$ may, by Fourier's theorem, be expressed in the form

$$U = \frac{1}{\pi}\int_0^\infty (F_q \cos qt + G_q \sin qt)\, dq \quad\text{..............(912)},$$

where the coefficients F_q, G_q are given by

$$\left.\begin{aligned} F_q &= \int_{t'=0}^{t'=\tau} U_{t=t'} \cos qt'\, dt' \\ G_q &= \int_{t'=0}^{t'=\tau} U_{t=t'} \sin qt'\, dt' \end{aligned}\right\} \quad\text{...................(913)}.$$

The general solution of equation (911) is readily obtained. Let us assume that at any instant $t=0$, the values of ϕ and $\dot{\phi}$ are the same as if the coordinate ϕ were describing the free harmonic vibration expressed by equation (910). The impulse Udt, acting from $t=0$ to $t=dt$, sets up an additional free vibration of initial displacement zero and velocity $\frac{Udt}{a}$; the displacement of this additional vibration at any subsequent time is therefore $\frac{U\,dt}{ap}\sin pt$. Compounding all these vibrations with the original vibration expressed by equation (910), we obtain for the vibration being described at the instant $t=\tau$ the well-known solution*

$$\phi = A\cos pt + B\sin pt + \frac{1}{ap}\int_{t'=0}^{t'=\tau} U_{t=t'}\sin p(t-t')\,dt' \quad\text{......(914)},$$

and, by equations (913), this may be written in the form

$$\phi = \left(A - \frac{G_p}{ap}\right)\cos pt + \left(B + \frac{F_p}{ap}\right)\sin pt.$$

The energy of this vibration is

$$\tfrac{1}{2}(a\dot{\phi}^2 + b\phi^2) = \tfrac{1}{2}b\left[\left(A - \frac{G_p}{ap}\right)^2 + \left(B + \frac{F_p}{ap}\right)^2\right]$$

$$= \tfrac{1}{2}b(A^2+B^2) + p(BF_p - AG_p) + \frac{1}{2a}(F_p^2 + G_p^2).$$

The first term $\tfrac{1}{2}b(A^2+B^2)$ is of course the energy of vibration at time $t=0$, so that the remaining terms represent the energy gained by the resonator from the field of radiation in the interval from $t=0$ to $t=\tau$. At time $t=0$, $\phi=A$ and $\dot{\phi}=pB$. Both ϕ and $\dot{\phi}$ are as likely to be positive as negative, so that on averaging over a large number of resonators, the average values of both A and B will be zero. Thus the last term

$$\frac{1}{2a}(F_p^2 + G_p^2) \quad\text{............................(915)}$$

will represent the average amount of energy absorbed by the resonator from the field of radiation in the interval from $t=0$ to $t=\tau$.

Let the force U be supposed to originate in an electric force Z in any direction, then we may assume that $U=cZ$, where c is a constant. Let the radiant energy per unit volume, analysed into waves of different frequencies, be supposed to be

$$\int_{q=0}^{q=\infty} E_q\,dq.$$

Then the mean value of Z^2 will be $\frac{4}{3}\pi$ times this total radiant energy, and the mean value of U^2 from time 0 to τ will be equal to c^2 times the mean value of Z^2. We accordingly have

$$\frac{1}{\tau}\int_{t=0}^{t=\tau} U^2\,dt = \tfrac{4}{3}\pi c^2\int_{q=0}^{q=\infty} E_q\,dq \quad\text{..................(916)}.$$

But also, from a well-known theorem of Lord Rayleigh†, if U is given by the expansion (912), we have

$$\frac{1}{\tau}\int_{t=0}^{t=\tau} U^2\,dt = \frac{1}{\pi\tau}\int_{q=0}^{q=\infty}(F_q^2 + G_q^2)\,dq \quad\text{...........(917)}.$$

* Rayleigh, *Theory of Sound*, I. § 66. † *Phil. Mag.* [5] XXVII. p. 466.

Equations (916) and (917) both express the spectroscopic analysis of the mean value of U^2, so that the constituents must be the same in the two equations, giving

$$\frac{1}{\pi\tau}(F_q^2 + G_q^2) = \tfrac{4}{3}\pi c^2 E_q,$$

and we now have, for the average rate of absorption of energy by the resonator, from formula (915),

$$\frac{1}{2a}(F_p^2 + G_p^2) = \frac{2}{3}\frac{\pi^2 c^2}{a} E_p \quad \text{.....................(918).}$$

474. We next examine the average rate of radiation of energy from the resonator. Assuming the radiating mechanism to be electrical, we may suppose the average rate of emission of radiation in the interval from $t = 0$ to $t = \tau$ to be given by either of the equivalent formulae

$$\frac{1}{\tau}\int_0^\tau C\ddot{\phi}^2 dt = -\frac{1}{\tau}\int_0^\tau C\dot{\phi}\dddot{\phi}\, dt,$$

in which C is a constant. The left-hand formula gives the expression of Larmor for the radiation, while the right-hand formula gives that of Lorentz.

On substituting for $\dddot{\phi}$ from equation (911), the right-hand formula becomes

$$-\frac{C}{a\tau}\int_0^\tau \dot{\phi}(\dot{U} - b\dot{\phi})\, dt.$$

Using the values for ϕ and $\dot{U}$ given by equations (910) and (912), we readily find that, on averaging over all resonators,

$$\int_0^\tau \dot{\phi}\dot{U}\, dt = 0,$$

so that the average rate of emission of energy is

$$\frac{C}{a\tau}\int_0^\tau b\dot{\phi}^2 dt = \frac{C}{a}\overline{b\dot{\phi}^2} = \frac{Cp^2}{a}\overline{a\dot{\phi}^2}.$$

475. If there is a steady state of equilibrium between the resonators and the field of radiation, this rate of emission must be equal to the rate of absorption obtained in formula (918). Comparing these expressions, we obtain

$$E_p dp = \frac{3C}{2\pi^2 c^2}\overline{a\phi^2}\, p^2\, dp \quad \text{.......................(919),}$$

which gives the partition of radiant energy in the steady state.

Whatever the mechanism of the resonator may be, if only it obeys the laws of classical mechanics, we may put

$$\overline{a\dot{\phi}^2} = RT \quad \text{..............................(920),}$$

and the value of E_p as given by formula (919) is seen to depend only on the quantities C and c, which in turn depend on the structure, but not on the motion, of the resonator.

For a Hertzian oscillator, $C/c = 2/3V^2$, where V is the velocity of light, and formula (919) becomes

$$E_p dp = \frac{RT}{\pi^2 V^3} p^2 dp \quad \text{.........................(921).}$$

Transforming to wave-lengths by means of the relation $p\lambda = 2\pi V$, this becomes

$$E_\lambda d\lambda = 8\pi RT\lambda^{-4} d\lambda \quad \text{.......................(922).}$$

This is identical with the result which was obtained in § 470 on the supposition of the existence of a material ether. Indeed we can see that there must be agreement of this kind, since the forces at work adding energy to, or extracting energy from, any resonator, are precisely those which would be at work if a material ether existed, subject always to the supposition of the classical laws of dynamics being obeyed.

If the resonator consists of a single free electron capable of performing isochronous oscillations of frequency $p/2\pi$, we have $c = e$ and $C = \frac{2}{3}e^2/V^3$, so that the partition of energy is again given by equations (921) and (922).

Radiation from Free Electrons.

476. The partition of energy to be expected from other mechanisms of radiation can be worked out in detail in a similar way, but so long as this mechanism is governed by the laws of ordinary classical mechanics, only one result can possibly be obtained, namely that which is given by the general dynamical theory of § 470, and is expressed by equation (922). The following cases are of interest.

When free electrons thread their way through the interstices of a solid, the forces to which they are subjected result in accelerations which must in turn, according to the classical mechanics, be accompanied by an emission of radiation. A steady state will be attained when the radiation emitted in this way is exactly equal to that absorbed by the matter. The problem of determining the steady state was first considered by Lorentz*, who shewed that the partition of energy for waves of great wave-length must be that expressed by formula (922). Later, the same question was attacked by the present writer†, who confirmed Lorentz's result by a different method, and shewed that it could be extended to waves of all wave-lengths.

* "On the Emission and Absorption by Metals of Rays of great wave-length," *Koninklijke Akad. van Wetenschappen*, Amsterdam, April 24, 1903.

† "The Motion of Electrons in Solids," *Phil. Mag.* XVII. (1909), p. 773, and *Phil. Mag.* XVIII. (1909), p. 209.

A number of free electrons moving in a field of radiation will, quite apart from the presence of matter, both emit and absorb radiation. A steady state will be reached when the radiation of any wave-length absorbed by the electrons is statistically equal to that emitted. It can be shewn* that, if the classical mechanics are obeyed, a final state will be reached only when the partition of energy in the ether is that given by equation (920).

477. The results given in §§ 472—476 provide conclusive proof that, whether an ether is assumed to exist or not, a steady state of equilibrium between matter and radiant energy can only be attained when the law of partition of the energy of radiation in terms of wave-length is that given by formula (922), namely

$$8\pi RT\lambda^{-4}d\lambda \quad \text{.........................(923).}$$

It follows, as we have already seen, that the temperature of the matter must be zero: there can be no equilibrium between matter and radiant energy until the matter has lost all its energy by radiation.

This is the conclusion arrived at from a study of the radiation problem based on the classical system of dynamics: the state of things predicted is, however, so utterly different from that observed in nature that we are compelled to contemplate an abandonment, or at least a modification, of the classical mechanics.

* *Phil. Mag.* XXVII. (1914), p. 14, or J. H. Jeans, *Report on Radiation and the Quantum Theory* (1914), p. 14.

CHAPTER XVII

RADIATION AND THE QUANTUM THEORY

THERMODYNAMICS OF RADIATION.

478. IN the previous chapter it was found that the classical system of mechanics, when applied to the radiation problem, led to a solution which proved to be in violent disagreement with experience. We shall begin the present chapter by explaining a different line of attack on this problem. This will be based only on general thermodynamical principles, and it will be noticed that it does not require any assumption to be made as to the existence, or non-existence, of an ether.

Consider an enclosure of any shape, of which the walls are impervious to energy of all kinds, and therefore in particular to radiation. Let it contain a certain amount of heated matter which will of course fill the enclosure with radiant energy, and let a steady state finally be reached in which the matter is at a temperature T. Since there is no loss of energy in this state, each piece of matter inside the enclosure retains its temperature indefinitely, and therefore the amount of energy it gains by absorption of radiation must exactly balance the amount it loses by emission of radiation. Considering two pieces of matter, A and B, it is readily seen that the stream of energy which flows from A to B must be exactly equal to that which flows from B to A. Hence we arrive at the conception of a stream of energy appropriate to matter of temperature T, this depending only on T and not on other quantities involved in the structure of matter. It follows that the density of radiant energy inside the enclosure will be a function of T only*.

Stefan's Law.

479. Let it be supposed possible to increase or decrease the volume of the enclosure, as for instance through one of the walls being supposed to consist in part of a moveable piston. Consider a change in which the total volume of the enclosure is increased from v to $v + dv$, and suppose that the amount of matter inside the enclosure is so infinitesimal that its heat-energy may be disregarded.

* A fuller discussion of the matter of this section will be found in almost any text-book on Heat. See for example, Poynting and Thomson's *Heat*, Chapter xv, or Preston's *Heat* (Second Edition), Chapter vi, Section 5.

The equation of conservation of energy for the radiation inside the enclosure can be written down in the same way as that for a gas in § 224. If E is the energy of radiation per unit volume inside the enclosure, the equation is readily seen to be (cf. equation (447))

$$dQ = d(Ev) + p\,dv \quad \text{.......................(924)},$$

in which p now denotes the pressure exerted on the piston by the radiation inside the enclosure, and so is the pressure of radiation corresponding to radiant energy E per unit volume.

Assuming the second law of thermodynamics to hold, dQ/T must be a perfect differential $d\phi$, where dQ refers to the change of energy in any reversible process. Such a process will be one in which the volume v and the temperature T change, while the radiant energy remains always in equilibrium with the temperature T of the matter. For a change of this type,

$$d\phi = \frac{dQ}{T} = \frac{d(Ev)}{T} + \frac{p}{T}\,dv = \left(\frac{E+p}{T}\right) dv + \frac{v}{T}\,dE \quad \text{.........(925)},$$

and this will be a perfect differential if

$$\frac{\partial}{\partial E}\left(\frac{E+p}{T}\right) = \frac{\partial}{\partial v}\left(\frac{v}{T}\right) = \frac{1}{T}.$$

The value to be assigned to p, the pressure of radiation, is known both from theory and experiment. Since the radiation now under consideration must be supposed to be scattered equally in all directions in space, the pressure per unit area must be equal to one-third of the energy per unit volume. Putting $p = \frac{1}{3}E$, the above equation becomes

$$\frac{dE}{E} = 4\frac{dT}{T},$$

of which the integral is

$$E = aT^4 \quad \text{...............................(926)},$$

where a is a constant.

The law expressed by this equation is generally known as Stefan's Law, having been announced as an empirical law by Stefan* in 1879. A theoretical proof, similar to the one given above, was published by Boltzmann† in 1884, following a method previously developed by Bartoli‡. Stefan's Law is found to agree well with experiment§, although the experimental determination of the absolute value of the constant a presents a problem of some difficulty.

* *Wiener Sitzungsber.* LXXIX. (2) (1879), p. 391.

† *Wied. Ann.* XXII. (1884), p. 291.

‡ *Sopra i movimenti prodotti dalla luce e dal calore* (La Monnier, Florence, 1876).

§ See, for instance, Winkelmann's *Handbuch d. Physik*, III. (Wärme), p. 374.

If dS is any element of surface of a perfect radiator, the flow of radiation through this element in directions which lie inside a small cone of solid angle $d\omega$ whose axis makes an angle θ with the normal to the surface is easily seen to be

$$CaT^4 dS \cos\theta \frac{d\omega}{4\pi}.$$

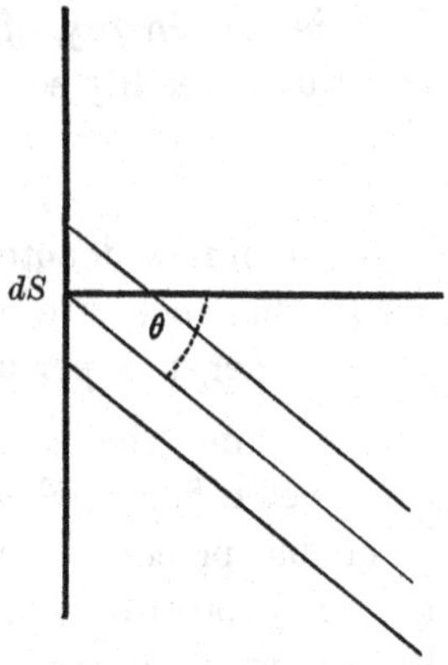

Replacing $d\omega$ by $\sin\theta\, d\theta\, d\phi$ and integrating with respect to ϕ from $\phi = 0$ to $\phi = 2\pi$ and with respect to θ from $\theta = 0$ to $\theta = \frac{1}{2}\pi$, the total radiation from the element of surface dS in unit time is found to be

$$CaT^4 dS \int_0^{\frac{1}{2}\pi}\int_0^{2\pi} \frac{\sin\theta\cos\theta\, d\theta\, d\phi}{4\pi} = \tfrac{1}{4} CaT^4 dS.$$

dS

θ

FIG. 24.

This may be put equal to σT^4, where

$$\sigma = \tfrac{1}{4} Ca \quad \text{.................................}(927).$$

The statement that the radiation from a perfect reflector at temperature T is equal to σT^4 per unit area per unit time provides an alternative form of Stefan's Law. It is usual to determine the constant σ by direct experiment, and then to deduce the value of a from relation (927).

Coblentz* and Millikan†, using widely different methods, agree in assigning to the constant a the value

$$\sigma = 5{\cdot}72 \times 10^{-5} \text{ erg. cm.}^{-2} \text{ degree}^{-4},$$

whence, using Michelson's value‡ $C = 2{\cdot}997 \times 10^{10}$, we find

$$a = 7{\cdot}64 \times 10^{-15} \text{ C.G.S. (centigrade) units §.}$$

The value of E at 0° C. ($T = 273{\cdot}1$) is accordingly

$$\sigma T_0^4 = 4{\cdot}25 \times 10^{-5} \text{ ergs per cu. cm.}$$

Wien's Displacement Law.

480. Wien ‖ has shewn how this thermodynamical argument can be extended so as to give not only the total energy E at temperature T, but also some knowledge of the partition of this energy according to wave-length.

* *Phys. Review*, VII. (1916), p. 694. See also *Scientific Papers of the Bureau of Standards*, Washington, Nos. 357 and 360 (1920).

† *Phil. Mag.* XXXIV. (1917), p. 16.

‡ *Astrophys. Journ.* LX. (1924), p. 260.

§ Earlier determinations of the value of a gave considerably lower values, such as $7{\cdot}06 \times 10^{-15}$ (Kurlbaum, *Wied. Ann.* LXV. (1898), p. 746), $7{\cdot}0 \times 10^{-15}$ (Bauer and Moulin, *Comptes Rendus*, 1910), $7{\cdot}10 \times 10^{-15}$ (Valentiner, *Ann. d. Physik*, XXXI. (1910), p. 275). For a general discussion of experimental values see W. W. Coblentz, *Journal of the Optical Society of America*, VIII. (1924), p. 11.

‖ *Berlin. Sitzungsber.* 9 Feb. 1893; *Wied. Ann.* LII. (1894), p. 156, and LVIII. (1896), p. 662. The present proof differs from that of Wien in many important respects.

Consider what may be called an adiabatic change, namely one in which the volume of the enclosure changes while no energy either enters or leaves. To study such a change we put $dQ = 0$ in equation (924) and obtain

$$d(Ev) = -p\,dv = -\tfrac{1}{3}E\,dv,$$

of which the integral is readily found to be

$$Ev^{\frac{4}{3}} = \text{constant},$$

or, since E is proportional to T^4,

$$Tv^{\frac{1}{3}} = \text{constant} \quad\text{..........................(928).}$$

This equation expresses the adiabatic law for radiation, just as equation (475) expressed the corresponding law for a gas.

Equation (928) shews that an alteration in the volume v must alter the temperature of the radiant energy. The partition of this energy according to wave-length must also be altered. To understand the mechanism by which this change is produced, let us fix our attention on a single beam of radiation of wave-length λ. This will be reflected round and round the enclosure, and the reflections must include a certain number from the piston whose motion changes the volume of the enclosure. At each such reflection the wave-length of the beam must be altered in accordance with Doppler's principle. If the motion of this piston is sufficiently slow, each beam will meet the piston the same number of times, and as the conditions of reflection will, on the average, be the same for every beam, the whole of the radiation which was initially of wave-length λ must ultimately be changed into radiation of some new wave-length λ'.

From Doppler's principle, it is clear that the ratio λ/λ' must depend solely on the motions of the walls of the enclosure, and must therefore be the same for all wave-lengths.

During the change in the volume of the enclosure there can be no direct transfer of energy between the different constituents of the original radiation, but each constituent is continually altering its energy by interaction with the walls of the enclosure. It can now be shewn that the final partition of radiant energy, after the change of volume has taken place, must itself be in thermodynamical equilibrium at some new temperature T'.

For if this were not so, it would be possible to allow the final radiation to interact with an infinitesimal amount of matter, and this would result in an interchange of energy between the various constituents of the radiation until thermodynamical equilibrium was attained. The entropy of the radiation would necessarily increase during this process. Let ϕ, ϕ' be the entropies before and after the change of volume, and let $\phi' + \delta$ be the entropy after thermodynamical equilibrium is attained, δ being positive. The volume of the enclosure may now be restored to its original value v. The change in entropy must by equation (925) be $\phi - \phi'$, so that the final entropy must be $\phi + \delta$.

The maximum entropy possible at the temperature is, however, by hypothesis, equal to ϕ, so that δ, which cannot be negative, must be zero. Thus the compressed radiation must itself have been in thermodynamical equilibrium at some new temperature T'.

It now follows that the ratio of change of wave-length λ/λ' must depend only on the volumes and temperatures before and after the change. The temperatures are however connected with the volumes by relation (928), so that λ/λ' may be thought of as depending only on v and v'. From a consideration of physical dimensions, we must have

$$\frac{\lambda}{\lambda'} = \left(\frac{v}{v'}\right)^n \quad \text{.......................... (929)},$$

but the determination of the value of n demands a detailed examination of the motion.

Let the radiation of wave-length between λ and $\lambda + d\lambda$ be of amount $E_\lambda d\lambda$ per unit volume. Then the energy of this wave-length which falls on an area dS of a moving piston in time dt, its direction lying within a small cone of angle $d\omega$, is

$$CE_\lambda d\lambda\, dS \cos\theta \frac{d\omega}{4\pi} dt,$$

where θ is the angle between the axis of this cone and the normal to the piston (cf. fig. 24). If u is the velocity of the piston, this energy, in accordance with Doppler's principle, experiences a change of wave-length $d\lambda$ such that

$$\frac{\delta\lambda}{\lambda} = \frac{2u\cos\theta}{C}.$$

Thus the average change of wave-length experienced in time dt by all the energy $vE_\lambda d\lambda$, whose wave-length lies between λ and $\lambda + d\lambda$, is

$$\frac{dt\,dS}{vE_\lambda d\lambda}\iint \frac{2\lambda u\cos\theta}{C} CE_\lambda d\lambda \cos\theta \frac{d\omega}{4\pi},$$

where the integration is with respect to $d\omega$ for all directions which represent radiation falling on to the area dS. Replacing $d\omega$ by $\sin\theta\, d\theta\, d\phi$, and integrating from $\phi = 0$ to 2π and from $\theta = 0$ to $\frac{1}{2}\pi$, we find that the average change of wave-length experienced by radiation of wave-length λ is

$$\frac{1}{3v}\lambda u\, dS\, dt.$$

This is the result produced by reflection from only one small area dS of a single moving piston. To obtain the total effect on the wave-length, we must integrate over the whole area of moving walls, and obtain for the average change of wave-length $\delta\lambda$

$$\delta\lambda = \frac{1}{3v}\lambda dt\, \Sigma \iint u\, dS.$$

If the change of volume in time dt is denoted by δv,

$$\delta v = dt\, \Sigma \iint u\, dS,$$

so that the average change of wave-length $\delta\lambda$ is given by

$$\delta\lambda = \frac{\lambda}{3v}\,\delta v.$$

We now see that the value of n in equation (929) must be $\frac{1}{3}$, so that the equation assumes the form

$$\frac{\lambda}{\lambda'} = \left(\frac{v}{v'}\right)^{\frac{1}{3}}.$$

Combining this with equation (928), we obtain

$$\lambda T = \lambda' T' \quad \text{..............................(930)},$$

so that λT remains constant for any constituent of the radiation through all changes of volume and temperature.

481. It follows that the law of partition of radiant energy at any temperature T can be deduced from that at any standard temperature T' by altering all wave-lengths in the ratio required by relation (930), while at the same time multiplying the energy throughout by the factor required to give the appropriate value to the total energy. For this reason, the law expressed by equation (930) is commonly called Wien's Displacement Law.

The law of partition of energy at temperature T must accordingly be of the form

$$E_\lambda d\lambda = f(T)\,\phi(\lambda T)\,d\lambda.$$

Integrating with respect to λ, the total energy per unit volume is

$$\begin{aligned} E &= \int_0^\infty E_\lambda d\lambda \\ &= \frac{1}{T} f(T) \int_{\lambda T=0}^{\lambda T=\infty} \phi(\lambda T)\, d(\lambda T) \\ &= \frac{1}{T} f(T) \times \text{a constant.} \end{aligned}$$

Hence for equation (926) to be satisfied, $f(T)$ must be proportional to T^5, and the partition of energy must be of the form

$$E_\lambda d\lambda = F(\lambda T)\, T \lambda^{-4} d\lambda \text{........................(931)}.$$

These laws are believed to be completely confirmed by observation.

482. So far as is known, equation (931) gives all the information which can be obtained about the law of partition of radiation from thermodynamical principles alone.

It will already have become evident from Chapter V that the second law of thermodynamics is very much more general than the particular set of dynamical laws from which it was there derived. Indeed, the second law appears to be deducible from almost any set of dynamical laws which imply the law of causation*. Thus equation (931) may be regarded as the limit of our knowledge until we assume some particular set of dynamical laws, when it must of course be possible fully to determine the function $F(\lambda T)$. Different forms of this function may be thought of as associated with different sets of dynamical laws.

483. For the classical dynamical laws, the function $F(\lambda T)$ was evaluated in the previous chapter, and was found to reduce to a constant, namely $8\pi R$, a value which was at once seen to be in contradiction with experience.

In general it may be noticed that $F(\lambda T)$ must be of the same physical dimensions as $8\pi R$, and so must be of the form†

$$F(\lambda T) = 8\pi R f\left(\frac{k}{RT\lambda}\right),$$

where k is a constant, of which the physical dimensions must be those of $RT\lambda$, so that f reduces to a pure numerical multiplier. It will be seen that the classical system of dynamics can provide no universal constant of the physical dimensions of $RT\lambda$ (*i.e.* energy × length), a circumstance which in itself shews the necessity of looking beyond the classical dynamics.

PLANCK'S RADIATION FORMULA.

484. According to Planck's radiation formula, the form of the function f is given by

$$f(x) = \frac{x}{e^x - 1} \quad \text{.........................(932)},$$

where x stands for $k/RT\lambda$, and k is an entirely new physical constant, to which no meaning has so far been assigned in terms of the older dynamics. Thus Planck's complete radiation formula is

$$E_\lambda d\lambda = \frac{k/RT\lambda}{e^{k/RT\lambda} - 1} 8\pi RT\lambda^{-4} d\lambda \quad \text{................(933)}.$$

Planck replaces k by hC, where C is the velocity of light, and h is a new constant commonly called Planck's constant. If ν is the number of vibrations per second of light of wave-length λ, so that $\lambda = C/\nu$, the value of x is

$$x = \frac{k}{RT\lambda} = \frac{h\nu}{RT}.$$

This formula is found to agree extremely well with experiment‡, the best

* *Phil. Mag.* xx. (1910), p. 943.

† Strictly speaking $f\left(\frac{k}{RT\lambda}\right)$ ought to be replaced by $f\left(\frac{k_1}{RT\lambda}, \frac{k_2}{RT\lambda}, \ldots\right)$, but this generalisation proves to be unimportant.

‡ Rubens and Michel, *Phys. Zeitsch.* XXII. (1921), p. 569, and Michel, *Zeitsch. für Physik*, IX. (1922), p. 285. See also, *La Théorie du Rayonnement et les Quanta* (Gauthier-Villars, Paris, 1912).

proof of this being that determinations of the unknown constant k/R made in all regions of the spectrum give substantially identical values*. The best determinations of the value of k/R (commonly denoted by the symbol C_2) give the value

$$C_2 = hC/R = 1\cdot4320 \text{ cm. degrees}\dagger.$$

The values C and R are known by other methods, so that a determination of C_2 leads at once to a value of Planck's constant h. The above value of C_2 leads to the value‡

$$h = 6\cdot554 \times 10^{-27} \text{ erg seconds},$$

which agrees well with the value of h determined in other ways§.

485. The simplest way of arriving theoretically at Planck's formula is perhaps the following.

Suppose that a dynamical system contains a great number of similar parts or components, each having only one degree of freedom, and having energy of the form

$$E = \tfrac{1}{2}(a\dot{\phi}^2 + b\phi^2).$$

These units can clearly describe isochronous vibrations of frequency $p/2\pi$, such that $ap^2 = b$. The values of ϕ and $\dot{\phi}$ for such a vibration will be of the forms

$$\phi = A\cos(pt + \eta), \quad \dot{\phi} = -Ap\sin(pt + \eta) \qquad (934).$$

From § 105, it appears that the law of distribution of the coordinates ϕ, $\dot{\phi}$ will be of the form

$$Ce^{-2hE}\, d\phi\, d\dot{\phi} \qquad (935).$$

From equations (934), it is easily seen that

$$d\phi\, d\dot{\phi} = Ap\, dA\, d\eta = dE\, d\eta/\sqrt{ab},$$

so that on integrating with respect to η, the law of distribution of values of E is found to be

$$\frac{2\pi C}{\sqrt{ab}} e^{-2hE} dE \qquad (936).$$

Imagine that equal small ranges of energy dE are marked out surrounding the values $E = 0$, $E = \epsilon$, $E = 2\epsilon$, etc. From formula (936), the numbers of units lying inside these ranges will stand in the ratio

$$1 : e^{-2h\epsilon} : e^{-4h\epsilon} : e^{-6h\epsilon} : \ldots,$$

* It is maintained by Nernst and Wulf (*Deutsch. Phys. Gesell.* xxi. (1919), p. 294) that k/R falls off for low values of x, but other investigators do not appear to have discovered any such falling off.

† W. W. Coblentz, *Bulletin* 15 *of the Bureau of Standards*, Washington (1920), and *Journal of the Optical Society of America*, viii. (1924), p. 11.

‡ See Coblentz, *Scientific Papers of the Bureau of Standards*, Washington, No. 360 (1920).

§ For a general discussion see R. T. Birge, *Phys. Rev.* xiv. (1919), p. 361; see also § 497 below.

so that if N have zero energy, the numbers having energies ϵ, 2ϵ, 3ϵ, ... will be $Ne^{-2h\epsilon}$, $Ne^{-4h\epsilon}$, $Ne^{-6h\epsilon}$, etc. The total number of systems whose energy lies within the specified small ranges will be

$$N(1 + e^{-2h\epsilon} + e^{-4h\epsilon} + e^{-6h\epsilon} + \dots) = \frac{N}{1 - e^{-2h\epsilon}} \dots\dots\dots\dots(937),$$

while the total energy of all these systems is

$$N(\epsilon e^{-2h\epsilon} + 2\epsilon \,.\, e^{-4h\epsilon} + 3\epsilon \,.\, e^{-6h\epsilon} + \dots) = \frac{N\epsilon}{e^{2h\epsilon}(1 - e^{-2h\epsilon})^2} \quad \dots(938).$$

On division, the mean energy of all the vibrations now under consideration is seen to be

$$\frac{\epsilon}{e^{2h\epsilon} - 1} = RT \frac{x}{e^x - 1} \quad \dots\dots\dots\dots\dots\dots\dots\dots(939),$$

where x stands for ϵ/RT.

486. If all values were possible for the energies of the units, the mean energy would, from the theorem of equipartition of energy, be RT. But when the energies are limited to the small equal ranges surrounding the values $E = 0, \epsilon, 2\epsilon, \dots$, we have found that the mean energy is only equal to $x/(e^x - 1)$ times the equipartition value. This multiplying factor is, however, exactly of the form demanded by Planck's formula (cf. equation (932)). To make the formulae agree completely the two values of x must be the same in equations (932) and (939), and this requires that

$$\epsilon = h\nu \quad \dots\dots\dots\dots\dots\dots\dots\dots\dots\dots\dots(940).$$

This last equation may be regarded as the fundamental equation of Planck's theory. The quantity ϵ measures the "quantum" of energy of frequency ν. It will be seen that the fundamental hypothesis on which the foregoing derivation of Planck's formula is based is the hypothesis that vibrational energy can occur only in complete integral quanta, each of amount ϵ given by equation (940).

487. In the foregoing analysis, it has not been necessary to specify the nature of the units under consideration.

Planck's original method* was to suppose that in all matter there were a great number of "resonators" of every possible period. The mean energy of any resonator of a definite frequency ν was shewn, by a method entirely different from that just given, to be that expressed by formula (939), so that the mean value of its kinetic energy $\frac{1}{2}a\overline{\dot{\phi}^2}$ was equal to half this, and the insertion of this value into equation (919) of the last chapter gave

$$E_\lambda d\lambda = 8\pi RT\lambda^{-4} \frac{x}{e^x - 1} d\lambda \quad \dots\dots\dots\dots\dots\dots(941),$$

which is Planck's formula.

* *Ann. d. Physik*, IV. (1901), p. 556.

This method of arriving at Planck's formula is open to the objection that in arriving at formula (939) it has been assumed that the energy of a resonator is limited to one of the values 0, ϵ, 2ϵ, ..., whereas in arriving at equation (919) it was supposed that the energy could, and did, vary continuously.

488. An alternative way of interpreting the analysis of § 485 is to suppose that the vibrations there considered are the vibrations of a material ether. The number of vibrations per unit volume within a range $d\lambda$ of wave-length is, from § 463, equal to $8\pi\lambda^{-4}d\lambda$; the mean energy of each of these vibrations is that given by formula (939), and on multiplication we arrive directly at formula (941) for the partition of energy in the ether.

With respect to this way of regarding the problem, it is to be noticed that the ether itself provides no mechanism for the interchange of energy between vibrations of different wave-lengths. This interchange can only occur through the intermediary agency of some kind of matter, and we are led to contemplate interchanges of energy between matter and ether taking place only by units of amount ϵ, each unit being of monochromatic radiation, the frequency ν being connected with ϵ by Planck's relation $\epsilon = h\nu$. We may if we wish go further and eliminate the hypothesis of an ether altogether, by considering simply interchanges of energy between matter and a field of radiation; to obtain Planck's law these must occur by whole units ϵ where $\epsilon = h\nu$.

489. Before discussing these or other physical interpretations of the quantum-theory, it will be convenient to consider another problem. We have seen that two physical schemes (§§ 487, 488) can lead to the equations of the quantum-theory, but neither of these schemes appears, at first sight at least, to be of great plausibility. The question arises whether schemes of any such revolutionary kind are necessary to explain the facts of radiation, or whether some simpler physical scheme, more in keeping with our well-established conceptions of physics, cannot be found to lead to the same result as the quantum-theory. We may in fact attack the converse problem: Given the observed laws of radiation, what systems of dynamics must be postulated in order to obtain these laws?

490. Let us return to the investigation of Chapter V (§ 78 *et seq.*), and examine to what extent it must be modified when the classical laws of mechanics are no longer supposed to hold.

The first step in the argument consisted in shewing (§ 85) that for a conservative system the swarm of representative points in the generalised space moved without any concentration taking place: the density of any part of the swarm remained perpetually the same.

When the classical laws are abandoned, this conclusion can no longer be established. And so long as the actual laws of motion are unknown the exact modification required cannot be ascertained.

It is, however, clear that the motion of the system under consideration must obey some law of determinism: if a state A is on one occasion succeeded by a state B, then the state A must always be succeeded by the state B. If this were not so, exactly similar experiments would not invariably lead to exactly similar results, and the uniformity of nature would disappear. The moving representative points in the generalised space must accordingly be supposed to follow definite tracks through this space. This being so, it will be possible to arrange the density of the initial swarm of particles in such a way that the density at any point of the space remains always the same. In Chapter V we had $\frac{d\rho}{dt}=0$ and $\frac{D\rho}{Dt}=0$ throughout the space, the former being a matter of arrangement, and the latter a consequence of our dynamical equations. In the present problem we can only arrange for the former equation to be true. This introduces the simplification that we need only consider one permanent swarm of points; the density varies at different points of the space, but does not vary with the time.

If the density of this swarm varied only by a finite amount from point to point, we should still be led to the theorem of equipartition of energy, and consequently to the radiation formula (923). For we have seen that equipartition of energy holds at every point of the generalised space except for small infinitesimal regions. It follows that the only way of escaping the equipartition formula is by supposing that the density of the swarm of representative points is zero at every point throughout the whole of the generalised space, except for small infinitesimal regions $R_1, R_2, \ldots$ in which equipartition does not hold, and at these there may be dense swarms of points. Further, in order to satisfy the hydrodynamical equation of continuity in the generalised space, zero density must be associated with infinite velocity, so that the motion of the representative points must consist of sudden jumps from one of the regions $R_1, R_2, \ldots$ to another.

In this way it appears that in seeking to avoid the equipartition formula (923), we are inevitably led to contemplate motion involving discontinuities of some kind*. We may next consider what special type of discontinuities must be postulated in order to arrive at Planck's law.

491. Let us return to the analysis given in §§ 91—98 of Chapter V, and suppose that E_1 is the energy of M vibrations each of frequency ν vibrations per second. According to Planck's formula, the value of E_1 must be

* See on this subject a very important paper by Poincaré, *Journ. de Phys.* [5] II. p. 5 (1912), and the same author's *Dernières Pensées*, Ch. VI, "L'hypothèse des Quanta." For a fuller discussion of the matter of §§ 490, 491 see J. H. Jeans, *Phil. Mag.* XX. (1910), p. 943.

given by

$$E_1 = \frac{M\epsilon}{e^{\frac{\epsilon}{RT}} - 1},$$

where $\epsilon = h\nu$, and on solving for T we obtain

$$\frac{1}{RT} = \frac{1}{\epsilon} \log \left(1 + \frac{M\epsilon}{E_1}\right) \quad \text{.......................(942).}$$

In equations (174) and (157) we had the relations

$$\frac{1}{RT} = \frac{d}{dE_1} \log F_1 (E_1) = \frac{d}{dE_1} \log W_1 \quad \text{..............(943),}$$

and these equations may be considered to be independent of any special system of dynamical laws if W is taken to represent, not the volume of the generalised space in which a certain partition of energy holds, but the total number of representative points in the space for which this partition holds. Comparing equations (942) and (943), we obtain

$$\frac{d}{dE_1} \log F_1 (E_1) = \frac{1}{\epsilon} \log \left(1 + \frac{M\epsilon}{E_1}\right),$$

giving on integration

$$\log F_1 (E_1) = \left(M + \frac{E_1}{\epsilon}\right) \log \left(M + \frac{E_1}{\epsilon}\right) - \frac{E_1}{\epsilon} \log \frac{E_1}{\epsilon} + \text{cons.} \quad \text{....(944).}$$

If we write P for E_1/ϵ, and use Stirling's formula (69), this equation becomes

$$F_1 (E_1) = \frac{(M+P)!}{P!} \times \text{a constant} \quad \text{..............(945).}$$

But $(M+P)!/P!$ is the number of ways in which P similar articles can be put into M similar pigeon-holes, or, more appropriately for our present investigation, is the number of ways in which P similar and indivisible units of energy can be distributed among M different similar vibrations. Since formula (945) gives the only possible value for $F_1 (E_1)$, it appears that Planck's formula can only be obtained by supposing that the total energy E_1 is made up of P similar indivisible units of energy, and that these are distributed indifferently between the M vibrations. Since P stands for E_1/ϵ, the unit of energy is

$$\frac{E_1}{P} = \epsilon = h\nu.$$

Thus we see that Planck's formula can only be arrived at by hypotheses which must be essentially identical with those already made in § 486. In the generalised space there are no representative points except in isolated regions which are such that the energy of every vibration is a multiple of h times its frequency. In the physical system the energy of each vibration must remain the same, and equal to a multiple of $h\nu$, until a sudden cataclysm of some kind results in a change by an amount which again must be a multiple of $h\nu$.

492. It is accordingly clear that the observed phenomena of black-body radiation point to an atomicity somewhere in the distribution of energy.

According to the original view of Planck (§ 487) this atomicity was an atomicity of the energy of the material system; vibrations of frequency ν in the system had energies which were confined to being integral multiples of $h\nu$. According to the alternative view put forward in § 488 the atomicity is one of radiant energy, the radiation of frequency ν in an enclosure being restricted to having energies which are multiples of $h\nu$. When there is no enclosure to prevent the escape of radiation, the hypothesis assumes the form that transfers of energy between the material system and radiation of frequency ν take place by units of amount $h\nu$. We may notice that Planck's original theory required also that transfers of energy between matter and radiation occurred in multiples of $h\nu$, but ν was here the frequency of the material vibration. Both views contemplate transfers by units $h\nu$, but in one case ν is the frequency of the radiation, and in the other case ν was the frequency of the radiator.

The advance of knowledge has made it possible to decide quite definitely between these two alternatives. As we shall see below, a material system, as for instance a hydrogen atom (§ 500), does in actual fact emit and absorb radiation by quanta of amount $h\nu$, but ν is the frequency of the radiation, and does not correspond to any vibration of the material system. Thus the proper interpretation of the analysis by which we obtained the formula for black-body radiation in § 485, is that ν represents the frequency of radiation, and that the energy of radiation of frequency ν changes only by whole units of amount $h\nu$.

493. From this point of view a short step leads directly to the hypothesis of "light-quanta," according to which all radiation consists of indivisible packets or "atoms" of monochromatic light, each of which travels through space like a bullet from a rifle until it hits a material target by which it is completely absorbed. This view was put forward as a working hypothesis by Einstein* in 1905, and at once enabled him to formulate the true law of photo-electric action (cf. § 497 below).

In spite of this success it appears fairly certain that the view must be regarded merely as a working hypothesis and not as a literal expression of actual fact. Against the supposition that radiation actually travels in indivisible quanta must be set practically all the evidence of the undulatory theory of light, and, in particular, that of the phenomena of diffraction and interference. If light-quanta were strictly indivisible, interference could not be obtained by splitting up a quantum into two beams: it could only occur at a point at which two or more quanta happened to exist simultaneously. The fainter the light in any experiment, the smaller the chance would be of two quanta coexisting in this

* *Ann. d. Phys.* XVII. p. 146.

way, so that if the light were sufficiently reduced in intensity the whole interference phenomenon ought to disappear. That this does not happen has been shewn by Taylor*, who reduced the intensity of illumination until an exposure of 2000 hours was necessary to affect a sensitive plate, and yet obtained photographs of diffraction patterns in which light and dark bands alternated with undiminished clearness. A simple calculation shews that, if light had existed only in indivisible quanta, no interference at all ought to have been observed in this experiment.

More evidence, equally adverse to the theory of light-quanta, can be obtained by considering what would have to be the size in space of light-quanta. It is possible to obtain interference over a path-difference equal to about a million wave-lengths, and this can hardly be interpreted except as meaning that a light-quantum must have a length in space comparable with a million wave-lengths, a length therefore of several feet. But it is hard to believe that a quantum as long as this can be indivisible.

Again, a telescope with a five-foot object-glass is found to have a greater resolving power than one with a five-inch object-glass. This, on the light-quantum theory, can only mean that the incident light-quanta must in some way be spread over the whole object-glass of the five-foot telescope. But it is difficult to think of quanta of five-foot cross-section as indivisible, and if they were, it is hard to see how any light at all could get into a five-inch telescope. If light-quanta were small and concentrated, the only difference in definition between a large and a small telescope would arise from the large telescope collecting a greater number of quanta than the small one. It would therefore shew itself as a mere difference in intensity, and a small telescope would resolve a pair of bright stars just as well as a large telescope would resolve an equally close pair of faint stars†. This is of course quite contrary to observation.

Various attempts have been made to reconcile these objections with different forms and modifications of the theory of light-quanta, but no success has so far been attained, and the general opinion of physicists seems to be that the theory cannot be regarded as an expression of physical reality.

Numerical Values.

494. Before turning to physical aspects of the quantum-theory, we may attempt, with the help of numerical data, to form some estimate of the magnitude of the unit of energy ϵ.

The value of Planck's constant h has been seen to be $6{\cdot}55 \times 10^{-27}$ erg secs., while the value of ν for yellow light (D_1 or D_2) is $5{\cdot}01 \times 10^{14}$. Thus for the unit of energy appropriate to light of this colour, we have

$$\epsilon = h\nu = 3{\cdot}28 \times 10^{-12} \text{ ergs}.$$

* *Proc. Camb. Phil. Soc.* xv. (1909), p. 114.

† These instances, and many others of an equally forcible nature, are given by Lorentz (*Phys. Zeitschrift*, xi. (1910), p. 349); see also *British Association Report*, Birmingham (1913), p. 376.

This may be compared with other amounts of energy met with in the Theory of Gases. In § 151 we found that at 0° C. the energy of translation of a molecule or atom (αT_0) is $5{\cdot}620 \times 10^{-14}$ ergs, so that the quantum ϵ for yellow light is about equal to the energy of translation of 60 atoms or molecules at 0° C. In accordance with Wien's displacement law, the value of λ, say λ_m, for which E_λ is a maximum, varies inversely as the temperature. Thus $\lambda_m T$ is a constant, and the value of this constant is found to be 0·2885 cm. deg.* The energy in a quantum of radiation of wave-length λ_m is

$$\epsilon = h\nu = \frac{hV}{\lambda_m} = 6{\cdot}81 \times 10^{-16} T.$$

The energy per atom of the solid at temperature T is however $3RT$ or $4{\cdot}12 \times 10^{-16} T$. Thus the quantum of light of any colour is equal to 1·65 times the energy of an atom of the solid at the temperature corresponding to this radiation (or, very approximately, to the energy of a molecule of a diatomic gas at this temperature).

495. The total radiant energy per unit volume in equilibrium with matter at temperature T is

$$\int_0^\infty \frac{8\pi}{C^3} \frac{h\nu^3 d\nu}{e^x - 1} = 6{\cdot}493 \left(\frac{RT}{Ch}\right)^3 8\pi RT,$$

while the number of quanta composing this radiation is

$$\int_0^\infty \frac{8\pi}{C^3} \frac{\nu^2 d\nu}{e^x - 1} = 60{\cdot}422 \left(\frac{RT}{Ch}\right)^3.$$

By division, the average energy per quantum at $T°$ abs. is found to be $2{\cdot}702RT$ or 1·801 times the energy of an atom at the same temperature.

At 0° C. the value of $(RT/Ch)^3$ is 6,935,000, so that there are about 420,000,000 quanta per cubic centimetre in radiation at 0° C., or about one quantum per cube of edge ·0013 cm.

The energy of bright sunlight is about 4×10^{-5} ergs per cubic centimetre, and if this is regarded as a collection of quanta of yellow light, there must be about ten million quanta per cubic centimetre. On the other hand the radiation from a sixth magnitude star, which is about the faintest star visible to normal eyesight, will contain only about one quantum per cubic metre.

496. According to the classical mechanics, the chance of a system possessing no energy at all is an infinitesimal one.

According to the quantum-theory, we saw in § 485 that out of a number

$$\frac{N}{1 - e^{-2h\epsilon}}$$

* Coblentz, *Scientific Papers of the Bureau of Standards*, Washington, Nos. 357, 360 (1920).

of vibrations, N must be supposed to possess no energy at all. The number which possess some energy must accordingly be

$$\frac{Ne^{-2h\epsilon}}{1-e^{-2h\epsilon}},$$

which is only a fraction $e^{-2h\epsilon}$ of the total number.

If ϵ is even moderately large compared with RT, the fraction $e^{-2h\epsilon}$ or $e^{-\epsilon/RT}$ will be very small. For instance, for matter at 0° C.,

$$RT = 3{\cdot}75 \times 10^{-14} \text{ ergs};$$

the quantum for yellow light is $3{\cdot}28 \times 10^{-12}$ ergs, so that in this case $\epsilon/RT = 87$ (about), and $e^{-2h\epsilon} = 10^{-38}$. Thus out of 10^{38} vibrators at 0° C. of frequency equal to that of yellow light, only one must be expected to have any energy, while it further appears from § 485 that the odds are 10^{38} to 1 that this one will only have one quantum of energy.

Corresponding to radiation of the wave-length λ_m, the quantum is found to be

$$\epsilon = 4{\cdot}965\,RT$$

so that even as regards vibrations of wave-length λ_m at any temperature, only one in $e^{4{\cdot}965}$, roughly 1 in 140, of the vibrations will have any energy. And of those vibrations which do possess energy, only about 1 in 140 will possess more than one quantum.

The Photo-Electric Phenomenon.

497. Since the early experiments of Hertz it has been known that the incidence of high-frequency light on the surface of a conductor results in its acquiring a positive charge if it was originally uncharged, or losing a negative charge if it originally had one. These phenomena are now known to be caused by the emission of negative electrons from the metal, the electrons being in some way set free by the incidence of the light.

In any one experiment, the velocities with which the electrons leave the metal are observed to have all values from zero to a certain clearly-defined maximum velocity v, the value of v depending on the conditions of the particular experiment. This is most naturally interpreted as meaning that in any one experiment, all the electrons are ejected out of their atoms with the same velocity v, but those which come from some distance below the surface lose part of their velocity in fighting their way out.

The maximum velocity v of the discharged electrons does not depend on the temperature* of the metal, or on the intensity of the incident light†, but solely on the nature of the metal and on the frequency of the light. For a

* Ladenburg, *Verhand. d. Deutsch. Phys. Gesell.* IX. (1907), p. 165; Lienhop, *Ann. d. Phys.* XXI. (1906), p. 281.

† Lenard, *Ann. d. Phys.* VIII. (1902), p. 149; Pohl and Pringsheim, *Verhand. d. Deutsch. Phys. Gesell.* XV. (1912), p. 974.

given metal this velocity increases as the frequency of the incident light is increased, but there is a critical frequency ν_0 below which the action does not occur at all. For any frequency ν above this, the velocity v is found to be given by*

$$\tfrac{1}{2}mv^2 = h(\nu - \nu_0) \quad \text{.........................(946)}.$$

An equation of this form was first suggested by Einstein†, as being the equation which ought to connect v and ν on the hypothesis of light-quanta. Making the simplest assumptions possible, it is clear that the kinetic energy $\tfrac{1}{2}mv^2$ of the projected electron ought to be equal to the energy of the radiation absorbed minus the work required to take the electron out of the field of force of its atom. The former amount of energy ought, on the light-quantum theory, to be $h\nu$, while the latter is of course eV, where V is the ionisation potential of the substance in question. Thus v and ν ought, on these very simple assumptions, to be connected by an equation of the form of (946) in which $h\nu_0$ ought to be equal to eV.

Experiment shews that the value of h determined from relation (946) is exactly equal, within the limits of experimental error, to the known value of h, the universal constant of Planck. For instance Millikan, in a series of very careful experiments, has obtained the following values for h:

from sodium‡, $h = 6{\cdot}561 \times 10^{-27}$,
from lithium§, $h = 6{\cdot}585 \times 10^{-27}$.

The value of h determined from a direct comparison of Planck's radiation formula with observed radiation is, as already mentioned,

$$h = 6{\cdot}554 \times 10^{-27}.$$

It is more difficult to compare the values of $k\nu_0$ and eV, largely because V is not known with great accuracy. What evidence is available suggests that $k\nu_0$ is very nearly equal to eV, but no very exact comparison is possible.

498. Einstein's conception of the photo-electric phenomenon, as expressed in his equation (946), appears to be valid over the whole range of frequencies of radiation for which the phenomenon can be observed, a range extending from the yellow light at which the alkaline metals are first affected up to the hardest of X-rays and even γ-rays.

In the case of rays of very high frequency equation (946) may assume a specially simple form. The value of ν_0, determined by the condition that $h\nu_0$ shall be the energy necessary to detach an electron from an atom, is generally of the order of the frequency of visible radiation, while the frequency of

* See in particular, A. Ll. Hughes, *Phil. Trans.* 212, A (1912), p. 205; Richardson and Compton, *Phil. Mag.* XXIV. (1912), p. 575; Millikan, *Phys. Review*, IV. (1914), p. 73, and *Phys. Review*, VII. (1916), pp. 18 and 355.

† *Ann. d. Phys.* XVII. (1905), p. 146.

‡ *Phys. Review*, IV. (1914), p. 73.

§ *Phys. Review*, VI. (1915), p. 55.

X-radiation is many thousands of times greater. Thus ν_0 may be neglected in comparison with ν and the equation assumes the simple form

$$\tfrac{1}{2}mv^2 = h\nu;$$

the whole quantum is now transformed into the kinetic energy of the ejected electron.

An allied phenomena may be mentioned here as an illustration of the wide applicability of the quantum-theory. When a stream of electrons all having the same high velocity v is caused to fall on a material target, X-radiation is emitted of which the spectrum is limited perfectly sharply to frequencies below a certain definite frequency ν. It has been shewn by Duane and Hunt*, as also by Hull† and others, that h and ν are connected by the relation

$$\tfrac{1}{2}mv^2 = h\nu,$$

where h again represents Planck's constants. Using this relation for the exact determination of h, Blake and Duane have obtained the value

$$h = 6{\cdot}555 \times 10^{-27},$$

which is probably very accurate and is in highly satisfactory agreement with the best values of h obtained by other methods.

The phenomena just mentioned as well as the simple photo-electric phenomenon, provide the strongest possible evidence for the view that Planck's quanta have a real physical meaning, and are not mere mathematical fictions introduced to explain an otherwise inexplicable radiation formula.

Bohr's theory of Line Spectra.

499. In 1913 Dr N. Bohr‡ put forward a theory of line-spectra according to which the emission of a line-spectrum is evidence of a phenomenon which is, roughly speaking, the converse of the photo-electric phenomenon. If the absorption of a quantum of energy of frequency ν results in the ejection of an electron from its orbit with a velocity v, then the consequence of an electron falling with initial velocity v into an atomic orbit may be expected to be the emission of a quantum of radiation of frequency ν. This, however, will by itself not explain the emission of isolated lines, for v can vary continuously, and so ν also ought to vary continuously. Bohr's theory accordingly finds it necessary to introduce a number of new additional hypotheses.

The theory is based on the Rutherford conception of the structure of the atom, according to which an uncharged atom of hydrogen consists of an electron of charge $-e$ revolving round a much heavier positive nucleus of charge $+e$; an uncharged atom of helium consists of two electrons revolving round a positive nucleus of charge $+2e$, and so on.

* *Phys. Review*, VI. (1915), p. 166.

† *Phys. Review*, VII. (1916), p. 157.

‡ *Phil. Mag.* XXVI. (1913), pp. 1, 476 and 857, XXVII. (1914), p. 506, and XXX. (1915), p. 394.

The inability of the classical mechanics to explain line-spectra becomes apparent on fixing our attention on the hydrogen atom. So long as we adhere to the classical mechanics, we are at a loss to understand how the two charges can continually go on rotating round one another at all: the analysis of § 84 would lead us to expect that the energy of the system would be continually dissipated by radiation until a state was reached in which no further radiation was possible, a state, therefore, in which the two charges had been reduced to rest. Moreover the total number of degrees of freedom of the two charges is only six, and of these three represent the freedom of the atom to move in space. Under the classical mechanics, it seems inconceivable that the remaining three degrees of freedom could produce the highly complicated line-spectrum of hydrogen.

Let us consider in detail the motion of an electron of mass m and charge $-e$ about a heavy nucleus of mass M and charge E. According to the classical mechanics it is possible for the two bodies to describe circular orbits about their common centre of gravity. If a is the distance apart of the two charges when ω revolutions are made per second, we have

$$\frac{eE}{a^2} = \frac{Mm}{M+m}(2\pi\omega)^2 a \quad \text{.......................(947)},$$

an equation in which the left-hand member represents the attractive force between the two charges, and the right-hand member represents the mass of either multiplied by its acceleration towards the centre of gravity of the two masses.

The potential energy of the orbit is $-eE/a$. The kinetic energy of the motion of the two masses relative to their centre of gravity is readily found to be

$$\frac{1}{2}\left(\frac{Mm}{M+m}\right)(2\pi\omega a)^2,$$

and this, from equation (947), is equal to $\frac{1}{2}eE/a$. Thus W, the work required to remove the electron out of its orbit to infinity, will also be equal to $\frac{1}{2}eE/a$.

The classical system of mechanics permits of W having any value from 0 to ∞. Bohr's primary assumption is that W must be of the form

$$W = \tfrac{1}{2}\tau h\omega \quad \text{.............................(948)},$$

where τ is a positive integer, which may have any value from 1 to ∞. This assumption is not one that can be deduced from the quantum-theory in the form in which this has so far been given in the present book, for this has required that the total energy, potential and kinetic, of a vibration of frequency ω shall be a multiple of $h\omega$, whereas Bohr's assumption is that the kinetic energy alone of a rotation of frequency ω shall be a multiple of $\frac{1}{2}h\omega$. We shall however see in the next Chapter that Bohr's assumption, as expressed in equation (948), is in full agreement with the general system of quantum-dynamics as at present developed.

Other physical interpretations can be given to Bohr's equation (948), of which perhaps the following is the simplest. Equation (948) requires that the angular momentum $2\pi m\omega a^2$ or $W/\pi\omega$ must be of the form $\tau h/2\pi$, and so is equivalent to requiring that the angular momentum must be atomic, and occur only in multiples of an "atom" of angular momentum $h/2\pi$, a conception due originally to Nicholson*. This interpretation of Bohr's equation will also be found later (§ 538) to possess a definite significance from the point of view of general quantum-dynamics.

Bohr's assumption at once prohibits the continuous variation of W, a and ω which is demanded by the classical mechanics. The possible values of W, a and ω are readily found from equations (947) and (948) to be

$$W = \frac{Mm}{M+m}\frac{2\pi^2 e^2 E^2}{\tau^2 h^2}, \quad 2a = \frac{M+m}{Mm}\frac{\tau^2 h^2}{2\pi^2 eE}, \quad \omega = \frac{Mm}{M+m}\frac{4\pi^2 e^2 E^2}{\tau^3 h^3} \quad \text{......(949)},$$

in which τ is restricted to integral values, so that W, a and ω are restricted to certain definite values. Thus a cannot gradually shrink, but is restricted to one of the values just found. It follows that there can be no oscillations of the electron in the plane of the orbit, so that the circumstance discovered by Nicholson, that such oscillations would be unstable, is no longer an objection to our present model of atomic structure.

500. We have next to consider what happens when, for reasons not at present specified, the atom suddenly shrinks from, say, the orbit $\tau = \tau_1$ to the orbit $\tau = \tau_2$. Formulae (949) shew that the atom must experience a loss of energy of amount ΔW given by

$$\Delta W = W_{\tau_2} - W_{\tau_1} = \frac{Mm}{M+m}\frac{2\pi^2 e^2 E^2}{h^2}\left(\frac{1}{\tau_2^2} - \frac{1}{\tau_1^2}\right) \quad \text{.........(950)}.$$

Bohr supposes that this amount of energy, suddenly set free from the atom, passes away into space in the form of one quantum of monochromatic radiation. The frequency of this radiation must, in accordance with Planck's equation (940), be determined by $\Delta W = h\nu$, so that ν must be given by

$$\nu = N\left(\frac{1}{\tau_2^2} - \frac{1}{\tau_1^2}\right) \quad \text{........................(951)},$$

where

$$N = \frac{Mm}{M+m}\frac{2\pi^2 e^2 E^2}{h^3} \quad \text{.......................(952)}.$$

According to Bohr's theory the frequencies of the different spectral lines of the element can be obtained by inserting different values of τ_1 and τ_2 in this formula. The lines can be sorted into spectrum series corresponding to different values of τ_2. We may now examine how far this formula is capable of giving the various observed spectral series.

* *Monthly Notices of the Royal Astron. Soc.* June, 1912.

The Hydrogen Spectrum.

501. The hydrogen spectrum is got by putting $E = e$, and so assigning to N the value N_H given by

$$N_H = \frac{Mm}{M+m} \frac{2\pi^2 e^4}{h^3} \quad \text{.........................(953).}$$

The value $\tau_2 = 1$ gives the series

$$\nu = N_H \left(1 - \frac{1}{n^2}\right) \dots (n = 2, 3, 4 \dots),$$

all the lines of which would lie in the ultra-violet. None of the lines of the series were known when Bohr's theory was originally published, but the series was subsequently discovered by Lyman*. For Lyman's series to agree with the series predicted by the theory, N_H must be equal to the Rydberg constant. We shall very soon see that the value of N_H given by equation (953) is exactly equal to this constant to within the limits of experimental error.

The value $\tau_2 = 2$ gives the series

$$\nu = N_H \left(\frac{1}{4} - \frac{1}{n^2}\right) \dots (n = 3, 4, 5 \dots),$$

which becomes the well-known Balmer series, when N_H is again supposed equal to the Rydberg constant.

The value $\tau_2 = 3$ gives the series

$$\nu = N_H \left(\frac{1}{9} - \frac{1}{n^2}\right) \dots (n = 4, 5, 6 \dots),$$

a series in the infra-red of which the first two lines ($n = 4, 5$) were discovered by Paschen† and the next three by F. S. Brackett‡. The remaining series, $\tau_2 = 4, 5, 6 \dots$, would be too far in the infra-red to be observed.

The Helium Spectrum.

502. The nuclear charge for helium is $E = 2e$, and in the normal uncharged atom there are two electrons revolving round the nucleus. But when, from ionisation or other cause, the atom loses one of its electrons, there will, in the remaining positively charged atom, be one electron revolving round a nucleus $2e$. The spectrum of the positively charged helium atom ought accordingly to be obtained on putting $E = 2e$ in the above formulae. Its frequencies will therefore be given by formula (951) if we assign to N a value N_{He} given by

$$N_{He} = \frac{M'm}{M'+m} \frac{8\pi^2 e^4}{h^3} \quad \text{.........................(954),}$$

where M' is the mass of the helium nucleus.

* *Nature*, XCIII. p. 241 (May 7, 1914).

† *Ann. d. Phys.* XXVII. (1908), p. 565.

‡ *Nature*, CIX. p. 209 (Feb. 16, 1922).

The mass of the hydrogen nucleus is about 1835 times that of the negative electron; that of the helium nucleus is about 7300 times that of the negative electron. Thus to a close approximation, equations (953) and (954) may be replaced by

$$N_H = \frac{2\pi^2 me^4}{h^3}, \quad N_{He} = \frac{8\pi^2 me^4}{h^3} \quad \text{.....................(955)},$$

giving the approximate relation

$$N_{He} = 4N_H \quad \text{..............................(956)}.$$

Thus the spectral series will be given by the formula

$$\nu = 4N_H\left(\frac{1}{\tau_2^2} - \frac{1}{\tau_1^2}\right) \quad \text{..........................(957)}.$$

The series $\tau_2 = 1$ is too far in the ultra-violet to be observed. Two lines of the series $\tau_2 = 2$ which is also in the ultra-violet have recently been discovered by Lyman*. The series $\tau_2 = 3$ may be divided into two parts according as τ_1 is even or odd, and so is equivalent to the two series

$$\nu = N_H\left(\frac{1}{(\frac{3}{2})^2} - \frac{1}{n^2}\right),$$

$$\nu = N_H\left(\frac{1}{(\frac{3}{2})^2} - \frac{1}{(n+\frac{1}{2})^2}\right).$$

Both these series were discovered by Fowler† in 1912 in a mixture of hydrogen and helium. At first the series were attributed to hydrogen, but after the appearance of Bohr's theory it was suspected that they might be due to helium. Subsequently E. J. Evans‡ succeeded in observing the first line ($\lambda = 4686$) of the first series in a tube which was believed to be free of hydrogen, and in which no trace of known hydrogen lines could be observed.

The series $\tau_2 = 4$ may again be regarded as falling into two parts according as τ_1 is even or odd, and these two parts give the series

$$\nu = N_H\left(\frac{1}{4} - \frac{1}{n^2}\right) \quad \text{.............................(958)},$$

$$\nu = N_H\left(\frac{1}{4} - \frac{1}{(n+\frac{1}{2})^2}\right) \quad \text{.......................(959)}.$$

The former of these is again the Balmer series, which now reappears as a helium series. The coincidence of the two series is, however, only approximate. If we replace the approximate equations (955) by the exact relations (953) and (954), it appears that the series are identical except for slight differences in the value of N. To these we shall return below (§ 503).

The second series (959) is the well-known "sharp" or Pickering series. This was observed by Pickering§ in the spectrum of the star ζ-Puppis, and

* *Nature*, CIV. (1919), p. 314.

† *Monthly Notices of the Royal Astron. Soc.* LXXIII. (1912), p. 62.

‡ *Phil. Mag.* XXIX. (1915), p. 284.

§ *Astrophys. Journ.* IV. (1896), p. 369, and V. (1897), p. 92.

was attributed to hydrogen simply on the grounds, now seen to be inadequate, that the analogous series (958) was a hydrogen series. Although there is no direct experimental evidence, there can be but little room for doubt that the lines of this series emanate from helium.

For a fuller study of the helium spectrum, reference should be made to a very full discussion by Paschen* of the helium spectrum in its relation to Bohr's theory.

Numerical Values.

503. On inserting the known numerical values

$$h = 6{\cdot}554 \times 10^{-27},$$
$$e = 4{\cdot}774 \times 10^{-10},$$
$$m = 9{\cdot}00 \times 10^{-28},$$

into formula (955), the value predicted for N_H is found to be

$$\frac{2\pi^2 me^4}{h^3} = 3{\cdot}284 \times 10^{15}.$$

The value observed for N_H is $3{\cdot}290 \times 10^{15}$. The numbers agree to well within the probable error resulting from our ignorance of the exact values of h, e and m, and the extra refinement of equation (953) is too slight to be worthy of consideration in determining the absolute value of N_H.

The case stands differently when we pass to a consideration of the relative values of N_H and N_{He}. In place of the approximate relation $N_{He} = 4N_H$, the exact equations (953) and (954) yield the exact relation

$$N_{He} = \frac{4(M+m)}{M + \frac{1}{4}m} N_H \qquad \text{.......................(960)},$$

where M is the mass of the hydrogen nucleus and that of the helium nucleus is assumed to be $4M$.

The value of M/m is known, from a large accumulation of direct experimental evidence, to be about 1835. On substituting this value, the ratio of N_{He} to N_H given by formula (960) is 4·001635, while the value observed spectroscopically is 4·001632†.

It is a consequence of Bohr's assumption, that the hydrogen atom, for which $E = e$, can have the diameter obtained by taking $\tau = 1$, namely

$$2a = \frac{h^2}{2\pi^2 me^2} \qquad \text{............................(961)},$$

and can also have diameters equal to 4, 9, 16, 25 ... times this. The normal hydrogen atom is that for which the loss of energy W has been greatest, and so is that for which $\tau = 1$. On inserting the numerical values just given for h, e and m, it is found that formula (961) predicts for the orbit in the normal hydrogen atom a diameter $2a = 1{\cdot}08 \times 10^{-8}$ cms.

* Bohr's "Heliumlinien," *Ann. d. Phys.* L. (1916), p. 901.
† Fowler, *Phil. Trans.* 214, A (1914), p. 258.

Absorption Spectra.

504. On this theory the absorption spectrum of a gas admits of a very simple interpretation. In an inert gas the vast majority of the atoms will be in the state $\tau = 1$. If radiation is supposed to pass through this gas in complete quanta, a quantum can be absorbed only if it is either just adequate to move the electron into some other orbit τ_1 or else to set the electron free altogether. Thus the absorption spectrum will consist simply of the series $\tau_2 = 1$ in formula (951) together with a continuous absorption band running upwards from the head ($\tau_1 = \infty$) of this series. The range covered by this band is exactly that for which the photo-electric effect occurs, and the presence of absorption shewn by this band would, on Bohr's theory, be evidence of the actual occurrence of photo-electric action.

R. W. Wood* has studied the absorption spectrum of sodium vapour, and has found it to be of exactly the type demanded by Bohr's theory. Fifty lines were observed in the absorption spectrum, their positions agreeing exactly with those of the principal sodium series, and in addition there was found to be a continuous absorption band beginning at the head of this series and extending to the extreme ultra-violet.

X-ray Spectra.

505. In 1913 Moseley† shewed that the frequencies in the X-ray spectrum of any element were given by the formula

$$\nu = N_H (n - \sigma)^2 \left(\frac{1}{\tau_2^2} - \frac{1}{\tau_1^2}\right),$$

in which N_H is the ordinary Rydberg constant for hydrogen, given by equation (953), τ_1 and τ_2 are integers, n is a constant for the element in question, namely its atomic number, and σ is small in comparison with n, being the same for all the lines which are believed to arise from the same ring of electrons.

Moseley's formula becomes identical with our previous formula (951) on taking

$$E = (n - \sigma) e,$$

and so would give the frequencies of radiation from an atom in which the orbits were determined by Bohr's mechanics if the force to the centre were $(n - \sigma) e/r^2$. In an element of atomic number n the charge on the nucleus is ne, but the force on an electron is the resultant of a force of attraction ne/r^2 to the nucleus and a force of repulsion from the other electrons of the atom. Moseley shewed that the small quantity σ has values such that the outstanding force $\sigma e/r^2$ may appropriately be attributed to these other electrons. Thus Bohr's theory appears to provide the key to X-ray, as well as to visual,

* *Physical Optics* (1911), p. 513, or Bohr, *l.c.* p. 17.

† *Phil. Mag.* xxvi. (1913), p. 1024.

spectroscopy. When we pass from optical spectra to X-ray spectra, with radiation of frequencies of from 1,000 to 100,000 times the frequencies of optical spectra, the emission of radiation still appears to be governed by the principles discovered by Bohr.

Band Spectra.

506. In recent years, the conceptions of the quantum-theory have been very successful in providing theoretical interpretation of band spectra. The ordinary band spectrum consists of a large number of lines which crowd together towards the "heads" of the bands in a strikingly orderly manner. In any single band the wave-numbers are obtained with high accuracy, by giving integral values to m in a formula of the type $A + Bm + Cm^2$, where A, B, C are constants for the particular band in question.

The band spectrum originates from complete molecules which may be either molecules of a single element or of a chemical compound. In addition to possessing the sub-atomic energy of the electrons of which its atoms are constituted, a molecule can possess also energy of rotation and energy of vibration of its atoms relative to one another.

As a preliminary to the discussion of the energy of a complete molecule, we may consider the energy of a simple Bohr atom in which a single electron of mass m describes a circular orbit of radius a about a nucleus of mass M.

Let K, k denote the radii of gyration of the nucleus and mass respectively. If the two masses were held at distance a apart and the rigid body so formed were made to rotate ω times per second about its centre of gravity, the kinetic energy of this motion would be

$$\frac{1}{2}\left(MK^2 + mk^2 + \frac{Mm}{M+m}a^2\right)(2\pi\omega)^2 \quad \text{..............(962).}$$

Numerically the terms MK^2 and mk^2 inside the bracket are very small in comparison with the remaining term $\frac{Mm}{M+m}a^2$. Thus the energy of the rotation may, to a close approximation, be expressed as

$$\frac{1}{2}\frac{Mm}{M+m}(2\pi\omega a)^2 \quad \text{............................(963).}$$

This is precisely the formula which was taken in § 499 to represent the kinetic energy of orbital motion in the Bohr atom. It now appears that the energy of description of an orbit in the plane of x, y may be thought of as the energy of a rigid-body rotation about the axis Oz. Thus if, as Bohr's theory supposes, orbital energy falls into quanta, then it is clear that rotational energy also must fall into quanta.

In general, if a body of mass M and radius of gyration k rotates at a rate of ω revolutions a second, its kinetic energy of rotation will be

$$\tfrac{1}{2}Mk^2(2\pi\omega)^2.$$

Bohr's supposition that double the kinetic energy will be equal to an integral number of quanta will be expressible by the equation

$$4\pi^2 Mk^2\omega = \tau h \qquad\qquad (964).$$

For a rigid body, the only values possible for ω will be a series of discrete values which will be in arithmetical progression. If k varies with ω, the values for ω are still discrete but follow a less simple law. Thus on the quantum-theory the dynamics of rotation differ widely from those of the classical mechanics.

The dynamics of a vibration of the atoms relative to one another may be treated in the same way; if ν is the frequency of a vibration, its kinetic energy must be of the form $\frac{1}{2}\tau h\nu$, where τ is an integer. The interpretation of band spectra is based on the supposition that when the total energy of a molecule changes by an amount ΔW, radiation is emitted of a frequency ν defined by $\Delta W = h\nu$, a supposition which clearly includes that of Bohr with regard to atomic radiation as a special case.

A summary of the methods and results of the theory will be found in *Nature* of June 14, 1924. The quantum-theory promises to be no less successful here than in the simpler problem of the line spectra of the elements.

507. We shall not go into the further development of Bohr's theory in the present chapter. Enough evidence has been given to shew that the hypothesis of quanta shews a capacity for interpreting, on broad lines at least, the phenomena of both line spectra and band spectra which have defied interpretation in terms of the classical mechanics. We shall conclude the present chapter by producing similar evidence in favour of the quantum-hypothesis from the phenomena of specific heats. The subsequent chapter will discuss the theoretical basis of quantum-dynamics in some detail, and we shall then be able to discuss the further theory of line spectra in a more systematic way than would be possible at the present stage.

Specific Heats.

508. From the brief sketch of Bohr's theory already given it will have become clear that this theory postulates the possibility of the internal energy of an atom being entirely independent of the temperature of the gas to which the atom belongs. Such independence, it need hardly be remarked, is entirely at variance with the principles of the classical system of mechanics. Quite apart from the theory of quanta, however, there is an overwhelming mass of evidence that such independence actually exists, perhaps the most convincing argument being that provided by the phenomenon of radioactivity. It is known beyond question that the radioactive process consists of a series of atomic disintegrations, so that the process would necessarily be influenced by changes in the internal energy of the atom, should any such occur. Thus if a rise of temperature produced a change in the internal energy of the atom this change would almost certainly shew itself in a change in the rate of the radioactive processes. Yet these processes are found to go on at a steady rate which is perfectly uninfluenced by changes of temperature. Just as many atomic explosions are found to occur at 5° abs. as at 500° C., from which we may infer that the internal atomic energy is not altered by the transition from one of these temperatures to the other.

It follows that the internal energy of the atom may not enter into the specific heats at all, so that, so far as evaluations of the specific heat are concerned, the atoms may be thought of as rigid bodies.

The Specific Heats of a Gas.

509. It will be found that many of the difficulties which have been encountered in connection with the specific heats of gases yield to the new conceptions which the quantum-theory has placed at our disposal. Let us consider the different kinds of gas in turn.

510. *Monatomic Gases.* The molecule of a monatomic gas is identical with the atom, and consists of a single massive positive nucleus surrounded by a number n of negative electrons. For such a structure, the value of Mk^2 is of the order of magnitude nma^2, where a is the mean distance from the nucleus to the electrons. The value of m is 9×10^{-28}; a is of the order of 10^{-8}, so that Mk^2 is of the order of $n \times 10^{-43}$. For the slowest rotation possible, other than $\omega = 0$, equation (964) shews that $Mk^2 \omega$ must be equal to $h/4\pi^2$ or $1{\cdot}64 \times 10^{-28}$. Thus ω must be of the order of $10^{15}/n$.

The numerical calculation of § 494 now makes it clear that a rotation which possesses only one quantum of energy, will still have energy many times that of the energy of translation of a molecule of the gas at all ordinary temperatures. It follows, as in § 494, that only an infinitesimal fraction of the whole system of molecules will have any rotation at all. Or, if we regard the motion of the electron in its orbit as constituting a rotation, then very few of the molecules will have any rotation beyond the one quantum which is necessary to keep the molecule in existence. Whichever way we choose to regard the matter, we are led to the same result, namely, that the energy of rotation will be very approximately independent of the temperature at all ordinary temperatures.

Thus the energy of rotation will not enter into the specific heats. We have already seen that the same is likely to be true of the internal energy, whence it appears that the specific heats of monatomic gases must be the same as though the molecules were hard structureless points.

This result is in full accord with experiment (cf. table on p. 190). It removes the difficulties which were found in § 250 to surround a discussion of specific heats in terms of the classical mechanics.

511. *Diatomic Gases.* The diatomic molecule consists of two positive nuclei surrounded by electrons in motion. The line joining the two nuclei may be called the nuclear axis; it forms an approximate axis of symmetry for the molecule.

The calculation of the last section is applicable to diatomic molecules as regards rotation round the nuclear axis only. It appears that these rotations cannot enter into the specific heats.

For rotations about axes perpendicular to the nuclear axis, the calculation assumes a different form. The value of Mk^2 may be replaced, very approximately, by $M_1a_1^2 + M_2a_2^2$, where M_1, M_2 are the masses of the nuclei and a_1, a_2 their distances from the centre of gravity of the molecule. For a molecule consisting of n electrons in addition to the two nuclei, the molecular weight will be approximately $2n$, so that $M_1 + M_2$ will be approximately equal to n times the mass of the hydrogen molecule, say $3n \times 10^{-24}$ grammes. If we suppose a_1, a_2 each of the order of 10^{-8} cms., Mk^2 will be of the order of $3n \times 10^{-40}$, and the slowest rotation permitted by the quantum-theory will now, from equation (964), have a frequency of the order of $10^{12}/2n$. The kinetic energy of this rotation, being equal to $\frac{1}{2}h$ times its frequency, will be of the order of $2 \times 10^{-15}/n$ ergs.

For all diatomic gases this is small compared with the energy of translation of the molecule at 0° C. Thus at 0° C. and all higher temperatures the falling of the rotation into quanta will not greatly affect the total energy of rotation, and the specific heats will be about equal to those predicted by ordinary mechanics on the supposition that rotation about the axis of symmetry may be neglected.

The energy of rotation corresponding to a single quantum, namely $2 \times 10^{-15}/n$ ergs, is, however, comparable with the energy of translation of the molecule at very low temperatures. For oxygen ($n = 15$) $2 \times 10^{-15}/n$ is equal to about $1{\cdot}3 \times 10^{-17}$ while the energy of one degree of freedom of translation at 1° abs. is about $0{\cdot}7 \times 10^{-17}$. Thus at the lowest temperatures only a few of the molecules will have even one quantum of energy of rotation, and the specific heats will approximate to those of a monatomic gas.

As has already been mentioned in § 252, this is precisely what is found experimentally to occur in the case of the simpler diatomic gases hydrogen, nitrogen, oxygen, carbon-monoxide and air. For these gases the value of γ, the ratio of the specific heats, is equal to $1\frac{2}{5}$ at ordinary temperatures and appears to approximate towards $1\frac{2}{3}$ at the lowest temperatures.

At the same time the specific heats of a number of diatomic gases* (*e.g.* Cl_2, B_2, I_2, BI, Cl I) seem to indicate more internal energy than can be accounted for by two degrees of freedom of rotation. This additional internal energy probably arises from vibratory motions of the atoms inside the molecule.

* Cf. § 253.

512. *Polyatomic Gases.* These differ from diatomic gases in that the molecules will in general have no nuclear axis. The rotational energy at ordinary temperatures will be that corresponding to three degrees of freedom; to this it may be necessary to add internal vibratory energy of unknown amount.

Specific Heats of Solids.

513. There are good reasons for supposing that a solid body must in general be regarded as a collection of atoms rather than of molecules; an individual atom in a solid cannot be said to be indefinitely associated with any other definite atom or atoms to form a molecule. To take a concrete instance, it has been shewn by W. H. Bragg and W. L. Bragg* that in a crystal of common salt each sodium atom is exactly equidistant from six atoms of chlorine, and is not associated with any one of these six more closely than with another.

Consider now a solid mass made up of exactly similar atoms, each of mass m. Let N atoms form a gramme of the solid, so that $Nm = 1$. If these atoms are regarded as points, the solid will have $3N$ degrees of freedom per gramme. Every possible displacement of the solid will alter the potential energy, so that each degree of freedom will have potential as well as kinetic energy associated with it, and the energy of one gramme of the solid will consist of $6N$ squared terms.

According to the classical mechanics, the energy of these terms will be $3NRT$, and we obtain as the specific heat of the substance at constant volume,

$$C_v = \frac{1}{J}\frac{d}{dT}(3NRT) = \frac{3NR}{J}.$$

Hence, since $Nm = 1$,

$$mC_v = \frac{3R}{J} \quad \text{..............................(965)},$$

which is constant. Let m_h denote the mass of the hydrogen atom, then m/m_h is the atomic weight of the element under consideration referred to the hydrogen atom as unity, and

$$\frac{m}{m_h} C_v = \frac{3}{J}\frac{R}{m_h} = 5{\cdot}92$$

on inserting the value $R/m_h = 8{\cdot}254 \times 10^7$ from the table on p. 119. If a is the atomic weight measured from the standard $O = 16$, the relation becomes

$$aC_v = 5{\cdot}96.$$

This expresses the law of Dulong and Petit, that *the product of the atomic weight of an element and its specific heat is constant.*

* *Roy. Soc. Proc.* LXXXVIII. A (1913), p. 428; see also W. L. Bragg, *Roy. Soc. Proc.* LXXXIX. A (1913), p. 468.

514. In a solid just as in a gas a distinction must be drawn between the specific heat at constant volume and that at constant pressure, the difference between the two depending on the amount of work required to compress the heated solid back to the volume it originally occupied. When determinations of specific heat are corrected so as to refer to constant volume, the product aC_v is found to be very nearly constant at high temperatures, and to have almost exactly the value 5·96 predicted by theory*.

Although the atomic heat aC_v is found to be consistently equal to 5·96 at high temperatures, a very remarkable falling off has been discovered at low temperatures, so that aC_v must be thought of as a function of the temperature. This is exactly what we should expect on the quantum-theory. The following investigation of the value of the atomic heat as a function of the temperature has been given by Debye†, and we shall find that the results obtained agree very closely with experiment.

Debye's Theory of Specific Heat.

515. The unit mass of solid containing N atoms must, as we have seen, possess $3N$ degrees of freedom, and therefore $3N$ independent vibrations. These vibrations may be regarded as different wave-motions in the solid, and so may be classified according to frequency by the method already used in the last chapter. Writing ν for the frequency ($\nu = p/2\pi$), the number of vibrations in the volume V occupied by unit mass, having frequencies within a range $d\nu$, will be (cf. § 463)

$$\frac{p^2 dp}{\pi^2}\left(\frac{1}{2a_1^3}+\frac{1}{a_2^3}\right)V = 8\pi V\left(\frac{1}{2a_1^3}+\frac{1}{a_2^3}\right)\nu^2 d\nu \qquad (966).$$

Clearly if we supposed that ν could have all values from 0 to ∞, the total number of vibrations would be infinite, whereas it is known to be $3N$. Just as in § 469, there is a limit set to ν by the coarse-grained structure of the medium, and, following Debye, we may make a simplifying assumption similar to that previously made. We may assume that formula (966) gives the number of vibrations accurately from $\nu = 0$ up to a limit ν_m, which is determined by the condition that the total number of vibrations within this range shall be equal to the required total number $3N$; there are supposed to be no vibrations of frequency greater than ν_m. The equation giving ν_m is accordingly

$$\tfrac{8}{3}\pi V\left(\frac{1}{2a_1^3}+\frac{1}{a_2^3}\right)\nu_m^3 = 3N \qquad (967).$$

Here a_1, a_2 are the velocities of compressional and distortional waves in the solid, and so are known in terms of the elastic constants of the substance, while $V = 1/\rho$ and $N = 1/m$.

* See Nernst and Lindemann, *Zeits. für Elektrochemie*, 1911, p. 817, also Nernst, *Ann. d. Phys.* xxxvi. (1911), p. 395, and a report in *La Théorie du Rayonnement et les Quanta*, p. 254.

† *Ann. d. Phys.* xxxix. (1912), p. 789.

According to the quantum-theory, the average energy of each of the vibrations of frequency ν which are enumerated in expression (967) must not be supposed to be RT, but

$$\frac{h\nu}{e^{\frac{h\nu}{RT}} - 1}.$$

The total energy E of the $3N$ vibrations of a gramme of the substance is accordingly

$$E = \int_0^{\nu_m} 8\pi V \left(\frac{1}{2a_1^3} + \frac{1}{a_2^3}\right) \frac{h\nu}{e^{\frac{h\nu}{RT}} - 1} \nu^2 d\nu \quad \text{...........(968)},$$

or, using relation (967),

$$E = 9N \int_0^{\nu_m} \frac{h\nu^3}{e^{\frac{h\nu}{RT}} - 1} \frac{d\nu}{\nu_m^3} \quad \text{.....................(969)}.$$

Unfortunately, it is not possible to evaluate this integral in finite terms except in special cases.

516. At high temperatures T is large, so that the index $h\nu/RT$ of the exponential in the denominator becomes a small quantity. The denominator may accordingly be replaced by $h\nu/RT$, and the integral now admits of direct integration yielding the value

$$E \doteq 3NRT \quad \text{...............................(970)},$$

which is identical with the value given by the classical mechanics, as of course it must necessarily be.

517. At low temperatures T is small, so that the index $h\nu/RT$ is large. The denominator in the integral of equation (969) now becomes a very large quantity except when ν is small. Thus the whole value of E is contributed by small values of ν and the value is very approximately the same as it would be if the integral were extended over the whole range from $\nu = 0$ to $\nu = \infty$.

This complete integral is easily evaluated. We have

$$\int_0^\infty \frac{x^3}{e^x - 1} dx = \int_0^\infty x^3 (e^{-x} + e^{-2x} + e^{-3x} + \ldots) \, dx.$$

The integral on the right can be evaluated term by term. Its value is

$$6\left(1 + \frac{1}{2^4} + \frac{1}{3^4} + \frac{1}{4^4} + \ldots\right) = \frac{\pi^4}{15} = 6{\cdot}495.$$

Thus when T is small, equation (969) assumes the form

$$E = 9N \frac{(RT)^4}{(h\nu_m)^3} \times 6{\cdot}495 \quad \text{.....................(971)}.$$

The specific heat at any temperature T is

$$C_v = \frac{1}{J} \frac{dE}{dT}.$$

When T is large, the specific heat assumes the constant limiting value, from equation (970),

$$C_v = \frac{3NR}{J} \equiv C_\infty \text{ say} \qquad \text{.......................(972).}$$

When T is small, the value of C_v is, from equation (971),

$$C_v = 36NR\left(\frac{RT}{h\nu_m}\right)^3 \times 6{\cdot}495 \qquad \text{.....................(973).}$$

Let us write Θ for $h\nu_m/R$, so that Θ is a characteristic temperature associated with each substance. From equation (967), Θ is given by

$$\Theta^3 = \frac{9Nh^3}{8\pi R^3 V\left(\dfrac{1}{2a_1^3} + \dfrac{1}{a_2^3}\right)} \qquad \text{.....................(974),}$$

so that it is possible to calculate Θ from the elastic constants of the substance.

Equation (973) now assumes the form

$$C_v = 77{\cdot}94\left(\frac{T}{\Theta}\right)^3 C_\infty \qquad \text{.......................(975).}$$

Thus at low temperatures C_v varies as T^3.

518. Still writing Θ for $h\nu_m/R$, let us put

$$\frac{h\nu}{RT} = \frac{\Theta}{T}\frac{\nu}{\nu_m} = x.$$

The general equation (969) can now be written in the form

$$E = 9N\int\left(\frac{RT}{h}\right)^4 \frac{h}{\nu_m^3}\frac{x^3 dx}{e^x - 1},$$

where the limits of integration are from $\nu = 0$ to $\nu = \nu_m$ and therefore from $x = 0$ to $x = \Theta/T$. The equation may further be re-written in the form

$$E = 9NRT\left(\frac{T}{\Theta}\right)^3 \int_{x=0}^{x=\Theta/T} \frac{x^3 dx}{e^x - 1},$$

and it is now clear that E is of the form

$$E = 9NRT \times (\text{a function of } \Theta/T).$$

Again, since

$$C_v = \frac{1}{J}\frac{dE}{dT},$$

it is clear that C_v will be of the form

$$C_v = \frac{9NR}{J} \times (\text{a function of } \Theta/T),$$

or, say,

$$C_v = \frac{3NR}{J} f\left(\frac{\Theta}{T}\right) = C_\infty f\left(\frac{\Theta}{T}\right).$$

Thus the ratio C_v/C_∞ is a function of Θ/T only. The atomic heat at any temperature T is now given by

$$aC_v = 5{\cdot}96 f\left(\frac{\Theta}{T}\right) \qquad \text{.........................(976).}$$

Details of the evaluation and computation of the function f will be found in Debye's paper*. The following table gives the values found for the function $f\left(\frac{\Theta}{T}\right)$:

$\frac{T}{\Theta}$	$f\left(\frac{\Theta}{T}\right)$	$\frac{T}{\Theta}$	$f\left(\frac{\Theta}{T}\right)$	$\frac{T}{\Theta}$	$f\left(\frac{\Theta}{T}\right)$
∞	1·000	·8	·926	·20	·369
4	·997	·7	·904	·15	·213
3	·994	·6	·872	·10	·0758
2	·988	·5	·825	·075	·0328
1·5	·978	·4	·745	·050	·00974
1·0	·952	·3	·607	·025	·00122
·9	·941	·25	·503	·000	·000

Comparison with Experiment.

519. A large amount of observational material is available for the comparison of Debye's theory with experiment. In fig. 25 the curve gives the theoretical value of $f\left(\frac{\Theta}{T}\right)$, while the marks +, o and × shew values of this function derived from the observed specific heats of aluminium, copper and silver respectively. It will be seen that the agreement between theory and observation is remarkably good.

In these comparisons between theory and experiment, Θ has been supposed to be an adjustable constant, and that value has been assigned to Θ which makes the observations fit the curve most closely. We may refer to this value of Θ as the observed value of Θ. The value of Θ can however be calculated directly from the elastic constants by the use of equation (974), so that Θ is in no sense an adjustable constant. The following table gives the observed values of Θ as used by Debye and also the values calculated from the elastic constants:

Element	Θ (observed)	Θ (calculated)
Aluminium.........	396	399
Copper..............	309	329
Silver	215	212
Lead.................	95	72

* *l.c.* p. 812.

The agreement between the observed values of Θ and those calculated from the elastic constants is seen to be considerably less good than the agreement exhibited in fig. 25. This, however, is not altogether surprising. The observed values of Θ and those calculated from the elastic constants will only agree if we lay great stress on just that part of Debye's theory which is

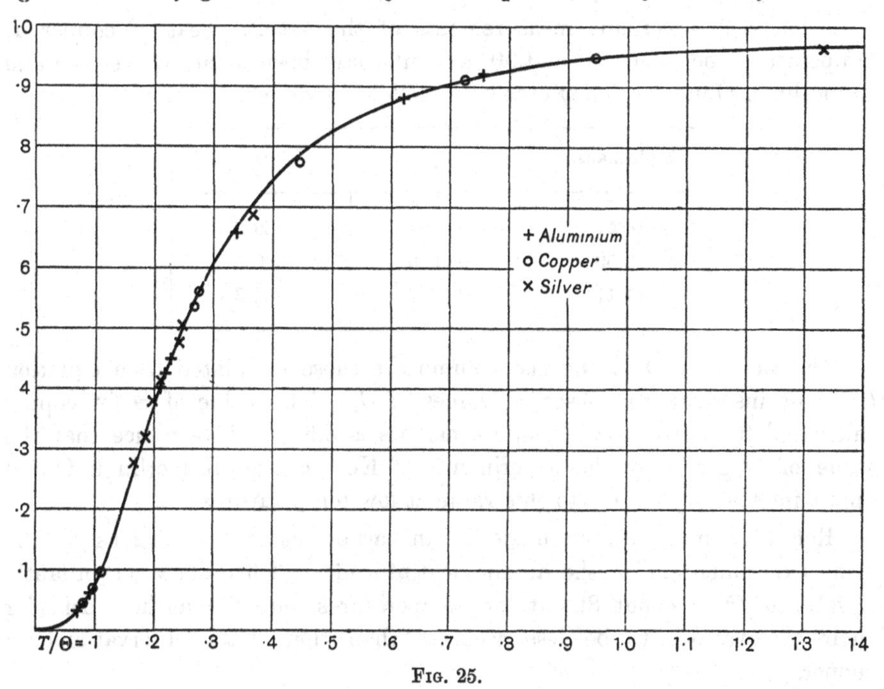

FIG. 25.

obviously weakest, namely the supposition that the vibrations of the solid fall off perfectly abruptly at a sharply defined frequency ν_m. As soon as this supposition is replaced by something more general the two values of Θ acquire different meanings.

This consideration does not, however, apply to measurements at low temperatures. Here, as we have already had occasion to notice, practically the whole heat-energy of the solid resides in vibrations of low frequency and the supposed existence of a maximum frequency ν_m becomes a matter of indifference. In accordance with what has already been said in § 516, the general integral (968) for the energy of heat-vibrations may be replaced by a similar integral in which the integration extends over all possible frequencies from $\nu = 0$ to $\nu = \infty$. Thus at low temperatures we have

$$E = \int_0^\infty 8\pi V\left(\frac{1}{2a_1^3} + \frac{1}{a_2^3}\right)\frac{h\nu}{e^{\frac{h\nu}{RT}} - 1}\nu^2 d\nu.$$

The integration can be effected as in § 517, and we finally obtain our previous equation (975) without making any use whatever of the supposed maximum frequency ν_m, Θ being defined by equation (974), and so being what we have referred to as the "calculated" value of Θ.

It follows that observations of specific heats at very low temperatures provide a test of those parts of Debye's theory which do not depend upon the assumption of a sharply defined limiting frequency.

Some very accordant measurements of the specific heat of copper at temperatures between 14° and 90° absolute have been made by Keesom and Kamerlingh Onnes*. They obtain the following values:

T (deg. abs.)	C_v	Θ
14·51	0·0396	329·6
15·59	0·0506	326·3
17·17	0·0687	324·6

The values of Θ in the last column are those calculated from equation (975) by inserting the observed values of C_v. The value of Θ for copper calculated from its known elastic constants is 329°, and we notice that the value of Θ given by the experiments of Keesom and Kamerlingh Onnes approximates very closely to this value at low temperatures.

Reference may also be made to an earlier set of observations by the same experimenters on the atomic heat of lead†. These shew concordantly a value of Θ of about 86° at low temperatures, but the authors consider their observations to be less accurate than their later observations on copper.

Extensions of Debye's Theory.

520. Attempts have been made to introduce greater precision into Debye's theory and also to extend it to substances other than chemical elements.

A solid mass of a compound substance cannot be treated as a collection of similar atoms, but may be thought of as a collection of similar molecules, each having internal vibrations of the kind already considered in § 510. On summing the contributions from the internal vibrations of the molecules and from the motions of the molecules in the elastic solid waves, we obtain for the total energy of N molecules of the solid an expression of the form

$$E = 9N \int_0^{\nu_m} \frac{h\nu^3}{e^{\frac{h\nu}{RT}} - 1} \frac{d\nu}{\nu_m{}^3} + N\Sigma \frac{h\nu_1}{e^{\frac{h\nu_1}{RT}} - 1} \quad \text{..............(977)}.$$

* *Verslag Amsterdam Akad.* June 26, 1915, p. 335, or *Commun. from the Phys. Lab. of Leiden*, No. 147 *a*.

† *Verslag Amsterdam Akad.* Oct. 31, 1914, p. 792, or *Commun. from the Phys. Lab. of Leiden*, No. 143.

Here the summation in the second term is over all the internal vibrations of the molecule. For a diatomic substance, this sum reduces to a single term. Such a term is commonly referred to as an Einstein-term, because Einstein at one time put forward a theory that the whole energy of a solid could be represented by terms of this form.

Nernst has shewn that an expression of the form (977) leads to approximately the right values of the specific heats of a number of chemical compounds. The formula contains no adjustable constants, for ν_m can be calculated from the elastic constants, and ν_1 is taken by Nernst to be the frequency of the infra-red absorption band as observed by Rubens. The following table, selected from a number given by Nernst*, will indicate the closeness of the agreement of this theory with experiment:.

VALUES OF $2C_p$ FOR KCl.

T (abs.)	Einstein term in $2C_v$	Debye term in $2C_v$	Correction term $2(C_p - C_v)$	Calculated $2C_p$	Observed $2C_p$
22·8	0·046	1·04	...	1·086	1·16
26·9	0·13	1·48	...	1·61	1·52
30·1	0·25	1·87	...	2·12	1·96
33·7	0·43	2·25	...	2·68	2·50
48·3	1·43	3·52	...	4·95	5·70
57·6	2·13	4·06	0·02	6·21	6·12
70·0	2·89	4·57	0·04	7·50	7·58
86·0	3·66	4·97	0·06	8·79	8·72
235	5·55	5·81	0·32	11·68	11·78
416	5·83	5·91	0·68	12·42	12·72
550	5·87	5·93	0·90	12·70	13·18

The agreement must not be expected to be perfect, for the theory is not perfect. The whole assembly of atoms in the solid interact with one another and we are not entitled to assume that the principal vibrations of the whole assembly fall into the two sharply divided classes of intra-molecular and inter-molecular motions. In confirmation of this remark it may be noticed that, when the various kinds of atom in the molecule are made to become all similar, formula (977) does not become identical with Debye's expression for the energy of an elementary substance.

521. This difficulty can be avoided by regarding the solid as a "space-lattice" of the type made familiar by the work of W. H. Bragg and W. L. Bragg on crystal structure. Attempts to improve the theory

* *Vorträge über die Kinetische Theorie der Materie* (1914), p. 81.

in this way have been made by Born and Karman* and by Thirring†. The formulae for the specific heat obtained in this way shew somewhat better agreement with observation than those of Debye, but the agreement is still by no means perfect‡.

The most marked success of these extensions of the theory occurs in the case of certain salts of comparative simple chemical structure. Thirring§ has given detailed calculations for the specific heats of NaCl, KCl, CaF_2 and FeS_2; he finds a mean error of 2·3 per cent., and a maximum error of 4 per cent. between his theory and observation.

522. We have now seen that the heat energy of a solid substance may be thought of as residing in its elastic vibrations, the atoms being treated as the ultimate particles of the solid, and each vibration having exactly the energy allotted to it by the quantum-theory. Thus the mean energy of any material vibration is exactly equal to that of a vibration of the same frequency in the ether. This was found in § 475 to be the condition, under the old mechanics, of equilibrium between matter and ether; it was one of the suppositions on which Planck's original theory was founded; the evidence of the specific heats of solids now shews it to be true in actual fact so far as the vibrations of solids are concerned.

523. The question suggests itself whether the mean energy of every material vibration must be put equal to that of the corresponding light-vibration. For instance, in § 468 we found that the random motions of molecules in a gas could be analysed into waves of sound in the gas, the frequencies forming a spectrum of a type similar to that formed by the elastic vibrations in a solid. The question arises whether these vibrations ought not to have allotted to them the mean energy predicted by the quantum-theory rather than the mean energy RT which corresponds to Maxwell's distribution of velocities. This suggestion has been put forward by Tetrode‖, and some of its consequences have been examined by Keesom¶, including especially its effect on the equation of state. If this point of view is accepted, Maxwell's law of distribution of velocities and the fundamental bases of the kinetic theory lose all claim to general validity: they must be relegated to the position of approximations which are true only in certain limiting conditions. But in a gas at ordinary temperature and pressure these limiting conditions are very nearly satisfied, at least so far as

* *Phys. Zeits.* XIII. (1912), p. 1, and XIV. (1913), pp. 15 and 65.

† *Phys. Zeits.* XIV. (1913), p. 867, and XV. (1914), pp. 127 and 180.

‡ See in particular Keesom and Kamerlingh Onnes, *Commun. from the Phys. Lab. of Leiden*, No. 143, p. 23.

§ *Phys. Zeits.* XV. (1914), p. 180.

‖ *Phys. Zeits.* XIV. (1913), p. 212.

¶ *Commun. from the Phys. Lab. of Leiden*, Suppl. No. 30 (1914), and *Phys. Zeits.* XIV. (1913), p. 665.

concerns the energy of molecular translation: the frequencies of the sound vibrations are so low that their energy has almost its equipartition value, Maxwell's law of distribution of velocities is very nearly true, and formulae based on Maxwell's law of distribution will be very nearly accurate. For instance, Keesom finds that for helium at 0° C. and at a pressure of 1 atmosphere, there would be a deviation of only 0·12 per cent. from the pressure calculated from Maxwell's law, while the error in the specific heats is shewn to be still less.

A somewhat different theory, based on the same fundamental conceptions, has been put forward by Lenz*, but weighty criticisms have been directed by Lorentz† against both this and the theory of Tetrode and Keesom just mentioned. Quite recently Einstein‡ has formulated a treatment of the question which seems to escape criticisms of this type.

524. Quite apart from the application of these ideas to a gas, it seems possible that they may be applicable to the quasi-gas formed by the free electrons in a solid§. In this case, owing to the small mass of the electron, the frequencies of the vibrations of the medium constituted by the electrons are very high, so that the old laws of partition of energy may not give anything even approaching a good approximation to the truth: a better approximation may even be obtained by regarding the wave-frequency as infinite, and so disregarding the energy of the free electrons altogether. To this approximation, we may have as many free electrons as we please in the solid without adding anything to the specific heat. It is quite in accord with observation that the energy of free electrons should add little or nothing to the specific heat. Nernst and Lindemann have found that the limiting value of the atomic heat has the theoretical value 5·96 for good and bad conductors equally, while Richter‖ has specially looked for the influence of free electrons on the specific heat in a series of Bi – Sn and Bi – Pb alloys, and has failed to find any. Clearly this conception can do a great deal towards removing the difficulties which have accumulated round the electron theory of metals (cf. § 408). Whether the theory can be completely re-established remains to be seen.

Let us consider from this point of view the problem of the electric and thermal conductivities, already discussed in § 404. For this the following method of procedure has been suggested by Königsberger¶, Hertzfeld**, and others.

* *Wolfskehl Kongress Vorträge*, p. 125.

† *Proc. Amsterdam Akad.* XIX. (1917), p. 737.

‡ *Preuss. Akad. Wiss. Berlin*, XXII. (1924), p. 261.

§ See papers by Keesom, *Communications from the Physical Laboratory of Leiden*, Suppl. No. 30 (1914), and *Phys. Zeits.* XIV. (1913), p. 670; also F. A. Lindemann, *Phil. Mag.* XXIX. (1915), p. 127.

‖ *Ann. d. Phys.* XXXIX. (1912), p. 1590.

¶ *Verh. d. Deutsch. Phys. Gesell.* XIII. (1911), p. 934.

** *Ann. d. Phys.* XLI. (1913), p. 27.

In § 404 we used the approximate formula

$$\vartheta = \frac{1}{3}\frac{\nu \bar{c} l}{J}\frac{d\bar{E}}{dT}$$

in which ϑ is the coefficient of conduction of heat and ν, $\bar{c}$, l and $\bar{E}$ refer to the heat-carriers which are believed to be the free electrons. In § 405 we obtained for the electric conductivity σ the corresponding formula (equation (819))

$$\sigma = \frac{\nu \bar{c} l e^2}{4\bar{E}}.$$

The conductivities ϑ and σ separately depend not only on E, but also on the unknown values of ν, $\bar{c}$ and l. The ratio of the conductivities, however, depends only on $\bar{E}$, being given by

$$\frac{\vartheta}{\sigma} = \frac{4}{3e^2 J}\bar{E}\frac{d\bar{E}}{dT} \quad \text{.........................(978).}$$

If the motion of the electrons is determined by the classical mechanics, $\bar{E}$ will be equal to $\frac{3}{2}RT$, and equation (978) reduces simply to the equation

$$\frac{\vartheta}{\sigma} = 3\frac{R^2 T}{e^2 J} \quad \text{..............................(979)}$$

already considered in § 406.

If, however, the energy of the electrons is supposed to be governed by the quantum-mechanics, then the value of $\bar{E}$ will be less at low temperatures than $\frac{3}{2}RT$, and ϑ/σ will have a value below that given by equation (979), as is found experimentally to be the case.

Hertzfeld assumes that $\bar{E}$ is given by an equation of the form

$$\bar{E} = \frac{1}{2}\frac{h\nu_0}{e^{\frac{h\nu_0}{RT}} - 1},$$

but this is open to the objection that it reduces at high temperatures to $\frac{1}{2}RT$ instead of to $\frac{3}{2}RT$. It seems more reasonable to assume for $\bar{E}$ the value

$$\bar{E} = \frac{3}{2}\frac{h\nu_0}{e^{\frac{h\nu_0}{RT}} - 1} \quad \text{..........................(980),}$$

this being the value which the quantum-theory would predict if each electron oscillated, in each direction in space, with a simple harmonic vibration of frequency ν_0. Equation (978) now assumes the form

$$\frac{\vartheta}{\sigma} = \frac{3R^2 T}{e^2 J}\frac{\left(\frac{h\nu_0}{RT}\right)^3 e^{\frac{h\nu_0}{RT}}}{\left(e^{\frac{h\nu_0}{RT}} - 1\right)^3} \quad \text{.......................(981).}$$

This reduces, as it ought, to equation (979) when T is large. At lower temperatures the last fraction on the right gives the diminution in $\vartheta/\sigma T$ from the value predicted by the classical mechanics.

In the following table, the first column gives the temperatures at which ϑ/σ has been determined for copper by Jäger and Diesselhorst and by Lees. The second column gives the observed values of $10^{-8}\times\vartheta/\sigma T$ already quoted in § 407, an addition of 0·03 being made, as by Hertzfeld, to the values observed by Jäger and Diesselhorst in order that the two sets of observations shall agree at 18° C. Finally the third column gives the values of $10^{-8}\times\vartheta/\sigma T$ calculated by Hertzfeld from equation (981). The value of ν_0 is such that $h\nu_0/R = 57{\cdot}77$, this value being selected so as to give the closest fit to the observed and calculated values.

T	$\frac{\vartheta}{\sigma T}\times 10^{-8}$ (observed)	$\frac{\vartheta}{\sigma T}\times 10^{-8}$ (calculated)
100° C.	2·35	2·363
18° C.	2·32	2·313
0° C.	2·30	2·296
− 50° C.	2·26	2·236
− 170° C.	1·85	1·865

The agreement is perhaps as good as could reasonably be expected to follow from the assumption of a single definite frequency ν_0 for the supposed electronic wave-motions. If, however, ν_0 has the assigned value, it will be found that at 0° C., $h\nu_0/RT$ has a value of about 0·212. From equation (980) it appears that at 0° C., $\bar{E} = 1{\cdot}35RT$. In other words, at 0° C. the electrons have 90 per cent. of their equipartition energy, an amount of energy which appears to be too great to reconcile with the observed value of the specific heat of copper.

The problem has been considered further by Hertzfeld*, Wien†, and Hauer‡, but it is difficult to find any explanation, which carries conviction in a physical sense, of the very long free paths which, as we saw in § 408, are required to account for the low resistances of metals at temperatures near to the absolute zero.

* *l.c.* ante.

† *Berliner Sitzungsber.* VII. (1913), p. 184.

‡ *Ann. d. Phys.* LI. (1916), p. 189.

CHAPTER XVIII

QUANTUM DYNAMICS

THE FOUNDATION OF THE NEW DYNAMICS.

525. IN preceding chapters we saw how the classical system of mechanics failed to account for certain phenomena of physics, and it became clear that these could only be explained in terms of a new system of dynamics which must be supposed to supersede the old classical system when we have to deal with phenomena conditioned by the fine-scale structure of matter.

The first instance of the failure of the classical mechanics was provided by the phenomenon of black-body radiation. On the assumption that black-body radiation represents a state of thermodynamical equilibrium, we found that the observed distribution of radiant energy in the spectrum of an ideal black body could be explained on the supposition that the energy of each part of the radiation-producing mechanism fell into "quanta." Corresponding to a simple harmonic vibration of frequency ν, there was supposed to be an energy-quantum of amount $h\nu$, where h is Planck's universal constant of which the value has already been given. Calling this energy ϵ, it was found that the energy of vibration could be $0, \epsilon, 2\epsilon, 3\epsilon, \ldots$ or any integral multiple of ϵ, but could not contain fractional parts of ϵ. The energy, after retaining one of these values for a certain time, had to be supposed to jump abruptly to some other value, which in turn was retained until another jump occurred*.

Debye's theory of specific heats is founded upon exactly similar conceptions. The atomic motions in a solid can be resolved, as in Fourier's analysis, into regular trains of waves, and the supposition that the energy of waves of frequency ν could occur only in whole quanta of amount $h\nu$ was found to lead to formulae for the specific heats of solids which proved to be in complete agreement with experiment.

Bohr's theory of line-spectra had a similar basis. The electron in the hydrogen atom was assumed to be capable of describing only certain orbits in which the energy had certain specified amounts, and the abrupt transition from one such orbit to another provided the origin of emitted radiation or the receptacle for absorbed radiation. These simple assumptions were found to give a very convincing explanation of the line-spectrum of hydrogen, and it was found that the helium spectrum could be similarly explained.

* Attempts have been made, particularly by Planck himself, to put the theory in such a form that the energy-jumps can be avoided. But definite energy-jumps seem to be essential to Bohr's theory of line-spectra, and if once they are admitted here there seems no valid reason for trying to avoid them in radiation theory.

Einstein has shewn* that Planck's radiation formula can be deduced from the supposition that the molecules of a substance have only a certain number of possible configurations, and that the emission of radiation takes place, in the way which had previously been imagined by Bohr (cf. § 500), when a molecule jumps from one of these states to another.

Continuous Motion and Discontinuities.

526. From the instances just given, it will be clear that the motion of a dynamical system, as contemplated by the quantum-theory, consists of spells of continuous motion punctuated by abrupt changes or jumps. Whether these jumps are actually instantaneous or not is so far unknown.

In every problem in which the quantum-theory has been successful, it has been assumed that the continuous motion is in accordance, in a certain restricted sense, with classical dynamics; the abrupt changes, so far as is known, have no relation at all to the classical dynamics. It is not possible to state with any sharpness or clearness the exact limitations which have to be placed on the conformity of the continuous motion with the classical dynamics, and it is by no means out of the bounds of possibility that this supposed conformity may ultimately prove to have been illusory.

In Planck's original theory of radiation, the motion of the assumed "vibrators" was supposed to be governed by classical mechanics, although it has since become clear that this supposition did not form an integral or essential part of the theory. In Debye's theory of specific heats it is not in any way necessary to suppose that the motions of the atoms conform to the classical mechanics. In the theory of line-spectra put forward by Bohr, as also in the brilliantly successful extensions of this theory developed by Sommerfeld, Epstein and others, it is assumed that the electron orbits described in the continuous parts of the motion are such as could be described, in conformity with the classical mechanics, under the electrostatic forces from the nucleus, but it is found necessary to disregard completely the electromagnetic reactions arising from the accelerations of the electron. These latter forces are, however, demanded by the classical mechanics just as imperatively as are the electrostatic forces. The only essential difference between the two sets of forces is a numerical one; the electromagnetic forces are of very small amount in comparison with the electrostatic forces.

Perhaps the best general statement that can be made with our present knowledge is that the continuous motions may be assumed to conform to the classical mechanics to a first approximation. The exact nature of the approximation cannot be stated in general terms. In the special problem of line-spectra the approximation consists in neglecting altogether the electromagnetic forces resulting from the accelerations of electric charges.

* *Phys. Zeitschrift.* XVIII. (1917), p. 122.

Allowed and Disallowed Motions.

527. The classical mechanics permit of a multiply infinite series of continuous motions for every dynamical system. The quantum-mechanics do not permit of all these motions being performed. In addition to the equations of the classical mechanics, the quantum-theory imposes equations of its own, and the effect of these equations is to restrict the number of possible continuous motions. The equations which are peculiar to the quantum-theory may accordingly be regarded as "equations of restriction." Following a usual terminology, we may speak of those continuous motions which conform to the classical mechanics and also to the equations of restriction as "allowed" motions. Motions which conform to the classical mechanics but violate the equations of restriction imposed by the quantum-theory may be spoken of as "disallowed" motions.

Thus in the radiation theory of Planck, as also in the theory of specific heats of Debye, the equations of restriction are contained in the single equation

$$E = \tau h\nu \qquad \text{.............................(982)},$$

where E is the energy of any vibration of frequency ν, and τ must be integral. In Bohr's theory of the hydrogen atom, the equation of restriction is our former equation (948), namely

$$W = \tfrac{1}{2}\tau h\omega \qquad \text{.............................(983)},$$

where τ must be integral.

528. So far as can at present be seen, the problem of discovering the complete laws of quantum-dynamics will consist of two parts:—

(i) The determination of the general equations of restriction in such a form as shall be applicable to all dynamical systems.

(ii) The determination of the mechanism of the supposed jumps from one allowed motion to another.

Practically no progress has been made with the second part of the problem; we turn at once to a discussion of attempts to solve the first part.

We have already knowledge of the form of the equation of restriction in two special cases, namely those given by equations (982) and (983) above. Both these equations refer to purely periodic motions having a single definite period. It is natural to try first whether they can be combined into a single more general equation, and it is found that this can readily be done.

Assuming that the motion of a simple harmonic vibrator is governed by the classical mechanics, it is known that the kinetic energy T and the potential energy V contribute, on the average, equal amounts to the total energy E. Thus if $\bar{T}$, $\bar{V}$ denote time-averages of T and V, taken over one or several complete periods, equation (982) can be put in the form

$$2\bar{T} = 2\bar{V} = \tau h\nu.$$

In Bohr's equation (983) T and V do not vary with the time, so that $\bar{T} = T$ and $\bar{V} = V$. We have also the relation (§ 499) $T = W = -\frac{1}{2}V$, so that Bohr's equation can be put in the form

$$2\bar{T} = -\bar{V} = \tau h\nu,$$

where ν is the frequency of description of the electron orbit.

It is now clear that the two equations of restriction (982) and (983) can be regarded as special cases of a single equation

$$2\bar{T} = \tau h\nu \quad \text{..............................(984)},$$

where $\bar{T}$ denotes the mean value of the kinetic energy T taken over a complete period, and ν is the frequency of the vibration or oscillation.

This single equation contains all that is dynamically necessary, in addition to the classical mechanics, for the theories of Planck, Debye and Bohr explained in the last chapter. The question naturally suggests itself whether we have in equation (984) the true equation of restriction for all cases of singly-periodic motion. Before discussing this question, it will be convenient to express equation (984) in a somewhat different form.

529. Suppose that a complete vibration occurs in the interval from $t = 0$ to $t = \sigma$, so that σ is the period of the vibration and $\nu\sigma = 1$. The value of $\bar{T}$ is given by

$$\bar{T} = \frac{1}{\sigma}\int_0^\sigma T\,dt = \nu\int_0^\sigma T\,dt,$$

so that

$$\frac{2\bar{T}}{\nu} = \int_0^\sigma 2T\,dt.$$

The right-hand member of the equation is of course the "action" $\int 2T\,dt$ taken over a complete period of the system, and equation (984) now assumes the form

$$\int_0^\sigma 2T\,dt = \tau h \quad \text{............................(985)}.$$

From the fundamental equation $\epsilon = h\nu$ it was clear that the universal constant h was of dimensions (energy) × (time). Its dimensions are therefore those of "action," and equation (985) expresses that the action is atomic, in the sense that the action of a simple vibration measured over a complete oscillation can be only an integral multiple of h.

Equation (985) can be put in still another form. Suppose as in § 78 that the dynamical system under consideration is specified by n generalised coordinates $q_1, q_2, \ldots q_n$ and their corresponding velocities $\dot{q}_1, \dot{q}_2, \ldots \dot{q}_n$. Let E be the total energy of the system and let momenta $p_1, p_2, \ldots p_n$ be introduced, defined by

$$p_s = \frac{\partial E}{\partial \dot{q}_s}, \text{ etc.}$$

The value of the "action" taken over any time-interval from 0 to t is

$$\int_0^t 2T dt,$$

and since T is necessarily a quadratic function of the velocities $\dot{q}_1, \dot{q}_2, \dots \dot{q}_n$, we have

$$2T = \Sigma \frac{\partial T}{\partial \dot{q}_s} \dot{q}_s = \Sigma \frac{\partial E}{\partial \dot{q}_s} \dot{q}_s = \Sigma p_s \dot{q}_s.$$

Hence

$$\int_0^t 2T dt = \Sigma \int_0^t p_s \dot{q}_s dt = \Sigma \int_0^t p_s dq_s$$

and equation (985) is seen to assume the form

$$\Sigma \int_{t=0}^{t=\sigma} p_s dq_s = \tau h \quad \text{.........................(986).}$$

The theory has so far been considered only in its application to systems having a definite single period of oscillation σ, in which case the integration is over one complete period.

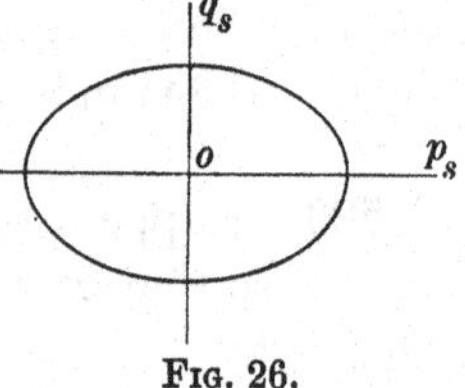

FIG. 26.

For such systems equation (986) admits of a simple interpretation in terms of the generalised space already introduced in § 80. The motion of the system, whatever it may be, can be represented by the motion of a single representative point in a space having

$$q_1, q_2, \dots q_n, \quad p_1, p_2, \dots p_n$$

as coordinates, and the projection of this point on the plane q_s, p_s will have q_s, p_s for its coordinates. If the motion has a single definite period σ all the coordinates will have returned to their initial values after an interval of time equal to σ*, so that the point q_s, p_s in the (q_s, p_s) plane will also have returned to its initial position. This point will have described a closed curve of which the area, according to the ordinary geometrical formula, will be

$$\int_{t=0}^{t=\sigma} p_s dq_s.$$

Thus equation (986) expresses that the sum of the areas described by all points such as p_s, q_s in a complete period of the system must be an integral multiple of h.

ADIABATIC INVARIANTS.

530. There is a simple and obvious condition to which all equations of restriction must conform.

Suppose that the system under consideration moves in continuous motion under the influence of an external field of force (or other external influence) which is capable of alteration. Let $a_1, a_2, \dots$ be parameters specifying this

* The case of azimuthal coordinates which increase by 2π in each period σ produces a slight complication in statement but none in fact.

field of force. Then $a_1, a_2, \ldots$ will enter into the total energy of the system, and therefore into the Hamiltonian equations of motion by which the continuous motion of the system is determined.

Of the various motions which are in accordance with the Hamiltonian equations, some will be "allowed" motions, and others not. Those which are allowed will be specified by the quantum equations of restriction, and since these motions will obviously depend on the values of $a_1, a_2, \ldots$, the quantum equations may also be expected to depend in general on $a_1, a_2, \ldots$. Thus a typical quantum equation may be supposed to be of the form

$$\Phi(a_1, a_2, \ldots) = \tau h \quad \text{.........................(987).}$$

Consider now a particular motion of a special system. Let it be specified by assigning particular integral values to τ in the various quantum equations, say

$$\left.\begin{array}{l}\Phi_1(a_1, a_2, \ldots) = \tau_1 h, \\ \Phi_2(a_1, a_2, \ldots) = \tau_2 h, \text{ etc.}\end{array}\right\} \quad \text{....................(988).}$$

While this continuous motion is in progress, let the values of $a_1, a_2, \ldots$ be gradually and slowly changed. The motion of the system required by the Hamiltonian equations will correspondingly change. Let $a_1, a_2, \ldots$ assume final values $a_1', a_2', \ldots$, then the final motion will depend upon $a_1', a_2', \ldots$ in the same way in which the initial motion depended upon $a_1, a_2, \ldots$. In particular $\Phi_1(a_1, a_2, \ldots)$ will have become changed into $\Phi_1(a_1', a_2', \ldots)$.

The motion which is finally taking place, since it is assumed actually to be in progress, must be supposed to be an "allowed" motion. It must therefore satisfy the quantum equations appropriate to the final values of $a_1, a_2, \ldots$ and these, in accordance with what has already been said, will be

$$\left.\begin{array}{l}\Phi_1(a_1', a_2', \ldots) = \tau_1 h, \\ \Phi_2(a_1', a_2', \ldots) = \tau_2 h, \text{ etc.}\end{array}\right\} \quad \text{....................(989).}$$

If we make the difference between $a_1, a_2, \ldots$ and $a_1', a_2', \ldots$ sufficiently small, we need not consider the possibility of any changes in the integers $\tau_1, \tau_2, \ldots$ so that we must have, by comparison of equations (988) and (989),

$$\Phi_1(a_1', a_2', \ldots) = \Phi_1(a_1, a_2, \ldots),$$
$$\Phi_2(a_1', a_2', \ldots) = \Phi_2(a_1, a_2, \ldots), \text{ etc.}$$

Since these equations must be true, no matter what final values are assigned to $a_1', a_2', \ldots$, provided only that these are adjacent to the values $a_1, a_2, \ldots$, it follows that $\Phi_1, \Phi_2, \ldots$ cannot vary with $a_1, a_2, \ldots$. In other words, the condition that equations (988) shall be a possible system of quantum equations is

$$\frac{\partial \Phi_1}{\partial a_1} = 0, \; \frac{\partial \Phi_1}{\partial a_2} = 0, \text{ etc., } \frac{\partial \Phi_2}{\partial a_1} = 0, \text{ etc.} \quad \text{..............(990),}$$

in which variations in Φ_1 are calculated on the supposition that the system moves continuously under the Hamiltonian equations while the change in Φ is taking place.

Following Einstein, Ehrenfest and Burgers, to whom the development of this set of ideas is due, we may say that a system is "adiabatically affected*" when the parameters $a_1, a_2, \ldots$ gradually have their values changed. Any quantity depending on the motion which does not change its value when the motion is "adiabatically affected" is spoken of as an "adiabatic invariant."

531. A simple illustration will perhaps elucidate the meaning of the criterion which has just been developed. For a number of mechanical systems it is possible to introduce or remove constraints in such a way that the period of the system can be altered without adding energy to, or removing energy from, the system. Let us test whether it is possible that Planck's original equation

$$E = \tau h\nu \qquad (991)$$

should provide the quantum equation of restriction for such systems. Assuming the adequacy of this equation let us suppose that the system is vibrating with total energy E equal to one quantum $h\nu$, and that while the motion is in progress a constraint is introduced so that the frequency ν is changed to some new value ν_1 greater than ν. The energy of the system is still E, but the frequency is ν_1 and we have, on putting $\tau = 1$ in equation (991),

$$E = \tau_1 h\nu_1 \qquad (992),$$

where $\tau_1 = \nu/\nu_1$, a fraction less than unity. Thus equation (991) with τ limited to integral values, cannot be the true quantum equation for such systems, since by a slight mechanical modification, we can pass to motions in which this equation is violated. We can express this concisely by saying that E/ν is not an adiabatic invariant for such systems.

532. Returning to the train of thought of § 529 we shall now prove that $2\overline{T}/\nu$ is an adiabatic invariant for all singly periodic mechanical systems which are liable to alteration through the changes of parameters $a_1, a_2, \ldots$ which enter into the specification of the potential energy, subject only to the condition that all the parameters change so slowly that the change of any one in the course of a single period of the system may be neglected. The proof which follows is substantially identical with the proof given by Ehrenfest† who in turn obtained the theorem by modification of an earlier theorem of Boltzmann‡. A somewhat different proof has been given by Bohr§.

* The terminology, which is not a particularly happy one, arose from analogy with the adiabatic transformation of radiation such as is considered in the proof of Wien's displacement law.

† *Verslag van de Gewone Vergaderingen der Wis- en Natuurkundige Afdeeling, Amsterdam Acad.* xxv. (1916), p. 412.

‡ Boltzmann, *Vorlesungen über Mechanik*, II. § 48, or *Wissenschaft. Abhand.* I. pp. 23, 229. See also Ehrenfest, *Verslag Amsterdam Acad.* xvi. (1913), p. 591.

§ "On the Quantum Theory of Line-Spectra," *Kgl. Danske Selsk. Skrifter, Naturvidensk. og Mathemat.* Afd. 8 (1918), IV. Part 1, p. 10.

We consider a system in which the specification of the potential energy as a function of the coordinates depends on slowly varying parameters $a_1, a_2, \ldots$.

Let L be the Lagrangian function of the system, equal to $T-W$, so that the equations of motion are

$$\frac{d}{dt}\left(\frac{\partial L}{\partial \dot{q}_s}\right)=\frac{\partial L}{\partial q_s} \quad\text{.............................(993).}$$

Let us compare the motion of the system before variation over a complete period extending from $t=t_1$ to $t=t_2$ with the motion after variation extending over a complete period from $t=t_1'$ to $t=t_2'$. As the parameters are supposed only to vary slowly, the motion after variation may be supposed to differ only to a small extent from that before variation. And, since the time will enter only through differentials, we may legitimately measure time in the varied motion from a different origin from that used for the original motion, and this may be chosen so that

$$t_1'=t_1+\Delta t_1, \quad t_2'=t_2+\Delta t_2,$$

where Δt_1, Δt_2 are both small quantities of the first order.

Let ϕ denote the value of any quantity at a time t in the first motion, and let the value of the same quantity at the same time t in the varied motion be denoted by $\phi+\delta\phi$, so that $\delta\phi$ is a small quantity of the first order.

The value of L integrated through a complete period of the motion before variation will be

$$I=\int_{t_1}^{t_2} L\,dt.$$

The corresponding value of the same quantity after variation may be supposed to be $I+\Delta I$, where ΔI is a small quantity, and where, by the ordinary rules of the calculus of variations,

$$\Delta I=\Delta\int_{t_1}^{t_2} L\,dt=\int_{t_1}^{t_2}\delta L\,dt+L_{t=t_2}\Delta t_2-L_{t=t_1}\Delta t_1 \quad\text{.........(994).}$$

Since L is a function of the parameters $a_1, a_2, \ldots$ as well as of the coordinates $q_s, \dot{q}_s, \ldots$, etc., we have

$$\delta L=\Sigma\frac{\partial L}{\partial q_s}\delta q_s+\Sigma\frac{\partial L}{\partial \dot{q}_s}\delta\dot{q}_s+\Sigma\frac{\partial L}{\partial a_r}\delta a_r.$$

On the right-hand side we can substitute the value of $\partial L/\partial q_s$ from equation (993) into the first term. In the second term we may replace $\delta\dot{q}_s$ by $\frac{d}{dt}(\delta q_s)$. The equation becomes

$$\delta L=\frac{d}{dt}\Sigma\left(\frac{\partial L}{\partial \dot{q}_s}\delta q_s\right)+\Sigma\frac{\partial L}{\partial a_r}\delta a_r.$$

On integrating from $t=t_1$ to $t=t_2$ this gives

$$\int_{t_1}^{t_2} \delta L\,dt = \left| \Sigma \frac{\partial L}{\partial \dot{q}_s} \delta q_s \right|_{t_1}^{t_2} + \Sigma \int_{t_1}^{t_2} \frac{\partial L}{\partial a_r} \delta a_r\,dt.$$

The variation in q_s at the lower limit $t=t_1$, say $(\Delta q_s)_1$, is given by

$$(\Delta q_s)_1 = (\delta q_s)_{t=t_1} + (\dot{q}_s)_{t=t_1} \Delta t_1,$$

so that the value of δq_s at $t=t_1$ will be given by

$$(\delta q_s)_{t=t_1} = (\Delta q_s)_1 - (\dot{q}_s)_{t=t_1} \Delta t_1,$$

and there is of course a similar relation for the time $t=t_2$. Using these relations, we find

$$\int_{t_1}^{t_2} \delta L\,dt = \left| \Sigma \frac{\partial L}{\partial \dot{q}_s} \Delta q_s \right|_{t_1}^{t_2} - \Sigma \left(\dot{q}_s \frac{\partial L}{\partial \dot{q}_s} \right)_{t=t_2} \Delta t_2 + \Sigma \left(\dot{q}_s \frac{\partial L}{\partial \dot{q}_s} \right)_{t=t_1} \Delta t_1 + \Sigma \int_{t_1}^{t_2} \frac{\partial L}{\partial a_r} \delta a_r\,dt.$$

The first term on the right-hand side vanishes, for, on account of the periodicity of the motion, both Δq_s and $\partial L/\partial \dot{q}_s$ have the same values at $t=t_2$ as at $t=t_1$. Substituting the remaining value for $\int_{t_1}^{t_2} \delta L\,dt$ into equation (994) we obtain

$$\Delta \int_{t_1}^{t_2} L\,dt = \Sigma \int_{t_1}^{t_2} \frac{\partial L}{\partial a_r} \delta a_r\,dt + \left(L - \Sigma \dot{q}_s \frac{\partial L}{\partial \dot{q}_s} \right)_{t=t_2} \Delta t_2 - \left(L - \Sigma \dot{q}_s \frac{\partial L}{\partial \dot{q}_s} \right)_{t=t_1} \Delta t_1 \quad \text{......(995).}$$

Let E denote the total energy $T+W$. We have, at any instant,

$$\Sigma \dot{q}_s \frac{\partial L}{\partial \dot{q}_s} = \Sigma \dot{q}_s \frac{\partial T}{\partial \dot{q}_s} = 2T,$$

so that

$$L - \Sigma \dot{q}_s \frac{\partial L}{\partial \dot{q}_s} = L - 2T = -E \quad \text{......(996).}$$

Further, since $a_1, a_2, \ldots$ enter only in W and not in T,

$$\frac{\partial L}{\partial a_r} = -\frac{\partial W}{\partial a_r} = -\frac{\partial E}{\partial a_r} \quad \text{......(997).}$$

Thus equation (995) reduces to

$$\Delta \int_{t_1}^{t_2} L\,dt = -\int_{t_1}^{t_2} \Sigma \frac{\partial E}{\partial a_r} \delta a_r - E_{t=t_2} \Delta t_2 + E_{t=t_1} \Delta t_1 \quad \text{......(998).}$$

We can evaluate $\Delta \int_{t_1}^{t_2} E\,dt$ by a much simpler calculation. The value of E, by the conservation of energy, remains constant throughout the description of a complete period of the motion, so that

$$\delta E = \Sigma \frac{\partial E}{\partial a_r} \delta a_r.$$

We now obtain, precisely as in equation (994),

$$\begin{aligned} \Delta \int_{t_1}^{t_2} E\,dt &= \int_{t_1}^{t_2} \delta E\,dt + E_{t=t_2} \Delta t_2 - E_{t=t_1} \Delta t_1 \\ &= \int_{t_1}^{t_2} \Sigma \frac{\partial E}{\partial a_r} \delta a_r + E_{t=t_2} \Delta t_2 - E_{t=t_1} \Delta t_1 \quad \text{......(999).} \end{aligned}$$

By comparison with equation (998), we see that

$$\Delta \int_{t_1}^{t_2} (L + E)\, dt = 0,$$

or

$$\Delta \int_{t_1}^{t_2} 2T dt = 0 \quad \text{......................(1000)},$$

which proves that $\int_{t_1}^{t_2} 2T dt$ or $2\overline{T}/\nu$ is an adiabatic invariant.

533. There are of course other adiabatic invariants besides $2\overline{T}/\nu$. The following examples are given by Ehrenfest*.

If a system possesses cyclic coordinates, then the corresponding momenta are adiabatic invariants.

If the rotation of a ring of electrons is affected by a changing magnetic field whose lines of force are perpendicular to the plane of rotation, then the sum of the moment of momentum and the electrokinetic momentum is an adiabatic invariant.

If a hydrogen atom of Bohr (§ 499) is affected by a changing electric field, then the moment of momentum about a line parallel to the lines of force is an adiabatic invariant.

If a particle of any kind is describing an orbit about a central field of force of varying intensity, the moment of momentum is an adiabatic invariant.

534. The theorem given in § 532 was applicable only to systems having a single definite period of vibration. To discover the proper extension of this theorem to systems of more general type, including non-periodic systems, forms one of the main problems of quantum-dynamics at the present time.

535. The energy of a certain class of dynamical systems can be put in the form

$$E = f_1(p_1, q_1) + f_2(p_2, q_2) + \dots + f_n(p_n, q_n) \quad \text{........(1001)},$$

and the motion of such systems, when calculated by the principle of least action, is found to be such that the changes in any one coordinate q_s are independent of the values of the other coordinates. For a certain number of the coordinates, say $q_1, q_2, \dots q_m$, the changes will be found to consist of regular oscillations, those of any one coordinate q_s being performed between fixed limits with a definite period σ_s. The kinetic energy falls into the sum of a number of contributions from the different coordinates and their velocities. If T_s is the contribution depending on q_s, $\dot{q}_s$, it is readily shewn that

$$\int_0^{\sigma_s} 2T_s dt$$

* *Verslag Amsterdam Acad.* xxv. (1916), p. 416, or *Phil. Mag.* xxxiii. (1917), p. 504.

is an adiabatic invariant, provided only that none of the periods $\sigma_1, \sigma_2, \ldots \sigma_m$ are equal or commensurable. In such a system it seems probable that the single equation (983) may be replaced by the system of s equations

$$\int_{t=0}^{t=\sigma_s} p_s dq_s = \tau_s h, \quad (s = 1, 2, \ldots m) \quad \text{.............. (1002)},$$

provided only that none of the periods $\sigma_1, \sigma_2, \ldots \sigma_m$ are commensurable.

An instance of such a system is provided by a solid body. The different wave-motions which represent the heat of the solid (§ 515) correspond to the different coordinates $q_1, q_2, \ldots q_m$ and equations (1002) simply express that the energy of the wave-motions of different periods must each be integral multiples of the corresponding energy-quanta.

536. In the most general class of systems the energy is of the general form

$$E = f(p_1, q_1, p_2, q_2, \ldots) \quad \text{.................... (1003)},$$

and is not separable into a sum of terms each of which depends only on one coordinate and its associated momentum. The momentum p_κ associated with the coordinate q_κ is given by

$$p_\kappa = \frac{\partial E}{\partial \dot{q}_\kappa} \quad \text{............................ (1004)}.$$

There will, however, in every problem be a certain number of integrals of the Hamiltonian equations of motion, say

$$\left.\begin{array}{r} \phi_1(p_1, q_1, p_2, q_2, \ldots) = \alpha_1 \\ \phi_2(p_1, q_1, p_2, q_2, \ldots) = \alpha_2 \\ \text{etc.} \end{array}\right\} \quad \text{................ (1005)},$$

where $\alpha_1, \alpha_2, \ldots$ are constants of integration which retain the same value throughout the motion so long as this remains in conformity with the classical dynamics. In certain problems it may be possible to eliminate all variables except p_κ and q_κ from equations (1004) and (1005), and so to express p_κ, for some coordinates at least, in the form

$$p_\kappa = f_\kappa(q_\kappa) \quad \text{............................ (1006)}.$$

The equation $f_\kappa(q_\kappa) = 0$ may have any number of roots; each root corresponds to a value of q_κ for which $p_\kappa = 0$, so that when this value of q_κ is reached p_κ in general changes sign and the direction of the momentum is reversed.

If $f_\kappa(q_\kappa) = 0$ has no roots, the momentum p_κ keeps the same sign throughout the motion (example: motion in a straight line under no forces).

If $f_\kappa(q_\kappa) = 0$ has only one root Q, the motion is one in which the momentum p_κ keeps the same sign until $q_\kappa = Q$. At this instant the sign of the momentum changes and after this the momentum keeps the same sign throughout the whole subsequent motion (example: motion in a hyperbolic orbit).

More generally $f_\kappa(q_\kappa) = 0$ has at least two roots, and if Q_1, Q_2 are two adjacent roots, there will be a motion possible in which q_κ oscillates between the values Q_1, Q_2. Systems of this type are called "conditionally periodic" systems. Each coordinate for which the momentum can be expressed in the form (1006) can perform a periodic libration between two fixed limits, just as though the coordinate belonged to a system possessing only one degree of freedom. For such a system, it has been assumed with great success by Sommerfeld, Epstein, and others that the equations of restriction are

$$\int p_s dq_s = \tau_s h \quad \text{.........................(1007)},$$

where the integral is taken over a complete libration of the q_s coordinate. There is one such equation of restriction for each coordinate whose momentum can be expressed in the form (1006).

This hypothetical supposition has been amply vindicated by success. Later it has been proved by Burgers* that the quantities on the left-hand side of equation (1007) are adiabatic invariants, subject only to the condition that none of the periods are equal or commensurable. It has further been shewn by Burgers† that if any r of the periods are equal or commensurable, say those corresponding to the coordinates $q_1, q_2, \ldots q_r$, then there is an adiabatic invariant of the form

$$\kappa_1 \int p_1 dq_1 + \kappa_2 \int p_2 dq_2 + \ldots + \kappa_r \int p_r dq_r \quad \text{...........(1008)},$$

where $\kappa_1, \kappa_2, \ldots$ are numerical multipliers.

It accordingly seems probable that the equation of restriction in this case is obtained by equating expression (1008) to some multiple of τh.

Applications to Spectroscopy.

The Bohr Atom.

537. The foregoing equations receive their simplest and most successful application in respect to electron orbits. Indeed the equations originated out of attempts by Sommerfeld to account for the fine structure of the hydrogen lines in terms of Bohr's theory of the hydrogen spectrum already given in § 499. The analysis which follows is substantially identical with that given by Sommerfeld‡.

* J. M. Burgers, *Amsterdam Academy Proc.* xxv. (1916), p. 849, or *Communications from the Phys. Laboratory of Leiden*, Supp. 41 *c*.

† J. M. Burgers, *Amsterdam Academy Proc.* xxv. (1916), p. 918, or *Communications from the Phys. Laboratory of Leiden*, Supp. 41 *d*.

‡ *Ann. d. Phys.* LI. (1916), pp. 1 and 125, or *Atombau und Spektrallinien* (Braunschweig, 1921), p. 263.

Consider the simplest case of a single electron of charge $-e$ describing a closed orbit about a positive nucleus of charge E. On taking $E=e$ we obtain the normal hydrogen atom. Taking $E=4e$ we obtain the positively charged helium atom, the spectrum of which has already been briefly considered in § 501.

If M, m are the masses of the nucleus and electron respectively, the kinetic and potential energies of the system are given by

$$T=\frac{1}{2}\frac{mM}{M+m}(\dot{r}^2+r^2\dot{\theta}^2);\quad V=-\frac{eE}{r} \quad \text{..............(1009)},$$

where r, θ are the polar coordinates of the electron relative to the nucleus. This expression for T assumes the centre of gravity of the system to be at rest. If the centre of gravity is in motion relative to the axes of reference, additional terms must of course be added to T, but these will not affect the internal motion of the atom.

For the hydrogen and helium atoms, m is very small in comparison with M, so that these equations become

$$T=\tfrac{1}{2}m(\dot{r}^2+r^2\dot{\theta}^2);\quad V=-\frac{eE}{r} \quad \text{..............(1010)}.$$

If we do not wish to neglect m/M we may still determine the motion from equations (1010), provided we replace m by $mM/(M+m)$ in the final result.

The two coordinates θ, r of the system will have corresponding momenta p_1, p_2 given by

$$p_1=mr^2\dot{\theta},\quad p_2=m\dot{r},$$

and the total energy $H=T+V$ expressed as a function of θ, r, p_1 and p_2 will be

$$H=\frac{1}{2}\left(\frac{p_2^2}{m}+\frac{p_1^2}{mr^2}\right)-\frac{eE}{r} \quad \text{.....................(1011)}.$$

538. The first of the Hamiltonian equations of motion is

$$\frac{dp_1}{dt}=-\frac{\partial H}{\partial \theta}=0 \quad \text{.......................(1012)},$$

and this has the integral $p_1=$ constant, or

$$mr^2\dot{\theta}=\text{constant},$$

expressing the constancy of angular momentum. In place of the second Hamiltonian equation, we may use the equation of energy. If W denote the negative energy of the orbit, this equation will be $H+W=0$, or from equation (1011),

$$\frac{1}{2}\left(\frac{p_2^2}{m}+\frac{p_1^2}{mr^2}\right)-\frac{eE}{r}+W=0 \quad \text{.................(1013)}.$$

Since p_1 is constant, this equation shews that p_2 is a function of r only Thus the system is seen to be of the general type discussed in § 536, as is of course otherwise known. The quantum equations of restriction will accordingly be assumed to be

$$\int p_1 d\theta = \tau_1 h \quad \text{..........................(1014)},$$

$$\int p_2 dr = \tau_2 h \quad \text{..........................(1015)},$$

the quantities on the left each being integrated through a complete cycle and so being adiabatic invariants.

In equation (1014) the limits for θ are from 0 to 2π, and since p_1 is constant the equation becomes

$$p_1 = \frac{\tau_1 h}{2\pi} \quad \text{..........................(1016)}.$$

Whatever orbit is described by the electron, the classical mechanics insist that its angular momentum shall remain constant. The quantum-mechanics insist that this constant value shall be an integral multiple of $h/2\pi$. Bohr's assumption for circular orbits, already introduced in § 499, now appears as a special case of this general law

539. From equations (1013) and (1016) we find

$$p_2^2 = \frac{2eEm}{r} - 2Wm - \left(\frac{\tau_1 h}{2\pi}\right)^2 \frac{1}{r^2}.$$

Let us write this in the form

$$p_2^2 = \left(\frac{\tau_1 h}{2\pi}\right)^2 \left(\frac{1}{r_1} - \frac{1}{r}\right)\left(\frac{1}{r} - \frac{1}{r_2}\right),$$

so that

$$\left(\frac{\tau_1 h}{2\pi}\right)^2 \frac{1}{r_1 r_2} = 2Wm; \quad \left(\frac{\tau_1 h}{2\pi}\right)^2 \left(\frac{1}{r_1} + \frac{1}{r_2}\right) = 2eEm \text{.........(1017)}.$$

The limits of the libration in r are from r_1 to r_2 and back again, so that equation (1015) may be replaced by

$$2\left(\frac{\tau_1 h}{2\pi}\right)\int_{r_1}^{r_2}\left[\left(\frac{1}{r_1} - \frac{1}{r}\right)\left(\frac{1}{r} - \frac{1}{r_2}\right)\right]^{\frac{1}{2}} dr = \tau_2 h.$$

To integrate we put $\frac{1}{r} = \frac{1}{r_1}\cos^2\theta + \frac{1}{r_2}\sin^2\theta$, and obtain

$$\int_{r_1}^{r_2}\left[\left(\frac{1}{r_1} - \frac{1}{r}\right)\left(\frac{1}{r} - \frac{1}{r_2}\right)\right]^{\frac{1}{2}} dr = 2\left(\frac{1}{r_1} - \frac{1}{r_2}\right)^2 \int_0^{\pi/2} \frac{\sin^2\theta\cos^2\theta\, d\theta}{\left(\frac{1}{r_1}\cos^2\theta + \frac{1}{r_2}\sin^2\theta\right)^2}$$

$$= \frac{\pi}{\sqrt{(r_1 r_2)}}(\sqrt{r_2} - \sqrt{r_1})^2.$$

Thus our equation becomes

$$\frac{\tau_1}{2\sqrt{(r_1 r_2)}} = \frac{\tau_2}{(\sqrt{r_2} - \sqrt{r_1})^2} = \frac{\tau_1 + \tau_2}{r_1 + r_2},$$

and on substituting for $r_1 r_2$ and $r_1 + r_2$ from equations (1017), this gives

$$W = \frac{2\pi^2 me^2 E^2}{(\tau_1 + \tau_2)^2 h^2} \quad \text{.........................(1018).}$$

From a well-known formula in the theory of central orbits, the eccentricity ϵ of the orbit is given by

$$1 - \epsilon^2 = \frac{2Wp_1^2}{me^2 E^2},$$

and this becomes, in the present problem,

$$1 - \epsilon^2 = \frac{\tau_1^2}{(\tau_1 + \tau_2)^2}.$$

It now appears that when the simple assumption of circular orbits is abandoned, the effect of the second quantum restriction is to limit the eccentricity of the orbit to certain definite values. The special value $\tau_2 = 0$ gives $\epsilon = 0$, and so corresponds to circular orbits. In general the only orbits allowed are those in which the ratio of the axes (b/a) is a commensurable fraction.

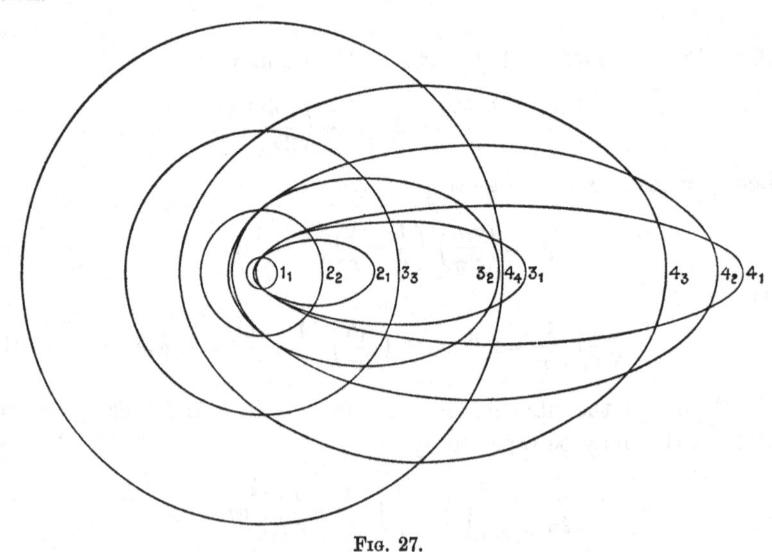

FIG. 27.

Fig. 27 shews the first few orbits which are possible for the simplest system of all, namely the hydrogen atom.

A comparison of equation (1018) with our former equation (949) shews that the values of W permitted by the present theory are precisely identical with those permitted by Bohr's simpler formula for circular orbits, namely,

$$W = \frac{2\pi^2 me^2 E^2}{\tau^2 h^2} \quad \text{.........................(1019).}$$

There is, however, the essential difference that, on the more general theory now under discussion, an orbit whose energy is given by equation (1019) can be described in a great number of ways, these ways corresponding to the number of ways in which τ can be expressed as the sum of two integers, τ_1 and τ_2. This circumstance appears at first sight to make the general theory of elliptic orbits more intelligible mechanically than was the simpler theory of Bohr. Under Bohr's simpler theory, a sudden change of energy required a sudden jump of the electron from one circular orbit to a second of different radius; thus a change of energy required the electron to jump from one point of space to another. Under the more general theory we have just examined, a jump of energy may require only a sudden jump in the velocity of the electron unaccompanied by a jump in space, so that the mechanical accompaniments of a sudden change in energy may, for certain values of τ_1, τ_2, τ_1' and τ_2', be merely such as might be produced by a sudden blow or impulse under ordinary mechanical laws. Nevertheless, Sommerfeld's work, which we shall discuss later, disposes of this conciliatory view of quantum-dynamics. Not only are spectral lines observed which, according to Sommerfeld's theory, are produced by actual jumps in space from one circular orbit to another, but the intensity of these lines greatly exceeds that of lines corresponding to mere changes of velocity.

A sudden jump of energy from a configuration τ_1, τ_2 of energy W to a second configuration τ_1', τ_2' of energy W' may be supposed to result in the emission (or absorption) of a quantum of radiant energy of frequency ν determined by

$$h\nu = W' - W = \frac{2\pi^2 m e^2 E^2}{h^2}\left[\frac{1}{(\tau_1+\tau_2)^2} - \frac{1}{(\tau_1'+\tau_2')^2}\right].$$

If, as in equation (952), we put

$$N = \frac{2\pi^2 m e^2 E^2}{h^3},$$

the system of frequencies is given by

$$\nu = N\left[\frac{1}{(\tau_1+\tau_2)^2} - \frac{1}{(\tau_1'+\tau_2')^2}\right] \quad \text{.................(1020)},$$

in which τ_1, τ_2, τ_1', τ_2' may have any integral values whatever. The value of ν depends, however, only on the two quantities $\tau_1+\tau_2$ and $\tau_1'+\tau_2'$. Each of these can have all possible integral values, but no others. Thus the spectrum given by equation (1020) is precisely identical with that given by the simpler equation (951) of Bohr, which, allowing for change of notation, may now be written in the form

$$\nu = N\left[\frac{1}{\tau_1^2} - \frac{1}{\tau_1'^2}\right] \quad \text{........................(1021)}.$$

This equation, it will be remembered, was deduced by assuming that all orbits were necessarily circular; indeed it is derivable at once from the general equation (1020) on taking $\tau_2 = \tau_2' = 0$.

540. Although the two equations (1020) and (1021) predict precisely the same spectrum, it is readily seen that their physical significance differs widely.

For instance, according to Bohr's simple formula (1021), the line H_a will be given by putting $\tau_1 = 2$, $\tau_1' = 3$. But according to the more complete formula (1020), the line H_a will be given by any set of integral values of $\tau_1, \tau_2, \tau_1', \tau_2'$ which make $\tau_1 + \tau_2 = 2$ and $\tau_1' + \tau_2' = 3$. The values $\tau_1 = 0$, $\tau_2 = 0$ are prohibited since these would make the electron and the nucleus coincide at their point of closest approach. Excluding these, there are two sets of values which make $\tau_1 + \tau_2 = 2$, namely

$$\tau_1 = 2,\ \tau_2 = 0 \ \text{ and } \ \tau_1 = 1,\ \tau_2 = 1;$$

and there are three sets of values which make $\tau_1' + \tau_2' = 3$, namely

$$\tau_1' = 3,\ \tau_2' = 0; \quad \tau_1' = 2,\ \tau_2' = 1; \quad \tau_1' = 1,\ \tau_2' = 2.$$

Thus there are six sets of values in all, shewing that, on the complete theory now under discussion, the line H_a can be generated in six distinct ways. In other words, it represents the superposition of six spectral lines.

A spectral line which represents the superposition of a number of coincident lines can be separated into its various constituent lines by various known physical methods, such as the superposition of a magnetic field (Zeeman effect) or the superposition of an electric field (Stark effect). Sommerfeld*, following up an early speculation of Bohr†, has suggested that the fine structure of the spectral lines may shew a permanent separation of the constituent lines resulting from the dependence of the mass of a moving electron upon its velocity. So long as we assume, as has so far been done, that the electron has a constant mass m which is independent of its velocity, the energies of all orbits for which $\tau_1 + \tau_2$ has the same value will be the same (cf. equation (1018)). As soon as the dependence of mass on velocity is taken into account, the energy depends, as we shall now see, on τ_1 and τ_2 separately, and the result must be a permanent separation of the constituents of a spectral line.

The Relativity Separation of Spectral Lines.

541. If c denote the velocity of an electron at any instant, and C is the velocity of light, the mass of the electron will be given, according to the theory of relativity, by the equation

$$m = m_0 (1 - \beta^2)^{-\frac{1}{2}} \quad \text{......................(1022)},$$

where β is written for c/C, and m_0 is the mass of the electron when at rest relative to the axes of reference.

* *Münchener Akad. Sitzungsber.* (1915), pp. 425, 459, and *Ann. d. Phys.* LI. (1916), p. 1. See also *Atombau und Spektrallinien*, Chap. v.

† *Phil. Mag.* XXIX. (1915), p. 332.

Let us, for the moment, simplify the problem by supposing the mass of the electron to be negligible in comparison with that of the nucleus round which it revolves. Then in place of equations (1010), the kinetic and potential energies will be given by

$$T = m_0 C^2 [(1-\beta^2)^{-\frac{1}{2}} - 1], \quad V = -\frac{eE}{r} \quad \text{............(1023)},$$

the former equation merely being the ordinary relativity formula for the mass of a moving particle. Except for terms of the order of $(c/C)^4$, it reduces to $\frac{1}{2}mc^2$.

From this we obtain as the equation of energy

$$W = \frac{eE}{r} - m_0 C^2 [(1-\beta^2)^{-\frac{1}{2}} - 1] \quad \text{..............(1024)},$$

where W is, as before, the amount of work required to move the electron from its orbit to infinity, and so is of course constant throughout the description of the orbit.

The principle of conservation of angular momentum gives, as before, the relation

$$mr^2\dot{\theta} = p_1,$$

where p_1 is constant and m is given by equation (1022). Using this relation, we may write

$$\frac{d}{dt} = \frac{p_1}{mr^2}\frac{d}{d\theta} \quad \text{..............................(1025)}.$$

If x, y are Cartesian coordinates in the plane of the orbit, such that $x = r\cos\theta$, $y = r\sin\theta$, we have

$$\dot{x} = \frac{d}{dt}(r\cos\theta) = \frac{p_1}{mr^2}\frac{d}{d\theta}(r\cos\theta) = \frac{p_1}{m}\left(\frac{1}{r^2}\frac{dr}{d\theta}\cos\theta - \frac{1}{r}\sin\theta\right),$$

so that, again using relation (1025),

$$\frac{d}{dt}(m\dot{x}) = \frac{p_1^2}{mr^2}\frac{d}{d\theta}\left(\frac{1}{r^2}\frac{dr}{d\theta}\cos\theta - \frac{1}{r}\sin\theta\right).$$

On putting $1/r = u$, this becomes

$$\frac{d}{dt}(m\dot{x}) = -\frac{p_1^2}{m}\left(\frac{d^2u}{d\theta^2} + u\right)\frac{\cos\theta}{r^2} \quad \text{.................(1026)}.$$

The equations of motion of the electron are, however, known to be*

$$\frac{d}{dt}(m\dot{x}) = -\frac{eE}{r^2}\cos\theta,$$

whence, on comparison with (1026),

$$\frac{d^2u}{d\theta^2} + u = \frac{eEm}{p_1^2} = \frac{eEm_0}{p_1^2}(1-\beta^2)^{-\frac{1}{2}} \quad \text{..............(1027)}.$$

* Cf. Jeans, *Electricity and Magnetism* (5th Edition), §§ 692, 694.

This is the differential equation of the orbit described when the relativity variation of mass with velocity is taken into account. The difference from the simpler equation obtained by neglecting the variation of mass consists solely in the presence of the factor $(1-\beta^2)^{-\frac{1}{2}}$ on the right-hand side.

542. The value of the factor $(1-\beta^2)^{-\frac{1}{2}}$ is at once given by the equation of energy (1024), and on using this value equation (1027) becomes

$$\frac{d^2u}{d\theta^2}+u\left[1-\left(\frac{eE}{p_1C}\right)^2\right]=\frac{eE}{p_1{}^2}\left(m_0-\frac{W}{C^2}\right) \quad\text{...........(1028).}$$

The general integral is of the standard form

$$u=A\cos\gamma\theta+B\sin\gamma\theta+D \quad\text{.................(1029),}$$

where, by substitution into the original equation, it is found that

$$\gamma^2=1-\left(\frac{eE}{p_1C}\right)^2,\quad D=\frac{eE}{\gamma^2p_1{}^2}\left(m_0-\frac{W}{C^2}\right) \quad\text{...........(1030),}$$

and A, B are constants of integration. By a suitable choice of the initial line $\theta=0$ we can make $B=0$. If we further put $A=\epsilon D$, the equation becomes

$$\frac{1}{r}=D(1+\epsilon\cos\gamma\theta) \quad\text{.......................(1031).}$$

This differs from the equation of the elliptic orbit which would be described by an electron of constant mass only in θ being replaced by $\gamma\theta$. Since γ differs only very slightly from unity, equation (1031) may be regarded as the equation of an elliptic orbit which undergoes a slow progression of the apses. Clearly ϵ is the eccentricity of this elliptic orbit.

We proceed to calculate the negative energy W of this orbit. Since W remains constant throughout the description of the orbit, it will suffice to evaluate it at the most convenient point and we shall select the apse at which $\theta=0$. At this point

$$\frac{1}{r}=D(1+\epsilon).$$

Also $\dot{r}=0$, so that

$$\beta=\frac{c}{C}=\frac{r\dot{\theta}}{C}=\frac{p_1}{mrC}=\frac{p_1D}{mC}(1+\epsilon),$$

whence, from equation (1022),

$$\frac{\beta}{(1-\beta^2)^{\frac{1}{2}}}=\frac{p_1D}{m_0C}(1+\epsilon)=\frac{eE}{\gamma^2p_1C}\left(1-\frac{W}{m_0C^2}\right)(1+\epsilon) \quad\text{......(1032).}$$

From the energy equation (1024) we obtain, on replacing $1/r$ by $D(1+\epsilon)$,

$$\frac{1}{(1-\beta^2)^{\frac{1}{2}}}=\left[1+\left(\frac{eE}{\gamma p_1C}\right)^2(1+\epsilon)\right]\left(1-\frac{W}{m_0C^2}\right) \quad\text{........(1033).}$$

Squaring equations (1033) and (1032) and subtracting corresponding sides, we find

$$1 = \left[\left\{1 + \left(\frac{eE}{\gamma p_1 C}\right)^2 (1+\epsilon)\right\}^2 - \left(\frac{eE}{\gamma^2 p_1 C}\right)^2 (1+\epsilon)^2\right]\left(1 - \frac{W}{m_0 C^2}\right)^2$$

$$= \left[1 + \left(\frac{eE}{\gamma p_1 C}\right)^2 (1-\epsilon^2)\right]\left(1 - \frac{W}{m_0 C^2}\right)^2 \quad \text{.......................(1034)},$$

an equation which determines the value of W.

543. Equations (1031) and (1034) give the orbit and the amount of its energy as predicted by the classical mechanics modified by the theory of relativity. On to these equations the quantum-theory imposes two additional equations of restriction, one corresponding to each of the coordinates θ and r.

The equation of restriction corresponding to the θ coordinate is the same as that of the simpler investigation of § 538 in which the relativity correction was neglected, namely

$$p_1 = \frac{\tau_1 h}{2\pi} \quad \text{..............................(1035)}.$$

To determine the second equation of restriction we notice that r performs a libration between values given by

$$\frac{1}{r} = D(1 \pm \epsilon).$$

In accordance with the ideas explained in § 536, Sommerfeld takes the second equation of restriction to be

$$\int p_2 dr_2 = \tau_2 h \quad \text{..........................(1036)},$$

where the integration is through a complete cycle of changes in r, say from $\theta = 0$ to $\theta = 2\pi/\gamma$. In this equation

$$p_2 = m\dot{r} = m\dot{\theta}\frac{dr}{d\theta} = \frac{p_1}{r^2}\frac{dr}{d\theta},$$

so that the equation becomes

$$\int_{\theta=0}^{\theta=2\pi/\gamma} \frac{p_1}{r^2}\left(\frac{dr}{d\theta}\right)^2 d\theta = \tau_2 h.$$

On using the value of r provided by equation (1031) and performing the integration, this gives

$$2\pi p_1 \gamma \left[\frac{1}{(1-\epsilon^2)^{\frac{1}{2}}} - 1\right] = \tau_2 h.$$

Combining this with equation (1035) it appears that the eccentricity of the orbit is now limited to values given by

$$1 - \epsilon^2 = \frac{\gamma^2 \tau_1^2}{(\tau_2 + \gamma\tau_1)^2},$$

where τ_1, τ_2 are integral. This differs from the value found in § 539 through the occurrence of $\gamma\tau_1$ in place of τ_1.

On substituting the value just obtained for $1-\epsilon^2$, and that for p_1 given by equation (1035), equation (1034) becomes

$$\left[1+\left(\frac{2\pi eE}{hC(\tau_2+\gamma\tau_1)}\right)^2\right]\left(1-\frac{W}{m_0C^2}\right)^2=1,$$

so that

$$\frac{W}{m_0C^2}=1-\left[1+\left(\frac{2\pi eE}{hC(\tau_2+\gamma\tau_1)}\right)^2\right]^{-\frac{1}{2}} \quad \text{..............(1037).}$$

The right-hand member can be expanded by the binomial theorem, giving

$$W=\frac{2\pi^2m_0e^2E^2}{(\tau_2+\gamma\tau_1)^2h^2}\left[1-\frac{3}{4}\left(\frac{2\pi eE}{hC(\tau_2+\gamma\tau_1)}\right)^2+\dots\right] \quad \text{......(1038).}$$

This is the equation which must replace equation (1018) when the relativity modification is taken into account. It is at once apparent that W no longer depends solely on the sum $\tau_1+\tau_2$.

From equations (1030) and (1035)

$$\gamma^2=1-\left(\frac{2\pi eE}{hC\tau_1}\right)^2,$$

so that

$$\gamma=1-\frac{2\pi^2e^2E^2}{h^2C^2\tau_1^2}+\dots$$

and

$$\frac{1}{(\tau_2+\gamma\tau_1)^2}=\frac{1}{(\tau_1+\tau_2)^2}\left[1+\frac{4\pi^2e^2E^2}{h^2C^2\tau_1(\tau_1+\tau_2)}+\dots\right].$$

Hence we find that the value of W given by equation (1038) can be expanded in the form

$$W=\frac{2\pi^2m_0e^2E^2}{(\tau_1+\tau_2)^2h^2}\left[1+\left(\frac{2\pi eE}{hC(\tau_1+\tau_2)}\right)^2\left(\frac{1}{4}+\frac{\tau_2}{\tau_1}\right)+\dots\right] \quad \text{...(1039).}$$

This is an expansion in powers of $(2\pi eE/hC)^2$, and since this quantity is numerically of the order of 5×10^{-5}, it is clear that terms beyond those written down, depending upon $(2\pi eE/hC)^4$, may be neglected.

544. The correction to the simple formula (1018) obtained by the neglect of the relativity correction is represented by the term in square brackets in equation (1039). It will be seen that it consists of two parts:

I. An increase of W by a factor

$$1+\frac{1}{4}\left(\frac{2\pi eE}{hC(\tau_1+\tau_2)}\right)^2 \text{ or } 1+\frac{W}{2m_0C^2}.$$

II. An increase of W by a further amount equal to $4\tau_2/\tau_1$ times the preceding.

For a circular orbit $\tau_2=0$, and the first correction alone appears. This correction is called by Sommerfeld "the relativistic correction for circular orbits."

The second correction depends on τ_2 and τ_1 separately and so is different for orbits of different eccentricities. It is this correction which gives rise to the splitting of the spectral lines.

If the electron in an atom suddenly passes from an orbit of energy $-W$ to one of energy $-W'$, the frequency of the emitted radiation is given as in § 500 by

$$h\nu = W' - W \quad \text{.........................(1040).}$$

If both W' and W can have a number of distinct but only slightly different values, the resulting value of ν will similarly have distinct and slightly differing values. In this way Sommerfeld accounts for the fine structure of a number of lines in the spectra of hydrogen and helium.

Fine Structure of Hydrogen and Helium Lines.

545. Let us consider first the ordinary Balmer series of hydrogen. As in § 501 this is obtained by taking $E = e$ and $\tau_1' + \tau_2' = 2$. If, as in equation (955), we write

$$\frac{2\pi^2 m_0 e^4}{h^3} = N_H,$$

we obtain from formula (1039)

$$W = N_H \frac{h}{(\tau_1 + \tau_2)^2}\left[1 + \frac{4\pi^2 e^4}{h^2 C^2 (\tau_1 + \tau_2)^2}\left(\frac{1}{4} + \frac{\tau_2}{\tau_1}\right)\right] \quad \text{......(1041),}$$

and, on putting $\tau_1' + \tau_2' = 2$,

$$W' = N_H \frac{h}{4}\left[1 + \frac{\pi^2 e^4}{h^2 C^2}\left(\frac{1}{4} + \frac{\tau_2'}{\tau_1'}\right)\right] \quad \text{.................(1042).}$$

On putting $\tau_1 + \tau_2 = n$, the frequencies are, from equation (1040), given by

$$\nu = N'\left(\frac{1}{4} - \frac{1}{n^2}\right) + N_H \frac{\pi^2 e^4}{h^2 C^2}\left(\frac{\tau_2'}{4\tau_1'} - \frac{4\tau_2}{n^4 \tau_1}\right) \quad \text{...........(1043),}$$

where

$$N' = N_H \left[1 + \frac{\pi^2 e^4}{h^2 C^2}\left(\frac{1}{4} + \frac{1}{n^2}\right)\right] \quad \text{.................(1044).}$$

The first term on the right-hand side of equation (1043) represents a Balmer series in which the value of the "Rydberg-constant" N' varies slightly from one member to the next. Measurements on the Balmer series by Paschen have in actual fact revealed a variation of this kind. Sommerfeld quotes the following values obtained by measurements on the lines H_α, H_β, H_γ, H_δ:

	H_α	H_β	H_γ	H_δ
$N' =$	109678·205	109678·164	109678·167	109678·198,

and gives the following table comparing the values of $10^6 \times \Delta N'/N$ given by these observations with the values predicted by equation (1044):

	$H_\alpha - H_\beta$	$H_\alpha - H_\gamma$	$H_\alpha - H_\delta$
Observed	0·37	0·35	0·06
Predicted by equation (1044)	0·61	0·89	1·04

There is not much agreement except as regards sign and order of magnitude, but Sommerfeld considers that the agreement is satisfactory in view of the extraordinary difficulty of making the measurements.

546. Let us pass now to the consideration of the fine structure of the lines of the Balmer series, represented by the last term on the right of equation (1043). For τ_1' and τ_2' there are only two sets of values (cf. § 540), namely

$$\tau_1' = 2, \quad \tau_2' = 0 \quad \text{and} \quad \tau_1' = 1, \quad \tau_2' = 1,$$

the former corresponding to a circular orbit in the final state of the atom and the latter to a final orbit of eccentricity $\frac{1}{2}\sqrt{3}$. Inserting these two sets of values, the series (1043) may be replaced by the two separate series

$$\nu = N'\left(\frac{1}{4} - \frac{1}{n^2}\right) - N_H \frac{\pi^2 e^4}{h^2 C^2} \frac{4\tau_2}{n^4 \tau_1} \quad \text{.....................(1045)},$$

$$\nu = N'\left(\frac{1}{4} - \frac{1}{n^2}\right) - N_H \frac{\pi^2 e^4}{h^2 C^2}\left(\frac{4\tau_2}{n^4 \tau_1} - \frac{1}{4}\right) \quad \text{...........(1046)}.$$

The corresponding lines in this series have a common frequency difference $\frac{1}{4}\frac{\pi^2 e^4}{h^2 C^2} N_H$, of which the numerical value is found to be 0·365. Thus the Balmer series ought to be a series of doublets of uniform frequency difference 0·365, each member of these doublets possessing its own fine structure.

This represents in general terms what is observed. Each line of the Balmer series is found to be a doublet and the frequency difference is either constant or approximately so. Michelson has determined the frequency differences of the lines H_α, H_γ as 0·32 and 0·42 respectively. It will be noticed that the mean of these is very close to 0·365, the value predicted by Sommerfeld's theory. The frequency difference of H_α has also been determined by Fabry and Buisson as 0·307. Sommerfeld considers that this value, which he regards as the most reliable, ought to be increased by one-quarter of its value on account of the details of fine structure of the two lines of which the doublet is composed. With this increase the value becomes 0·384 which is in very close agreement with the theoretical value of 0·365.

Merton*, as the result of a careful experimental study of the problem, has formed the opinion that observation is completely reconciled with Sommerfeld's theoretical value when allowance is made for the great effect of impurities on the appearance of the lines.

The value of this quantity can, however, be more accurately determined by indirect methods. The same doublet occurs in the spectrum of positively charged helium and also in the L-series of the X-ray spectra of all elements. From the former source Paschen has determined the value of the frequency difference as $0{\cdot}3645 \pm 0{\cdot}0045$; the various doublets in the L-series give a value approximating very closely to the theoretical value 0·365†. Thus there is little room for doubt that the agreement between theory and observation is complete.

* *Proc. Roy. Soc.* 97 A (1920), p. 307.

† Sommerfeld, *Atombau und Spektrallinien* (2nd Edition, 1921), pp. 343 and 359.

547. Sommerfeld considers in full detail the fine structure predicted by theory for the lines both of the Balmer series and of other hydrogen and helium series. In fig. 28 an example is given of the type of agreement with observation given by the theory. The top line shews the relative positions of the lines which on Sommerfeld's theory ought to constitute the helium line $\lambda = 4686$. The two lower lines shew the observed structure of this line, representing current and spark spectra respectively.

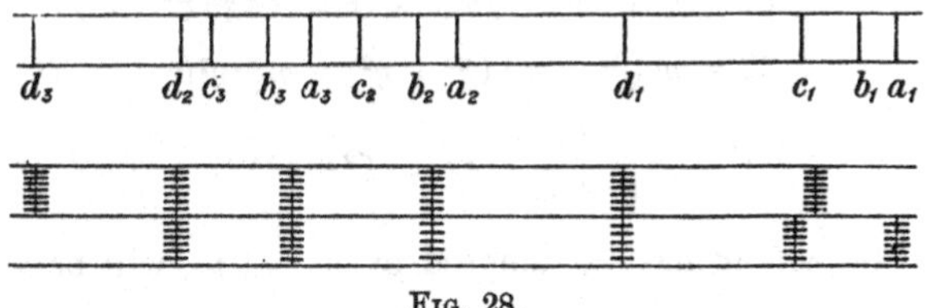

Fig. 28.

It will be noticed that certain lines predicted by theory do not appear at all in the observed spectrum. Sommerfeld introduces into his theory certain conjectures as to the relative intensity to be expected theoretically in the different constituent lines, and also a conjectural limitation ($\tau_1 \nless \tau_2$ and $\tau_1' \nless \tau_2'$ in our notation) which would prevent certain lines from appearing at all. For details of this part of Sommerfeld's work, as well as for a fuller study of the fine structure of individual lines, reference should be made to Sommerfeld's original paper*.

The Stark Effect.

548. It has been found by Stark† that spectral lines are split up in the presence of an electrostatic field, the degree of separation being proportional to the intensity of the field. The theoretical investigation of this effect by Epstein‡ provides a further instance of a successful application of quantum-dynamics.

We consider first the general motion of an electron around a massive nucleus in the presence of a uniform field of electrostatic force. Let the nucleus be taken for origin and the direction of the force be taken for axis of x. The kinetic and potential energies are given by

$$2T = m(\dot{x}^2 + \dot{y}^2 + \dot{z}^2), \quad V = -\frac{Ee}{r} - eXx \quad \text{...........(1047)},$$

where X is the electric intensity. The problem forms a special case of the general problem of the motion of a particle under two centres of force, and the solution of Epstein, apart from the final introduction of quantum-dynamics, is a special case of the known solution first given by Euler and Jacobi.

* *Ann. d. Phys.* LI. (1916), p. 1.

† *Ann. d. Phys.* XLIII. (1914), p. 965, and XLVIII. (1915), p. 193. See also *Ann. d. Phys.* XLIII. (1914), p. 983, and the two succeeding papers.

‡ *Ann. d. Phys.* L. (1916), p. 489.

549. Let us first change to cylindrical coordinates x, ϖ, θ having Ox as axis. Then equations (1047) become

$$2T = m(\dot{x}^2 + \dot{\varpi}^2 + \varpi^2\dot{\theta}^2), \quad V = -\frac{Ee}{(x^2+\varpi^2)^{\frac{1}{2}}} - eXx.$$

We next change further to coordinates defined by the transformation

$$x + i\varpi = \tfrac{1}{2}(\xi + i\eta)^2.$$

We have $\quad x = \frac{1}{2}(\xi^2 - \eta^2), \quad \varpi = \xi\eta, \quad x^2 + \varpi^2 = \frac{1}{4}(\xi^2 + \eta^2)^2,$

whence

$$2T = m(\xi^2 + \eta^2)(\dot{\xi}^2 + \dot{\eta}^2) + m\xi^2\eta^2\dot{\theta}^2,$$

$$V = -\frac{2Ee}{\xi^2 + \eta^2} - \tfrac{1}{2}eX(\xi^2 - \eta^2).$$

The momenta, which may be denoted by p_1, p_2, p_3, will be given by

$$p_1 = \frac{\partial T}{\partial \dot{\xi}} = m(\xi^2 + \eta^2)\dot{\xi}, \quad p_2 = \frac{\partial T}{\partial \dot{\eta}} = m(\xi^2 + \eta^2)\dot{\eta}, \quad p_3 = \frac{\partial T}{\partial \dot{\theta}} = m\xi^2\eta^2\dot{\theta},$$

and the Hamiltonian function H, which is the total energy $T + V$, expressed as a function of p_1, p_2, p_3, ξ, η and θ, assumes the form

$$H = \frac{1}{2m(\xi^2+\eta^2)}\left[p_1^2 + p_2^2 + \frac{\xi^2+\eta^2}{\xi^2\eta^2}p_3^2 - 4mEe - meX(\xi^4 - \eta^4)\right] \text{...(1048).}$$

The equations of motion of the electron are the usual canonical equations

$$\frac{dp_1}{dt} = -\frac{\partial H}{\partial \xi}, \text{ etc.}$$

Since $\partial H/\partial\theta = 0$, the third of these equations shews that $dp_3/dt = 0$. In other words p_3 has a constant value which we may denote by α. This measures the angular momentum about the axis of x.

The two remaining equations do not admit of direct integration but it is known from the classical solution of Euler and Jacobi that the variables are separable when ξ, η coordinates are used, so that p_1, p_2 are of the forms

$$p_1 = f_1(\xi), \quad p_2 = f_2(\eta) \quad \text{......................(1049).}$$

Assuming these forms, the actual values of the functions f_1 and f_2 are readily obtained.

Throughout the motion the energy $T + V$ has a constant value which we may as before denote by $-W$. Thus the equation of energy is $H + W = 0$, where H is given by equation (1048), and this may be written in the form

$$2m(\xi^2+\eta^2)W + [f_1(\xi)]^2 + [f_2(\eta)]^2 + \left(\frac{1}{\xi^2} + \frac{1}{\eta^2}\right)\alpha^2 - 4mEe - meX(\xi^4 - \eta^4) = 0$$

$$\text{......(1050).}$$

In this equation W is regarded as a constant, so that the equation is of the form

(a function of ξ) + (a function of η) = 0.

This can only be satisfied if the function of ξ has a constant value C, and the function of η has the constant value $-C$. Thus equation (1050) is equivalent to the two equations:

$$[f_1(\xi)]^2 + 2m\xi^2 W + \frac{\alpha^2}{\xi^2} - 2mEe - meX\xi^4 - C = 0 \quad \text{......(1051)},$$

$$[f_2(\eta)]^2 + 2m\eta^2 W + \frac{\alpha^2}{\eta^2} - 2mEe + meX\eta^4 + C = 0 \quad \text{......(1052)}.$$

These equations are first integrals of the equations of motion. The complete integration is effected, and the orbit discussed, in Epstein's original paper, but the above equations contain all that is essential for our present purpose.

550. The equations just obtained will represent all orbits which are consistent with the classical mechanics. We now introduce the quantum equations of restriction to select those which are allowed by the quantum-mechanics.

Since $m(\xi^2 + \eta^2)\dot{\xi} = p_1 = f_1(\xi)$, it follows that $\dot{\xi}$ will vanish when $f_1(\xi) = 0$; *i.e.* when

$$meX\xi^6 - 2m\xi^4 W + (2mEe + C)\xi^2 - \alpha^2 = 0 \quad \text{.........(1053)}.$$

The electrostatic fields available in the laboratory correspond to values of X such that the coefficient of ξ^6 in this equation is small compared with the remaining coefficients. Thus the equation will have one root for which ξ^2 is very large and two other roots in ξ^2 which will approximate to those of the equation

$$2m\xi^4 W - (2mEe + C)\xi^2 + \alpha^2 = 0.$$

This latter equation of course determines the points at which $\dot{\xi} = 0$ in the orbit described when $X = 0$. This orbit may be elliptic or hyperbolic, but at present we are only interested in elliptic orbits, in which case the equation will have two real roots and equation (1053) may be supposed to have two real roots to correspond.

Similarly the equation $f_2(\eta) = 0$ which determines values of η at which $\dot{\eta} = 0$ may be supposed to have two real roots for which η^2 is not large of the order of $1/X$.

The coordinates ξ, η will each perform librations between values of ξ and η, say ξ_1, ξ_2 and η_1, η_2, determined by the equations just discussed. Hence, in accordance with the principles explained in § 536, the quantum equations of restriction may be taken to be

$$\int_{\xi_1}^{\xi_2} p_1 d\xi = \tau_1 h, \quad \int_{\eta_1}^{\eta_2} p_2 d\eta = \tau_2 h, \quad \int_0^{2\pi} p_3 d\theta = \tau_3 h \quad \text{.........(1054)}.$$

Since p_3 has the constant value α, the third equation may at once be written in the form

$$\alpha = \frac{\tau_3 h}{2\pi} \quad \text{............................ (1055)},$$

while the two former equations become

$$\int_{\xi_1}^{\xi_2} f_1(\xi)\,d\xi = \tau_1 h, \quad \int_{\eta_1}^{\eta_2} f_2(\eta)\,d\eta = \tau_2 h \quad \ldots\ldots\ldots\ldots(1056),$$

in which $f_1(\xi)$ and $f_2(\eta)$ are given by equations (1051) and (1052).

The integrations involved in evaluating the left-hand members of these equations prove to be extremely laborious. Clearly the first equation will give τ_1 in terms of the three constants W, α and C which occur in the specification of $f_1(\xi)$ in equation (1051). The second equation will give τ_2 in terms of the same three constants. From these two equations and (1055), it is possible to eliminate α and C, and we are left with an equation giving W in terms of τ_1, τ_2, τ_3. Epstein carries the calculation only as far as first powers of X, since it is readily shewn that for all electric fields available in the laboratory, the terms in X^2 will be negligible. Details of the integration will be found in Epstein's original paper*; the final result proves to be

$$W = \frac{2\pi^2 m e^2 E^2}{h^2}\left[\frac{1}{(\tau_1+\tau_2+\tau_3)^2}\right] + \frac{3h^2 X}{8\pi^2 m e E}(\tau_1-\tau_2)(\tau_1+\tau_2+\tau_3) \ldots(1057),$$

and this equation expresses the energy of all "allowed" orbits as a function of the three integers τ_1, τ_2, τ_3.

551. Suppose that the electron falls from an orbit of energy $-W$ specified by the integers τ_1, τ_2, τ_3 to one of energy $-W'$ specified by the integers $\tau_1', \tau_2', \tau_3'$. In accordance with Bohr's supposition explained in § 500, we assume that the energy which is set free appears as a single quantum of monochromatic radiation. The frequency ν of this radiation will, as in § 500, be given by

$$h\nu = W' - W.$$

Let us put $\tau_1+\tau_2+\tau_3 = n$, $\tau_1'+\tau_2'+\tau_3' = n'$, so that n, n' are integers; then, by the use of relation (1057), we find that the frequency of the emitted radiation will be given by

$$\nu = \frac{2\pi^2 m e^2 E^2}{h^3}\left(\frac{1}{n'^2} - \frac{1}{n^2}\right) + \frac{3hX}{8\pi^2 m e E}\left[n'(\tau_1'-\tau_2') - n(\tau_1-\tau_2)\right] \ldots(1058).$$

On putting $X = 0$ this formula reduces naturally to the formula of Bohr already given in § 500. Thus the first term on the right of equation (1058) represents the undisplaced line, while the second term represents the splitting produced by the electrostatic field.

552. Fortunately the amount of this splitting under fields available in the laboratory is so large as to render comparison with experiment comparatively easy.

We notice first that $n'(\tau_1'-\tau_2') - n(\tau_1-\tau_2)$ is necessarily an integer. Call this Z, then the displacement produced by an electric field X is

$$\delta\nu = \frac{3h}{8\pi^2 m e E} XZ.$$

* *l.c.* pp. 503—509.

Thus for a given field X, every displacement is an integral multiple of the same quantity $3hX/8\pi^2meE$, so that the components of any line ought to form a regular series of equidistant lines. The unit of separation $\Delta\nu = 3hX/8\pi^2meE$ is easily calculated in terms of X, so that the prediction of theory is a quite definite one, in which adjustable constants do not figure at all. That this prediction is very closely confirmed by observation will appear in the detailed observations given below.

The quantities τ_1 and τ_2 have entered symmetrically into our fundamental equations, so that any negative value of $\tau_1 - \tau_2$ ought to be precisely as possible and as likely as the equal positive value. In other words an electrostatic field ought so to split up each spectral line that its components are arranged symmetrically about the position of the original line. This is found to be the case experimentally, but does not provide a verification of the theory since it must obviously be the case with any splitting in which the displacement is proportional simply to X.

553. In the two former problems of orbital motion under discussion, the angular momentum fell into quanta, and the possibility of a zero number of quanta could be disregarded from the outset, since the corresponding orbits would be such that the positive nucleus and negative electrons would coalesce at intervals and then separate out again. In the present problem also, the angular momentum p_3 falls into quanta, but the value $\tau_3 = 0$ cannot be dismissed for the reasons which previously held. An orbit for which $\tau_3 = 0$ is, in the present problem, merely one described permanently in a plane through the nucleus containing the direction of the electric field. If, however, as Epstein notes, both τ_3 and one of the integers τ_1 and τ_2 vanish, then the orbit reduces to a straight line through the nucleus. Such orbits are described by Epstein as "pendelbahnen," and it would appear that they ought to be rejected for the same reasons as held before.

Epstein retains the "pendelbahnen," although describing them as extremely improbable. He first tries the effect of rejecting values of τ_1, τ_2, τ_3 which are excluded by the conjectural principle (Auswahlprinzip) of Sommerfeld to which reference has already been made (§ 547). This principle is found to be too drastic, since certain lines which it rejects are actually observed. Epstein accordingly replaces it tentatively by a modified and less drastic principle of selection according to which all lines must be rejected, and so ought not to be observed, except those for which

$$\tau_1' \leqslant \tau_1, \quad \tau_2' \leqslant \tau_2, \quad \tau_3' \leqslant \tau_3 + 1,$$

the difference from Sommerfeld's principle lying in the replacement of τ_3 by $\tau_3 + 1$.

Even with this modified principle, it still appears that a number of lines are predicted which are not observed, and one line appears to be observed which is not predicted*. Moreover the lines which correspond to "pendelbahnen" are observed, although faintly. A detailed comparison with observations on the lines H_α, H_β, H_γ, H_δ will be found in Epstein's original paper.

554. An alternative plan which does not appear to have been considered by Epstein is to discard Sommerfeld's Auswahlprinzip altogether, and reject purely and simply those orbits for which $\tau_3 = 0$. This at once rejects the impossible "pendelbahnen." It rejects also, for less apparent reasons, elliptic orbits in which there is no angular momentum about the lines of electric force. The principle of rejection is, however, no more arbitrary than that of Epstein, while it has the advantage of introducing an agreement between theory and observation which is practically complete.

555. As instances of the good agreement of Epstein's theory with observation, we may consider in detail the lines H_β and H_γ. The observations are those of Stark, the calculated lines are those given in Epstein's paper†, except that the lines have been selected in accordance with the alternative principle of choice just explained.

For these two lines $\tau_1 + \tau_2 + \tau_3 = n = 2$. Our principle of selection rejects the value $\tau_3 = 0$, so that $\tau_1 - \tau_2$ can be 1, 0 or -1. For H_β,

$$\tau_1' + \tau_2' + \tau_3' = n' = 4,$$

so that $\tau_1' + \tau_2'$ can be 0, 1, 2 or 3 and $\tau_1' - \tau_2'$ can be 3, 2, 1 or 0. Similarly for H_γ, $n' = 5$, so that $\tau_1' - \tau_2'$ can be 4, 3, 2, 1 or 0.

For H_β the values of Z predicted are 14, 12, 10, ... 2, 0. All these lines, except $Z = 14$, are observed as follows:

	$Z=$	12	10	8	6	4	2	0
$\Delta\lambda$	calculated	19·4	16·1	12·9	9·7	6·5	3·2	0
	observed (p)	19·4	16·3	13·2	10·0	6·7	3·3	0
	" (s)	19·3	16·4	13·2	9·7	6·6	3·4	0

Here (p) refers to light polarised so that the electric force is parallel to the lines of the external electric field, while (s) refers to light polarised perpendicular to this.

* The line in question is $Z = 22$ in H_γ. The calculated value of $\Delta\lambda$ for this line is 29·3, for $Z = 21$ it is 28·0. Stark observed a line for which $\Delta\lambda = 29{\cdot}4$, and Epstein, having rejected the value $Z = 22$, identifies the line with $Z = 21$, thereby introducing the one and only discrepancy between theory and observation which is equal to, or even comparable with, an Ångström unit.

† *l.c.* p. 512.

For H_γ the values of Z predicted are those given below, all the lines being observed:

	$Z=$	22	20	18	17	15	13	12
$\Delta\lambda$	calculated	29·3	26·6	23·9	22·7	20·0	17·3	16·0
	observed (p)	29·4	—	23·9	—	19·9	—	15·9
	„ (s)	—	26·3	—	22·8	—	17·3	—
	$Z=$	10	8	7	5	3	2	0
$\Delta\lambda$	calculated	13·3	10·6	9·3	6·7	4·0	2·7	0
	observed (p)	—	10·6	—	6·6	—	2·6	—
	„ (s)	13·3	—	9·7	—	3·9	—	0

Other applications of Quantum Dynamics.

556. In the foregoing pages we have given a very brief account of what are probably the most brilliant triumphs of the quantum-mechanics—Bohr's original theory of the spectrum of the mono-electronic atom, Sommerfeld's extension to the fine structure of the spectral lines, and Epstein's investigation of the Stark effect. It may be well to state explicitly what the reader will probably have conjectured, that want of space has prevented anything like full justice being done to the investigations in question. Many points of interest, some of them of primary importance, have of necessity been omitted. For a full study of the problems, the reader should refer to the original papers.

Beyond the investigations which we have briefly summarised are others upon which we have not been able even to touch. Foremost among these should be mentioned investigations of the Zeeman effect by Sommerfeld*, Debye† and Bohr‡. In this connection also should be mentioned papers by Kramers§ and Burgers‖. Epstein¶ has extended his study of the Stark effect to a discussion of Fowler's helium series, and Sommerfeld, in a sequel to the paper from which we have so largely quoted, has given a theory of the X-ray spectrum, which proves to be in striking agreement with observation**.

557. Although the quantum-dynamics has been triumphantly successful in these applications, it must still, as a system of dynamics, be regarded as in an unsatisfactory state. It is only competent to deal with an excessively small range of dynamical phenomena, and even here progress has been rather by

* *Phys. Zeits.* XVII. (1916), p. 491 and *Atombau und Spektrallinien* (2nd edn, 1921), p. 416.

† *Phys. Zeits.* XVII. (1916), p. 507, or *Göttingen Nachrichten* (1916).

‡ "On the Quantum Theory of Line-Spectra," *Danish Academy* (1918), Ser. 8, iv, 1, p. 79.

§ *Danish Academy* (1919), Ser. 8, iii, 3, p. 287.

‖ "Het Atoommodel van Rutherford-Bohr," Dissertation, Haarlem, 1918.

¶ *Ann. d. Phys.* LVIII. (1919), p. 553.

** *Ann. d. Phys.* LI. (1916), p. 125. See also *Atombau und Spektrallinien* (2nd edn, 1921), pp. 115 and 354.

guesswork than by the utilisation of any established principles. The laws which have been discovered are no doubt cases of wide general laws, but so far only meagre success has followed attempts to discover these laws by generalisation from what is already known.

Consequently a very special value attaches to all attempts to extend the present known quantum laws into a complete or even a partially complete system of mechanics. One line of attack will be to examine the exact relation of what is known of quantum-dynamics to the old classical dynamics. The work of Bohr in this field is of great importance*. Another line of attack will be to try to generalise Sommerfeld's phase-integral equation

$$\int p\,dq = \tau h$$

so as to apply to any dynamical system. The main trouble is to find a generalisation which shall be invariant for all coordinates by which the motion of the system can be specified. Considerable discussion of this question will be found in papers by Sommerfeld, Ehrenfest and Burgers to which reference has already been made. Mention must also be made of a valuable paper by Planck†. But it cannot be said that any great progress has yet been made in bringing quantum-dynamics into its proper relation to general statistical mechanics.

Questions of fundamental importance to the progress of physics still remain untouched, for the reason that the tools for the work have not yet been created. Among such questions may be mentioned the dynamics of the interaction between radiation and electrons, and the dynamics of the apparent sudden electronic jumps of which the spectral lines are believed to be evidence. The field which remains unexplored is incomparably vaster and more important than that which has so far been conquered.

* "On the Quantum Theory of Line-Spectra," Parts I. and II. Parts III. and IV. which will deal specially with the problem mentioned are not yet published.

† *Ann. d. Phys.* L. (1916), p. 385.

APPENDIX A

INTEGRALS INVOLVING EXPONENTIALS

A TYPE of integral which occurs very frequently in the mathematics necessary to the Kinetic Theory is

$$\int u^n e^{-\lambda u^2} du \quad \text{.......................................(i)},$$

where n is integral. This can be evaluated in finite terms when n is odd, and can be made to depend on the integral

$$\int_0^u e^{-\lambda u^2} du \quad \text{.......................................(ii)},$$

when n is even. In each case the reduction is most quickly performed by successive integrations by parts with respect to u^2. Tables for the evaluation of the integral (ii) will be found in Appendix B.

When, as is generally the case, the limits of integration are from $u=0$ to $u=\infty$, the results of integration are expressed by the formulae

$$\int_0^\infty u^{2\kappa} e^{-\lambda u^2} du = \frac{1.3...(2\kappa-1)}{2^{\kappa+1}} \sqrt{\frac{\pi}{\lambda^{2\kappa+1}}},$$

$$\int_0^\infty u^{2\kappa+1} e^{-\lambda u^2} du = \frac{\kappa\,!}{2\lambda^{\kappa+1}}.$$

The following cases of the general formulae are of such frequent occurrence that it may be useful to give the results separately:

$$\int_0^\infty e^{-hmu^2} du = \frac{1}{2}\sqrt{\frac{\pi}{hm}}, \qquad \int_0^\infty e^{-hmu^2} u\,du = \frac{1}{2hm},$$

$$\int_0^\infty e^{-hmu^2} u^2 du = \frac{1}{4}\sqrt{\frac{\pi}{h^3m^3}}, \qquad \int_0^\infty e^{-hmu^2} u^3 du = \frac{1}{2h^2m^2},$$

$$\int_0^\infty e^{-hmu^2} u^4 du = \frac{3}{8}\sqrt{\frac{\pi}{h^5m^5}}, \qquad \int_0^\infty e^{-hmu^2} u^5 du = \frac{1}{h^3m^3},$$

$$\int_0^\infty e^{-hmu^2} u^6 du = \frac{15}{16}\sqrt{\frac{\pi}{h^7m^7}}, \qquad \int_0^\infty e^{-hmu^2} u^7 du = \frac{3}{h^4m^4}.$$

Each integral can be obtained by differentiating the one immediately above it with respect to hm. In this way the system can be extended indefinitely.

APPENDIX B

THE following tables will be found of use for various numerical calculations in connection with the Kinetic Theory. The values of $\psi(x)$ are from a table by Tait in the paper already referred to (p. 255).

x	x^2	e^{-x^2}	$\frac{2}{\sqrt{\pi}}\int_0^x e^{-x^2}dx$	$\psi(x)$ Defined by equation (712), p. 254
·1	·01	·99905	·11246	·20066
·2	·04	·96080	·22270	·40531
·3	·09	·91393	·32863	·61784
·4	·16	·85214	·42839	·84200
·5	·25	·77880	·52050	1·08132
·6	·36	·69768	·60386	1·33907
·7	·49	·61263	·67780	1·61819
·8	·64	·52729	·74210	1·92132
·9	·81	·44486	·79691	2·25072
1·0	1·00	·36788	·84270	2·60835
1·1	1·21	·29820	·88021	2·99582
1·2	1·44	·23693	·91031	3·41448
1·3	1·69	·18452	·93401	3·86538
1·4	1·96	·14086	·95229	4·34939
1·5	2·25	·10540	·96611	4·86713
1·6	2·56	·07730	·97635	5·41911
1·7	2·89	·05558	·98379	6·00570
1·8	3·24	·03916	·98909	6·62715
1·9	3·61	·02705	·99279	7·28366
2·0	4·00	·01832	·99532	7·97536
2·1	4·41	·01215	·99702	8·70234
2·2	4·84	·00791	·99814	9·46467
2·3	5·29	·00504	·99886	10·26236
2·4	5·76	·00315	·99931	11·09547
2·5	6·25	·00197	·99959	11·96402
2·6	6·76	·00116	·99976	12·86798
2·7	7·29	·00068	·99987	13·80734
2·8	7·84	·00039	·99992	14·78225
2·9	8·41	·00022	·99996	15·79255
3·0	9·00	·00012	·99998	16·83830

INDEX OF SUBJECTS

The numbers refer to pages

INDEX OF NAMES

Printed in the United States
By Bookmasters